FUNDAMENTALS
OF FORENSIC SCIENCE

D0068844

FUNDAMENTALS
OF FORENSIC SCIENCE

Max Houck
Director, Forensic Science Initiative, Research Office
Manager, Forensic Business Research and Development
College of Business & Economics
West Virginia University
Morgantown, West Virginia

Jay Siegel
Director of Forensic and Investigative Sciences Program
Indiana University and Purdue University
Indianapolis, Indiana

AMSTERDAM • BOSTON • HEIDELBERG • LONDON
NEW YORK • OXFORD • PARIS • SAN DIEGO
SAN FRANCISCO • SINGAPORE • SYDNEY • TOKYO
Academic Press is an imprint of Elsevier

Academic Press is an imprint of Elsevier
84 Theobald's Road, London WC1X 8RR, UK
Radarweg 29, PO Box 211, 1000 AE Amsterdam, The Netherlands
30 Corporate Drive, Suite 400, Burlington, MA 01803, USA
525 B Street, Suite 1900, San Diego, CA 92101-4495, USA

First edition 2006
Reprinted 2006

British Library Cataloguing in Publication Data
Houck, Max M.
  Fundamentals of forensic science / Max Houck, Jay Siegel.
  p. cm.
  Includes bibliographical references and index.
  ISBN 0-12-356762-9 (hardcover: alk. paper)   1. Forensic sciences.
  2. Criminal investigation.   I. Siegel, Jay A.   II. Title.
  HV8073.H77 2006
  363.25—dc22

Library of Congress Catalog Number: 2005028127

ISBN–13: 978-0-12-356762-8
ISBN–10: 0-12-356762-9

For information on all Academic Press publications
visit our website at books.elsevier.com

Printed and bound in *China*

06 07 08 09 10   10 9 8 7 6 5 4 3 2

*Dedicated to Lucy,
the love of my life,
and to my parents,
who are with me always.
MMH*

*Dedicated to Maggie:
You have made my life's journey worth traveling
and this book possible.
JAS*

CONTENTS

FOREWORD

We live in an era where most professions favor multi-skilling, flexibility, and the ability to adapt to rapid changes. Forensic science is a perfect example, and there is little doubt that forensic science disciplines are currently evolving at a rapid pace. In recent years, forensic science has also become increasingly popular worldwide. This has resulted in many challenges, but especially the importance of sound and relevant education and training in forensic science. Recent reviews of forensic science education and training in the US, UK, and Australia also point to a similar conclusion.

Textbooks continue to be essential learning tools in areas of knowledge. However, writing a fundamental textbook on forensic science is not an easy task because of the multi-disciplinary nature of the field. It is often difficult to go beyond a basic introductory level due to the amount of background information needed in each discipline. Another challenge is to introduce and develop concepts that are specific to the forensic context (e.g., what is evidence? Locard's exchange principle, etc.) while keeping enough rigorous scientific content to assist the student to become a sound scientist and stay technologically current. This balance is crucial for forensic scientists who need to apply good science to solve the questions asked by the justice system in a logical and relevant manner, and ultimately to provide valuable expert evidence. This book succeeded in this endeavor by introducing crime scene investigation and the nature of evidence with references to Kirk and Locard up front, and also by including a chapter on reports and testimony at the end of the book.

In addition to traditional forensic disciplines (e.g., DNA, fingerprints, drugs, fibers, etc.), *Fundamentals of Forensic Science* also includes topics that are not commonly covered by other textbooks: pathology, anthropology and odontology, and entomology. This gives a more complete and realistic view of forensic science as applied in real casework.

A prime task of forensic science educators is to prompt critical thinking amongst students to eventually generate high-level forensic scientists that are able to critically review results, data, circumstantial information, legal propositions, and cases. The format of this book, with feature boxes and lists of questions in each chapter, will undoubtedly assist in reaching this goal.

There is no doubt that *Fundamentals of Forensic Science* fills a need in forensic science education. I have known the authors for almost a decade. Both are internationally-acclaimed forensic scientists with significant casework and academic experience. This unique combination has allowed them to deliver a fine book that will rapidly become an essential reference not only for students and educators, but also for forensic science professionals and lawyers. I hope you will enjoy reading and using this book for whatever your application might be.

Claude Roux, PhD
Professor of Forensic Science
University of Technology, Sydney

ACKNOWLEDGEMENTS

Little is written about how to write a textbook, probably because no one wants to be responsible for telling the truth about such a long and painful process. Therefore, appreciation and thanks are due to those who provided support to us when we had nowhere else to turn.

For reviewing early drafts of chapters: Jeff Wells, Suzanne Bell, Michael Bell, James Amrine, Richard Ernest, David Foran, Chris Bommarito, and Lucy Davis Houck. Additional thanks to Chris Tindall, Larry Quarino, and Bruce McCord for doing such a great job of reviewing the manuscript. Special thanks to Bruce for his help with the DNA chapter. We would also like to thank the Honorable David Lawson, US District Court Judge for the Eastern District of Michigan for his excellent review of the manuscript. His legal/scientific perspective added valuable insight to this project.

For photographs and figures, Mark Sandercock, Silvana Tridico, Terry McAdam, Brad Putnam, John Black, Richard Ernest, Chris Bommarito, Cheryl Lozen, John Lentini, Bob Kallman, Marilyn Bagley, Jamie Downs, Sarah Jones, Eric Dahlberg and Frans Janssen.

For additional support, whether they knew they were giving it or not (in no particular order), the WVU Forensic and Investigative Science students, Mary Holleran, Colleen Lankford, Nick Fallon (for getting us into this mess in the first place), Melissa Schoene, Anjali Swienton, Thompson Cigar Company, David Corbett, and the crew at Café Bacchus, Sheralee Brindell, Maureen Bottrell, Maria Yester. Also, students in the Master of Science Program at Michigan State for reading and commenting on many of the chapters.

Special thanks go to Robin Bowen, Keith Morris, Mark Dale, Barry Fisher, and Mark Listewnik and the staff at Academic Press including Pam Chester and Jenn Soucy.

PREFACE

Fundamentals of Forensic Science represents a different, albeit more realistic, view of the field of forensic science than is found in other textbooks. This view includes areas that are central to criminal investigations but fall outside the typical definition of "criminalistics." From the beginning, we decided to make *Fundamentals of Forensic Science* reflect how professional forensic scientists work and not how forensic science academicians teach. This enabled us to include the "-ologies" (pathology, entomology, anthropology, etc.) that many instructors don't traditionally teach—but that's probably because the chapters don't exist in other books. We felt that many instructors would like to teach these topics but don't have the fundamental resource materials to do so; additionally, students may want to read about a discipline that interests them but isn't covered in the course. The instructor may have local experts lecture on these specialties but, without these chapters, the students don't have any foundation to appreciate what the expert presents. If the instructor uses a video of a case, in the absence of a local expert, the students can be even more lost—the application of the methods in the case are key and the background information may be glossed over. In this regard, *Fundamentals of Forensic Science* provides the basis for the integration of these critical topics into the overall course. Our hope is that *Fundamentals of Forensic Science* fills this need.

We also offer a new perspective on the nature of forensic evidence. In his *Science* article, "Criminalistics" from 1963, Kirk opines that the principles that bind the various disciplines into the whole of forensic science "center on identification and individualization of persons and of physical objects". But this is only part of the larger nature of the discipline: The binding principles relate to relationships between people, places, and things as demonstrated by transferred evidence. It doesn't matter so much that this ceramic shard came from a particular lamp— it *does* matter, however, that the shard was found in the dead person's head and the suspect's fingerprints are found on the lamp. It is not merely the identification or individualization of the objects but it is the *context* of those people, places, and things and their relationship or interrelatedness within that context that provides its value in the justice system. A crime scene is a set of spatial relationships and/or properties; all evidence is spatial in that sense. Even an item of evidence discarded a distance from the scene by the perpetrator has meaning. A crime scene can also be viewed as a piece of recent history. It has a story to tell and the various pieces of evidence carry the facts of the story within them. In that sense, forensic scientists are auditors and storytellers.

In *Fundamentals of Forensic Science*, we stress these associations and how they relate the evidence to the facts of the crime. We also emphasize that *all* evidence is *transfer* evidence (à la Locard), even evidence that may not have been characterized as such, like DNA (semen transferred by sexual contact in a sexual assault), pathology (the pattern of a weapon transferred and recorded in the wound of a victim), or entomology (the number and kinds of maggots that have accumulated—transferred from the environment—on a decomposing body). Locard's Exchange Principle, then, is *the* binding principle in forensic science because it focuses on reconstructing relationships in the commission of a crime through the analysis of transferred information.

Forensic science education is entering an exciting era, ushered in largely by the work of the Technical Working Group on Education and Training in Forensic Science (TWGED). This group, sponsored by the National Institute of Justice (NIJ) and West Virginia University, generated guidelines for building careers in forensic science, curricula for undergraduates and graduates, and continuing education for professional forensic scientists. These guidelines led the American Academy of Forensic Sciences (AAFS) to form the Forensic Education Program Accrediting Commission (FEPAC), an accrediting body for forensic science educational programs. New forensic science educational programs appear weekly, it seems, and, because the quality of education goes to the heart of any profession, standards are a necessary component to assure that they prepare students properly for careers in our field.

The teaching of forensic science has spread from graduate and four-year programs to community colleges and high schools. While writing a book targeted for one end of that spectrum most likely makes it unsuitable for the other end, we see *Fundamentals of Forensic Science* as being appropriate across that spectrum. Educators teaching a forensic science course for the first time will find the supplemental course materials helpful in getting started. Experienced educators will find these resources helpful as well but will also appreciate the breadth and depth of the chapters of this text. Despite its broad applicability, our intent in writing *Fundamentals of Forensic Science* was for students who have already taken basic science courses.

Fundamentals of Forensic Science is organized roughly along the timeline of a real case. It begins with an introduction and history of forensic science as background to the discipline and the structure of a modern forensic science laboratory. Chapter 2 covers the processing of crime scenes and Chapter 3 covers the nature of forensic evidence. In Chapters 4 (Microscopy), 5 (Spectroscopy), and 6 (Chromatography), we cover the basic methods of analysis used in most, if not all, forensic science examinations. The biological sciences are then presented: Pathology (Chapter 7), anthropology and odontology (Chapter 8), entomology (Chapter 9), serology and blood pattern analysis (Chapter 10), DNA (Chapter 11), and finally hairs (Chapter 12). The next chapters address the chemical sciences, drugs (Chapter 13), toxicology (Chapter 14), fibers (Chapter 15), paints (Chapter 16), soils and glass (Chapter 17), and arson/explosives (Chapter 18). The third section covers physical evidence, including friction ridges (Chapter 19), questioned documents (Chapter 20), firearms and toolmarks (Chapter 21), shoeprints, tire treads, and other impression evidence (Chapter 22). The final chapter in the book looks at the intersection of forensic science and the law (Chapter 23).

Feature boxes throughout the book emphasize resources on the World Wide Web ("On the Web"), historical events in forensic science ("History"), practical issues in laboratory analysis ("In the Lab"), and topics for further reading or interest ("In More Detail"). Each chapter ends with two types of questions to help with chapter review and discussion: "Test Your Knowledge" questions target key terms and information from the chapters while the questions under "Consider This . . ." offer topics and issues that should challenge the students knowledge and understanding of the chapter contents.

With a project like writing a textbook (we submit that *no* project is like writing a textbook!), compromises must invariably take place. Our aim was to yield only where necessary and to dig in when we felt our vision of the book was in jeopardy. We feel that the decisions we made have resulted in a better product and hope that you do as well.

MMH
JAS

PART
ONE

Criminal Justice and
Forensic Science

Introduction

KEY TERMS

Anthropometry

American Academy of Forensic Sciences (AAFS)

American Society of Crime Laboratory Directors (ASCLD)

American Society for Testing and Materials, International (ASTM)

ASCLD Laboratory Accreditation Board (ASCLD-LAB)

Behavioral sciences

Bertillonage

Chain of custody

Criminalistics

Criminalists

Forensic anthropology

Forensic engineering

Forensic odontology

Forensic pathology

Forensic science

Forensic Science Education Program Accreditation Commission (FEPAC)

International Organization for Standardization (ISO)

Medical examiner

Questioned documents

Technical Working Group on Education and Training in Forensic Science (TWGED)

Toxicology

WHAT IS FORENSIC SCIENCE?

The *Oxford English Dictionary* lists one of the first uses of the phrase "forensic science" to describe "a mixed science" (Oxford English Dictionary, 2005). The early days of forensic science could certainly be called mixed, when science served justice by its application to questions before the court. Forensic science has grown as a profession and into a science in its own right. Given the public's interest in using science to solve crimes, it looks as if forensic science has an active, if hectic, future.

Forensic science describes the science of associating people, places, and things involved in criminal activities; these scientific disciplines

3

assist in investigating and adjudicating criminal and civil cases. The discipline divides neatly into halves, like the term that describes it. Science is the collection of systematic methodologies used to increasingly understand the physical world. The word "forensic" is derived from the Latin *forum* for "public" (Oxford English Dictionary, 2005). In ancient Rome, the Senate met in the forum, a public place where the political and policy issues of the day were discussed and debated; even today, high school or university teams that compete in debates or public speaking are called "forensics teams." More technically, "forensic" means "as applied to public or legal concerns." Together, "forensic science" is an apt term for the profession of scientists whose work answers questions for the courts through reports and testimony.

AREAS OF FORENSIC SCIENCE

CRIMINALISTICS

The term **criminalistics** is sometimes used synonymously with "forensic science." Criminalistics is a word imported into English from the German *kriminalistik*. The word was coined to capture the various aspects of applying scientific and technological methods to the investigation and resolution of legal matters. In some forensic science laboratories, forensic scientists may be called **criminalists**. Criminalistics is generally thought of as the branch of forensic science that involves collection and analysis of physical evidence generated by criminal activity. It includes areas such as drugs, firearms and toolmarks, fingerprints, blood and body fluids, footwear, and trace evidence. "Trace evidence" is a term of art that means different things to different people. It might include fire and explosive residues, glass, soils, hairs, fibers, paints, plastics and other polymers, wood, metals, and chemicals. These items are generally analyzed by forensic science or forensic science laboratories. To avoid confusion, unnecessary terminology, and regionalism, the phrases "forensic science" and "forensic scientist" will be used instead of "criminalistics" and "criminalist."

FORENSIC PATHOLOGY

Back in the days when the *Quincy* television show was popular, many people thought of **forensic pathology** and forensic science as the same thing; this misperception persists today. Forensic pathology is conducted by a **medical examiner**, who is a physician, specially trained in clinical and anatomic pathology, whose function is to determine the cause and manner of death in cases in which death occurred under suspicious or unknown circumstances. This determination often involves a teamwork approach with the autopsy or postmortem examination of the body as the central function. Other team members may include toxicologists, anthropologists, entomologists, and radiologists. Medical examiners are often called to death scenes to make some preliminary observations including an estimate of the time since death.

FORENSIC ANTHROPOLOGY

Forensic anthropology is a branch of physical anthropology, the study of humans and their ancestors. Forensic anthropology deals with identifying people who cannot be identified through soft tissue features, such as fingerprints or photographs. Typically, forensic anthropologists analyze skeletal remains to determine if they are human and, if so, the age, sex, height, and other characteristics, such as socio-economic status, of the deceased. If the characteristics of the remains compare favorably with those of the missing person in question, then further methods (such as X-rays) are employed to positively identify (individualize) the remains.

Forensic anthropologists figure prominently in the reconstruction and identification of victims in mass fatalities, such as bombings and airplane crashes. Working closely with pathologists, dentists, and others, forensic anthropologists aid in the identification of people who otherwise might never be identified.

FORENSIC ODONOTOLOGY

Sometimes called forensic dentistry, **forensic odontology** has a large number of applications to the forensic sciences. They include identification of human remains in mass disasters (enamel is the hardest material produced by the body and intact teeth are often found), postmortem X-rays of the teeth can be compared to antemortem X-rays, and the comparison of bitemarks. One of the most famous of all serial killers in the United States, Theodore Bundy, was brought to justice in part on evidence of bitemarks. He bit his last victim after her death. The forensic pathologist was able to obtain a plaster impression of the bitemark, which was compared to a known impression of Bundy's teeth. Lowell Levine, a forensic odontologist, testified at Bundy's trial that the bitemarks on the victim's body were made by Bundy. This was important evidence that the jury used to convict him of the murder. As a consequence of this conviction, Bundy was executed (Rule, 1980).

FORENSIC ENGINEERING

In 1980, a balcony lining the inside of the lobby of a large hotel in Kansas City collapsed and many people were injured and some died. Forensic engineers investigated the site and determined that the concrete support used in construction of the balcony was made of substandard materials. This led to criminal charges against the contractor. This case illustrates the value that a forensic engineer has in helping to investigate situations involving failure analysis of materials and constructions. Forensic engineers are also heavily involved in reconstruction of traffic accidents. They can determine path, direction, speed, the person driving, and the type of collision, from what may seem to the layperson as scant evidence.

TOXICOLOGY

Toxicology involves the chemical analysis of body fluids and tissues to determine if a drug or poison is present. Toxicologists are then able to determine how much and what effect, if any, the substance might have had on the person. Forensic toxicologists often work hand in hand with forensic pathologists. More than half of the cases that forensic toxicologists receive involve drunk driving cases and the determination of the level of alcohol in blood or breath.

BEHAVIORAL SCIENCES

Forensic psychiatrists and psychologists have long been involved in the forensic sciences in the determination of a person's competency to stand trial and to aid in one's own defense. Although each state has its own standards for determining insanity, the question usually revolves around whether or not the defendant had the mental capacity to form an intent to commit the crime and/or whether he or she knew right from wrong.

In recent years, behavioral forensic scientists have been called upon to assist law enforcement agents and forensic pathologists in the investigation of serial crimes by creating psychological profiles of the criminals. Such profiling has provided useful information about the person that the police should look for as they investigate serial crimes. People generally act in predictable, reproducible ways when they commit crimes, and the discovery of these behavioral patterns can provide clues to the personality of the offender. Behavioral scientists may also be called upon to help in interviewing or interrogating suspects in crimes or to develop profiles of likely airplane hijackers and possible terrorists.

QUESTIONED DOCUMENTS

Questioned document examination is a complicated and broad area of study; a trainee may study with an experienced examiner for several years before being qualified. This field has many facets including the comparison of handwritten or typewritten documents to determine their source or authenticity. In addition, questioned document examiners may be called upon to detect erasures or other obliterations, forgeries, altered documents, charred documents, and counterfeit currencies. Questioned document examiners analyze papers and inks to determine their source and age.

A BIT OF FORENSIC SCIENCE HISTORY

Some forms of what we would now call forensic medicine were practiced as far back as the fifth century. During the next thousand years, there were many advances in science, but only forensic medicine was practiced to any great extent. The science of toxicology, the study of poisons, was one of the first "new" forensic sciences to emerge. In an early case, a Mr. Lefarge died under mysterious circumstances, and his wife fell under suspicion. In 1840 the famous French scientist Mathcu Orfilia examined Lafarge's remains and determined that he had ingested arsenic. Orfilia further showed that the source of the arsenic could only have been poisoning, and Lefarge's wife was subsequently convicted of the crime (Wilson and Wilson, 2003).

The eighteenth and nineteenth centuries saw considerable advances in the science of personal identification. Because police photography had not been developed and fingerprints weren't being used, there needed to be methods of reliably tracking a person either through the police process or during incarceration (Thorwald, 1964). Enter Alphonse Bertillion, a French criminologist, who developed a method of recording physical features of a person in such a way that the record would be unique to that person. This method was referred to as **anthropometry** or **Bertillonage**, after its creator. Bertillion developed a set of precise measuring instruments to be used with his method. The Bertillionage system became very popular throughout Europe and the United States. It became widely used in U.S. prisons, which needed a way to track the prisoners. The Bertillion system was plagued by problems of reproducibility and was finally discredited in the United States at Leavenworth Federal Prison in Kansas. In 1903 a prisoner named William West was admitted to the prison to serve a sentence. When he was measured using the Bertillion system, it was found that a man with the name William West with virtually the same set of measurements was already at the prison! This sounded the death knell for Bertillionage and opened the door for the study of fingerprints. Bertillion used fingerprints in his system but didn't have a good way to organize them for mass searches (Wilson and Wilson, 2003). Dr. Juan Vucetich, a Croatian who lived in Argentina and worked for the La Plata police force, conceived of a method of fingerprint classification in 1894 that provided for 1,048,576 primary classifications of fingerprints. As history and culture would have it, his work was largely unheard of in Europe until much later. Sir William Herschel, a British officer in India, and Henry Faulds, a Scottish medical doctor, both studied fingerprints as a scientific endeavor to see whether they could be used reliably for identification. In 1901, Sir Edward Henry devised a fingerprint classification system, still used today to categorize sets of fingerprints and store them for easy retrieval (Thorwald, 1964).

Modern blood and body fluid typing got its start around 1900 when Karl Landsteiner showed that human blood came in different types; his work led to the ABO blood typing system. This work, in turn, led to the discovery of other blood antigen systems such as Rh, MnSs, and the Lewis systems. White blood cell antigen systems were also discovered. From these discoveries came the forensic typing of blood to help distinguish one individual from another (Nuland, 1988).

After Watson and Crick discovered the structure and functions of DNA in the early 1950's, it wasn't until Sir Alec Jeffries developed the first forensic DNA typing method, which he coined, regrettably, "DNA fingerprinting," in 1984 that forensic DNA technology was born. The work of Kary Mullis in the 1980's led to the discovery of the polymerase chain reaction (PCR), the way our bodies reproduce DNA. This discovery led to Mullis being awarded the 1993 Nobel Prize in Chemistry (Malmstrom, 1997).

In the early part of the twentieth century, Goddard popularized the comparison microscope, which is two standard microscopes joined by an optical bridge. This tool revolutionized the comparison of bullets, cartridges, toolmarks, hairs, and fibers. Microscopy is the mainstay of forensic science laboratories and includes newer methods, such as the scanning electron microscope.

Several professional forensic organizations exist, and membership can convey many benefits, not the least of which is meeting other forensic scientists and developing contacts. Many of these organizations have journals associated with them. Refer to "On the Web: Professional forensic organizations" for more information about these groups.

FORENSIC SCIENCE LABORATORY ORGANIZATION AND SERVICES

Although it may seem contradictory, there is no one structure for the organization of a forensic science laboratory. The organization of laboratories varies by jurisdiction, agency, and history. The variation becomes more pronounced when laboratories in the United States are compared with those in other countries. The examinations and services that a forensic science laboratory offers also vary, depending on budget, personnel, equipment, and crime statistics. This section will focus on laboratories in the United States and answer two questions: First, how is the laboratory administered, and second, what services does the laboratory provide?

ON THE WEB: Professional forensic organizations

Some of these organizations have regional groups affiliated with them; check the websites for contact information.

American Academy of Forensic Sciences
www.aafs.org

International Association for Identification
www.theiai.org

Association of Firearms and Toolmarks Examiners
www.afte.org

American Society of Questioned Document Examiners
www.asqde.org

Society of Forensic Toxicologists
www.soft.org

FORENSIC SCIENCE LABORATORY ADMINISTRATION

The vast majority of forensic science laboratories in the United States are public; that is, they are financed and operated by a federal, state, or local unit of government. They number something over 470 today. There are also an undetermined number of private forensic science laboratories, and some estimates put this number at 50–100.

Private Forensic Science Laboratories

Most private laboratories serve a niche by performing only one or two examinations, such as drugs, toxicology, or questioned documents. Many are "one-person" operations, often a retired forensic scientist providing services in the specialties practiced when employed in a public laboratory. Today a significant number of the private laboratories are devoted to DNA analysis in either criminal cases or in the civil area, chiefly in paternity testing. Private laboratories serve a necessary function in our criminal justice system in that they are able to provide forensic science services directly to persons accused of crimes. Most public laboratories can provide forensic science services only to police or other law enforcement departments and will not analyze evidence requested by an accused person except under a court order. Some public laboratories, however, will accept evidence from private citizens, and the fee is subsidized by the jurisdiction where the laboratory operates.

Public Forensic Science Laboratories

Public forensic science laboratories are administered and financed by a unit of government that varies with the jurisdiction. Different states have different models, and the federal government has its own collection of laboratories. Laboratories administered by the federal government, typical state systems, and local laboratories will be discussed separately.

FEDERAL GOVERNMENT FORENSIC SCIENCE LABORATORIES

When most people think of federal forensic science laboratories, the only name that usually pops up is the Federal Bureau of Investigation (FBI) laboratory. While this is certainly the most famous forensic science laboratory in the United States if not the world, it is far from being the only one in the federal government. A surprising number and type of laboratories are administered by several departments of the U.S. government.

THE DEPARTMENT OF JUSTICE

The Federal Bureau of Investigation (FBI) is a unit of the Department of Justice. It has one laboratory, in Quantico, Virginia, near its training academy. It also maintains a research laboratory, the Forensic Science Research and Training Center in Quantico. The FBI laboratory supports investigative efforts of the FBI and will, upon request, analyze certain types of evidence for state and local law enforcement agencies and forensic science laboratories.

The Drug Enforcement Administration (DEA) is responsible for investigating major illicit drug enterprises and helping interdict shipments of drugs from other countries. In support of this function, the DEA maintains a network of seven drug laboratories throughout the United States. They are in Washington, D.C., Miami, Chicago, Dallas, San Francisco, New York, and San Diego. There is also a research and support laboratory, the Special Testing and Research Laboratory, in Chantilly, Virginia. The DEA laboratories support not only the efforts of the DEA investigators but will also work with local law enforcement in joint operations.

The Bureau of Alcohol, Tobacco and Firearms (BATF) has three regional laboratories: Washington, D.C., Atlanta, and San Francisco. There is also a fire research laboratory in conjunction with the Washington, D.C., laboratory. Although the primary responsibilities of the BATF are embodied in its name—the regulation of alcohol, tobacco, and firearms—the laboratories have particular expertise in fire scene analysis and explosives. They also have the capability of questioned document and fingerprint analyses as well as trace evidence.

THE DEPARTMENT OF THE TREASURY

Although one wouldn't usually think of looking at the Treasury Department for forensic science laboratories, it has the Internal Revenue Service Laboratory in Chicago. This laboratory specializes in the various disciplines of questioned document analysis including inks and papers. A good deal of its work includes authentication of signatures on tax returns, fraudulent documentation relating to taxation, and other forms of fraud in the name of avoiding federal taxation.

THE DEPARTMENT OF HOMELAND SECURITY

The relatively new Department of Homeland Security (DHS) acquired the Secret Service Laboratory in Washington, D.C. This laboratory has two major functions. The first is in the area of counterfeiting and fraud—including counterfeit currency, fraudulent credit cards, and related documents. One of the world's largest libraries of ink standards is located here, and questioned document analysis is also a major function. The second major component of the Secret Service Laboratory supports its function of executive protection. This laboratory engages in research and development of countermeasures and protection of the president and other officials.

THE DEPARTMENT OF THE INTERIOR

The Department of the Interior has a unique laboratory: The U.S. Fish and Wildlife Service operates a forensic science laboratory in Ashland, Oregon. One of the few animal-oriented forensic science laboratories in the world, its mission is to support the efforts of the Service's investigators who patrol the national parks. Among other duties, these agents apprehend poachers and people who kill or injure animals on the endangered species list. Thus, the laboratory does many examinations involving animals and has particular expertise in the identification of hooves, hairs, feathers, bones, and other animal tissues. The laboratory also provides consulting services for other countries in their efforts to track people who traffic in animal parts such as bear gall (in certain parts of Asia bear gallbladders are thought to improve sexual potency) and elephant ivory. The laboratory maintains some of the most sophisticated instrumentation and has some of the world's leading experts in animal forensic science.

THE U.S. POSTAL SERVICE

Although the Postal Service is not strictly a federal agency nor is it managed by one, it is considered to be a quasi-federal agency. The service maintains a laboratory in the Washington, D.C., area that supports the Service's efforts to combat postal fraud. This effort mainly involves questioned document analysis although the laboratory also has fingerprint and trace evidence capabilities.

Additional federal laboratories include the Department of Defense's Army Criminal Investigation Division Laboratory in Georgia;

the Navy Drug laboratories in Norfolk, Long Beach, Honolulu, and Japan; and Air Force Drug Laboratory in San Antonio.

STATE AND LOCAL FORENSIC SCIENCE LABORATORIES

Every state in the United States maintains at least one forensic science laboratory. Historically, no nationwide effort has been made to standardize laboratory organization or function, so each state has developed a system that meets its particular needs. These forensic science laboratories have arisen from two sources. The most prevalent is law enforcement: The majority of forensic science laboratories are administered by a unit of a state or local police or other law enforcement agency. The other source of forensic science laboratories is health departments or other scientific agencies. In Michigan, for example, the modern Michigan State Police Laboratory system developed from the merger of a smaller MSP laboratory and the state Health Department laboratory. The Michigan State Police laboratory had expertise in firearms, questioned documents, and fingerprints, whereas the Health Department laboratory had expertise in drugs, toxicology, and trace evidence, such as hairs and fibers. The state police in Michigan now operate a network of seven regional laboratories. In all states a statewide laboratory or laboratory system is operated by the state police, state department of justice, or as an independent state laboratory system, such as in Virginia. In California, for example, the state department of justice operates an extensive network of state-financed laboratories, whereas West Virginia has a single laboratory that serves the whole state.

Besides the statewide laboratory system, most states also have one or more laboratories operated by a local governmental unit. For example, in Maryland some counties have laboratories under the jurisdiction of the county police department separate from the state system. In Texas, some police or sheriff's departments in major cities operate city laboratories, as in Fort Worth, and in California, Los Angeles has a county and a city laboratory. In Michigan, the Detroit City Police Department has its own forensic science laboratory, but the rest of Wayne County surrounding Detroit is serviced by the state police laboratories. This patchwork of political, geographical, and historical jurisdictions can be confusing but is usually maintained because of real societal needs, such as population levels, crime rates, and geography.

FORENSIC SCIENCE LABORATORY SERVICES

Forensic science laboratories offer different levels of service. In a statewide system, for example, at least one laboratory will offer a full range of forensic science services (typically at the headquarters laboratory), whereas the regional laboratories may offer only limited services (say, fingerprints and drugs) and then send the other evidence to the headquarters laboratory. In Michigan, for example, the headquarters laboratory of the Michigan State Police in Lansing offers a complete set of forensic science services, and the other six regional laboratories offer the more high-volume services. This section discusses the capabilities of a typical full-service forensic laboratory. Keep in mind that the designation of "full service" may mean different things in different states: A state may not offer gunshot residue analysis in even its best-equipped laboratory but would still describe it as "full service."

STANDARD LABORATORY SERVICES

EVIDENCE INTAKE

All forensic science laboratories have a system for receiving evidence. The laboratory may have one employee assigned to manage this unit full time and may employ several additional people, depending on the volume of evidence and casework. The evidence intake unit will have a secured area for storing evidence, the size of which depends again on the volume of work; it may be a room or a warehouse. A police officer or investigator will bring evidence to the laboratory and fill out a form that describes the evidence and the types of examinations requested. A unique laboratory number will be assigned to the case, and each item of evidence will be labeled with this number, along with other identifying information, such as the item number. This continues the **chain of custody** of the evidence, the documentation of the location of evidence from the time it is obtained to the time it is presented in court. The chain of custody begins at the crime scene when the evidence is collected. The job of the evidence intake unit is like that of inventory control for a business.

Modern intake systems use computerized systems that generate barcodes that are placed on each item of evidence or its container. The barcode is scanned by each unit of the laboratory that takes possession of that item so the evidence can be easily traced by computer as it makes its way through the laboratory. Paperwork accompanies the evidence, either in hard copy or electronic form, as each analyst signs or accepts possession of the evidence.

ANALYTICAL SECTIONS

Once the evidence has been received by the laboratory, it will be assigned to one or more forensic units for analysis; each unit, in turn, assigns a scientist to take charge of the evidence, and its analysis. Many times more than one scientist will be asked to analyze an item of evidence, and then arrangements must be made to get the evidence to each scientist in a logical order. For example, a gun may have to be test fired, but also may contain fingerprints and suspected blood. The examinations must be performed in an order that will not disrupt or destroy any of the evidence on the gun. An example of a full-service laboratory analytical sections might contain

Photography
Biology/DNA
Firearms and Toolmarks
Footwear and Tire Treads
Questioned Documents
Friction Ridge Analysis (fingerprints)
Chemistry/Illicit Drugs
Toxicology
Trace Evidence

What all these analyses have in common is that a microscope is used in some fashion because the items examined are small. In some laboratories, one forensic scientist may be certified to examine several of these evidence types; in larger laboratories that have the luxury of specialization, a scientist may examine only one or two.

OTHER LABORATORY SERVICES

Some laboratories offer services in addition to those listed in the preceding section depending on the need for such services and the availability of qualified scientists. Laboratories that have an occasional need for these services may submit the evidence to the FBI laboratory, a private laboratory, or a local specialist. Specialists areas include polygraph (so-called lie detectors), voiceprint and speaker identification, bloodstain pattern analysis, entomology, odontology, and anthropology.

ADMINISTRATIVE ISSUES WITH FORENSIC SCIENCE LABORATORIES

Forensic science laboratories are faced with ever-increasing demands and workloads. Courts have come to expect more and higher quality expert testimony and speedier turnaround times from forensic laboratories. More scrutiny also has been placed on the forensic science systems around the world by the public, the media, and government officials. This has caused a number of administrative issues to assume greater importance, two of the major ones are accountability and access.

Accountability

Virtually every hospital and clinic in the United States has to be accredited by a responsible agency. Environmental and pharmaceutical companies, among others, also have accreditation procedures. Thus, it might come as a surprise to many people to find out that there is no mandatory accreditation process for the nation's forensic science laboratories. Considering the impact that forensic science can have on trials, this is a disturbing situation.

Arguably, the major reason for this state of affairs is that forensic science laboratories historically have arisen within police agencies whose focus is not science. Movements in the United States and worldwide to accredit forensic science laboratories have had some success: Some states, such as New York, make it mandatory for forensic laboratories to be accredited, but many seek accreditation voluntarily. In the United States, the **American Society of Crime Laboratory Directors (ASCLD)** has formed a subsidiary, the **ASCLD Laboratory Accreditation Board (ASCLD-LAB)**, which provides accreditation services for public and private laboratories worldwide. The accreditation process is rigorous and involves a self-study process, an extensive checklist of requirements, and an on-site evaluation by trained members of the accrediting board. It should be stressed that ASCLD accreditation does not directly address the competence of the individual forensic scientists who work at the laboratory. It does mean that the laboratory meets certain minimum criteria for the physical plant (facilities, heating cooling, etc.), security, training, equipment, quality assurance and control, and other essential features. Reaccreditation is required every five years to maintain the laboratory's status.

Standards play a major role in helping laboratories become accredited. The **American Society for Testing and Materials, International (ASTM)** publishes voluntary, consensus standards for a wide variety of sciences, including forensic science (Committee E30, Volume 14.02). They are voluntary because individuals and agencies independently choose to adhere to them. The standards are written through a consensus process, meaning that everyone on the subcommittee, committee, and the Society has had a chance to read, comment, and vote on the standard.

Other accreditation processes, such as the **International Organization for Standardization (ISO)**, are gaining headway, and it is hoped that someday soon forensic science laboratory accreditation will become mandatory, and every laboratory will become accredited. More information about forensic standards and accrediting agencies can be found on the websites listed in "On the Web: Accreditation."

Access to Laboratory Services

The majority of forensic science laboratories in the United States are funded by the public and administered by a unit of federal, state, or local government. These laboratories support the law enforcement functions of the parent agency or the government. Police officers, detectives, crime scene investigators, and prosecutors generally have open access to the services of the laboratory, including expert testimony by its forensic sciences at no cost to the agency. Considering that the public pays for these services, it might seem obvious that a person accused of a crime should also have access to these services. That, however, is not the case. Very few public forensic science laboratories will permit accused persons access to forensic science services even if that person is willing and able to pay for them.

ON THE WEB:
Accreditation

American Society of Crime Laboratory Directors (ASCLD)   **www.ascld.org**

ASCLD Laboratory Accreditation Board (ASCLD-LAB)   **www.ascld-lab.org**

American Society for Testing and Materials, International (ASTM) **www.astm.org**

International Standards Organization (ISO) **www.iso.org**

How then do criminal defendants gain access to forensic science services? The options are limited. Private laboratories serve defendants (and anyone, really), but the cost is generally high and often courts will not authorize enough money for indigent defendants to cover the costs of analysis and testimony. If an accused person has a public defender for an attorney, most public defenders' offices do not have sufficient funds to pay for analyses of evidence. Even people willing and able to pay may not have a qualified forensic science laboratory available. This results is an imbalance in the resources available to the prosecution and defense in a criminal case. It is interesting to note that the British justice system, faced with this same problem, is addressing this imbalance by requiring public laboratories to charge all agencies for scientific analysis and testimony. This solution has not been universally embraced (see "In More Detail: Public or Private?"), however, and only time will tell if it succeeds or creates new problems.

THE FORENSIC SCIENTIST

Forensic scientists have two major duties: performing scientific analysis of evidence and offering expert testimony in criminal and civil proceedings. There are sometimes other responsibilities such as offering training in evidence collection and preservation, doing research, or performing other studies such as validation procedures for new methods, but the major duties take up most of the forensic scientists' time.

EDUCATION AND TRAINING OF FORENSIC SCIENTISTS

Science is the heart of forensic science. Court decisions, such as *Daubert v. Merrill Dow* (1993), have reinforced this fact. A forensic scientist must be well versed in the methods and requirements of good science in general and in the specific techniques used in the particular disciplines being practiced. Additionally, the forensic scientist must be familiar with the rules of evidence and court procedures in the relevant jurisdictions. The knowledge, skills, and aptitudes needed in these areas are gained by a combination of formal education, formal training, and experience.

IN MORE DETAIL:

Public or private?

The majority of forensic science laboratories in the United States are public; that is, they are funded by and operated within government agencies, such as local or state police departments. Police and prosecutors submit evidence for analysis, and they don't need to pay for this service. The costs are covered by taxes. A few private commercial laboratories charge for their services. Anyone with sufficient money can walk in and request examinations.

In the United Kingdom, the Forensic Science Service (FSS) is a quasi-governmental forensic laboratory. It is a governmental agency, but it also charges police and agencies for its services. If it gets to make a profit, it keeps that money: In 2003, the FSS made a profit of £10 million (US $17,475,937 in 2003), which was reinvested in public service. A movement to fully privatize the FSS is afoot, and forensic agencies around the world are watching to see what happens. The implications strike to the core of what constitutes forensic investigations and analyses.

Police need accurate information and they need it quickly. A profit-oriented FSS would, in essence, charge by the hour and the examination. Thus, dollars and not necessarily solution would drive the investigation. At £1375 (about US $2400 in 2003) per DNA sample, an investigator might think twice about indiscriminately submitting evidence that would clog up the laboratory. The cost should lead police to investigate, collect, and submit smarter rather than out of fear.

But the police might also find all kinds of ways to cut costs. What if, as an example, five cigarette butts were found at a crime scene, and the victim was known not to smoke. The most effective method for determining if the suspect's DNA is on the cigarette butts (offering probable cause to arrest him) is to test them all. This is costly—almost $10,000 by the prices quoted previously. The cost-effective approach, however, would be to test one and see if a result is obtained; if that one didn't yield a result, the next one could be tried, and

so on. This testing could take, literally, months and give the suspect time to flee or, worse, commit more crimes. This has been referred to as the "staging problem," in which evidence is analyzed in stages.

Selectivity and screening can also be problems. In the case of a rape, a vaginal swab is collected to check for the rapist's semen but, for the sake of argument, let's say the results are inconclusive. Paying a scientist to pore over the victim's bedding in search of semen stains or other trace evidence wouldn't be cheap. If the police are pressured to save money, the cheap-and-easy method wins—although it isn't effective.

Another method to reduce costs would be to quickly screen evidence and not continue an analysis if the initial presumptive test is negative. Sensitivity of presumptive tests is an issue, however; so-called screening tests may not be as sensitive as other "in lab" tests. Major cases or those with a high media profile would probably get nearly unlimited budgets, but what about the "average" rapes, burglaries, and death investigations? DNA isn't always found or effective; in some cases, like spousal rape, it can't prove a thing. Hours and hours of searching for microscopic evidence may not be cost effective but it can be *results* effective. And in some instances, it could be cost effective (see Table 1–1, Cost Effectiveness of Microscopical Hair Examinations).

For example, in three hypothetical cases involving hairs in three laboratories with differing sampling protocols, the laboratory employing microscopical hair examinations has a more efficient and cost-effective supply chain than laboratories that do not. Therefore, while hair examiners may be considered to be a "waste of time and money" by some laboratory managers, they, in fact, can save both time and money.

Regardless of how the FSS proceeds with privatization, the argument about the best way to provide proper access to forensic sciences for all citizens has yet to be concluded.

Table 1.1. Cost Effectiveness of Microscopical Hair Examination.

($1500 PER SAMPLE FOR MTDNA)	LAB #1 ANALYZES DNA ON ALL Q HAIRS, PLUS ALL K SAMPLES		LAB #2 ANALYZES ONLY TWO HAIRS FROM VICTIM'S UNDERWEAR AND SAK, PLUS ALL K SAMPLES		LAB #3 ANALYZES ONLY ONE HAIR AFTER MICROSCOPICAL EXAM FROM EACH ITEM, PLUS 2 K SAMPLES	
	COST	ACCURACY	COST	ACCURACY	COST	ACCURACY*
Small Case 2 positive/5 total 2 Known samples	$10,500	100%	$9,000	80%	$6,000	90%
Medium Case 5 positive/15 total 4 Known samples	$36,000	100%	$12,000	27%	$9,000	90%
Large Case 15 positive/50 total 6 Known samples	$84,000	100%	$15,000	8%	$12,000	90%

*Based on rates published in Houck, M.M. and Budowle, B. *JFS*, V47, N5, 2002; these rates are not applicable to any one particular case, set of samples, or examiner.

Table taken from Houck, M.M. & Walbridge, S. (2004, February). *Could Have, Should Have, Would Have: The Utility of Trace Evidence*. Presented at the American Academy of Forensic Sciences annual meeting, Dallas, TX.

Education

Historically, forensic scientists were recruited from the ranks of chemistry or biology majors in college. Little or no education was provided in the forensic sciences themselves—all of that was learned on the job. Since the middle of the twentieth century, undergraduate and then graduate programs in forensic science have been offered by a handful of colleges and universities in the United States. The early bachelor's degree programs provided a strong chemical, mathematical, biological, and physical science background, coupled with applied laboratory experience in the analysis of evidence, with classes in law and criminal procedure mixed in. These programs also offered opportunities for a practicum in a functioning forensic science laboratory to see how science was applied in forensic laboratories. The American Academy of Forensic Sciences website (**www.aafs.org**) lists many programs that offer a bachelor's degree with some level of forensic emphasis.

In the past 20 years or so, graduate degrees, particularly at the master's level, have become the norm. About two dozen graduate (MA,

MS, or PhD) programs in the United States are listed on the American Academy of Forensic Sciences website. They typically require a bachelor's degree in a science and then teach the applications of the science to forensic work, as well as relevant aspects of law, criminal investigation, and criminal justice classes. A research component is also generally required. For more information about forensic science educational accreditation standards, see "In More Detail: TWGED and FEPAC."

IN MORE DETAIL:
TWGED and FEPAC

In *Forensic Science Review of Status and Needs*, a published report from the National Institute of Justice in 1999, it was noted that the educational and training needs

> of the forensic community are immense. Training of newcomers to the field, as well as providing continuing education for seasoned professionals, is vital to ensuring that crime laboratories deliver the best possible service to the criminal justice system. Forensic scientists must stay up to date as new technology, equipment, methods, and techniques are developed. While training programs exist in a variety of forms, there is need to broaden their scope and build on existing resources. (5)

Forensic Science Review of Status and Needs made a number of recommendations, including seeking mechanisms for

- Accreditation/certification of forensic academic training programs/institutions,
- Setting national consensus standards of education in the forensic sciences, and
- Ensuring that all forensic scientists have professional orientations to the field, formal quality-assurance training, and expert witness training.

The **Technical Working Group on Education and Training in Forensic Science (TWGED)** was created in response to the needs expressed by the justice system, including the forensic science and law enforcement communities, to establish models for training and education in forensic science. West Virginia University, in conjunction with the National Institute of Justice, sponsored TWGED, which was made up of over 50 forensic scientists, educators, laboratory directors, and professionals. TWGED drafted a guide addressing qualifications for a career in forensic science, undergraduate curriculum in forensic science, graduate education in forensic science, training and continuing education, and forensic science careers outside the traditional forensic science laboratory.

Seeing this as an opportunity, the **American Academy of Forensic Sciences (AAFS)** initiated the **Forensic Science Education Program Accreditation Commission (FEPAC)** as a standing committee of the Academy. The FEPAC drafted accreditation *standards* for forensic education programs based on the TWGED guidelines. The mission of the FEPAC is to maintain and to enhance the quality of forensic science education through a formal evaluation and recognition of college-level academic programs. The primary function of the Commission is to develop and to maintain standards and to administer an accreditation program that recognizes and distinguishes high-quality undergraduate and graduate forensic science programs.

A pilot program of six forensic educational institutions was run in 2003; as of September 2005, nine programs have been accredited.

For more information about FEPAC, visit **www.ncjrs.gov** and **www.aafs.org**.

Educational programs are not, however, designed to provide training so that graduates can start working cases on their first day in a forensic science laboratory.

Formal Training

Once scientists are employed by a forensic science laboratory, they begin formal training. New scientists are normally hired as specialists; they will learn how to analyze evidence in one or a group of related areas. Thus, someone may be hired as a drug analyst, a trace evidence analyst, or a firearms examiner. Training requires a period of apprenticeship in which the newly hired scientist works closely with an experienced scientist. The length of time for training varies widely with the discipline and the laboratory. For example, a drug chemist may train for three to six months before taking cases, while a DNA analyst may train for one to two years, and a questioned document examiner may spend up to three years in apprenticeship. Training usually involves mock case work as well as assisting in real cases. Ideally, it will also include proficiency testing at intervals and mock trials at the end of the training.

On-the-Job Training: Experience

Some people say that you don't really learn to drive a car until after you get your license. That is when you begin to experience all of the everyday driving situations that make you a good driver—or a bad one. The same could be said for forensic science. After you have been educated and trained and begin to work your own cases, only then do you learn how to be an effective forensic scientist. You learn how to manage time and resources, and you experience the pressure of testifying in court. You learn "how to hurry up and wait" to testify, how to handle the media (or not), and how to deal with harried attorneys. These are aspects of the career that are difficult to convey to someone who hasn't experienced them.

ANALYSIS OF EVIDENCE

The reason someone wants to become a forensic scientist is to analyze evidence. The science and method of this process fills much of the rest of this book. But besides the routine analysis of evidence, many important aspects other than science affect how evidence is analyzed:

- *Chain of custody.* The forensic scientist must be constantly aware of the requirements of the chain of custody. Evidence can be rendered inadmissible if the chain of custody is not properly constructed and maintained.

- *Turnaround time.* Federal and state "speedy trial" laws require that an accused person be brought to trial within a specified window of time after arrest; this is usually 180 days but may vary with the jurisdiction. If the forensic science laboratory cannot analyze and report evidence out in a timely manner, the accused may be released for failure of the government to provide a speedy trial.

- *Preservation and spoilage.* Forensic scientists have a duty to preserve as much of the evidence as is practical in each case and to ensure that the evidence is not spoiled or ruined. In some cases, so little evidence exists that there is only one chance for analysis. In such cases, the prosecutor and defense attorney should be apprised before the analysis takes place.

- *Sampling.* In many cases there is so much evidence that sampling becomes an issue. This often happens with large drug cases that may have hundreds or thousands of similar exhibits; it can also be true of blood stains, fibers, or any type of evidence. The opposite may also be true: insufficient sample for complete or repeat analysis. Finally, in some cases any type of analysis is destructive and there is no opportunity for reanalysis.

- *Reports.* Every laboratory has protocols for writing laboratory reports, but a surprising lack of uniformity exists from laboratory to laboratory. Some laboratories mandate complete reports for each case, whereas others have bare-bones reports with a minimum of description and explanation. Reports of forensic science analysis are scientific reports and should be complete like any other scientific report.

EXPERT TESTIMONY

Being a competent analytical scientist is only half the battle in a forensic science laboratory. The forensic scientist must also be able to explain his or her findings to a judge or jury in a court of law. This is one of the key factors that distinguish careers in forensic science from those in other sciences.

CHAPTER SUMMARY

There are a number of definitions of an expert. For forensic science purposes, an expert may be thought of as a person who possesses a combination of knowledge, skills, and abilities in a particular area that permit him or her to draw inferences from facts that the average person would not be competent to do. In short, an expert knows more about something than the average person and has the credentials to prove it. An expert does not have to possess a PhD. Many experts have accumulated expertise over many years of experience and may not have much education. For example, suppose that a person is killed while driving his car because the brakes failed, and he crashed into a tree. If an average group of people were to inspect the brakes of the car, they would not be competent to determine why the brakes failed or even if they did. This would require the services of an expert mechanic to examine the brake system and then make conclusions about if, why, and how the brakes failed. A difference exists, however, between an expert and a forensic scientist—a mechanic is not a forensic scientist. That difference is what this book is about.

Forensic science is a wide-ranging field with a rich, if untapped, history. In many ways, the discipline has suffered from that lack of historical knowledge and our ignorance of it—not knowing the past dooms one to repeat it, and so forth. Forensic science also occupies what may be a unique niche between law enforcement and the courts. The pressures from either side color much of what is accepted as forensic science and yet practitioners must adhere to the tenets of science. As a growth industry, one would be hard pressed to find another discipline with so much rich material to mine or such promise in the dazzling future of technology.

Test Your Knowledge

1. What is forensic science?

2. How is forensic science different from other sciences, such as biology and physics?

3. What does the word "forensic" mean?

4. Name four disciplines within the forensic sciences.

5. What are the two kinds of forensic science laboratories?

6. What is the main difference between these two types of forensic laboratories?

7. Name three federal agencies that have forensic science laboratories.

8. What is a chain of custody?

9. Who accredits forensic science laboratories?

10. What is TWGED?

11. What is FEPAC?

12. What is ASCLD?

13. What is ASCLD-LAB?

14. To whom are forensic laboratories accountable?

15. What is a forensic anthropologist?

16. Who was Bertillion?

17. What laboratory analyzes wildlife samples in criminal cases?

18. Why would the Department of Defense need forensic laboratories?

19. What is a forensic toxicologist? How would this differ from a regular toxicologist?

20. What is an expert?

Consider This . . .

1. Why do you think a mechanic is or is not a forensic scientist?

2. Is privatization a good way to ensure every citizen has access to forensic science services?

3. Why is formal training necessary when someone is hired by a forensic science laboratory?

BIBLIOGRAPHY

Beavan, C. (2001). *Fingerprints: The origins of crime detection and the murder case that launched forensic science.* New York: Hyperion.

Bodziak, W.J. (1995). *Footwear impression evidence.* Boca Raton, FL: CRC Press.

Butler, J. (2001). *Forensic DNA typing: Biology and technology behind STR markers.* San Diego, CA: Academic Press.

Casey, E. (Ed.) (2002). *Handbook of computer crime investigation.* San Diego, CA: Academic Press.

Daubert v. Merrill Dow Pharmaceuticals, Inc. (1993). 509 U.S. 579, 113 S.Ct 2786, 125 L. Ed2d 469 (1993).

Douglas, J.E., Burgess, A.W., Burgess, A.G., & Ressler, R. (1992). *Crime classification manual,* New York: Lexington Books.

Fisher, B.A.J. (2004). *Techniques of crime scene investigation,* 7th ed. Boca Raton, FL: CRC Press.

Gerber, S.M. (1983). *Chemistry and crime.* Washington, D.C.: American Chemical Society.

Goff, M.L. (2000). *A fly for the prosecution.* Cambridge, MA: Harvard University Press.

Hilton, O. (1982). *Scientific examination of questioned documents.* New York: Elsevier Science Publishing Company.

Houck, M.M. (Ed.) (2000). *Mute witnesses: Trace evidence analysis.* San Diego, CA: Academic Press.

Houck, M.M. (Ed.) (2003). *Trace evidence analysis: More cases from mute witnesses.* San Diego, CA: Academic Press.

Malmstrom, B.G. (Ed.) (1997). *Nobel lectures, chemistry 1991–1995.* Singapore: World Scientific Publishing Co.

Mohr, J.C. (1993). *Doctors and the law: Medical jurisprudence in nineteenth-century America.* Baltimore, MD: The Johns Hopkins University Press.

Nuland, S.B. (1988). *Doctors and the biography of medicine.* New York City, Alfred A. Knopf, Inc.

National Institute of Justice (1999). *Forensic science: review of status and needs,* Washington, D.C., Department of Justice, Office of Justice Programs, National Institute of Justice.

Oxford English Dictionary (2005). Oxford, United Kingdom. Oxford University Press.

Rule, A. (1980). *The stranger beside me.* New York, W.W. Norton.

Taylor, K.T. (2001). *Forensic art and illustration.* Boca Raton, FL: CRC Press.

Thorwald, J. (1964). *The century of the detective.* New York, Harcourt, Brace & World.

Ubelaker, D. (1992). *Bones: A forensic detective's casebook.* New York: Edward Burlingame Books.

Wilson, C., & Wilson, D. (2003). *Written in blood: A history of forensic detection.* New York, Carroll and Graf Publishers.

Crime Scene Investigation

KEY TERMS

Artifact	Datum	Matrix
Blood-borne pathogens	Evidence	Organic or environmental remains
Chain of custody	Feature	
Context	Material Safety Data Sheet (MSDS)	Provenience
Crime scene investigator (CSI)		Universal Precautions

INTRODUCTION

An argument becomes heated, old psychological wounds are reopened, lingering hate resurfaces, there is a struggle, finally a heavy ceramic vase is hefted and crashes down on someone's head—and a crime is committed. Suddenly, normal household items are transformed into evidence, their importance changed forever. Processing a crime scene, collecting these items, this evidence, appears deceptively simple at first. But this perception comes from the investigations we read in novels and see on television and the movies in which we *know* what's important (the camera lingers on the crucial evidence), *who* the short list of suspects are (they wouldn't be on the show if they weren't involved), and that the crime will be wrapped up in an hour (with commercials). We see the murder weapon being collected and bagged, and the next time it appears, it's presented in court to the witness! The reality of crime scene processing is more involved and detailed than what we read or see in the media.

Without a crime scene, nothing would happen in a forensic laboratory. The scene of the crime is the center of the forensic world, where everything starts, and the foundation upon which all subsequent analyses are based. Normal household items are transformed from the mundane to that special category called "evidence." The importance of a properly processed crime scene cannot be overemphasized—and yet, it is where devastating mistakes occur that affect an entire case. Many agencies have recognized the significance of the crime scene and employ specially and extensively trained personnel to process them. The processing of a crime scene is a method of "careful destruction": It is a one-way street, and you can never go back and undo an action. Standard operating procedures and protocols guide the crime scene investigator, providing a framework for comprehensive and accurate evidence collection, documentation, and transmittal to the laboratory.

This chapter will focus on the scene itself and the collection of evidence. Because the nature of evidence and the way things become evidence can be complex, evidence itself will be discussed separately and in depth in the next chapter.

The scene of the crime is the focus of the forensic laboratory. It is where everything starts and is the foundation upon which all forensic analyses are based. The importance of a properly processed crime scene cannot be overemphasized—and yet, it is where even simple mistakes occur that affect an entire case and many people's lives.

OF ARTIFACTS AND EVIDENCE

The goal of an archaeological excavation is to carefully collect and record all the available information about a prehistoric or historic site of human activity. The goal of processing a crime scene is to collect and preserve evidence for later analysis and reporting. What these two processes have in common is that they are one-way: an action is taken, an artifact moved, a piece of evidence collected, you can't undo it any more than you can unring a bell. Crime scenes and archaeological sites are made up of the physical remains of past human activity and, in a sense, are snapshots of the "leftovers" of a completed process. When a scene is processed or a site dug, the procedure is one of "careful destruction": The scene or site will never exist in exactly the same way as it did before the process started. All the information, the relationships, the **context** of the items must be documented as they are destroyed to allow for some level of reconstruction in the laboratory or museum. It is an awesome responsibility to work a scene or excavate a site, and neither should be taken lightly.

Several technical terms that are used in archaeology may be of use in crime scene processing. The first is the idea of a **datum**, a fixed reference point for all three-dimensional measurements. The datum should be something permanent, or nearly so, like a light switch (pick a corner!), a tree, or a post. If no datum easily suggests itself, an artificial one, such as a post, nail, or mark, can be made. Ultimately, all measurements must be able to be referenced to the datum.

Other terms that can be borrowed from archaeology suggest the nature of what is found. An **artifact** is a human-made or modified portable object. A **feature** is a non-portable artifact, such as a fire pit, a house, or a garden. **Organic** or **environmental remains** (non-artifactual) are natural remnants that nonetheless indicate human activity, such as animal bones or plant remains but also soils and sediments. An archaeological site, then, can be thought of as a place where artifacts, features, and organic remains are found together. Their location in relation to each other sets the internal context of the site. In order to reconstruct this context once the site or scene has been processed, it is necessary to locate the position of each item. Thus, the **provenience** is the origin and derivation of an item in three-dimensional space, in relation to a datum and other items. When an artifact is uncovered at a site, it is measured to the reference points for that excavation unit, including its depth. A similar process occurs at a crime scene when evidence is located. As the noted archaeologists Colin Renfrew and Paul Bahn (2000) put it,

> In order to reconstruct past human activity at a site it is crucially important to understand the *context* of a find, whether artifact, feature, structure or organic remain. A find's context consists of its immediate **matrix** (the material surrounding it), its provenience (horizontal and vertical position within the matrix), and its association with other finds (occurrence together with other archaeological remains, usually in the same matrix) (50).

The similarities between archaeology and crime scene processing are numerous and deep. Serious crime scene students would do well to study archaeological methods to enhance their forensic skills.

Evidence can be defined as information, whether personal testimony, documents, or material objects, that is given in a legal investigation to make a fact or proposition more or less likely. Most of the evidence discussed in this chapter relates to physical evidence—that is, things involved in the commission of the crime under investigation. The nature of evidence will be discussed in more depth in the next chapter.

CRIME SCENE INVESTIGATION

As Paul Kirk, the noted Californian forensic science pioneer described it, forensic science is interested in the unlikely and the unusual (Kirk, 1963, p. 368). This is certainly true of crime scenes: Each one is unique. The crime committed, the location, the items used, the people involved—all vary from scene to scene. Although nearly every police and forensic agency has written protocols about processing a crime scene, these protocols may be trumped by the circumstances of the crime scene. As Barry Fisher (2000), Director of the Los Angeles County Sheriff's Department Crime Laboratory, noted, "There are few absolute rules in crime scene investigations . . . There are always cases where guidelines cannot be followed . . . Situations demand that investigators be flexible and creative when necessary" (2004; page 49). That is, **crime scene investigators** (or **CSIs**, for short) must know and follow their agency's protocols but must be ready to improvise, within accepted limits, to protect and/or preserve evidence, as shown in the example in Figure 2.1.

FIRST ON THE SCENE

The success of any crime scene investigation depends in large part on the actions taken in the first few minutes after the first officer or CSI arrives (or just FO, for short). This sounds odd, to be sure. "How can a few minutes matter to a crime scene that's just been sitting there for hours or days?" you may ask. But crime scenes are a complex mix of static and dynamic information, a scene fixed in time like a photograph but slowly degrading, much like poorly archived historical photographs. The majority of the physical evidence will be generated by the processing of the crime scene, and the relationships among the people, places, and things (the context) will tell the story of what happened. Remember, facts alone are not sufficient—by themselves, they explain nothing. Facts must be interpreted in light of the circumstances or context surrounding the crime. Once an item is moved,

Figure 2.1. *This little fellow, along with three of its littermates, was at the scene of a triple homicide in the northwest United States. Before they could be rounded up, they tracked blood around the crime scene; this photo was taken as a reminder. The cats were adopted by various agency personnel (Anonymous by submitter's request).*

it can never be placed back exactly as it was: The context is diturbed, and the subsequent interpretation may be biased and inaccurate.

The primary task of the FO at a crime scene is to *secure the scene and prevent destruction or alteration of the critical and sometimes fragile context of a crime scene.* The assumption is that the perpetrator has left physical evidence at the crime scene. Therefore, the FO's duties are simple in concept but complex in execution:

1. Detain any potential suspects.
2. Render medical assistance to those who need it.
3. Do not destroy, alter, or add any evidence at the scene.
4. Prevent others, even superiors, from doing the same.

But not all crime scenes are equal. A homicide in a small house's bedroom is certainly easier to seal and guard than a body found in the middle of a wooded park or a busy highway. The FO should not simply rush into a scene but approach it carefully, thoughtfully. Sometimes the best thing to do is just prevent further entry until additional agency staff arrive.

Once the immediate scene is secured, the lead investigator further defines and evaluates the scene. The scene may be large or small, extensive or discrete, made up of several locations or centered in one area. With the crime scene defined and its borders identified, the initial survey begins to develop an overview and devise a plan of action.

PLAN OF ACTION

Prepare

The officers or investigators assigned to the scene should have obtained a search warrant, if necessary, by the time the crime scene processing begins. If there is time, the search should be discussed with involved personnel before arriving at the scene. A command station for communication and decision making should be established in an area away from the scene but still within the secured perimeter. If personnel task assignments don't already exist, they should be made before arrival at the scene. Depending on the number of personnel available, each may be assigned multiple responsibilities.

Optimally, the person in charge of the scene is responsible for scene security, evidence or administrative log, the preliminary survey, the narrative description, problem resolution, and final decision making. The person in charge of photography arranges, takes, and coordinates photography and keeps the photograph log. The person assigned to prepare the sketch does so in coordination with other methods of documentation; for complex scenes, multiple personnel may be assigned to this task for reconstructive purposes. An evidence custodian takes charge of items collected as evidence, logs them in, and assures that the packaging is labeled properly and sealed.

Communication between the various agencies' representatives, such as medical examiners, laboratory personnel, emergency medical technicians, and attorneys, is crucial to a smooth and successfully executed crime scene process. Questions that arise during the crime scene search can be resolved more easily (with less administrative backlash later) by involving and engaging the proper individuals.

Think Ahead

Fifteen minutes of thought can save hours, and possibly lives, later on. Prepare the paperwork to document the search *before* searching. Agree upon terminology: If everyone refers to it as the "living room," then there will be less confusion afterward if questions come up ("Did we collect that from the *front* room?" "Do you mean the *living* room?").

Arrange for protective clothing, communication, lighting, shelter, transportation, equipment, food, water, medical assistance, and security for personnel. Processing crime scenes can be tedious, physically demanding work, and people, even professionals, perform poorly when they are tired. In prolonged searches, use multiple shifts or teams. If one doesn't exist, develop a transfer mechanism for paperwork and responsibility from one team to the next.

SECURE THE SCENE

If the FO hasn't done so, take control of the scene *immediately*. Determine the extent to which the scene has, or has not, been protected. Talk to personnel who have knowledge of the original condition. Keep out unauthorized personnel. Record who enters and leaves. Now, and throughout the processing of the scene, you cannot take too many notes.

Regarding note taking, it is important to remember the central nature of crime scene notes. They are the documentation of who did what when, contemporaneous with those activities. The adage from quality assurance, "if isn't written down, it didn't happen," is a good guide on what to record. This means that if a supervisor tells a CSI to "process the front bedroom," the supervisor makes a note of that and the time in her notes—as does the employee in *his own notes*. Later, the two sets of notes should correspond, and if a question arises (say, in court), then the activities can be corroborated. Taking contemporaneous notes is crucial to a successful crime scene investigation.

PRELIMINARY SURVEY

The survey is an organizational stage to plan for the search. A few minutes planning and discussion can be of great value later on. Cautiously walk the scene. Crime scenes can be emotional experiences, but you need to stay professional and calm. Take preliminary photographs to establish the scene and delineate the extent of the search area. The initial perimeter may be expanded later if more evidence is found. Make note of special "problem" areas, such as tight spaces, complex evidence arrangements, or environments with transient physical evidence (blood in a running shower, for example). Take extensive notes to document the scene, physical and environmental conditions, and personnel tasks and activities.

Evaluate what physical evidence collection requirements you may have. Make sure you have enough supplies. You don't want to get halfway through and run out of packaging or gloves! Focus first on evidence that could be lost or damaged; leave the more robust evidence for last. All personnel should consider the variety of possible evidence, not just evidence within their specialties.

Collection of evidence is more than just "bagging and tagging." Process the easily accessible areas first, of course, but then move on to out-of-the-way locations, like cupboards, under rugs or carpeting, or drawers. Look for hidden items, secret compartments, and false fronts. Things may not be what they seem, and crime scene personnel must evaluate whether evidence appears to have been moved or altered. Remember, things at a crime scene are just things until they are designated as evidence and then recorded and collected. The scene may not even be *the* scene. The scene may be contrived to look like an accident or some other type of crime.

PHOTOGRAPHY

The photography of the crime scene should begin as soon as possible. The photographic log documents all of the photographs taken and includes a description and location of what's in the photograph. A progression of establishing (overall or perspective views), medium (within 6′), and close-up (within 12″) views of the crime scene should be collected. Multiple views, such as eye level, top, side, and bottom, help to represent what the scene or a piece of evidence looked like in place. Start with the most fragile areas of the crime scene first; move through the scene as evidence is collected and processing continues. Document the process itself, including stages of the crime scene investigation, discoveries, and procedures. Photographs must be taken *before* the evidence is recovered.

Photographs should be taken with and without a scale. Photographs that include a scale should also have the photographer's initials and the date. This is easily accomplished by using a disposable plastic ruler and writing the pertinent information (case number, item number, etc.) on it with a permanent marker. Scales allow photographs to be reproduced at defined scales (1:1, 1:2, 1:10, etc.). Photograph the crime scene in an overlapping series using a wide-angle lens, if possible; 50 mm lenses are the standard issue for cameras. Use both and lots of film. It's almost impossible to take too many photographs. All of these images can help later on with reconstruction questions.

When the exterior crime scene is photographed, establish the location of the scene by a series of overall photographs, including one or more landmarks, with 360° of coverage. Photograph entrances and exits. Prior photographs, blueprints, or maps of the scene may be of assistance, and they should be obtained, if available.

SKETCH

Crime scene sketches may look crude at times, but they contain one very important element for reconstruction: numbers. Distances, angles, time, temperature—all of these elements make the crime scene sketch, an example of which is shown in Figure 2.2, central to all subsequent work. Sketches complement photographs and vice versa. Items of evidence can be located on the sketch as it is made to help establish locations later. Although they are quantitative, sketches are normally not drawn to scale. However, sketches should have measurements and details for a drawn-to-scale diagram. Sketches should include

- Case identifier
- Date, time, and location
- Weather and lighting conditions
- Identity and assignments of personnel

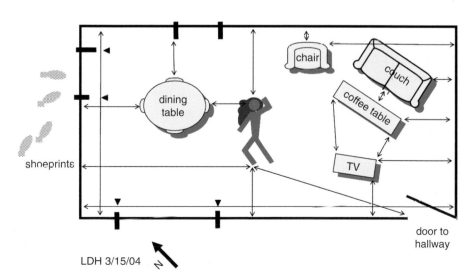

Figure 2.2. A typical crime scene sketch; measurements would accompany all of the arrows and descriptions when the scene is completed.

- Dimensions of rooms, furniture, doors, and windows
- Distances between objects, persons, bodies, entrances, and exits
- An arrow pointing toward magnetic north

CHAIN OF CUSTODY

Arguably, the single most important piece of paper generated at a crime scene is the **chain of custody**. This form, an example of which is shown in Figure 2.3, documents the movement of evidence from the time it is obtained to the time it is presented in court. The most compelling evidence in the world can be rendered useless if inaccuracies or gaps exist in a chain of custody. Where was the evidence? Who had control of it? When? Who last had this item? Could it have been tampered with during this gap in time? Documenting each exchange of an item from person, to evidence locker, to person, to agency may seem to be a nuisance, but it is the foundation that permits forensic science results to enter a courtroom.

CRIME SCENE SEARCH AND EVIDENCE COLLECTION

The crime scene search should be methodical and performed in a specific pattern. The choice of pattern may be dictated by the location, size, or conditions of the scene. Typical patterns are spiral, strip or lane, and grid and are shown in Figure 2.4. Adhering to the selected pattern prevents "bagging and tagging" random items with no organization or system. Measurements showing the location of evidence should be taken with each object located by two or more measurements from non-movable items, such as doors or walls. These measurements should be taken from perpendicular angles to each other to allow for triangulation.

Be alert for all evidence: The perpetrator had to enter or exit the scene. Mark evidence locations on the sketch and complete the evidence log with notations for each item of evidence. If possible, having one person serving as evidence custodian makes collection more regular, organized, and orderly. Again, if possible, two persons should observe evidence in place, during recovery, and being marked for identification. Use tags, or if feasible, mark directly on the evidence.

| Bakersfield Forensic Laboratory 123 Main Street Bakersfield, WV 26501 | | Agency Number | 72204 |
| | | Laboratory number | 615243 |

Chain of Custody

Received From	Delivered to	Date/Time	Items
D. Green Print Name *David Green* Signature	B. Putnam Print Name *Bradford Putnam* Signature	7/22/04 2:14p	1-26, 28
Received From	Delivered to	Date/Time	Items
D. Green Print Name *David Green* Signature	B. Schneckster Print Name *B Schneckster* Signature	7/22/04 2:45p	27, 29
Received From	Delivered to	Date/Time	Items
B. Putnam Print Name *Bradford Putnam* Signature	D. Green Print Name *David Green* Signature	7/29/04 9:16 am	1-26, 28
Received From	Delivered to	Date/Time	Items
Print Name Signature	Print Name Signature		
Received From	Delivered to	Date/Time	Items
Print Name Signature	Print Name Signature		
Received From	Delivered to	Date/Time	Items
Print Name Signature	Print Name Signature		

Figure 2.3. *The chain of custody form documents the movement of evidence from the time it is obtained to the time it is presented in court. The most compelling evidence in the world can be rendered useless if inaccuracies or gaps exist in a chain of custody.*

Figure 2.4. *It is best to have an organized systematic search of a crime scene. The strip (or lane), the spiral, and the grid are three of the most common patterns.*

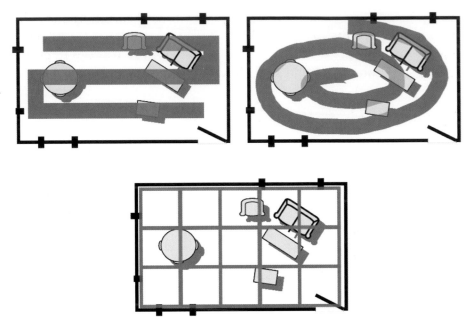

Wear gloves to avoid leaving fingerprints, but be aware that after about 30 minutes, it is possible to leave fingerprints *through* latex gloves! Evidence should not be handled excessively after recovery. Seal all evidence packages with tamper-evident tape at the crime scene. An important activity often overlooked is the collection of known standards from the scene, such as fiber samples from a known carpet or glass from a broken window. Monitor the paperwork, packaging, and other information throughout the process for typographic errors, clarity, and consistency.

Simple geometry can help locate and reconstruct where things were in a sketch. Always take measurements from at least two locations. This will help with checking distances and triangulating "untaken" measurements later. In trigonometry and elementary geometry, triangulation is the process of finding a distance to a point by calculating the length of one side of a triangle, given measurements of angles and sides of the triangle formed by that point and two other reference points. In many ways, measuring a crime scene is surveying, the art and science of accurately determining the position of points and the distances between them; the points are usually on the surface of the earth. Surveying is often used to establish land boundaries for ownership (such as buying a house) or governmental purposes (geographic surveys). Large crime scenes may require standard surveying (and the prepared CSI would do well to learn a bit of surveying), but processing an *indoors* crime scene is much the same except for issues of points (guns, not mountains) and scale (inches, not miles).

FINAL SURVEY

When the crime scene is finished, there is still work to be done. A final survey is recommended to review all aspects of the search. Discuss the search and ask questions of each other. Read over the paperwork for a final check for completeness. Take photographs of the scene showing the final condition. Secure all evidence and retrieve all equipment. A final walk-through with at least two people from different agencies (if possible) as a check on completeness is a must.

The crime scene can be released after the final survey; this event should be documented in the paperwork, including the time and date of release, and to whom and by whom it was released. Remember that other specialists, such as a bloodstain pattern analyst or a medical examiner, may need access to the scene before it is released. Once the scene has been released, reentry may require a warrant; therefore, the scene should be released only after all personnel are satisfied that the scene was searched correctly and completely.

SUBMISSION OF EVIDENCE TO THE LABORATORY

The collected evidence may be submitted to the laboratory by that agency's personnel (that is, laboratory personnel) or by CSIs or law enforcement officers. A form is typically filled out or a letter written detailing what is submitted, under what criminal circumstances, who is submitting the items, and what laboratory examinations are requested.

SAFETY

Walking into a crime scene is one of the most hazardous activities a forensic scientist or crime scene investigator can do. Chemical and biological threats abound, not to mention knives, firearms, explosives . . . the list goes on. Worse, coming in at or near the end of the action,

Figure 2.5. Bystanders protect themselves seconds after a second explosion detonated outside the Atlanta Northside Family Planning Services building in Atlanta, Georgia, in this Thursday, January 16, 1997, Associated Press file photo. (below) Eric Rudolph, the longtime fugitive charged in the 1996 Olympic Park bombing in Atlanta and in attacks at an abortion clinic and a gay nightclub, was arrested June 1, 2003, in the mountains of North Carolina by a local sheriff's deputy.

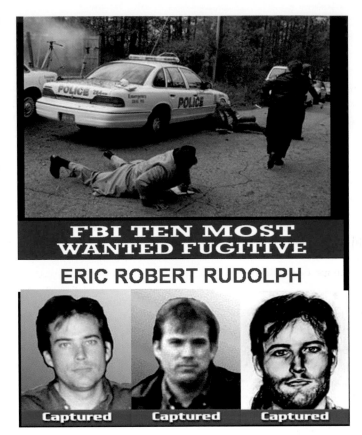

crime scene personnel have little or no foreknowledge of what's in store for them. Add in the prospect of intentional manufacture or use of chemical or biological agents or explosives by terrorists, and the issue of safety for crime scene personnel becomes of paramount concern, as shown in Figure 2.5.

The increase in blood-borne pathogens (AIDS and hepatitis, for example) and other pathogens that may be encountered at crime scenes (like the Hanta virus) has made law enforcement and CSIs more aware of personal protection when responding to crime scenes. Although the risk of infection to crime scene responders is exceedingly low, precautions are typically mandated by individual agencies' protocols. Additionally, federal laws or regulations from one of several health agencies may be applicable to crime scene personnel (see "On the Web: Safety").

ON THE WEB:
Safety

Occupational Safety and Health Administration, www.osha.gov

The mission of the Occupational Safety and Health Administration (OSHA) is to save lives, prevent injuries, and protect the health of America's workers. To accomplish this, federal and state governments must work in partnership with the more than 100 million working men and women and their 6.5 million employers who are covered by the Occupational Safety and Health Act of 1970.

The Centers for Disease Control, www.cdc.gov

The Centers for Disease Control and Prevention (CDC) is recognized as the lead federal agency for protecting the health and safety of people at home and abroad, providing credible information to enhance health decisions, and promoting health through strong partnerships. The CDC serves as the national focus for developing and applying disease prevention and control, environmental health, and health promotion and education activities designed to improve the health of the people of the United States.

The Morbidity and Mortality Weekly Report, www.cdc.gov/mmwr

The *Morbidity and Mortality Weekly Report* (*MMWR*) Series is prepared by the Centers for Disease Control and Prevention. The data in the weekly *MMWR* are provisional, based on weekly reports to CDC by state health departments. The reporting week concludes at close of business on Friday; compiled data on a national basis are officially released to the public on the succeeding Friday. An electronic subscription to MMWR is free.

National Institute for Occupational Safety and Health, www.cdc.gov/niosh

The National Institute for Occupational Safety and Health (NIOSH) is the federal agency responsible for conducting research and making recommendations for the prevention of work-related disease and injury. The Institute is part of the Centers for Disease Control and Prevention. NIOSH is responsible for conducting research on the full scope of occupational disease and injury ranging from lung disease in miners to carpal tunnel syndrome in computer users. In addition to conducting research, NIOSH investigates potentially hazardous working conditions when requested by employers or employees; makes recommendations and disseminates information on preventing workplace disease, injury, and disability; and provides training to occupational safety and health professionals. Headquartered in Washington, D.C., NIOSH has offices in Atlanta, Georgia, and research divisions in Cincinnati, Ohio; Morgantown, West Virginia; Bruceton, Pennsylvania; and Spokane, Washington.

SOURCES AND FORMS OF DANGEROUS MATERIALS

Inhalation

At a crime scene, airborne contaminants can occur as dust, aerosol, smoke, vapor, gas, or fume. Immediate respiratory irritation or trauma might ensue when these contaminants are inhaled. Some airborne contaminants can enter the bloodstream through the lungs and cause chronic damage to the liver, kidneys, central nervous system, heart, and other organs. Remember that some of these inhalants may be invisible!

Skin Contact

Because processing a crime scene requires the physical collection of items, skin contact is a frequent route of contaminant entry into the body. Direct effects can result in skin irritation or trauma at the point of contact, such as a rash, redness, swelling, or burning. Systemic effects, such as dizziness, tremors, nausea, blurred vision, liver and kidney damage, shock, or collapse, can occur when the substances are absorbed through the skin and circulated throughout the body. The use of appropriate gloves, safety glasses, goggles, face shields, and protective clothing can prevent this contamination.

Ingestion

Ingestion is a less common route of exposure. Ingestion of a corrosive material can cause damage to the mouth, throat, and digestive tract. When swallowed, toxic chemicals can be absorbed by the body through the stomach and intestines. To prevent entry of chemicals or biological contaminants into the mouth, wash hands before eating, drinking, smoking, or applying cosmetics. Also, do not bring food, drink, or cigarettes into areas where contamination can occur.

Injection

Needle sticks and cuts from contaminated glass, hypodermic syringes, or other sharp objects can inject contaminants directly into the bloodstream. Extreme caution should be exercised when handling objects with sharp or jagged edges.

UNIVERSAL PRECAUTIONS

The Occupational Safety and Health Administration (OSHA) issued regulations regarding occupational exposure to **blood-borne pathogens** (BBP) in December 1991. Those occupations at risk for exposure to BBPs include law enforcement, emergency response, and forensic laboratory personnel (Title 29 CFR, 1991).

Fundamental to the BBP standard is the primary concept for infection control called **Universal Precautions**. It requires employees to treat all human blood, body fluids, or other potentially infectious materials as if they *are* infected with diseases such as hepatitis B virus (HBV), hepatitis C virus (HCV), and human immunodeficiency virus (HIV). The following protective measures should be taken to avoid direct contact with these potentially infectious materials (Title 29 CFR, 1991):

- Use barrier protection such as disposable gloves, coveralls, and shoe covers when handling potentially infectious materials. Gloves should be worn, especially if there are cuts, scratches, or other breaks in the skin.

- Change gloves when torn, punctured, or when their ability to function as a barrier is compromised.

- Wear appropriate eye and face protection to protect against splashes, sprays, and spatters of infectious materials. Similar precautions should be followed when collecting dried bloodstains.

- Place contaminated sharps in appropriate closable, leak-proof, puncture-resistant containers when transported or discarded. Label the containers with a BIOHAZARD warning label. Do not bend, recap, remove, or otherwise handle contaminated needles or other sharps.

- Prohibit eating, drinking, smoking, or applying cosmetics where human blood, body fluids, or other potentially infectious materials are present.

- Wash hands after removing gloves or other personal protective equipment (PPE). Remove gloves and other PPE in a manner that will not result in the contamination of unprotected skin or clothing.

- Decontaminate equipment after use with a solution of household bleach diluted 1:10, 70 percent isopropyl alcohol, or other disinfectant. Non-corrosive disinfectants are commercially available. Allow sufficient contact time to complete disinfection.

(Source: Title 29 CFR, 1991)

In addition to Universal Precautions, prudent work practices and proper packaging serve to reduce or eliminate exposure to potentially infectious materials. Packaging examples include puncture-resistant containers used for storage and disposal of sharps.

Chemical Safety

A wide variety of health and safety hazards can be encountered at a crime scene. Some of those hazards are listed in Table 2.1. This awareness comes from the information contained in a **Material Safety Data Sheet** (MSDS) (for example, http://www.msdssolutions.com or

Table 2.1. *Numerous chemical safety hazards can be encountered at crime scenes. Source: National Research Council, 1981.*

MATERIAL	EXAMPLES
Flammable or combustible materials	Gasoline, acetone, ether, and similar materials ignite easily when exposed to air and an ignition source, such as a spark or flame.
Explosive materials	Over time, some explosive materials, such as nitroglycerine and nitroglycerine-based dynamite, deteriorate to become chemically unstable. In particular, ether will form peroxides around the mouth of the vessel in which it is stored. All explosive materials are sensitive to heat, shock, and friction, which are employed to initiate explosives.
Pyrophoric materials	Phosphorus, sodium, barium, and similar materials can be liquid or solid and can ignite in air temperatures less than 130° F (540° C) without an external ignition source.
Oxidizers	Nitrates, hydrogen peroxide, concentrated sulfuric acid, and similar materials are a class of chemical compounds that readily yield oxygen to promote combustion. Avoid storage with flammable and combustible materials or substances that could rapidly accelerate its decomposition.

`http://siri.uvm.edu`) and appropriate training. The MSDS provides information on the hazards of a particular material so that personnel can work safely and responsibly with hazardous materials; an MSDS is typically available through a vendor's website.

Remember, when working with chemicals, be aware of hazardous materials, disposal techniques, personal protection, packaging and shipping procedures, and emergency preparedness.

PERSONAL PROTECTIVE EQUIPMENT

Hand Protection

Hand protection should be selected on the basis of the type of material being handled and the hazard or hazards associated with the material. Detailed information can be obtained from the manufacturer. Nitrile gloves provide protection from acids, alkaline solutions, hydraulic fluid, photographic solutions, fuels, aromatics, and some solvents. It is also cut resistant. Neoprene gloves offer protection from acids, solvents, alkalies, bases, and most refrigerants. Polyvinyl chloride (PVC) is resistant to alkalies, oils, and low concentrations of nitric and chromic acids. Latex or natural rubber gloves resist mild acids, caustic materials, and germicides. Latex will degrade if exposed to gasoline or kerosene and prolonged exposure to excessive heat or direct sunlight. Latex gloves can degrade, losing their integrity. Some people are allergic to latex and can avoid irritation by wearing nitrile or neoprene gloves.

Gloves should be inspected for holes, punctures, and tears before use. Rings, jewelry, or other sharp objects that can cause punctures should be removed. Double-gloving may be necessary when working with heavily contaminated materials; double-gloving is also helpful if "clean" hands are needed occasionally. If a glove is torn or punctured, replace it. Remove disposable gloves by carefully peeling them off by the cuffs, slowly turning them inside out. Discard disposable gloves in designated containers and, it should go without saying, do not reuse them.

Eye Protection

Safety glasses and goggles should be worn when handling biological, chemical, and radioactive materials. Face shields can offer better protection when there is a potential for splashing or flying debris. Face shields alone are not sufficient eye protection; they must be worn in combination with safety glasses. Contact lens users should wear safety glasses or goggles to protect the eyes. Protective eyewear is available for those with prescription glasses and should be worn over them.

Foot Protection

Shoes that completely cover and protect the foot are essential—*no sandals or sneakers!* Protective footwear should be used at crime scenes when there is a danger of foot injuries due to falling or rolling objects or to objects piercing the sole and when feet are exposed to electrical hazards. In some situations, shoe covers can provide protection to shoes and prevent contamination to the perimeter and areas outside the crime scene.

Other Protection

Certain crime scenes, such as bombings and clandestine drug laboratories, can produce noxious fumes requiring respiratory protection. In certain crime scenes, such as bombings or fires where structural damage can occur, protective helmets should be worn.

TRANSPORTING HAZARDOUS MATERIALS

Title 49 of the Code of Federal Regulations codifies specific requirements that must be observed in preparing hazardous materials for

ISSUES IN . . .
Crime scene investigation: Gloria Ramirez

The strange and still unsolved case of Gloria Ramirez, unfairly dubbed "The Toxic Lady," stands as a precautionary tale about the unpredictability of events during an investigation.

Ramirez, diagnosed with advanced cervical cancer, was having nausea, heartbeat, and breathing difficulties at her home in Riverside, California, on the evening of February 19, 1994. She was rushed to Riverside General Hospital by paramedics who administered oxygen along the way. At the hospital, she was in respiratory and cardiac distress, which caused a critically low blood pressure. Despite this, Ramirez responded to questions but acted lethargic and vomited in the emergency room (ER). The doctors gave her drugs to calm her down (Valium, Versed, and Ativan) and to restore a normal heart rate (lidocaine and other antiarrhythmic drugs). Shortly after her arrival, Ramirez went into full cardiac arrest.

The ER staff applied a breathing tube to provide oxygen, and Ramirez was defibrillated (an electric shock delivered to restore a normal heart rhythm). After this, accounts vary as to what exactly happened.

As a nurse drew blood for routine tests, one of the doctors smelled ammonia and felt dizzy. The nurse keeled over. The senior medical resident checked on the nurse to make sure she was not hurt; the doctor took the syringe and smelled ammonia. She noticed the blood had funny manilla-colored crystals in it; then she passed out and went into convulsions. Other ER staff were also affected, and the ER was cleared. The fire department's Hazardous Materials team was called in, and incoming patients were rerouted to other hospitals. Attempts to revive Ramirez failed, and she was pronounced dead. As many as 23 staff were affected, reporting nausea and headaches. The senior medical resident was in the hospital for two weeks with breathing disorders, hepatitis, and pancreatitis, in addition to other maladies.

An autopsy was later conducted by professionals wearing protective suits with respirators. Ramirez had been suffering from a urinary blockage as well as the cervical cancer; she had died of kidney failure. No known toxic chemicals were found, neither in Ramirez nor the hospital's plumbing or ventilation systems.

Various agencies have come up with explanations of what happened in the Riverside ER, from the laughable (the ER staff were overcome by the "smell of death") to the bizarre (the hospital was running a secret methamphetamine lab). Scientists at Livermore National Laboratory have provided the most scientific explanation to date, but it still has weaknesses. Dimethyl sulfone ($DMSO_2$) was found in Ramirez's blood. $DMSO_2$ is a metabolic product of dimethyl sulfoxide (DMSO), a solvent sometimes used by cancer patients and athletes as a pain remedy. The theory runs like this: Due to Ramirez's use of DMSO and her urinary blockage, DMSO accumulated in her bloodstream and the oxygen the paramedics gave her in the ambulance converted the DMSO in her blood into a high concentration of $DMSO_2$. Some unknown catalyst, perhaps the electric defibrillation, converted the $DMSO_2$ into $DMSO_4$ and induced the unhealthy effects. When the nurse drew the blood and it cooled to room temperature, the straw-colored crystals formed. The $DMSO_4$ evaporated leaving no clues behind.

Beyond this theory, no credible explanation has ever been offered for the strange case of Gloria Ramirez.

Sources: Adams, 1996; Pilkington, 2004.

shipment by air, highway, rail, or water. All air transporters follow these regulations, which describe how to package and prepare hazardous materials for air shipment. Title 49 CFR 172.101 (http://hazmat.dot.gov) provides a Hazardous Materials table that identifies items considered hazardous for the purpose of transportation, special provisions, hazardous materials communications, emergency response information, and training requirements. Training is required to properly package and ship hazardous materials employing any form of commercial transportation.

CHAPTER SUMMARY

The crime scene is the center of the forensic world. The importance of a carefully processed crime scene cannot be overstated. The processing of a crime scene is a one-way street—there is no going back. Standard operating procedures and protocols guide the crime scene investigator, but training, experience, and education all play a role in adapting to each unique crime scene.

Test Your Knowledge

1. What is a chain of custody?
2. What is a crime scene?
3. What should the first officer or CSI at the crime scene do?
4. Name four safety issues for CSIs.
5. Is it okay to only take photographs or only draw sketches? Why not?
6. How many photographs should you take at a crime scene?
7. Name three agencies that regulate worker safety.
8. What is a datum?
9. What is provenience?
10. When is it acceptable to release a crime scene?
11. What is a MSDS?
12. Who should be involved in the final walk-through of a crime scene?
13. What should be included in a crime scene sketch?
14. Should you take photographs with or without a scale?
15. What does "BBP" stand for?
16. What does the Ramirez case teach us about safety?

17. What is a "Universal Precaution"?

18. Why is processing a crime scene considered "careful destruction"?

19. What's involved in making a plan for a crime scene?

20. Why is it important to have a plan for a crime scene?

Consider This . . .

1. How would you process an underwater crime scene? A homicide scene on a beach? Outside during a thunderstorm? What protocols would change? How would you process and package evidence? How would you maintain the integrity of the evidence?

2. How would you process a crime scene (use Figure 2.3 as a basis) with two people? Assign tasks and duties. How would you process the same scene with 10 people? What would you do the same or do differently? Would the quality of the scene processing be the same?

3. Take one of the extreme examples in *Consider This* #1. How would you explain to a jury that you followed your agency's protocols—but also did not?

BIBLIOGRAPHY

Adams, C. (1996). "What's the story on the 'toxic lady'?" *The Straight Dope*, March 22, retrieved from **www.straightdope.com**.

American National Standards Institute. (1989). *American national standard practice for occupational and educational eye and face protection.* New York: Author.

Casey, E. (ed.) (2002). *Handbook of computer crime investigation.* San Diego: Academic Press.

Douglas, J., Burgess, A.W., Burgess, A.G., & Ressler, R. (1992). *Crime classification manual.* New York: Lexington Books.

Federal Bureau of Investigation Laboratory Division. (1999). *Forensic science handbook.* Washington, D.C.: Author.

Fisher, B. (2000). *Techniques of crime scene investigation.* 7th ed., Boca Raton, FL: CRC Press.

Kirk, P.L. (1963). "Criminalistics," *Science* V140, N3565: 367–370.

C. Gorman. (ed.). Hazardous Waste Handling Pocket Guide. Schenectady, New York: Genium.

Laboratory survival manual. (2005). Environmental Health and Safety Office, University of Virginia, Charlottesville, Virginia. Retrieved from: **www.keats.admin.virginia.edu/lsm/home.html**

National Institute of Justice. (2000). *A guide for explosion and bombing scene investigation.* Washington, D.C.: Author.

National Institute of Justice. (2000). *Fire and arson scene evidence: A guide for public safety personnel.* Washington, D.C.: Author.

National Research Council, Committee on Hazardous Substances in the Laboratory. (1981). *Prudent practices for handling hazardous chemicals in laboratories.* Washington, D.C.: National Academy.

Osterburg, J. (2000). *Criminal investigation.* Cincinnati: Anderson Publishing.

Pilkington, M. (2004). "Deadly miasma," *The Guardian,* Thursday July 1.

Renfrew, C., & Bahn, P. (2000). *Archaeology: Theories, methods, and practice,* 3rd ed. New York: Thames and Hudson.

Technical Working Group on Crime Scene Investigation. (2000). *Crime scene investigation: A guide for law enforcement.* Washington, D.C.: National Institute of Justice.

Title 29 CFR Section 1910.134. (1991). *Respiratory protection.* Washington, D.C.: U.S. Department of Labor, Occupational Safety and Health Administration.

Title 29 CFR Section 1910.136. (1991). *Foot protection.* Washington, D.C.: U.S. Department of Labor, Occupational Safety and Health Administration.

Title 29 CFR Section 1910.1030. (1991). *Occupational exposure to bloodborne pathogens: Final rule.* Washington, D.C.: U.S. Department of Labor, Occupational Safety and Health Administration. Retrieved from: **www.osha-slc.gov/OshStd_data/1910_1030.html**

Title 29 CFR Section 172.101. (1991). *Purpose and use of hazardous materials table.* Washington, D.C.: U.S. Department of Transportation, Office of Hazardous Materials Safety. Retrieved from: **http://hazmat.dot.gov**.

Upfal, M. J. (1991). *Pocket guide to first aid for chemical injuries.* (J. R. Stuart, Ed.) Schenectady, New York: Genium.

The Nature of Evidence

CONTENTS

KEY TERMS

Class

Coincidental associations

Common source

Comparison

Contamination

Demonstrative evidence

Direct transfer

Evidence

False negative (Type II error)

False positive (Type I error)

Hypothesis

Identification

Indirect transfer

Individualization

Known evidence

Locard

Locard Exchange Principle

Negative control

Persistence

Positive control

Printer analysis

Probative value

Proxy data

Questioned evidence

Repeatability

Scientific method

Testable

Testability

Trier-of-fact

INTRODUCTION

John Adams, in his *Argument in Defense of the Soldiers in the Boston Massacre Trials*, in December 1770, said that, "Facts are stubborn things; and whatever may be our wishes, our inclinations, or the dictates of our passion, they cannot alter the state of facts and evidence" (Zobel, 1996, page 293). Evidence is critical to a trial; it provides the foundation for the arguments the attorneys plan to offer. It is viewed as the impartial, objective, and even stubborn information that leads a judge or jury to their conclusions. As you will see, evidence is complicated, and much goes into getting evidence ready before it can go into court.

WHAT IS EVIDENCE?

In a trial, the jury or judge hears the facts or statements of the case to decide the issues; whoever determines guilt or innocence is called the **trier-of-fact**. During the trial, the trier-of-fact must decide whether the statements made by witnesses are true. This is done mainly through the presentation of information or evidence. **Evidence** can be defined as information, whether in the form of personal testimony, the language of documents, or the production of material objects, that is given in a legal investigation, to make a fact or proposition more or less likely. For example, someone is seen leaving the scene of a homicide with a gun, and it is later shown by scientific examination that bullets removed from the body of the victim were fired from that gun. This could be considered evidence that the accused person committed the homicide. Having the association of the bullets to the gun makes the proposition that the accused is the perpetrator more probable than it would be if the evidence didn't exist. In this chapter we will explore the nature of evidence, how it is classified, and how we decide what value the evidence has in proving or disproving a proposition.

KINDS OF EVIDENCE

Most evidence is real; that is, it is generated as a part of the crime and recovered at the scene or at a place where the suspect or victim had been before or after the crime. Hairs, fingerprints, paint, blood, and shoeprints are all real evidence. Sometimes, however, items of evidence may be created to augment or explain real evidence. For example, diagrams of hair characteristics, a computer simulation of a crime scene, or a demonstration of blood stain pattern mechanics may be prepared to help the trier-of-fact understand complex testimony. Such **demonstrative evidence** was not generated directly from the incident but is created later. Because it helps explain the significance of real evidence, it does help make a proposition more or less probable and is, therefore, evidence. "In More Detail: Kinds of Evidence" lists other varieties of evidence.

IN MORE DETAIL:
Kinds of evidence

Circumstantial evidence: Evidence based on inference and not on personal knowledge or observation.

Conclusive evidence: Evidence so strong as to overbear any other evidence to the contrary.

Conflicting evidence: Irreconcilable evidence that comes from different sources.

Corroborating evidence: Evidence that differs from but strengthens or confirms other evidence.

Derivative evidence: Evidence that is discovered as a result of illegally obtained evidence and is therefore inadmissible because of the primary taint.

Exculpatory evidence: Evidence tending to establish a criminal defendant's innocence.

Foundational evidence: Evidence that determines the admissibility of other evidence.

Hearsay: Testimony that is given by a witness who relates not what he or she knows personally, but what others have said, and that is therefore dependent on the credibility of someone other than the witness.

Incriminating evidence: Evidence tending to establish guilt or from which a trier-of-fact can infer guilt.

Presumptive evidence: Evidence deemed true and sufficient unless discredited by other evidence.

Prima facie (prī-mə **fay**-shə) *evidence*: Evidence that will establish a fact or sustain a judgment unless contradictory evidence is produced.

Probative evidence: Evidence that tends to prove or disprove a point in issue.

Rebuttal evidence: Evidence offered to disprove or contradict the evidence presented by an opposing party.

Tainted evidence: Evidence that is inadmissible because it was directly or indirectly obtained by illegal means.

Source: Garner, 2000.

LEVELS OF EVIDENCE

Not all evidence is created equal: Some items of evidence have more importance than others. The context of the crime and the type, amount, and quality of the evidence will dictate what can be determined and interpreted. Most of the items in our daily lives are produced or manufactured *en masse*, including biological materials (you have thousands of hairs on your body, for example). This has implications for what can be said about the relationships between people, places, and things surrounding a crime.

FORENSIC SCIENCE IS HISTORY

Forensic science is an historical science: The events in question have already occurred and are in the past. Forensic scientists do not view the crime as it occurs (unless they're witnesses); they assist the inves-

Table 3.1. *Forensic science is an historical science because it reconstructs past events from the physical remnants (proxy data) of those events. In this way, forensic science is similar to other historical sciences such as geology, astronomy, paleontology, and archaeology.*

	FORENSIC SCIENCE	ARCHAEOLOGY	GEOLOGY
Time Frame	Hours, days, months	Hundreds to thousands of years	Millions of years
Activity Level	Personal; Individual	Social; Populations	Global
Proxy Data	Mass-produced	Hand-made	Natural

tigation through the analysis of the physical remains of the criminal activity. Many sciences, such as geology, astronomy, archaeology, paleontology, and evolutionary biology, work in the same way: No data is seen *as it is created*, but only the *remnants,* or **proxy data**, of those events are left behind. Archaeologists, for example, analyze cultural artifacts of past civilizations to interpret their activities and practices. Likewise, forensic scientists analyze evidence of past criminal events to interpret the actions of the perpetrator(s) and victim(s). Table 3.1 compares differences between some historical sciences.

Just as archaeologists must sift through layers of soil and debris to find the few items of interest at an archaeological site, forensic scientists must sort through all of the items at a crime scene (think of all the things in your home, for example) to find the few items of evidence that will help reconstruct the crime. The nature and circumstances of the crime will guide the crime scene investigators and the forensic scientists to choose the most relevant evidence and examinations. Many methods may seem "forensic," but the definition may occasionally be stretched. See "In More Detail: But is it *forensic* science?" for a discussion of this issue.

THE BASIS OF EVIDENCE: TRANSFER AND PERSISTENCE

When two things come into contact, information is exchanged. Seems pretty simple, doesn't it? And yet it is the central guiding theory of forensic science. Developed by Edmund **Locard**, a French forensic microscopist in the early part of the twentieth century, this theory

IN MORE DETAIL:

But is it *forensic* science?

Many people identify forensic science as "science applied to law," but in truth it isn't that simple. If structural engineers are consulted to determine why a bridge failed, write a report, testify once, and then never work on a legal case again, are they *forensic* engineers? Most people wouldn't think so, but what if they did it 3, 9, or even 21 times in their career? Many forensic scientists don't work at government forensic laboratories, so forensic science can't be defined that way. At what point does the *application* of science in the legal arena shift to *forensic* science?

Reconstructing events to assist the justice system happens all the time without being forensic science proper. A good example is the case of a Florida dentist who unwittingly passed on his HIV infection to several of his patients (Ou et al., 1992). Ou's group reported in 1990 that a young woman with AIDS had probably contracted her HIV infection during an invasive dental procedure. The dentist had been diagnosed with AIDS in 1986 and continued to practice general dentistry for two more years. The dentist went public for the safety of his patients, requesting that they all be tested for HIV infection. Out of 1,100 people who were tested, seven patients were identified as being HIV-positive.

HIV is genetically flexible and changes its genetic makeup during its life cycle, resulting in a variety of related viral family lines or strains (called quasi-species). Investigators used the degree of genetic similarity among the HIV strains in the seven infected patients, along with epidemiologic information, to evaluate whether the infections originated with the dentist or were from other sources. The investigators used genetic distance, constructed "family tree" diagrams, and developed amino acid "signature patterns."

Of the seven patients, five had no other identified HIV risk other than visiting the dentist. These five patients were infected with HIV strains that were closely related to those of the dentist's infection; moreover, these strains were different from the strains found in the other two HIV infected patients and 35 other HIV infected people in the same geographical area. As the authors of the paper note:

> In the current investigation, the divergence of HIV sequences within the Florida background population was sufficient to identify strain variation . . . this investigation demonstrates that detailed analysis of HIV genetic variation is a new and powerful tool for understanding the epidemiology of HIV transmission. (Ou et al., 1992, p. 1170)

They call it an "investigation," they're doing DNA analysis, they're reconstructing the transfer of something from one person to others . . . is this *forensic* science?

Don't be confused simply because a science is *historical*, because it uses proxy data to represent past events, or because it uses the same techniques or methods as a forensic science. Forensic science is the demonstration of relationships between people, places, and things involved in legal cases through the identification, analysis, and, if possible, individualization of evidence. Because nothing legal is at issue in the dentist "case," it isn't forensic. With the increased popularity of forensic science, students and professionals must be cautious about the use of "forensic" as a buzzword in the media and professional publications.

Sources: Ou, C., Ciesielski, C.A., Myers, G., Bandea, C.I., Luo, C., Korber, B.T.M., et al. (1992).

Table 3.2. In a sense, all evidence is transfer evidence *in that it has a source and moves or is moved from that source to a target/location. Note that there are levels to various types of evidence, from the fundamental (striations on the barrel-cutting tool) to the specific (the bullet in the victim's body identified by the striations).*

ITEM	TRANSFERRED *FROM* (SOURCE)	TRANSFERRED *TO* (TARGET/LOCATION)
Drugs	Dealer	Buyer's pocket or car
Blood stains	Victim's body	Bedroom wall
Alcohol	Glass	Drunk driver's bloodstream
Semen	Assailant	Victim
Ink	Writer's pen	Stolen check
Handwriting	Writer's hand/brain	Falsified document
Fibers	Kidnapper's car	Victim's jacket
Paint chips/smear	Vehicle	Hit-and-run victim
Bullet	Shooter's gun	Victim's body
Striations	Barrel of shooter's gun	Discharged bullet
Imperfections	Barrel-cutting tool	Shooter's gun's barrel

posits that this exchange of information occurs, even if the results are not identifiable or are too small to be found (Locard, 1930). The results of such a transfer would be proxy data—not the transfer itself, but the remnants of that transaction. Because forensic science demonstrates associations between people, places, and things through the analysis of proxy data, essentially *all evidence is transfer evidence.* Table 3.2 lists some examples in support of this concept.

The conditions that affect transfer amounts include

- The pressure applied during contact
- The number of contacts (six contacts between two objects should result in more transferred material than one contact)
- How easily the item transfers material (mud transfers more readily than does concrete)

- The form of the evidence (solid/particulate, liquid, or gas/aerosol)
- How much of the item is involved in the contact (a square inch should transfer less than a square yard of the same material)

Evidence that is transferred from a source to a location with no intermediaries is said to have undergone **direct transfer**; it has transferred from A to B. **Indirect transfer** involves one or more intermediate objects; the evidence transfers from A to B to C, as shown in Figure 3.1.

Indirect transfer can become complicated and poses potential limits on interpretation. For example, say you own two dogs, and before you go to work each day, you pet and scratch them. At work, you sit in your desk chair and talk on the phone. You get up to get a cup of coffee. When you come back, a colleague is sitting in your desk chair waiting to tell you some news. You have experienced a *direct transfer* of your dogs' hairs from them to your pants. Your chair, however, has received an *indirect transfer* of your dogs' hairs; your dogs have never sat in your office desk chair. The colleague who sat in your chair has also experienced an indirect transfer of anything on the chair, except for any fibers originating from the chair's upholstery. How would you interpret finding your dogs' hairs on your colleague if you didn't know she had sat in your chair? As you can see, while direct transfer may be straightforward to interpret, indirect transfers can be complicated and potentially misleading. It may be more accurate to speak of direct and indirect *sources*, referring to whether the originating source of the evidence is the transferring item, but the "transfer" terminology has stuck. This leads to unsupportable statements regarding certain types of indirect transfer (secondary, tertiary, etc.); in almost no cases can a forensic scientist tell the difference between secondary (one intermediary) and tertiary (two intermediaries).

The second part of the transfer process is **persistence**. Once the evidence transfers, it will remain, or persist, in that location until it further transfers (and, potentially, is lost), degrades until it is unusable or unrecognizable, or is collected as evidence. How long evidence persists depends on

- What the evidence is (such as hairs, blood, toolmarks, accelerants)
- Location of the evidence

A ⇒ B
Direct transfer
(source)

A ⇒ B ⇒ C
Indirect transfer
(source)

A ⇒ B ⇒ C ⇒ D ⇒ E
Indirect transfers
(source)

Figure 3.1. Direct transfer describes the movement of items from the source to the recipient (A to B), whereas indirect transfer involves an intermediate object that conveys the items to the recipient (A to B to C). Sometimes, direct transfer is referred to as primary transfer, and indirect transfers are listed as secondary, tertiary, etc., but this terminology becomes clumsy after several exchanges. It may be more accurate to speak of direct and indirect sources.

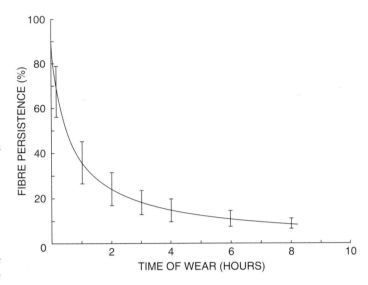

Figure 3.2. Trace evidence, such as fibers, tends to be lost at a geometric rate with normal activity. For example, numerous fiber transfer studies demonstrate that, from the time of transfer with normal activity, after about four hours 80% of the transferred fibers are lost. Transfer and persistence studies with other evidence types have shown similar loss rates. This graph shows a typical fiber loss curve (for acrylic and wool fibers) showing one standard deviation limit (Pounds & Smalldon, 1975, p. 34).

- Environment around the evidence
- Time from transfer to collection
- "Activity" of or around the evidence location

For example, numerous fiber transfer studies demonstrate that, from the time of transfer with normal activity, after about four hours 80% of the transferred fibers are lost. Transfer and persistence studies with other evidence types have shown similar loss rates, as depicted in Figure 3.2.

CONTAMINATION

Once the activity surrounding the crime has stopped, any transfers that take place may be considered **contamination**—that is, an undesired transfer of information between items of evidence. You would not want to package, for example, a wet bloody shirt from the victim of a homicide with clothes from a suspect; in fact, *every* item of evidence (where practical) should be packaged *separately*. Contamination is itself evidence of a kind; this is why it is so difficult to falsify a case or plant evidence. Based on Locard's Principle, *every* contact produces some level of exchange, including contamination. It is nearly impossible to completely prevent contamination, but it can be severely minimized through properly designed facilities, adequate protective clothing, and quality-centered protocols that specify the handling and packaging of evidence.

IDENTITY, CLASS, AND INDIVIDUALIZATION

All things are unique in space and time. No two (or more) objects are absolutely identical. Take, for example, a mass-produced product like a tennis shoe. Thousands of shoes of a particular type may be produced in any one year. The manufacturer's goal, to help sell more shoes, is to make them all look and perform the same—consumers demand consistency. This consistency is a help and a hindrance to forensic scientists because it makes it easy to separate one item from another (this red tennis shoe is different from this white one), but these same characteristics make it difficult to separate items with many of the same characteristics (two red tennis shoes). Think about two white tennis shoes that come off the production line one after the next. How would you tell them apart? You might say, "this one" and "that one," but if they were mixed up, you probably couldn't sort them again. You'd have to label them somehow, like numbering them "1" and "2."

Now consider if the two shoes are the same except for color—one's white and one's red. Of course, you could tell them apart, but would you put them in the same category? Compared with a brown dress shoe, the two tennis shoes would have more in common with each other than with the dress shoe. All the shoes, however, are more alike than any of them are compared to, say, a baseball bat. Forensic scientists have developed terminology to clarify the way they communicate about these issues.

Identification is the examination of the chemical and physical properties of an object and using them to categorize the object as a member of a group. What is the object made of? What is its color, mass, and/or volume? The process of examining a white powder, performing one or two analyses, and concluding it is cocaine is identification. Determining that a small colored chip is automotive paint is identification. Looking at debris from a crime scene and deciding it contains hairs from a black Labrador retriever is identification (of those hairs). All of the characteristics used to identify an object help to refine that object's identity and its membership in various groups. The debris has fibrous objects in it, and that restricts what they could be—most likely hairs or fibers rather than bullets, to use an absurd example. The microscopic characteristics indicate that some of the fibrous objects are hairs, that they are from a dog, and the hairs are most like those from a specific breed of dog. This description places the hairs into a group of objects with similar characteristics, called a **class**. All black Labrador retriever hairs would fall into a class; these belong to a larger class of items called *dog hairs*. Further, all dog hairs can be included in the class of *non-human hairs* and, ultimately, into a more inclusive class called *hairs*. Going in the other direction, as the process of identification of evidence becomes more specific, it permits the analyst to classify the evidence into successively smaller classes of objects.

Class is a movable definition. It may not be necessary to classify the evidence beyond *dog hairs* because you are looking for human hairs or textile fibers. Although it is possible to define the dog hairs more completely, you may not need to do so in the case at hand. Multiple items can be classified differently, depending on what questions need to be asked. For example, an orange, an apple, a bowling ball, a bowling pin, and a banana could be classified, as shown in Figure 3.3, by *fruit v. non-fruit, round things v. non-round things*, and *organic v. inorganic*. Notice that

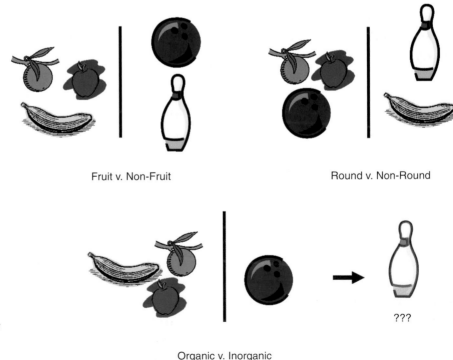

Fruit v. Non-Fruit   Round v. Non-Round

Organic v. Inorganic

Figure 3.3. A class is a group of things with similar characteristics. The size of the class can vary widely depending on the characteristics used for definition, such as the class "all oranges" versus the class "all oranges in your refrigerator."

the bowling pin doesn't fit into either of the classes in the last example because it is made of wood (which is organic) but is painted (which has inorganic components).

Stating that two objects share a class identity may indicate they come from a **common source**. What is meant by a "common source" depends on the material in question, the mode of production, and the specificity of the examinations used to classify the object. A couple of examples should demonstrate the potential complexity of what constitutes a common source. Going back to the two white tennis shoes, what is their common source—the factory, the owner, or where they are found? Because shoes come in pairs, finding one at a crime scene and another in the suspect's apartment could be considered useful to the investigation. The forensic examinations would look for characteristics to determine whether the two shoes were owned by the same person (the "common source"). If the question centered on identifying the production source of the shoes, then the factory would be the "common source."

Another example is fibers found on a body left in a ditch; the fibers are determined to be from an automobile. A suspect is developed, and fibers from his car are found to be analytically indistinguishable in all tested traits from the crime scene fibers. Is the suspect's car the "common source"? For investigative and legal purposes, the car should be considered as such. But certainly it is not the only car with that carpeting. Other models from that car manufacturer or even other car manufacturers may have used that carpeting, and the carpeting may not be the only product with those fibers. But given the context of the case, it may be reasonable to conclude that the most logical source for the fibers is the suspect's car. If the fibers were found on the body but no suspect was developed, part of the investigation may be to determine who made the fibers and to track what products those fibers went into in an effort to find someone who owns that product. In that instance, the "common source" could be the fiber manufacturer, the carpet manufacturer, or the potential suspect's car, depending on what question is being asked.

If an object can be classified into a group with only one member (itself), it has been **individualized**. An individualized object has been associated with one, and only one, source: It is unique. The traits that allow for individualization depend, in large part but not exclusively, on the raw materials, manufacturing methods, and history of use. Sometimes, sufficiently increasing class traits can lead toward individualization; for example, John Thornton's article on the classification of firearms evidence is an excellent, if overlooked, treatment of this issue (1986).

INDIVIDUALIZATION OF EVIDENCE

To individualize evidence means to be able to put it into a class with one member. It is the logical extension of the classification of evidence discussed previously. If a forensic scientist can discover properties (normally physical) of two pieces of evidence that are unique—that is, they are not possessed by any other members of the class of similar materials—then the evidence is said to have been individualized. An example would be the broken turn signal lens in a hit-and-run case: If the broken pieces of plastic found at the crime scene can be fit back to the lens from the suspected automobile, then it is reasonable to conclude that those pieces of plastic were previously one continuous piece of plastic. This conclusion implies that there is no other lens in the entire world that those broken pieces could have come from. Clearly, no one has tested these pieces of plastic against all of the other, similar broken turn signal lenses to see if they could fit.

Think about the scenario this way. Take ten of the same model cars that have the same type of turn signal lights and break them. Common sense and experience dictate that each breaking of the lenses would result in different size and shape fragments being produced. It would not be reasonable to predict or assume that two breakings would yield exactly the same number and shape of broken pieces. The innumerable variables, such as force of the blow, the shape of the lens, microstructure of the lenses, chemical nature of the material, and direction of the blow, cannot be exactly duplicated; therefore, the number and shapes of the fragments produced are, essentially, random. The probability of two (or more) breaks exactly duplicating the number and shape of fragments is unknown but generally considered to be zero.

Because of the limitations of time, space, and the sheer number of things in the world that could potentially become evidence, far more evidence is at the class level than individual level. The fact that *every single* item of a specific make, model, and type cannot be accounted for in an analysis (think of the number of golf balls in existence, for example), most interpretations of forensic evidence are relegated to statistical interpretation. Even DNA scientists, as will be discussed later, interpret their results through statistics. If an item of evidence cannot be individualized, some form of qualitative or quantitative statistics is used in evaluating its evidentiary significance.

KNOWN AND QUESTIONED ITEMS

Continuing with the hit-and-run example, a motorist strikes a pedestrian with his car and then flees the scene in the vehicle. When the pedestrian's clothing is examined, small flakes and smears of paint are found embedded in the fabric. When the automobile is impounded and examined, fibers are found embedded in an area that clearly has been damaged recently. How is this evidence classified? The paint on the victim's coat is **questioned evidence** because the original source of the paint is unknown. Likewise, the fibers found on the damaged area of the car are also questioned items. The co-location of the fibers and damaged area and the wounds/damage and paint smears are indicative of recent contact. When the paint on the clothing is analyzed, it is compared to paint from the car; this is **known evidence** because it is known where the sample originated. When the fibers on the car are analyzed, they are compared to fibers taken from the clothing, which makes them known items as well. Thus, the coat *and* the car are sources of *both* kinds of items, which allows for their reassociation, but it is their *context* that makes them questioned or known.

Back at the scene where the body is found, there are some pieces of yellow, hard, irregularly shaped material. In the lab, the forensic scientist will examine this debris and will determine that it is plastic, rather than glass, and further it is polypropylene. This material has now been put in the class of substances that are yellow and made of polypropylene plastic. Further testing may reveal the density, refractive index, hardness, and exact chemical composition of the plastic. This process puts the material into successively smaller classes. It is not just yellow polypropylene plastic but has a certain shape, refractive index, density, hardness, etc. In many cases this may be all that is possible with such evidence. The forensic scientist has not been able to determine the exact source of the evidence, but only that it could have come from any of a number of places where this material is used—class evidence.

Suppose that the car that is suspected to be involved in the hit and run has a turn signal lens that is broken and some of the plastic is missing. The pieces are too small and the edges too indefinite for a physical match. Pieces of this plastic can be tested to determine if it has the same physical and chemical characteristics as the plastic found at the crime scene (color, chemical composition, refractive index, etc.). If so, we could say that the plastic found at the scene could have come from that broken lens. This is still class evidence because there is nothing unique about these properties that would be different from similar plastic turn signal lenses on many other cars.

RELATIONSHIPS AND CONTEXT

The relationships between the people, places, and things involved in crimes are critical to deciding what to examine and how to interpret the results. For example, if a sexual assault occurs and the perpetrator and victim are strangers, more evidence may be relevant than if they live together or are sexual partners. Strangers are not expected to have ever met previously and, therefore, would not have transferred evidence before the crime. People who live together would have some opportunities to transfer certain types of evidence (head hairs and carpet fibers from the living room, for example) but not others (semen or vaginal secretions). Spouses or sexual partners, being the most intimate relationship of the three examples, would share a good deal more information. The interaction of these evidence environments is shown in Figure 3.4.

Stranger-on-stranger crimes beg the question of **coincidental associations**; that is, two things which previously have never been in contact with each other have items on them which are analytically indistinguishable at a certain class level. Attorneys in cross examination may ask, "Yes, but couldn't [insert evidence type here] really have come from *anywhere*? Aren't [generic class level evidence] very *common*?" It has been proven for a wide variety of evidence that coincidental matches are extremely rare. The variety of mass-produced goods, consumer choices, economic factors, and other product traits create a nearly infinite combination of comparable characteristics for the items involved in any one situation. Some kinds of evidence, however, are either quite common, such as white cotton fibers, or have few distinguishing characteristics, such as white architectural paint.

It is important to establish the context of the crime and those involved early in the investigation. This sets the stage for what evidence is significant, what methods may be most effective for collection or

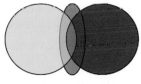

Victim and Criminal only
interact at a Crime Scene
unfamiliar to both

Ex. Sexual assault in an alley

Victim and Criminal interact at a
Crime Scene familiar to both

*Ex. Spouse kills co-habitating
spouse*

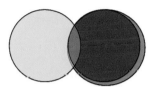

Victim and Criminal interact
at a Crime Scene familiar
only to the Criminal

*Ex. Kidnapping and assault in
Criminal's house*

Victim and Criminal interact
at a Crime Scene familiar only
to the Victim

Ex. Home invasion

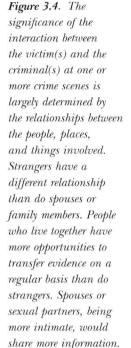

Figure 3.4. *The significance of the interaction between the victim(s) and the criminal(s) at one or more crime scenes is largely determined by the relationships between the people, places, and things involved. Strangers have a different relationship than do spouses or family members. People who live together have more opportunities to transfer evidence on a regular basis than do strangers. Spouses or sexual partners, being more intimate, would share more information.*

Victim

Criminal

Crime Scene

analysis, and what may be safely ignored. Using context for direction prevents the indiscriminate collection of items that clog the workflow of the forensic science laboratory. Every item collected must be transferred to the laboratory and cataloged—at a minimum—and this takes people and time. Evidence collection based on intelligent decision making, instead of fear of missing something, produces a better result in the laboratory and the courts.

COMPARISON OF EVIDENCE

There are two fundamental processes in the analysis of evidence. The first has already been discussed: identification. Recall that identification is the process of discovering physical and chemical characteristics of evidence with an eye toward putting it into successively smaller classes. The other process is **comparison**. Comparison is done in order to try to establish the source of evidence. The questioned evidence is compared with objects whose source is known. The goal is to determine whether or not sufficient common physical and/or chemical characteristics between the samples exist. If they do, it can be concluded that an association exists between the questioned and known evidence. The strength of this association depends on a number of factors, including

- Kind of evidence
- Intra- and inter-sample variation
- Amount of evidence
- Location of evidence
- Transfer and cross-transfer
- Number of different kinds of evidence associated to one or more sources

Individualization occurs when at least one unique characteristic is found to exist in both the known and the questioned samples. Individualization cannot be accomplished by identification alone.

CONTROLS

Controls are materials whose source is known and which are used for comparison with unknown evidence. Controls are often used to determine whether a chemical test is performing correctly. They may also be used to determine if a substrate where evidence may be found is interfering with a chemical or instrumental test. There are two types of controls: positive and negative.

Consider a case in which some red stains are found on the shirt of a suspect in a homicide. The first question that needs to be answered about these stains is: Are they blood? A number of tests can determine whether a stain may be blood. Suppose one of these tests is run on some of the stains and the results are *negative*. There are a number of reasons why this might happen:

- The stain isn't blood.
- The stain is blood, but the reagents used to run the test are of poor quality.
- There is something in the shirt that is interfering with the test.

Before concluding that the stain isn't blood, a number of additional steps could be taken. One might be to run a different presumptive test and see if the results change. Another is to run the first test on a sample that is known to be blood and that should yield a positive test. This known blood is a **positive control**. It is a material that is expected to give a positive result with the test reagents and serves to show that the test is working properly. In this case, if the positive control yields a correct result, then it can be presumed that the reagents are working properly and another cause must be sought for the negative result on the blood-soaked shirt. The shirt fibers may contain some dye or other material, for example, that deactivates the blood test so that it will fail to react with blood. To test this hypothesis, some fibers from the shirt that have absolutely no stains on them could be tested. This would be a **negative control** for the shirt: the results of the test are expected to come out negative. If the test results are negative, it could still mean that the shirt contains something that interferes with the test. This could be verified only by running a different test on the stain. Other negative controls can be run on "blank" samples—that is, those prepared similarly to the test materials being used but without any sample present.

If the initial test for blood was done on the stained shirt and came out positive, it should not immediately be assumed that the stain is definitely blood—a sample of the unstained shirt fibers should be tested as a negative control. A negative result here would mean that the positive result on the stain most likely means that the stain is blood.

What is the consequence of not running a positive or negative control? If a negative control is not used, a **false positive** result may be obtained; that is, it may be concluded that the stain is blood when it is not. This gives rise to what statisticians call a **Type I error**. Type I errors are serious because they can cause a person to be falsely incriminated in a crime.

Failure to run a positive control can cause a **false negative** result. This can give rise to what is called **Type II error**. This type of error means that a person may be falsely exonerated from a crime that he or she really did commit. Any error is problematic, but from the criminal justice standpoint, a Type II error is less serious than a Type I error. The criminal justice system would rather have someone falsely released than falsely accused. Positive and negative controls are usually easy to obtain and should be used to minimize the chance of errors.

ANALYSIS OF EVIDENCE: SOME PRELIMINARY CONSIDERATIONS

Science is a way of examining the world and discovering it. The process of science, the **scientific method**, is proposing and refining of plausible explanations about any unknown situation. It involves asking and answering questions in a formal way and then drawing conclusions from the answers. Science, through its method, has two hallmarks. The first is the questions that are asked must be **testable**. It is not scientific to ask, "How many angels can dance on the head of a pin?" or "Why do ghosts haunt this house?" because a test cannot be constructed to answer either of these questions. The second hallmark of science is **repeatability**. Science is a public endeavor, and its results are published for many reasons, the most important of which is for other scientists to review the work and determine if it is sound. If nobody but you can make a particular experiment work, it isn't science. Other scientists must be able to take the same kinds of samples and methods, repeat your experiments, and get the same results for it to be science (see "History: The Method of Science" for a discussion of scientific models).

In the language of science, the particular questions to be tested are called **hypotheses**. Suppose hairs are found on the bed where a victim has been sexually assaulted. Are the hairs those of the victim, the suspect, or someone else? The hypothesis could be framed as: "There is a significant difference between the questioned hairs and the known hairs from the suspect." Notice that the hypothesis is formed as a neutral statement that can be either proven or disproven.

After the hypothesis has been formed, the forensic scientist seeks to collect data that sheds light on the hypothesis. Known hairs from the suspect are compared with those from the scene and the victim. All relevant data will be collected without regard to whether it favors the hypothesis. After it is collected, the data will be carefully examined to determine what value it has in proving or disproving the hypothesis; this is its **probative value**. If the questioned hairs are analytically indistinguishable from the known hairs, then the hypothesis is rejected. The scientist could then conclude that the questioned hairs could have come from the suspect.

But suppose that *most* of the data suggests that the suspect is the one who left the hairs there, but there is not enough data to associate the hairs to him. It cannot be said that the hypothesis has been *dis-*

HISTORY:
The method of science

[An important person in the history of science] was not a scientist at all, but a lawyer who rose to be Lord Chancellor of England in the reign of James I, Elizabeth's successor. His name was Sir Francis Bacon, and in his magnum opus, which he called *Novum Organum*, he put forth the first theory of the scientific method. In Bacon's view, the scientist should be a disinterested observer of nature, collecting observations with a mind cleansed of harmful preconceptions that might cause error to creep into the scientific record. Once enough such observations have been gathered, patterns will emerge from them, giving rise to truths about nature.

Bacon's idea, that science proceeds through the collection of observations without prejudice, has been rejected by all serious thinkers. Everything about the way we do science—the language we use, the instruments we use, the methods we use—depends on clear presuppositions about how the world works. At the most fundamental level, it is impossible to observe nature without having some reason to choose what is worth observing and what is not worth observing.

In contrast to Bacon, [Sir Karl] Popper believed all science begins with a prejudice, or perhaps more politely, a theory or hypothesis. Popper was deeply influenced by the fact that a theory can never be proved right by agreement with observation, but it can be proved wrong by disagreement with observation. Because of the asymmetry, science makes progress uniquely by proving that good ideas are wrong so that they can be replaced by even better ideas. Thus, Bacon's disinterested observer of nature is replaced by Popper's skeptical theorist.

Popper's ideas . . . fall short in a number of ways in describing correctly how science works. Although it may be impossible to prove a theory is true by observation or experiment, it is nearly just as impossible to prove one is false by these same methods. Almost without exception, in order to extract a falsifiable prediction from a theory, it is necessary to make additional assumptions beyond the theory itself. Then, when the prediction turns out to be false, it may well be one of the other assumptions, rather than the theory itself, that is false.

It takes a great deal of hard work to come up with a new theory that is consistent with nearly everything that is known in any area of science. Popper's notion that the scientist's duty is then to attack that theory at its most vulnerable point is fundamentally inconsistent with human nature. It would be impossible to invest the enormous amount of time and energy necessary to develop a new theory in any part of modern science if the primary purpose of all that work was to show that the theory was wrong.

Another towering figure in the twentieth century theory of science is Thomas Kuhn. A paradigm, for Kuhn, is a sort of consensual world view within which scientists work. Within a given paradigm, scientists make steady, incremental progress, doing what Kuhn calls "normal science."

As time goes on, difficulties and contradictions arise that cannot be resolved, but one way or another, they are swept under the rug, rather than be allowed to threaten the central paradigm. However, at a certain point, enough of these difficulties have accumulated so that the situation becomes intolerable. At that point, a scientific revolution occurs, shattering the paradigm and replacing it with an entirely new one.

If a theory makes novel and unexpected predictions, and those predictions are verified by experiments that reveal new and useful or interesting phenomena, then the chances that the theory is correct are greatly enhanced. [However, science] does undergo startling changes of perspective that lead to new and, invariably, better ways of understanding the world. Thus, science does not proceed smoothly and incrementally, but it is one of the few areas of human endeavor that is truly progressive. [Science] is, above all, an adversary process. The scientific debate is very different from what happens in a court of law, but just as in the law, it is crucial that every idea receive the most vigorous possible advocacy, just in case it might be right.

Excerpted from: Goodstein, D. How science works. In *Reference manual on scientific evidence* (2nd ed) (pp. 67–82). Washington, D.C.: Federal Judicial Center.

proved (there are some similarities), but neither can it be said that it has been *proved* (some differences exist but are they significant?). Although we would like to be able to prove unequivocally that someone is or is not the source of evidence, it is not always possible. As we have previously learned, not all evidence can be individualized. The important thing to note here is that evidence analysis proceeds by forming many hypotheses and perhaps rejecting some as the investigation progresses.

Some preliminary questions must be answered before we even begin to formulate hypotheses. Is there sufficient material to analyze? If the amount of the evidence is limited, then choices have to be made about which tests to perform and in what order. The general rule is to perform non-destructive tests first because they conserve material. Most jurisdictions also have evidentiary rules that require that some evidence be kept for additional analyses by opposing experts; if the entire sample will be consumed in an analysis, then both sides must be informed that not enough evidence will be available to have additional analyses performed.

If extremely large amounts of material are submitted as evidence, how are they sampled? This often happens in drug cases in which, for example, a 50-pound block of marijuana or several kilograms of cocaine are received in one package. The laboratory must have a protocol for sampling large quantities of material so that samples taken are representative of the whole. In the other kinds of cases where this occurs, many exhibits appear to contain the same thing—100 0.5-ounce packets of white powder. The laboratory and the scientist must decide how many samples to take and what tests to perform. This is especially important because the results of the analyses will ascribe the characteristics of the samples to the whole exhibit, such as identifying a thousand packets of powder as 23% cocaine based on analysis of a fraction of the packets.

What happens in cases in which more than one kind of analysis must be done on the same item of evidence? Consider a handgun received into evidence from a shooting incident with red stains and possible fingerprints on it. This means that firearms testing, serology, latent print, and possibly DNA analysis must be performed on the handgun. These analyses should be put into an order so that one exam does not spoil or preclude the subsequent exam(s). In this case, the order should be first serology, then latent print, and finally firearms testing.

Figure 3.5. *Even one small item of evidence can be subjected to multiple examinations and may travel through most of a forensic laboratory. A threat letter, like this one, could pass through bacterial diagnosis, trace evidence, DNA, questioned documents, latent print analysis, and content analysis (© Yahoo News, with permission).*

It is important to note that one seemingly small piece of evidence can be subjected to many examinations. Take the example of a threatening letter, as depicted in Figure 3.5, one that supposedly contains anthrax or some other contagion. The envelope and the letter could be subjected to the following exams, in order:

- *Disease diagnosis,* to determine if it really contains the suspected contagion
- *Trace evidence,* for hairs or fibers in the envelope or stuck to the adhesives (stamp, closure, tape used to seal it)
- *DNA,* from saliva on the stamp or the envelope closure
- *Questioned documents,* for the paper, lettering, and other aspects of the form of the letter
- *Ink analysis,* to determine what was used to write the message, address, etc.
- *Handwriting, typewriter,* or ***printer analysis,*** as appropriate
- *Latent fingerprints*
- *Content analysis,* to evaluate the nature of the writer's intent and other investigative clues

In this example, the ordering of the exams is crucial not only to ensure the integrity of the evidence, but also the safety of the scientists and their co-workers. Other evidence can also be very, very large—ocean currents, for example (see "In More Detail: Rubber Duckies and Human Remains"). It is important to realize that *anything* can become evidence, and forensic scientists must keep open minds if they are to solve the most difficult of crimes.

IN MORE DETAIL:
Rubber duckies and human remains

In January 1992, a container ship en route from Hong Kong to America encountered a storm, and several containers broke free from their moorings and dropped into the water. At least one, containing 29,000 plastic bath toys, split open. Drifting at the whim of the wind and ocean currents are ducks, along with red beavers, green frogs, and blue turtles. Scientists estimate the toys moved up the western coast of North America, crossed the waters of the North Pole, and were—as of 2003—headed toward the U.K, as shown in Figure 3.6.

Oddly, very little is known about how winds and currents move drifting objects. Two scientists, Curtis Ebbesmeyer, an oceanographer in Seattle, and James Ingraham, a scientist at the National Marine Fisheries Service, have been carefully recording each reported sighting of the plastic toys to better understand the phenomena. Beachcombers report sightings of finds to **www.beachcombers.org** and the data is entered into Ingraham's ocean modeling program, OSCUR (Ocean Surface Currents Simulation). OSCUR uses air pressure metrics dating back to 1967 to calculate wind speed, direction, and surface currents. The floating toy finds help the scientists to check and improve the performance of OSCUR.

Ebbesmeyer and Ingraham have tracked the journeys of everything from toy cars, balloons, ice hockey gloves, even five million pieces of Lego, all lost from ships over the years. Recently, they've been processing data from 33,000 Nike shoes that fell off a ship near California. OSCUR estimated a landing for about 1,600 of the shoes (roughly 2% of the dunked shoes)—this is as accurate an estimate as that of oceanographers who deliberately release objects to measure currents.

But are cute bathtoys and scientific ingenuity *forensic?* Using OSCUR, Ebbesmeyer predicted the final resting place of George Karn, a crewman lost from the *Galaxy*, a freezer long-liner that caught Pacific cod with miles of baited hooks, which sunk in the Bering Sea in 2002. Starting from the location of the *Galaxy's* sinking, the model ran forward in time and came up with a location—an island 430 nautical miles southwest of the disaster. On June 9, 2003, while working at Portage Bite, a seldom-visited site on Tanaga Island far west in the Aleutians and 1,400 nautical miles due north of Hawaii's Midway Island, a beachcomber spotted a lower jawbone; the extensive dental work told him it was human. Upon subsequent search of the area, an orange survival suit was discovered. State troopers traced the suit's serial number to the *Galaxy*. Karn's body drifted in an unusual way, possibly leading to his delayed discovery. The two calculated where Karn would have drifted if lost on the same day (October 20) of each year from 1967 to 2002. These drifts terminate after 3.5 months, the time interval between the disaster and Tanaga Island. All but three of the 36 drifts headed west toward Siberia, nearly the opposite direction of where Karn drifted. If George had perished in most years except 2002, he would have drifted west toward Kamchatka, then south into the wide North Pacific, never to be found.

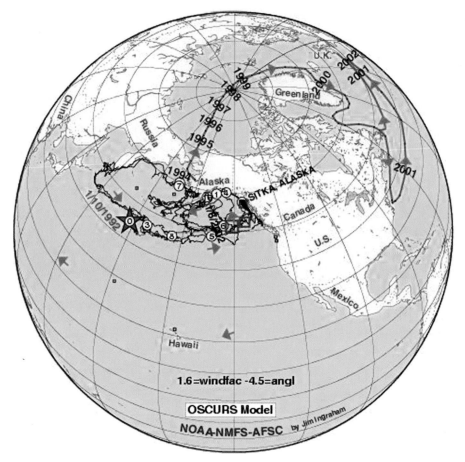

Figure 3.6. *Calculated drifts of bathtub toys lost at sea. Even seemingly obscure information like this can be of use in solving crimes and finding victims (C. Ebbesmeyer, with permission).*

CHAPTER SUMMARY

Anything can be submitted for scientific analysis in an investigation, becoming the samples that yield data for the forensic scientist to interpret. As evidence, however, these samples and data follow different rules than in other scientific, non-forensic laboratories. The context of the evidence is central to how it is analyzed and interpreted in the reconstruction of the criminal events. The scientific method still applies, however, and forensic scientists still employ that approach as do other non-forensic scientists. These differences and similarities will follow the forensic scientist into the courtroom and either support, if done well, or weaken, if done poorly, the fruits of their scientific labors.

Test Your Knowledge

1. What is a "trier-of-fact"?

2. What is evidence?

3. Name four kinds of evidence.

4. What is exculpatory evidence?

5. What is "proxy data"?

6. How is direct transfer different from indirect transfer? Give an example.

7. What is persistence in relation to evidence?

8. Is contamination evidence? Why or why not?

9. What is class-level evidence?

10. What does it mean to identify something?

11. What is a "common source"?

12. If you have individualized two pieces of evidence, how many common sources could they have come from?

13. What is the difference between questioned and known evidence?

14. What is a control? How is it different from known evidence?

15. What is the probative value of an item of evidence?

16. What is the difference between a Type I and a Type II error?

17. What are the two hallmarks of science?

18. What is a cross-transfer?

19. Name three ways an association between a questioned and known item can be *strengthened*.

20. Name three ways an association between a questioned and known item can be *weakened*.

Consider This . . .

1. How do transfer and persistence relate? How would this affect the collection of evidence? What would be the difference in processing a crime scene one hour after the crime and 48 hours afterward?

2. Why is context important to forensic science? How does this determine what evidence should be collected and analyzed?

3. Why is forensic science a historical science? Does this make it inferior to non-historical sciences? What are the limits of historical sciences?

BIBLIOGRAPHY

Collins, H.M., & Pinch, T. (1998). *The Golem: What you should know about science.* Cambridge, UK: Cambridge University Press.

Federal Judicial Center. (2000). *Reference manual on scientific evidence* (2nd ed.). Washington, D.C: Author.

Garner, B.A. (Ed.) (2000). *Black's Law Dictionary.* 7th ed., St. Paul, MN: West Group.

Klotter, J., & Ingram, J.L. (2003). *Criminal evidence* (8th ed.). Philadelphia: Anderson Publishing.

Locard, E. (1930). Analysis of dust traces. *American Journal of Police Science,* 276–298.

Ou, C., Ciesielski, C.A, Myers, G., Bandea, C.I., Luo, C., Korber, B.T.M., Mullins, J.I., Schochetman, G., Berkelman, R.L., Economou, A.N., Witte, J.J., Furman, L.J. Satten, G.A., MacInnes, K.A., Curran, J.W., & Jaffe, H.W. (1992). Laboratory Investigation Group, Epidemiologic Investigation Group, "Molecular Epidemiology of HIV Transmission in a Dental Practice," *Science* 256, 1165–1171.

Pounds, C.A., & Smalldon, K.W. (1975). The transfer of fibres between clothing materials during simulated contacts and their persistence during wear, Part II—fibre persistence. *Journal of the Forensic Science Society* 15, 29–37.

Thornton, J.I. (1986). Ensembles of subclass characteristics in physical evidence examination. *Journal of Forensic Sciences* 31(2), 501–503.

Zobel, H.B. (1996). *The Boston Massacre.* New York City: W.W. Norton.

PART
TWO

Analytical Tools

Microscopy

KEY TERMS

Achromatic objectives

Analyzer

Anisotropic

Apochromats

Astigmatism or spherical aberration

Binocular

Birefringence

Chromatic aberration

Compound magnification system

Condenser

Condenser diaphragm

Coverslips

Critical illumination

Curvature of field

Empty magnification

Eyepiece or ocular

Field diaphragm

Field of view

Fluorescence

Fluorites or semi-apochromats

Fluorophores

Focal length

Focus

Infinity-corrected lens systems

Interference colors

Isotropic

Köhler illumination

Lens

Mechanical stage

Monocular

Mounting media or mountants

Normal

Numerical aperture

Objective

Phosphorescence

Plan achromats

Polarization colors

Polorized light

Polarizer

Polarizing light microscope

Real image

Resolution

Rotating stage

Simple magnification system

Snell's Law

Spherical aberration

Stage

Tube length

Virtual image

INTRODUCTION

The microscope is a nearly universal symbol of science, representing our ability to explore the world below the limits of our perception. Forensic science is equally well represented by the microscope; illustrations in Sir Arthur Conan Doyle's Sherlock Holmes stories show the great detective peering through a microscope at some minute evidence. As Dr. Peter DeForest (2002, 217) has stated, "Good criminalistic technique demands the effective use of the microscope."

The microscope may seem to be a relic of an antiquated age of science when compared with some of today's advanced instrumentation. But, as the life's work of Dr. Walter McCrone and others has shown, microscopy is applicable to every area of forensic science (see Table 4.1). Microscopy can be as powerful as many current technologies and, in some cases, more powerful. For example, microscopy can easily distinguish between cotton and rayon textile fibers, whereas to an infrared spectrometer they both appear to be cellulose.

Forensic microscopy is more than simply looking at small things. It requires the student (and the expert) to know a great deal about

Table 4.1. *Microscopy has nearly unlimited application to forensic sciences.*

ART FORGERIES	ASBESTOS	BUILDING MATERIALS
Bullets	Hairs	Product tampering
Chemistry	Handwriting	Questioned documents
Drugs	Minerals	Serology
Dust	Paint	Soil
Fibers	Paper	Tapes
Fingerprints	Photographic analysis	Toolmarks
Food poisoning	Pollen	Wood
Glass	Polymers	

many things, how they are made, how they are used, and their physical and chemical natures. Chamot and Mason (1940, 1958), in their classic text *Handbook of Chemical Microscopy, Volume 1*, succinctly describe the role of the forensic (or technical, in their words) microscopist:

> The technical microscopist is concerned with form, but also with formation and function. He needs to know, as completely as possible, the existing structure of the specimen, but he frequently has to investigate or at least postulate how that structure developed or was produced, how it can be controlled, and how it affects performance. The correlation of these three aspects of his studies is too specific to the material involved to be dealt with here .... But even descriptive microscopy often requires more than superficial observation, or the ordinary arts of varying focus and illumination that experience makes habitual. And there are many properties closely governing non-microscopical behavior that can be usefully explored, as a background for understanding it and as an adjunct to tests on a larger scale. (p. 173)

A full explanation of microscopy and the optical principles involved is beyond the scope of this book—the physics and geometry get complicated. Additional details will be listed throughout this chapter, but only the core information necessary for an understanding of the fundamentals of microscopy will be presented. For a fuller treatment of the optical theory of microscopy, see DeForest (2002) or McCrone, McCrone, and Delly (1978).

MAGNIFICATION SYSTEMS

If you want to see more detail in an object—a postage stamp, for example—you need to magnify the image, as shown in Figure 4.1. The easiest way to do this is with a common pocket magnifier or hand lens; this is a **simple magnification system**, a single lens used to form an enlarged image of an object. A similar system is used to project the image of a 35 mm slide or transparency in a lecture hall. If you removed the screen where the focused image is projected and put, say, the hand lens in its place, you would produce a second, larger image, as shown in Figure 4.2. This is the basic principle of all microscopy—a **compound magnification system**, where magnification occurs in two stages and the total magnification is the product of the magnification of the first lens and the second lens. The observer looks at the first image

Figure 4.1. *To see more detail in an object, here a postage stamp, the observer must magnify the image. This is accomplished with a simple hand lens, which enlarges the image 10 times, or 10×.*

Figure 4.2. *In a compound magnification system, magnification occurs in two stages, and the total magnification is the product of the first lens and the second lens. So, a 10× lens and a 4× lens would produce a 40× image (10 × 4 = 40), or one that has been magnified 40 times. The observer looks at the first image with a lens that produces an enlarged image called a virtual image. This is the image the eye perceives and is visible only as a result of the compound magnification system.*

with a lens that produces an enlarged image called a **virtual image**. This is the image the eye perceives—a real, projectable image does not exist where the virtual image appears to be—and is visible only as a result of the compound magnification system. A more commonplace example of a virtual image is that seen in a mirror: If you are standing 2′ away from the mirror, your image in the mirror looks as if it is standing 2′ away from the other side of the mirror. Were a white screen or glass plate substituted for the mirror, no image would be visible. By contrast, a **real image** is one that could be seen *on* the screen—that is, projected *onto* the screen.

THE LENS

Most of us are familiar with the lenses in our daily lives: eyeglasses, reading magnifiers, and the like. In microscopy, a **lens** means a very specific thing: a translucent material that bends light in a known and predictable manner. For example, an ideal converging lens causes all light entering the lens from one side of the lens to meet again at a point on the other side of the lens, as shown in Figure 4.3. In doing so, an image of the original object is produced.

The size and position of an image produced by a lens can be determined through geometry based on the **focal length** of the lens, which is the distance between the two points of focus on either side of the lens, as shown in Figure 4.4. Focal length is important in microscopy because it determines much of the image quality. Think of it this way: If one eye is too far or too close, we have difficulty seeing a clear image of an object. Why? Our eyes, being curved, cannot maintain a clear point of focus for all distances: About 10 inches or 25 cm is the distance that a human eye can easily distinguish between two objects next to each other. Lenses are made using this "ideal" viewing distance, or focal length.

At 25 cm, the **resolution**, or the minimum distance two objects can be separated and still be seen as two objects, of the human eye is between 0.15 and 0.30 mm. Therefore, this is the limit of our eyes without assistance. If we want better resolution (that is, to see more detail in the postage stamp), we must magnify the image. If the hand lens we use magnifies an image four times (the shorthand for this is "4×"), then we will be able to resolve two objects that are about 0.05 mm apart (for the math used to obtain this

Figure 4.3. An ideal converging lens causes light entering it from one side to meet again (converge) at a point on the other side of the lens.

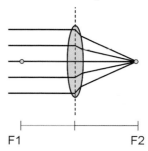

F1   F2

value, see "In More Detail: Why Resolution Is More Important Than Magnification"). Magnification with one lens cannot continue indefinitely, however. As magnification *increases*, lens diameter *decreases* to bend the light more to make a larger image. A simple lens that magnifies 1000× would be only 0.12 mm in diameter! Therefore, about 10× to 15× is the practical limit of magnification for simple lenses.

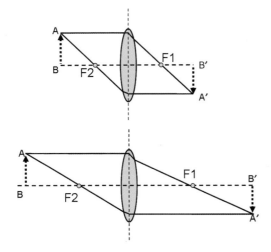

Figure 4.4. The distance from the two focal points (F1 and F2) is the focal length of a lens. F2 sits on a plane where the image will appear to be in focus when the object being viewed is at F1.

COMPOUND MAGNIFYING SYSTEMS

A compound microscope, as the name implies, employs a magnification system that exceeds the limits imposed by simple lenses. A second lens is placed in line with the first lens, and this further enlarges the image. The total magnification of the microscope is the product of the two lenses. A 10× lens and a 4× lens would produce a 40× image (10 × 4 = 40), or one that has been magnified 40 times. Lenses of up to 40× can be used in a compound microscope, and higher magnifications are possible with special lenses.

Even lenses in compound microscopes have resolution limits, however, and it is possible to continue to magnify an image but not improve its resolution; this is called **empty magnification**. The result of empty magnification is a larger but fuzzier looking image, as shown in Figure 4.5.

Figure 4.5. Empty magnification results from an increase in magnification without an increase in resolution.

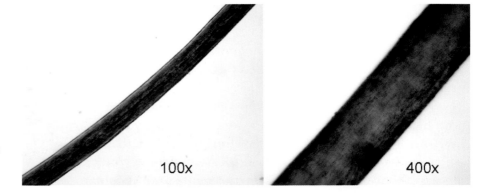

100x   400x

THE MICROSCOPE

The fundamental design of the microscope has not changed much since its original invention; improvements to nearly every component, however, have made even the most inexpensive microscopes suitable for basic applications. In this section, refer to Figure 4.6 for a diagram of the important parts of a microscope.

Starting at the top, the **eyepiece** or **ocular** is the lens that you look into when viewing an object microscopically. A microscope may be **monocular**, having one eyepiece, or **binocular**, having two eyepieces; most microscopes found in laboratories today are binocular. Many microscopes today are trinocular; they have an eyepiece that accommodates a video or digital camera. Typically, the eyepiece(s) will have a magnification of 10× and may be focusable; this enables the viewer to adjust the eyepieces if one eye is stronger than the other. The area seen when looking through the eyepieces is called the **field of view** and will change if the specimen is moved or the magnification is changed.

The next lens in the microscope is called the **objective** lens (or just "the objective"), because it is closest to the object or specimen being studied. The objective is the most important part of the microscope. Objectives come in many types (see "In More Detail: Lens Corrections") and magnifications (typically, 4×, 10×, 15×, 20×, and 25×; higher magnifications are possible). Each objective will have information about it engraved into its body in a specific format, as shown in Figure 4.7. Although the information may vary by manufacturer, objectives

Figure 4.7. *The objective lens, called so because it is closest to the object or specimen being viewed. The objective is the most important part of the microscope and comes in many types and magnifications. The information on the lens is very specific: "10×" is the magnification, "0.25 na" is the numerical aperture, "170 mm" is the tube length (some objectives are now infinity-corrected and are labeled "∞"), and "0.17 mm" is the recommended thickness of cover slip to use.*

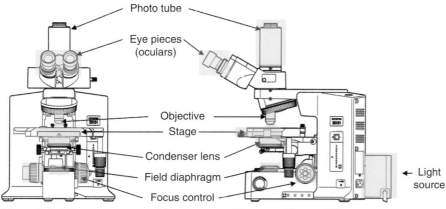

Figure 4.6. *The various parts of the microscope (Courtesy: Olympus USA).*

IN MORE DETAIL:
Why resolution is more important than magnification

In the card game of microscopy, numerical aperture always trumps magnification. This short aside should give you an understanding of why that is so.

The minimum distance d, which must exist between two separate points in the specimen in order for them to be seen as two distinct points, is shown in equation 4.1:

Equation 4.1 $d = \lambda/2NA$

or the wavelength of light divided by twice the numerical aperture (NA). The numerical aperture is further defined as shown in equation 4.2:

Equation 4.2 $NA = n\ sine\ u$

where n represents the refractive index of the medium between the coverslip and the front lens, and u is half the angle of aperture of the objective (refer to Figure 4.8). The refractive index of air is 1.0; practically speaking, this means the NA of any lens system with air as the intermediate medium (so-called dry systems; other systems use oil as the intermediate medium, improving their NA) will be less than 1 because half of the angle u in air cannot be more than 90°.

The resolving power of the human eye or the objective lens is not enough for a magnification of, say, 10,000×, because two points on the object can be seen as separate only if the distance between them is within the limit of the resolving power. If the distance is below the resolving power, then you wouldn't see two objects; if it were higher, you would see only two (and not several) points with no more detail than before. The maximum useful magnification available is about 1,000 times the NA of the objective.

Source: Davidson and Abromowitz, 2005.

Figure 4.8. The numerical aperture is an angular measure of the lens' light-gathering ability. It is an indication of the lens' resolving power.

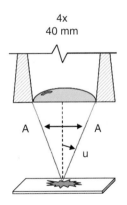 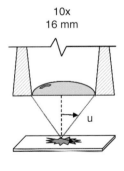 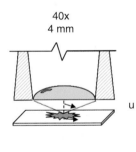

will usually have the magnification, the numerical aperture, the tube length, and the thickness of coverslip that should be used with the objective. The **numerical aperture** is an angular measure of the lens' light-gathering ability and, ultimately, its resolving quality, as shown in Figure 4.8 (see "In More Detail: Why Resolution Is More Important Than Magnification"). The **tube length** is the distance from the lowest

IN MORE DETAIL:
Lens corrections

Achromatic objectives are the least expensive objectives, and they are found on most microscopes. These objectives are designed to be corrected for **chromatic aberration**, where white light from the specimen is broken out into multiple colored images at various distances from the lens. Achromats are corrected for red and blue only, and this can lead to substantial artifacts, such as colored halos. Because of this, it may be necessary to use a green filter and employ black-and-white film for photomicrography.

A simple lens focuses a flat specimen on a microscope slide onto the lens, a rounded surface. This results in an aberration called **curvature of field** and results in only part of the image being in focus. Regular achromats lack correction for flatness of field, but recently most manufacturers have started offering flat-field corrections for achromat objectives, called **plan achromats**.

Astigmatism or **spherical aberration** results from a lens not being properly spherical. As a result, specimen images seem to be "pulled" in one direction when you are focusing through the lens. Most modern microscope objectives are corrected for spherical aberration.

Fluorites or **semi-apochromats** are a step up in corrected lenses because the mineral fluorite was the original method used for correction. Fluorites are also corrected for **spherical aberration**, where the light passing near the center of the lens is less refracted than the light at the edge of the lens. Fluorite objectives are now made with advanced glass formulations that contain fluorspar or synthetic substitutes. These materials give fluorites a higher numerical aperture, better resolution, and higher contrast. The cost for fluorite objectives, of course, is higher than that for achromats.

The most highly corrected objectives are the **apochromats**, which contain several internal lenses that have different thicknesses and curvatures in a specific configuration unique to apochromats. Apochromats are corrected for three colors (red, green, and blue) and, thus, have almost no chromatic aberration. They are very costly but provide even better numerical aperture and resolution than fluorites.

In the last decade, major microscope manufacturers have all migrated to **infinity-corrected lens systems**. In a typical microscope, the tube length (distance from the top of the eyepiece to the bottom of the objective) is set to 160 mm, but in these systems, the image distance is set to infinity, and a lens is placed within the tube between the objective and the eyepieces to produce the intermediate image. Infinity-corrected lens systems produce very high quality images and allow for the addition of a variety of analytical components to the microscope. More information on infinity-corrected lenses and microscopy can be found on the Web at **www.microscopyu.com**.

part of the objective to the upper edge of the eyepiece; this has been standardized at 160 mm in modern microscopes. Because the tube length determines where the in-focus image will appear, objectives must be designed and constructed for a specific tube length (however, read about "infinity-corrected" lenses in "In More Detail: Lens Corrections"). **Coverslips**, the thin glass plates that are placed on top of mounted specimens, protect the specimen and the objective from damage. They come in a range of thicknesses measured in millimeters

(0.17 mm, for example). All of this information is important to the microscopist's proper use of a particular objective.

The microscope **stage** is the platform where the specimen sits during viewing. The stage can be moved up or down to **focus** the specimen image, meaning that portion of the specimen in the field of view is sitting in the same horizontal plane; typically, stages are equipped with a coarse and fine focus. Stages may be **mechanical** (that is, having knobs for control of movement), **rotating** (able to spin 360° but not move back and forth), or both.

The **condenser** is used to obtain a bright, even field of view and improve image resolution. Condensers are lenses below the stage that focus or condense the light onto the specimen field of view. Condensers also have their own **condenser diaphragm** control to eliminate excess light and adjust for contrast in the image. The condenser diaphragm is different from the **field diaphragm**, a control that allows more or less light into the lens system of the microscope.

The illumination of the microscope is critical to a quality image and is more complicated than merely turning on a light bulb. Two main types of illumination are used in microscopy: critical and Köhler. **Critical illumination** concentrates the light on the specimen with the condenser lens; this produces an intense lighting that highlights edges but may be uneven. **Köhler illumination**, named after August Köhler in 1893, sets the light rays parallel throughout the lens system, allowing them to evenly illuminate the specimen. Köhler illumination is considered the standard setup for microscopic illumination (Davidson and Abromowitz, 2005).

REFRACTIVE INDEX

The refraction of visible light is an important characteristic of lenses that allows them to focus a beam of light onto a single point. Refraction (or bending of the light) occurs as light passes from one medium to another when there is a difference in the index of refraction between the two materials, and is responsible for a variety of familiar phenomena such as the apparent distortion of objects partially submerged in water.

Refractive index is defined as the relative speed at which light moves through a material with respect to its speed in a vacuum. By convention, the refractive index of a vacuum is defined as having a value

Table 4.2. *The refractive indices of several materials.*

MATERIAL	REFRACTIVE INDEX
Air	1.0008
Ice	1.310
Water	1.330
Glass, soda-lime	1.510
Ruby	1.760
Diamond	2.417

of 1.0. The index of refraction, N (or n), of other transparent materials is defined through equation 4.3:

$$\text{Equation 4.3} \qquad N = C/v$$

where C is the speed of light, and v is the velocity of light in that material. Because the refractive index of a vacuum is defined as 1.0 and a vacuum is devoid of any material, the refractive indices of all transparent materials are therefore greater than 1.0. For most practical purposes, the refractive index of light through air (1.0008) can be used to calculate refractive indices of unknown materials. Refractive indices of some common materials are presented in Table 4.2.

When light passes from a less dense medium (such as air) to a more dense medium (such as water), the speed of the wave decreases. Alternatively, when light passes from a more dense medium (water) to a less dense medium (air), the speed of the wave increases. The angle of refracted light is dependent on both the angle of incidence and the composition of the material into which it is entering. We can define the **normal** as a line perpendicular to the boundary between two substances. Light will pass into the boundary at an angle to the surface and will be refracted according to **Snell's Law**, as shown in equation 4.4:

$$\text{Equation 4.4} \qquad N_1 \times sin(q_1) = N_2 \times sin(q_2)$$

where N represents the refractive indices of material 1 and material 2, and q are the angles of light traveling through these materials with

Figure 4.9. Samples to be viewed microscopically must be mounted in a material that has an RI near their own; for example, Permount has an RI of 1.525 and hair has an RI around 1.5. If the RI of the sample and the mountant are too different, like this hair in water (RI = 1.33), then optical distortion results. The RI of glass is about 6° more than that for water, meaning that a light ray gets bent more passing through glass than water.

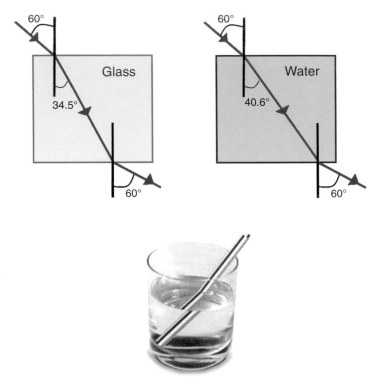

respect to the normal. Several important points can be drawn from this equation. When $N(1)$ is greater than $N(2)$, the angle of refraction is always smaller than the angle of incidence. Alternatively, when $N(2)$ is greater than $N(1)$, the angle of refraction is always greater than the angle of incidence. When the two refractive indices are equal ($N(1) = N(2)$), then the light is passed through without refraction. The concept of refractive index is illustrated in Figure 4.9 for the case of light passing from air through both glass and water. Notice that while both beams enter the more dense material through the same angle of incidence with respect to the normal (60°), the refraction for glass is almost 6° more than that for water due to the higher refractive index of glass.

Samples to be viewed in transmitted light must be in a material with a refractive index that is close to their own. Numerous materials are commercially available to use as **mounting media** or **mountants**. The RI of water is about 1.33 and therefore makes a poor mounting medium because it refracts the light so much less than a hair, which has a refractive index of about 1.5.

POLARIZED LIGHT MICROSCOPY

One of the most powerful tools forensic scientists have at their disposal is the **polarizing light microscope**, a tool of nearly infinite uses and applications. Sadly, in this age of computerized instrumentation, few scientists routinely use a polarized light microscope, or PLM. Something can be learned about almost every kind of sample, from asbestos to zircon, by using PLM. The PLM exploits optical properties of materials to discover details about the structure and composition of materials, and these discoveries lead to its identification and characterization.

Materials fall into one of two categories. The first are materials that demonstrate the same optical properties in all directions, such as gases, liquids, and certain glasses and crystals. These are **isotropic** materials. Because they are optically the same in all directions, they have only one refractive index. Light, therefore, passes through them at the same speed with no directional restrictions.

The second category is **anisotropic** materials, which have optical properties that vary with the orientation of the incoming light and the optical structure of the material. About 90% of all solid materials are anisotropic. The RIs vary in anisotropic materials depending both on the direction of the incident light and on the optical structure. Think of anisotropic materials as having a "grain," like wood, with preferential orientations, as illustrated in Figure 4.10.

Because of their inhomogeneous internal structure, anisotropic materials divide light rays into two parts. PLM uses this to cause the light rays to interact in a way that yields information about the material. Light is emitted from a source in all directions; in the wave model of light, all directions of vibration are equally possible. If the light passes through a special filter, called a **polarizer**, then the only light that passes is that which vibrates in that "preferred" direction; light that vibrates in only one direction is called **polarized light** (see Figure 4.11). Our eyes are "blind" to the vibrational direction of light; it can be seen only by a color effect or by intensity. This may sound complicated, but chances are good that you have seen polarized light—through polarized sunglasses! They reduce the glare, like off a car hood on a sunny day, by filtering out all the light except for that

Figure 4.10. Isotropic materials have the same optical properties in all directions, whereas anisotropic ones have differing properties based on the incident light and the internal structure of the material. Anisotropic materials can be envisioned as having a "grain."

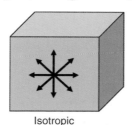
Isotropic

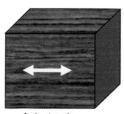
Anisotropic

Figure 4.11. *Because of the orientation of the polarizing filter, only light rays that are in line with its orientation can pass through. This is how polarized sunglasses work, by filtering out scattered light rays and allowing only certain ones through. When the preferred orientations of the filters, sometimes called polars, are at right angles to each other, no light can pass through. Varying degrees of rotation will allow progressively more light through until the polars are aligned. (Source: Wikipedia,* www.wikipedia.com, *with permission).*

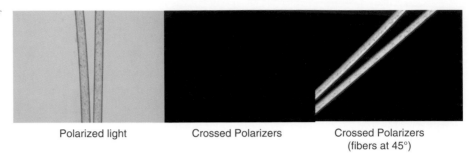

Polarized light   Crossed Polarizers   Crossed Polarizers
(fibers at 45°)

which is traveling in the direction preferred by the orientation of the treated sunglass lens.

All light that reflects off a flat surface is at least partially polarized. The easiest way to visualize polarization is to imagine a wave vibrating perpendicular to the direction in which it's traveling. The light can move in two directions or vectors (the *x* and *y* components). In this simple example, assume the two components have exactly the same frequency (occurrence over time). The *x* and *y* components can differ in two other ways. The two components may differ in amplitude, and the two components may not have the same phase (they may not hit their peaks and troughs at the same time). By tracing the shape as the light wave, the light's polarization state can be described as illustrated in Figure 4.11.

A PLM uses two polarizing filters (or **polarizers**, sometimes called **polars**, for short): one called the polarizer (that's obvious, isn't it?) and the other called the analyzer (for reasons that will become obvious). The polarizer sits beneath the stage and has its preferred vibration direction set left-to-right (sometimes called the "east-west" direction, like on a map). The **analyzer**, aligned opposite that of the polarizer (that is, north-south), is located above the objectives; the analyzer can be manually slid into or out of the light path. If the analyzer is inserted with its orientation opposite that of the analyzer (at right angles), what should be seen? Nothing. The filters are said to be crossed, and no light can pass through the microscope to the viewer's eyes. The field of view appears black or very, very dark, as shown in Figure 4.12. Information can be obtained both in plane-polarized light (only the polarizer in place) or with crossed polarizers (polarized *and* analyzer in place).

Anisotropic materials split light into component light rays. Birefringence is the result of this division of light into at least two rays (the ordinary ray and the extraordinary ray) when it passes through certain

Figure 4.12. *When an anisotropic material is placed under crossed polarizers and rotated on the optical axis of the microscope, polarization colors result. (left) A grain of sillimanite, a mineral component found in a soil sample from a crime scene (Academic Press, by permission). (middle, right) A section of brass metal with a fracture in polarized light and under cross-polarizers (Carl Zeiss, with permission).*

types of material, depending on the polarization of the light. Two different refractive indices are assigned to the material for different polarization orientations (rotating the sample under the polarizing filter). Birefringence is quantified as shown in equation 4.5:

$$\text{Equation 4.5} \qquad \Delta n = n_e - n_o$$

where n_o is the refractive index for the ordinary ray, and n_e is the refractive index for the extraordinary ray.

The difference in velocity of the ordinary and extraordinary rays is called retardation and increases linearly with both the thickness of a specimen and with the birefringence. The greater the thickness, the greater the retardation (the thicker the fiber, the farther one ray lags behind the other); and the greater the difference between the refractive indices (that is, the higher the birefringence) to begin with, the greater the retardation. This can be related as shown in equation 4.6:

$$\text{Equation 4.6} \qquad r = t(n_2 - n_1)$$

where r is retardation, t is thickness, and $n_2 - n_1$ is birefringence.

When these out-of-phase waves of light strike the analyzer, it diffracts them into various colors depending on the wavelengths being added and subtracted through interference; they are called **interference colors**. These colors are caused by the interference of the two rays of light split by the anisotropic material interfering destructively with each other; that is, they cancel each other out to a greater or lesser degree. The colors produced are indicative of the fiber's polymer type and molecular organization. The birefringence of a fiber can be determined with the polarizing microscope by examining the fiber between crossed polars. The characteristic birefringence of a given substance is the numerical difference between the maximum and minimum refractive indices. Birefringence will be greatest when the polymers in the fiber are lined up parallel to the longitudinal axis of the fiber and will be zero if they are randomly organized. A chart of diameter, birefringence, and retardation, pictured in Figure 4.13, is called a Michel-Levy Chart, after its inventor, Auguste Michel-Lévy (1844–1911). Michel-Lévy, a French geologist, was born in Paris and became inspector-general of mines and director of the Geological Survey of France. He was distinguished for his research into the microscopic structure and origin of eruptive minerals; importantly, Michel-Lévy was a pioneer in the use of the polarizing microscope for the determination of minerals. The chart assists in the identification of birefringent materials. One of the ingenious things about the chart is that if two of the parameters are known, the third can be calculated (using equation 4.6). For more information about the Michel-Levy Chart, see Delly (2003); for more information on microscopy, see "On the web: Microscopy."

FLUORESCENCE MICROSCOPY

Fluorescence is the luminescence of a substance excited by radiation. Luminescence can be subdivided into **phosphorescence**, which is characterized by long-lived emission, and fluorescence, in which the emission stops when the excitation stops. The wavelength of the emitted fluorescence light is longer than that of the exciting radiation. In other words: A radiation of relatively high energy falls on a substance. The substance absorbs and/or converts (into heat, for example) a certain, small part of the energy. Most of the energy that is not absorbed by the substance is emitted again. Compared with the exciting radiation, the fluorescence radiation has lost energy, and its wavelength will be longer than that of the exciting radiation. Consequently, a fluorescing substance can be excited by near-UV invisible radiation, and its fluorescent components (**fluorophores**) are seen in the visible range.

In a florescence microscope, the specimen is illuminated with light of a short wavelength, for example, ultraviolet or blue. Part of this light is absorbed by the specimen and reemitted as fluorescence. To enable the comparatively weak fluorescence to be seen, despite the strong

---

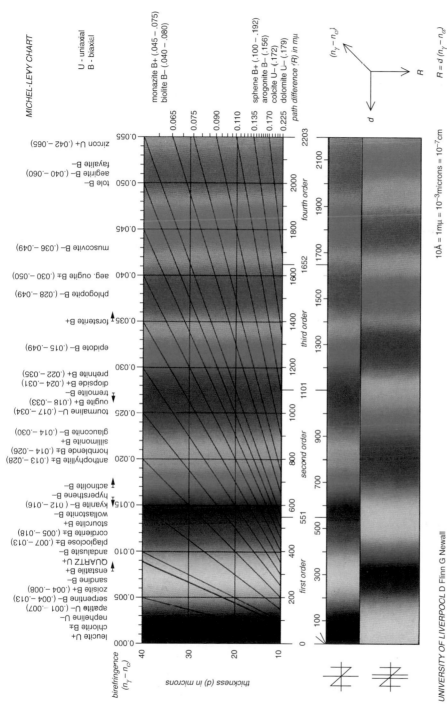

UNIVERSITY OF LIVERPOOL D Flinn G Newall

Figure 4.13. *The Michel-Lévy Chart devised in 1888 by a French geologist, August Michel-Lévy.*

illumination, the light used for excitation is filtered out by a secondary (barrier) filter placed between the specimen and the eye. This filter, in principle, should be fully opaque at the wavelength used for excitation and fully transparent at longer wavelengths so as to transmit the fluorescence. The fluorescent object is therefore seen as a bright image against a dark background.

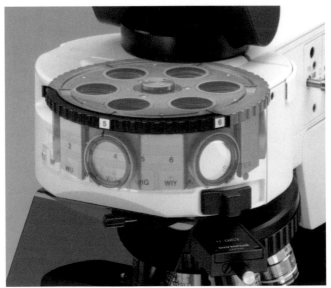

Figure 4.14. A *fluorescence microscope uses various filters to exclude and excite specific wavelengths of light to induce fluorescence. Most microscope companies now package filters in sets, or cubes, to make choosing combinations easier (Courtesy: Olympus USA).*

It follows that a fluorescence microscope differs from a microscope used for conventional light microscopy mainly in that it has a special light source and a pair of complementary filters. The lamp should be a powerful light source, rich in short wavelengths; high-pressure mercury arc lamps are the most common. A primary or excitation filter is placed somewhere between the lamp and the specimen. The filter, in combination with the lamp, should provide light over a comparatively narrow band of wavelengths corresponding to the absorption maximum of the fluorescent substance. The secondary, barrier, or suppression filter prevents the excitation light from reaching the observer's eye and is placed anywhere between the specimen and the eye. A fluorescence microscope and filter sets are shown in Figure 4.14.

CHAPTER SUMMARY

The microscope is a nearly universal symbol of science, and forensic science is equally well represented by the microscope. For all its power and simplicity, microscopy is sometimes neglected in modern laboratories in favor of expensive and complicated instrumentation. Microscopy provides fast, low-cost, and definitive results to the trained scientist. The wise forensic scientist would learn and develop strong microscopy skills to ensure successful scientific investigations.

Test Your Knowledge

1. What is a simple magnification system? How is it different from a compound magnification system?
2. What is a virtual image?
3. What is focal length?
4. What is resolution?
5. Why is resolution more important than magnification?
6. What are the main parts of a microscope?
7. If you saw "10×/0.54/170/0.17" on an objective, what would it mean?
8. What is astigmatism?
9. What does a condensing lens do?
10. What is the difference between a real image and a virtual image?
11. What is the refractive index of air? Of water? Of a diamond?

12. "Angle in, angle out" is a shorthand way of characterizing what principle in microscopy?

13. What is a mounting medium?

14. Name three materials that are isotropic—besides air.

15. What does a polarizing filter do?

16. Why do crossed polarizing filters create a black field of view?

17. If you had a pair of polarizing sunglasses, how could you tell their polarization direction?

18. What is birefringence?

19. A fluorescing substance contains _____.

20. What materials can be accurately analyzed by microscopy?

Consider This . . .

1. Why do you see polarization (interference) colors *only* when the analyzer is placed into the light path?

2. What do you think Chamot and Mason meant when they said, "But even descriptive microscopy often requires more than superficial observation"?

3. Polarizing sunglasses cancel out the glare from surfaces. All polarizing filters have a preferred orientation, even the sunglasses. What is the preferred orientation of polarizing sunglasses? Why?

BIBLIOGRAPHY

Chamot, E.M., & Mason, C.W. (1940, 1958). *Handbook of chemical microscopy, Volumes 1 and 2.* New York: John Wiley and Sons, Inc.

Davidson, M.W., & Abromowitz, M. (2005). *Optical microscopy*, retrieved from **http://micro.magnet.fsu.edu**, Olympus America, Inc. and Florida State University.

DeForest, P.R. (2002). Foundations of forensic microscopy. In R. Saferstein (Ed.), *Forensic science handbook, Volume I* (2nd ed.). Englewood Cliffs, NJ: Prentice-Hall.

Delly, J.G. (2003, July). The Michel-Levy Interference Color Chart: Microscopy's magical color key. *Modern Microscopy*. Retrieved from **www.modernmicroscopy.com**

McCrone, W.C., McCrone, L.B., & Delly, J.G. (1978). *Polarized light microscopy.* Ann Arbor, MI: Ann Arbor Science Publications, Inc.

Spectroscopic Techniques

KEY TERMS

Atomic absorption spectroscopy (AA)

Absorption spectrum

Attenuated total reflectance (ATR)

Chemical shifts

Conjugated bond

Conjugated carbon/carbon double bone

Covalent bond

Cuvette

dcfinitely

Diamond cell

Diffuse reflectance

Elastic/inelastic

Electromagnetic radiation

Electron

Emission

Energy (E)

Emission spectroscopy (ES)

Excitation

Frequency (ν)

Gamma ray

Hertz

Inductively coupled plasma mass spectrometery (ICPMS)

Infrared

Ionic bond

Matrix assisted laser desorption ionization (MALDI)

Microwave

Monochromator

Nernst glower

Neutron

Nucleus

Orbital

Photocell

Photon

Proton

Quantized

Radiowaves

Reflectance

Sine waves

Spectrum

Thermocouple

Transmission spectrum

Ultraviolet

Valence electron

Wavelength (λ)

Wavenumber

X-rays

INTRODUCTION

In this chapter, we will discuss some physical properties of matter, namely the ways that substances and objects react when exposed to various kinds of light. Light waves contain energy that can be transferred to matter when exposed to the light. Some of the energy

THE CASE

A woman is abducted at gunpoint from a main street in town. Her abduction happens so fast that no one gets a good look at the license plate of the car, but witnesses note that the abductor was wearing a plaid shirt and blue jeans. He takes the woman to nearby woods and assaults her and then leaves her there. She flags down a passing motorist, who calls the police. She is taken to the hospital. An investigator sends her clothing to the crime lab, where a forensic scientist finds some small, blue fibers on her clothing. These appear to be blue denim fibers of the type used to make blue jeans.

Meanwhile the police find a car matching the description of the one used in the abduction. The driver is wearing clothing similar to that described by witnesses. He is arrested. His jeans are sent to the lab where the forensic scientist removes specimen (known) fibers from the jeans. The scientist will subject these fibers as well as the ones taken from the woman's clothing to a number of tests including visible and infrared spectrophotometry. These tests can help determine if the fibers taken from the woman's clothes could have come from the suspect's blue jeans.

can be absorbed by the matter, which will, in turn, undergo chemical and physical changes as a result. Matter is ultimately made up of individual molecules that are made up of atoms connected by bonds between electrons. Depending on the energy content of the light (or electromagnetic radiation as it is also called), the nucleus or the electrons in certain atoms can undergo changes. Each molecule will absorb different wavelengths of light, and this array of absorbed wavelengths of light, called a spectrum, is characteristic of each type of molecule. The field of spectroscopy can be very useful to forensic scientists as they compare items of evidence from a crime scene to those taken from a victim or suspect in a crime.

In this chapter we will look at the different types of electromagnetic radiation and the different types of effects they can have on substances. Those types of radiation that reveal the most information for the comparison of evidence will be discussed.

LIGHT AND MATTER

PROPERTIES OF LIGHT

Light, or **electromagnetic radiation**, can be visualized in a number of ways. One way is to think of light as existing in **sine waves** irradiating out from a source in all directions. There are many types of waves, each one described as a **wavelength (λ)**, which is the distance between any two waves, measured at the same points on the waves. This is shown in Figure 5.1.

Light can also be described in terms of **frequency (ν)**, which is the number of waves that pass a given point in one second (cycles per second or **Hertz**). There is a simple inverse relationship between wavelength and frequency, as shown in equation 5.1:

$$\text{Equation 5.1} \qquad c = \lambda \nu$$

In equation 5.1, c is the speed of light (3×10^8 meters/sec or about 186,000 miles/sec). This equation shows that, as frequency increases, wavelength decreases and vice versa. Given the speed of light and either the frequency or wavelength, the other variable can be calculated. For example, your favorite FM radio station might be located at 90.5 on the dial. This is shorthand for a broadcast frequency of 90.5 Megahertz (millions of Hertz) or 9.05×10^7 Hertz. Using equation 5.1, we can determine the wavelength of this light to be about 3.1 meters, which is about 10 feet. Radio waves are very long waves compared to other types. This means they contain very little energy and are not harmful to living things.

Frequency can also be expressed in other units. Most important of these is **wavenumbers**. A wavenumber is the inverse of the wavelength measured in centimeter. Thus, a wavenumber is $1\,\text{cm}^{-1}$. For example, another way of expressing the broadcast frequency of the radio station would be in wavenumbers. To convert, change the wavelength from 3.1

Figure 5.1.
Electromagnetic radiation can be viewed as a sine wave. The wavelength is the distance between two corresponding peaks or valleys and is denoted by the Greek letter λ. The number of waves that pass a given point in one second is referred to as the frequency of the light and is denoted by the Greek letter ν (Courtesy: William Reusch, 1999. www.chemistry. msu.edu/~reusch/ virtualtext/intro1. htm#contnt*).*

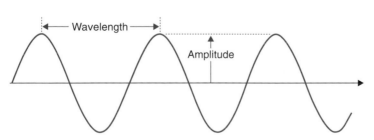

meters to 310 centimeters and then take the reciprocal; the result is $3.2 \times 10^{-2}\,cm^{-1}$.

Likewise, wavelength can be expressed in any unit of length. Usually, units are chosen for a given region of the electromagnetic spectrum such that they describe relatively small, whole numbers. Thus, in the ultraviolet region, nanometers (nm, 10^{-9} meters) are used, and the ultraviolet comprises about 200–450 nm.

Light can also be thought of as tiny packets of **energy (E)** called **photons**. The energy of a photon can be related to wavelength or frequency, as shown in equations 5.2 and 5.3:

$$\text{Equation 5.2} \qquad E = h\nu$$
$$\text{Equation 5.3} \qquad E = hc/\lambda$$

In these equations, h is a constant of proportionality called Planck's constant. It is there to ensure that the units are the same on both sides of the equation. These equations show that, as the frequency of light goes up, so does its energy, and as the wavelength goes up, the energy goes down.

The various types of light encountered in our world can be described by arranging them in order of decreasing frequency, as shown in Figure 5.2.

At the far left of this electromagnetic spectrum are **gamma rays**. These rays are very energetic and can pass through matter. They can be dangerous to life in that they can damage or destroy cells. Next lower in energy are **X-rays**. These rays can also pass through most matter but are deflected by dense matter such as bones. This is the principle behind X-ray cameras that are used to show the insides of

Figure 5.2. The electromagnetic spectrum. Along the top of the chart are the frequencies of electromagnetic radiation or light in decreasing order. Scientists divide the spectrum into regions. Within each region, electromagnetic radiation has different effects on matter that it comes in contact with. For forensic science purposes, the most important regions are the ultraviolet/visible (UV/visible) and the infrared (IR) (Courtesy: William Reusch, 1999. www.chemistry. msu.edu/~reusch/ virtualtext/intro1. htm#contnt).

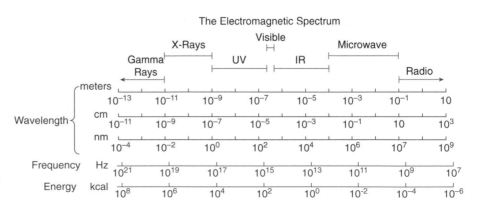

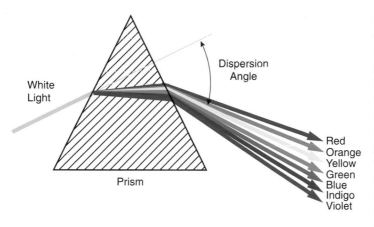

White Light

Prism

Dispersion Angle

Red
Orange
Yellow
Green
Blue
Indigo
Violet

Figure 5.3. The color spectrum. When white light is refracted by a prism, it breaks up into various colors. Light of these wavelengths is called the visible spectrum because when photons reach our eyes, our optic nerves send images to our brain that register as a color. The highest frequency (lowest wavelength) light is violet, and the lowest is red. Frequencies higher than violet are in the ultraviolet region. We do not see this light as colored. Frequencies below the red are in the infrared region. We do not see this light as being colored either (Courtesy: William Reusch, 1999. www.chemistry. msu.edu/~reusch/ virtualtext/intro1. htm#contnt*).*

bodies. The X-rays reflect off bone and other dense tissue and are detected while the others pass through soft tissue.

Lower in frequency and energy than X-rays is ultraviolet (UV) and visible radiation. Visible light has important applications in forensic science. Light in this region is not energetic enough to pass through matter. Instead, when a molecule absorbs this light, electrons are shifted from one **orbital** to another. An orbital is an energy level where an electron resides. The ultraviolet region is so-called because it borders on the violet area of the visible region, which is light that human eyes can detect and see as color. As frequencies of visible light decrease, the light changes from violet down to red at the lowest frequencies. Figure 5.3 shows the color spectrum.

Below the red region of visible light is the **infrared** region (*infra* means "below"). When absorbed by matter, this type of light causes bonds between atoms in a molecule to vibrate like two weights on either end of a spring. The infrared region is also very important in the analysis of chemical evidence in forensic science.

At still lower frequencies than infrared light is the **microwave** region. These light waves cause molecules to rotate or spin.*

At the lowest end of the light spectrum are **radio waves**. These waves have very long wavelengths and very small frequencies. Some of these waves are meters long! They carry radio and TV signals. Remember the example of the radio station at 90.5 Megahertz. Its wavelength is more than 10 feet long!

---

*The microwave region also has some practical uses such as microwave ovens. This is discussed below in more detail.

IN MORE DETAIL:
The microwave region

Microwaves serve a number of purposes. They are used as carrier waves for some transmissions of audio devices. Their major effect chemically is to cause molecules to rotate or spin. The practical effect of this is that, when adjacent molecules absorb microwave radiation and spin, they rub against each other and cause friction. This friction, in turn, generates heat. This is the principle behind microwave ovens. When food is put in a microwave oven and the oven is turned on, the food is bombarded with microwaves. The water molecules in the food absorb the microwaves and begin to spin, creating heat that cooks the food. Because of the energy of the microwaves, not enough heat is produced to caramelize or char the sugars in the food, so the food doesn't "brown" like it would in a conventional radiative or convection oven. When microwave ovens first came out, people thought that they weren't really cooking food, especially meats, because they didn't turn brown. This spawned the production of "browning sauces" that would cause the food to turn brown when being cooked in a microwave oven. Some people believed that these ovens cooked food from the inside out. This cannot be true, of course, because the microwaves will be absorbed by the first molecules they encounter, which would be on the outside surface.

INTERACTIONS OF MATTER AND LIGHT

One of the most important properties of matter is how it acts when exposed to light of various frequencies. In some cases these interactions can be used to gain a great deal of information about the substance being studied and can even be used to identify it, to the exclusion of other substances.

When a substance is exposed to electromagnetic radiation, it undergoes changes that may or may not be reversible and that depend on the energy of the radiation. For example, light of the highest frequencies and energies can cause matter to lose electrons and undergo irreversible changes. In the case of living matter, mutations in the tissue may result in cancer or some other disease. Forensic science, however, is more concerned with what happens to a substance when it is exposed to light of much lower frequencies and energies. There are two principal regions of light that are most important in characterizing evidence: the ultraviolet/visible (UV/visible) and the infrared (IR).

Remember, forensic scientists are most interested in comparing evidence from a crime scene (unknown evidence) to some object or material (known evidence) to see to what extent they may be associated with each other. This is accomplished principally by comparing as many physical and chemical properties as possible. The more characteristics the known and unknown samples have in common, the higher the degree of association. Behavior of matter when exposed to light is a very important physical property and is greatly exploited by forensic scientists.

UV/VISIBLE SPECTROPHOTOMETRY

ULTRAVIOLET LIGHT AND MATTER

All matter consists of atoms that are made up of negatively charged electrons that inhabit orbitals that exist in approximately concentric spheres around the **nucleus**, which is made up of positively charged **protons** and neutral **neutrons**. A neutral atom has equal numbers of electrons and protons, so there is not net positive or negative charge. When atoms combine to make molecules (the building blocks of compounds, materials, or substances), they do so by sharing or donating/accepting electrons to form **covalent** or **ionic bonds**. The electrons that are shared are those that are farthest from the nucleus, called the **valence electrons**. Valence electrons in atoms and in molecules can be promoted to a higher energy level by absorbing energy from light or other energy sources. This process is said to be **quantized**, because the atom or molecule can only absorb the exact amount or quantum of energy that corresponds to the difference in energy between the occupied and unoccupied energy level. In other words, a molecule will absorb energy and promote an electron if it is exposed to a photon of the proper energy. You can visualize these energy levels to be like stairs on a staircase; you can be on one stair or the next stair but cannot occupy the space between stairs. Figure 5.4 illustrates this concept.

A photon that causes electron promotion in atoms and molecules is in the UV/visible region of the electromagnetic spectrum. When a substance is exposed to UV/visible radiation, it will absorb certain photons of particular energy (and thus particular frequencies or wavelengths). If the amount of each wavelength of light that is absorbed by a substance throughout the UV/visible region is plotted, a **spectrum** is generated. See Figure 5.5 for a UV/visible spectrum of heroin.

Note that the peaks in the UV/visible spectrum of a typical substance tend to be few in number and quite broad in shape. This is due to the nature of the absorbance of this type of energy; there are not

Figure 5.4. *An orbital is like a stair on a staircase. An electron can be on a stair but not between stairs.*

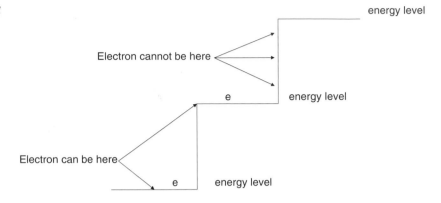

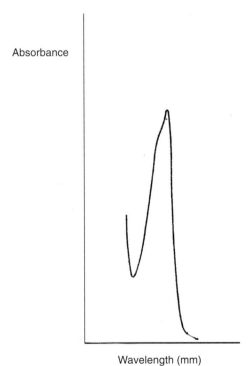

Figure 5.5. *The ultraviolet spectrum of heroin.*

too many electrons that can be promoted in a typical molecule, so there are not very many different wavelengths where an appreciable number of photons is absorbed. The broadness of the peaks is due to the temperature; at very low temperatures, UV absorptions are narrower. The practical effect of these characteristics of UV/visible spectra is that they are not commonly used for absolute identification of a pure chemical substance. Closely related substances would exhibit UV/visible spectra that are practically indistinguishable. For example, morphine and heroin (which is derived from morphine and is similar in structure) have very similar UV/visible spectra.

Not every substance will absorb energy in the UV/visible range. Certainly any substance that appears to the human eye as possessing a color will absorb in this region because the sensation of color is caused by light reflection from a substance that is received by our optic nerves, which, in turn, send a signal to the brain that is registered as the quality of color. Many organic substances will also have a UV/visible spectrum, but they usually possess a number of **conjugated carbon/carbon double bonds**. These are alternating single and double (or triple) bonds in the molecule. Any compound that is based on the benzene ring, for example, will absorb strongly in the UV/visible region. It is conceivable that several substances could have the same chromophore and thus the same UV/visible spectrum. This is one reason why UV/visible spectra

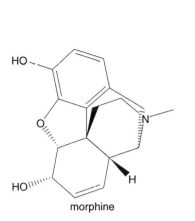

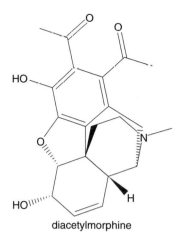

morphine

diacetylmorphine

Figure 5.6. *Structures of morphine and diacetylmporphine (heroin). Although the structures are different, the UV/visible spectra are virtually the same.*

cannot be used for unequivocal identification of a substance. This can be seen in Figure 5.6, which shows the structures of morphine and diacetylmorphine (heroin).

OBTAINING A UV/VISIBLE SPECTRUM

The UV/visible spectrum of a substance is obtaincd by using a UV/visible Spectrophotometer. A simplified diagram of a spectrophotometer is shown in Figure 5.7.

The source is a light that emits all of the wavelengths of UV or visible light. Most often, this is a deuterium (heavy hydrogen) lamp for UV light and an incandescent light for visible wavelengths. Next, there is a **monochromator**. This devise selects one particular wavelength (or a small packet of wavelengths) to be exposed to the sample. The monochromator is a prism or grating (see Figure 5.3) that is rotated, thus exposing the sample to steadily increasing or decreasing wavelengths of light throughout the entire spectrum. The monochromator is coupled to the detector so that the amount of transmitted light at each wavelength is known.

The sample holder comes next. Most samples are run as solutions in a solvent that does not absorb light in the UV/visible range. Methyl alcohol is a popular solvent for many substances such as drugs. The holder is a rectangular, quartz glass **cuvette**, usually with a 1 cm square base. Spectra of gases can also be obtained using a special gas-tight holder.

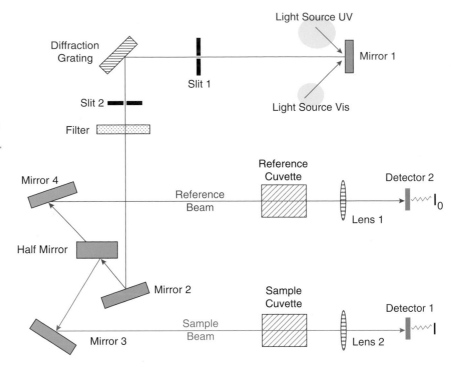

Figure 5.7. A *simplified diagram of an ultraviolet/visible spectrophotometer (Courtesy: William Reusch, 1999.* www.chemistry. msu.edu/~reusch/ virtualtext/intro1. htm#contnt).

Finally, there is the detector. The detector must be sensitive to changes in the intensity of UV/visible light that reaches it. Most often, these detectors are **photocells**. A photocell is a device that converts UV or visible light into an electric current. The more intense the light, the more current will be created. At a wavelength where the sample is not absorbing any light, the detector will be producing maximum electric current because all of the light from the source reaches the detector. When the sample absorbs some or all of a wavelength of light, less light reaches the detector, and thus, less current is created.

Modern spectrophotometers are computer controlled. The computer stores the wavelength and corresponding electric current and then, when the entire spectrum has been obtained, will construct a graph of wavelength (or frequency) versus intensity of transmitted light. This is called the **transmission spectrum**. The computer can also convert this to the wavelength versus amount of light absorbed by the sample, called the **absorption spectrum**. In practice, a spectrum of the analyte will be obtained, and then a spectrum of just the solvent will be obtained. The latter will be subtracted from the former, yielding the transmission spectrum.

UV/VISIBLE MICROSPECTOPHOTOMETRY

Many types of evidentiary materials are too small to be accommodated by an ultraviolet spectrophotometer. For example, a scientist might want to determine the exact color of a tiny paint chip and compare it to paint taken from a car suspected of being involved in a hit-and-run case. This could be accomplished by determining the exact wavelengths of light that are absorbed or reflected by the paint. A tiny paint chip could not be made to fit properly into a large spectrophotometer. The solution is to combine a powerful microscope with a UV/visible spectrophotometer. Light that travels through the microscope onto the paint chip is detected by a photocell, just as in a normal spectrophotometer. The ultraviolet and visible spectra are thus obtained and comparisons can be made. Figure 5.8 shows a UV/visible microspectrophotometer.

APPLICATIONS OF UV/VISIBLE SPECTROSCOPY IN FORENSIC SCIENCE

The major application of this type of spectroscopy is the determination of the exact color of an object or substance. The paint chip described previously illustrates when this would be needed. Another example is in the case cited at the beginning of this chapter. One of the major comparisons that the forensic scientist would attempt to accomplish would be the color of the fibers found on the victim's clothes with the fibers taken from the suspect's blue jeans. This is especially critical since so many pairs of jeans appear to be the same color to the naked eye.

Other applications include the characterization of organic substances such as illicit drugs. As was mentioned previously, the opiates such as heroin and morphine have characteristic UV/visible spectra. Also, polymers such as plastics, fibers, and paints have UV/visible spectra that help identify them.

MOLECULAR FLUORESCENCE

Fluorescence is a phenomenon whereby a substance absorbs light of one wavelength and then emits light of a longer wavelength (lower energy). Everyday objects, such as fluorescent tape worn by bicyclists on their clothes, take advantage of fluorescence. If a bicyclist is wearing such tape at night, and a car headlight reaches it, it will "glow" green or some other color. This is fluorescence.

Figure 5.8. *A UV/visible microspectrophotometer.*

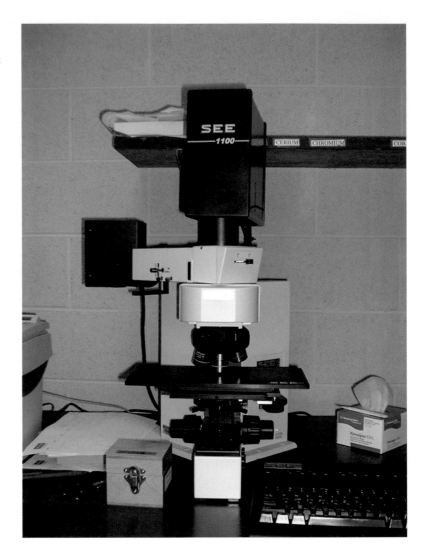

Not all substances fluoresce, and those that do generally exhibit UV/visible fluorescence. That is, they absorb light in the UV or visible region and emit light in that same region, although always at a longer wavelength. A fluorescence spectrophotometer looks a lot like a basic UV/visible spectrophotometer with some important differences. Figure 5.9 is a schematic of a fluorescence spectrophotometer.

The detector is set at a right angle to the source, with the sample at the apex, so that the detector does not see any light that leaves the source and is directly transmitted by the sample. The detector sees only light that is fluoresced by the sample. In addition, there is one monochro-

IN MORE DETAIL:
Fluorescence spectroscopy

Comparatively few substances exhibit luminescence, which is fluorescence or phosphorescence. In organic chemistry the types of molecules that luminesce are generally those that contain conjugated double bonds with carbon. These are double bonds that alternate with single bonds as shown here:

$$C=C-C=C-C=C$$

The conjugation allows the Π bonds to be delocalized over several carbon atoms. This has the effect of lowering the energy of these orbitals. As a result, the energy needed to promote electrons from their ground state into these orbitals is in the UV/visible range. When an electron is promoted, it will eventually fall back to its lowest, ground state. The particular mechanism by which this happens determines whether the molecule will fluoresce or phosphoresce. In both cases, some of the energy absorbed by the molecule to promote the electrons into excited Π orbitals is lost by vibrations or some other mechanism. Then the electron drops back to lower levels, emitting photons of light as it does so. These photons are of lower energy than the light that the molecule absorbed in the first place. This is why a molecule may absorb yellow light and then emit red light, the latter being of lower energy than the former.

Measurement of luminescence can be a powerful tool in characterizing those substances that have this property. Fluorescence is very efficient and thus very sensitive. Small amounts of such substances can be detected by fluorescence. It is also a good quantitative technique.

Fluorescence is a short-lived phenomenon, lasting just fractions of a second. Phosphorescence, on the other hand, is a longer-lasting process. It may persist for seconds or even minutes.

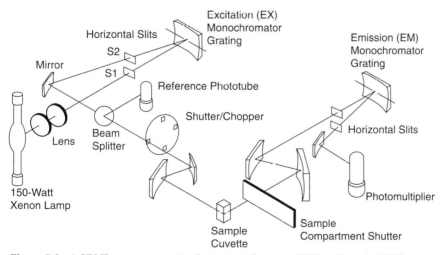

Figure 5.9. *A UV Fluorescence spectrophotometer (Courtesy: William Reusch, 1999.* www.chemistry.msu.edu/~reusch/virtualtext/intro1.htm#contnt*).*

mator between the source and the sample, so the wavelengths of light that reach the sample can be selected; and there is another monochromator between the sample and the detector, so the wavelengths of light that reach the detector can be selected. When the two monochromators are used, the specific wavelengths the sample absorbs **(excitation)** and which ones it gives off **(emission)** can be determined. The excitation and emission wavelengths are characteristic of a particular substance. The process by which fluorescence occurs is discussed in more detail below.

In forensic science, only a few substances are commonly analyzed by fluorescence. They include drugs such as LSD, as well as certain pigments and dyes that are found in paints, inks, fibers, and other materials. Fingerprints can be visualized by using a strong light source and a fluorescent dye. Optical brighteners are used in some fabrics, and they can be characterized by fluorescence.

Figure 5.10. A molecule can be visualized as a set of weights connected by springs. The springs are the chemical bonds made up of shared electrons. This diagram shows a molecule of water (H_2O) and a model using weights and springs. When the springs are pulled or bent, they contract and expand or wag back and forth. When the bonds between two atoms absorb energy, they vibrate back and forth at characteristic frequencies.

INFRARED (IR) SPECTROSCOPY

INFRARED LIGHT AND MATTER

Infrared light photons contain less energy than UV/visible photons, and thus, when this light strikes a substance, different things happen. For example, electrons do not change their orbitals when exposed to IR light. Instead, bonds between electrons start to vibrate. In this interaction, it helps to visualize a molecule as two weights connected by a spring, as shown in Figure 5.10.

The chemical bond is the spring, and the atoms on either side of the bond are the weights. When the weights attached to a spring are pulled apart, the system will vibrate back and forth at a frequency that depends on the strength of the spring and the amount of weight on either side. When infrared radiation of the proper energy strikes a molecule, one of the bonds can absorb it and cause a vibration to take place. Like UV/visible interactions with matter, infrared absorptions are also quantized. The energy of a photon must exactly match the proper energy of vibration of one of the bonds in the molecule.

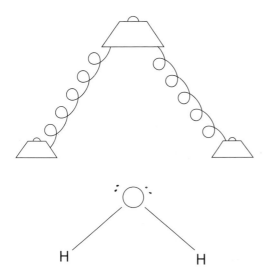

Each different chemical bond in the molecule has its own characteristic vibrations, and each bond can undergo a number of different kinds of vibrations. Some of the most common vibrations of molecules are shown in Figure 5.11.

The result is that, unlike UV/visible absorptions, there are many infrared absorptions in each type of molecule. Even the slightest change in the composition of a molecule will result in a different infrared spectrum. Thus, the infrared spectrum of a pure substance may be unique and might be used to unequivocally identify that substance. Infrared spectrophotometry is one of the two analytical techniques, along with gas chromatography-mass spectrometry, that can be used for identification of pure substances, such as drugs.

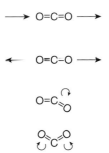

Figure 5.11. Some of the vibrations that a simple molecule can undergo. More complicated molecules can undergo many more complex vibrations. The result is that the infrared spectrum of complex molecules contains dozens of absorbances.

OBTAINING AN INFRARED SPECTRUM

The basic infrared spectrophotometer is somewhat similar in appearance to the basic UV/visible spectrophotometer. There are some differences, which are explained in this section. Modern IR instruments called Fourier Transform Infrared Spectrophotometers (FTIRs) are based on somewhat different principles, but the results are the same as with basic instruments. The concept of FTIR is discussed in more detail in the "In the Lab" box below.

The light source for infrared radiation is different than for UV or visible. It emits light of all wavelengths in the infrared region simultaneously. The basic instrument also has a monochromator that performs the same function as in the UV/visible spectrophotometer.

Sampling for IR spectrophotometry is somewhat more flexible than for UV/visible spectrophotometry. Solid spectra can be easily obtained by preparing potassium bromide (KBr) pellets. KBr, which is a white powder, becomes a hard, solid, transparent pellet when subjected to high pressure. It also does not absorb infrared radiation in the same regions where organic compounds commonly absorb. If a scientist has a solid, pure drug sample, such as heroin, an IR spectrum can be obtained by making a 10% solid solution with KBr and then forming a small pellet using high pressure and a die. This pellet contains small particles of the drug dispersed throughout the solid KBr. The pellet can be put directly in the beam of the IR, and an excellent spectrum can be obtained. Pure liquid spectra can be obtained by putting a thin film between two premade KBr plates. Vapor spectra can be made using a special gas-tight cell with KBr windows.

IN THE LAB:

FTIR

Modern analytical infrared spectrophotometers represent a great improvement over classical instruments. In the classical setup, the sample is exposed to each frequency or wavelength of light individually. This is accomplished using a rotating prism that refracts the light, breaking it up into individual wavelengths. A series of mirrors is used to focus the light on the sample, and then the transmitted light is further focused on the detector. The light must be attenuated (reduced) by a slit before reaching the detector to avoid overloading it. The result is that obtaining a spectrum takes a long time and involves a lot of moving parts and a degradation of sensitivity because of all of the mirrors and slits.

Modern instruments take advantage of a mathematical concept called the Fourier Transform. The light from the source is split into two beams by a beam splitter. A moving mirror is used to change the path length of one of the beams. When the beams recombine, they are in phase or out of phase, depending on the position of the moving mirror. This apparatus is called a Michaelson Interferometer. The resultant beam is called an interferogram. The entire panoply of wavelengths of light from the source are put through the interferometer and then exposed to the sample, which absorbs some of the light as usual. The interferogram is sent to the detector, where it is subjected to a mathematical reconversion to individual wavelengths of light using a Fourier Transform. This process has only one moving part, and it takes around one second to obtain a complete infrared spectrum.

If the scientist has a very small sample that is capable of transmitting light, a **diamond cell** can be used. In this case, the sample is squeezed between two tiny diamond chips (diamond is practically transparent in the IR where most substances absorb). This flattens out the sample, making it easier to obtain a spectrum. The light from the source is focused on the diamond windows to get a high-quality spectrum.

Even some materials that are opaque—they do not transmit light—can be analyzed by IR. Two techniques are available to obtain **reflectance** spectra. One, called **diffuse reflectance**, uses an apparatus that causes the IR source light to bounce off the surface of the material, which then absorbs some of the light. The other method, **attenuated total reflectance (ATR)**, uses a special crystal in contact with the sample, which causes multiple reflections so that the sample has several opportunities to absorb IR light.

Detectors are also different from those used in UV/visible spectrophotometry. IR detectors are usually some type of **thermocouple**, a device that converts heat into electricity. To increase the sensitivity of these detectors, they are often housed in a flask that has liquid nitrogen circulating around it to make it very cold.

One of the advantages of having computers that control infrared spectrophotometers is the ability to create libraries of infrared spectra. Most infrared spectra are unique, and the ability of computerized instruments to compare spectra with library entries may be a good method of identification. This type of analysis is tempered, however, by the fact that some IR spectra of similar substances are almost indistinguishable. Modern computers are capable of searching a library containing hundreds of spectra in just a few seconds. They will return a list of the substances whose spectra most closely match the unknown.

INFRARED MICROSPECTOPHOTOMETRY

Suppose a forensic scientist receives some single fibers in the lab and wishes to obtain their visible spectra to determine their exact colors and the IR spectra to determine their chemical makeup. None of the conventional sampling methods for either technique are engineered to accept a single, thin fiber. To solve these types of problems, the forensic scientist can join a microscope with a spectrophotometer to make a microspectrophotometer. A microspectrophotometer can be thought of as a microscope with an appropriate light source and a monochromator and detector. The object (fiber in this case) is mounted on the stage of the microscope, and all of the light is focused on the object. The detector detects the light transmitted by the object. These valuable tools extend these techniques to evidence that couldn't be analyzed this way in the past. Figure 5.12 shows an infrared spectrophotometer with attached microscope.

APPLICATIONS OF IR IN FORENSIC SCIENCE

IR spectrophotometry is very versatile, and there are a number of applications. Solid materials, such as purified drug exhibits, are easily confirmed using IR. Paint samples can be ground up or analyzed by diamond cell or ATR. Liquid hydrocarbons such as gasoline can be easily characterized by IR. Plastics and other polymers can be analyzed by transmitted or reflected light. Figure 5.13 shows how a suspected drug can be identified by comparing its infrared spectrum with a known sample.

Figure 5.12. A Fourier Transform Infrared Spectrophotometer with attached microscope. Small objects such as single fibers can be mounted under the microscope, and high-quality infrared spectra can be obtained.

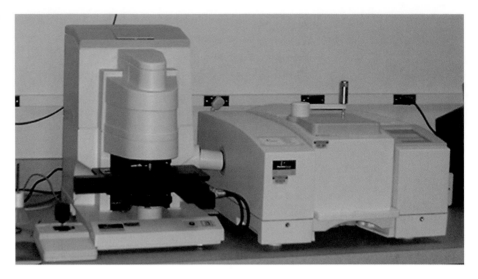

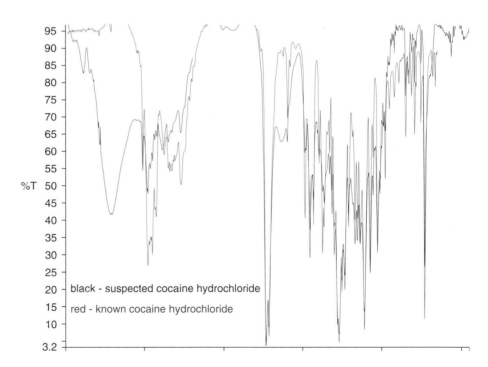

black - suspected cocaine hydrochloride

red - known cocaine hydrochloride

Figure 5.13. Infrared spectra of known and suspected cocaine.

RAMAN SPECTROSCOPY

Raman spectroscopy is a companion technique to infrared spectrophotometry. It also depends on the vibrations that bonds undergo within molecules. Some types of vibrations do not result in the absorption of infrared radiation, and no peak will be seen in the IR spectrum. These vibrations may, however, give rise to Raman bands.

In the Raman effect, absorption of photons in the ultraviolet or visible range induces the scattering of photons of light rather than simple absorption/emission. The vast majority of the scattering incidents are **elastic**; that is, the energy of the scattered photon is the same as that of the incident or absorbed photon. Approximately one out of a million of the scatterings is **inelastic**; the energy of the scattered photon can be greater (anti-Stokes line) or less (Stokes line) than that of the incident one. The Raman spectrum is measured as the **chemical shifts** of the emitted photons, which is the difference in energy between the incident photon and the inelastically scattered emitted photon.

MASS SPECTROMETERY

Mass spectrometry is the analysis of unknown substances by measuring the masses of ions created by bombarding the substance with energy. Strictly speaking, the energy used for ionization of materials is not normally what we consider light. It can be a stream of electrons, a stream of small molecules, or a laser. Once ionized, the molecule may survive intact, or it may fragment into daughter ions, which themselves may further decompose or react with other daughter ions. If the ionization conditions are kept constant, the relative amounts and compostion of the daughter ions will be very reproducible. After the ions are formed, they are separated in various ways and then detected. The separators and detectors actually measure the mass-to-charge ratio of the ions. Sometimes a species will lose two electrons, giving it a +2 charge. This would be detected as having a similar "mass" as an ion that weighed half as much with a single charge. The basic components of a mass spectrometer are shown in Figure 5.14.

SAMPLE INTRODUCTION

Mass spectrometers are very versatile. Samples can be introduced in almost any form. Solids and liquids and even gases can be directly injected into the instrument. A mass spectrometer can be designed as a detector for a gas chromatograph or HPLC. As each analyte component is separated, it is introduced into the ionization chamber of

Figure 5.14. *A mass spectrometer.*

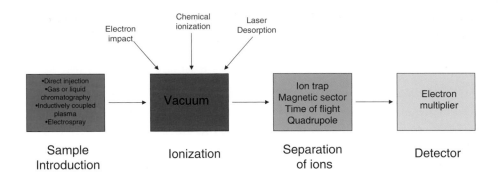

the mass spectrometer. In the case of HPLC, the mobile phase liquids are stripped off before the analyte is ionized.

Even intractable or insoluble materials such as glass can be introduced into a mass spectrometer. The elemental composition of glass samples can be identified and quantified by **inductively coupled plasma mass spectrometery (ICPMS)**. In ICPMS the glass is digested and is transformed into an aerosol by a nebulizer, which breaks up the sample into very small droplets. The aerosol is heated to 8000°C by argon plasma, in which it is vaporized, molecular bonds are broken, and the resulting atoms are ionized.

IONIZATION

Once the sample has been introduced, it is sent to an ionization chamber that has been evacuated so that there is a vacuum of approximately 10^{-5} Torr. There are a number of mechanisns for ionizing the sample molecules. One method, ICPMS, was discussed in the preceding section.

Electron Impact

In the most common method of mass spectrometry, the analyte is bombarded by a stream of energetic electrons. The molecules absorb the energy of the electrons and then lose an electron to form a cation. Depending on the nature of the analyte and the energy of the electrons, the parent ion may undergo decomposition into smaller ions. These, in turn, may further decompose or may react with other energetic species to form new daughter ions. If the ionization conditions are kept constant, the fragmentation pattern formed by a given substance will be highly reproducible, both qualitatively and quantitatively.

Except for very similar substances such as enantiomers, each substance will form a unique mass spectrum by electron impact.

Chemical Ionization

Electron impact ionization tends to treat molecules harshly, and they decompose extensively. Sometimes it is important for the scientist to know what the molecular weight of the original analyte is. In such cases small molecules such as methane or butane can be used to ionize the analyte. These molecules are much less energetic than electrons and do not impart as much energy to the analyte; therefore, decomposition is much less extensive, and more of the molecular ions remain intact.

Laser Desorption

Laser desorption mass spectrometry (LDMS) uses a laser to ionize the analyte. In some cases the laser can be applied directly to the sample. It will remove molecules from the surface of the material and then ionize them. Very little decomposition of the parent ion takes place, so molecular ions are always prominent. In some situations the laser is unable to directly desorb substances. In such cases, a matrix is used. The analyte is embedded in the solid or liquid matrix. The matrix absorbs the laser energy and transfers it to the analyte. This technique is called **matrix assisted laser desorption ionization** or MALDI.

SEPARATION OF IONS

After the ions have formed, they must be separated. Since they are all positive ions, they are affected by magnetic fields, and their separation is based on this property.

Magnetic Sector

In the magnetic sector mass spectrometer, the ions are accelerated through a curved magnetic field toward the detector. Smaller ions are deflected to a greater extent as they pass through the field.

Quadrupole

The quadrupole mass spectrometer has a set of four rods arranged in the corners of a square. Diagonally opposite rods have positive charges, and the others, negative. As ions pass through the rods, the strength of the fields created by the rods is altered. This allows only certain mass/charge ratio ions to get through. Continual changing of the voltages across the rods permits all ions within a given range to be analyzed.

Ion Trap

In the ion trap instrument, ions are focused by a quadrupole into an ion trap where they are collected. They are then ejected toward a conventional detector.

Time of Flight

In the time of flight mass analyzer, the ions are accelerated by a magnetic field of known strength. The time it takes for a given ion to reach the detector is then used to determine the mass/charge ratio.

DETECTION

Detectors in mass spectrometry are based on electron multipliers. After being separated, the ions from the mass analyzer are put through an amplifier to boost their signal. They then strike the surface of the detector, kicking loose electrons, which are measured. The more ions of a given mass/charge ratio, the more electrons are released and the stronger the signal.

ATOMIC SPECTROSCOPY

All of the spectroscopic methods discussed thus far are for the characterization of molecules. Sometimes it is important to be able to measure certain elements in analytes. Most often these include trace metals, but other elements may also be measured. The two major methods for measuring elemental species are **atomic absorption spectroscopy (AA)** and **emission spectroscopy (ES)**.

ATOMIC ABSORPTION

In atomic absorption spectroscopy, a solution of the analyte is introduced into a flame or furnace and heated until vaporized. Each individual element is analyzed separately. A lamp is chosen that emits light at the proper wavelengths of absorption for that element. The amount of light that is absorped is measured. This is a function of the concentration of that element. AA is very sensitive but requires a separate lamp for each element that is sought.

EMISSION SPECTROSCOPY

Emission spectroscopy has the advantage over AA of being able to analyze multiple elements simultaneously. The analyte solution is introduced into a flame, discharge, or plasma to vaporize it. The high temperature atomization of the analyte drives the atoms to high energy levels. As they return to ground states, they emit photons of characteristic wavelengths. A high resolution spectrometer is used to determine the emission wavelengths and thus the elements present.

BACK TO THE CASE

At the beginning of this chapter, a case described the abduction of a woman. One of the major pieces of evidence that had the potential of linking the suspect to the crime scene was blue denim fibers. Some of these fibers had apparently been transferred from the suspect's blue jeans to the victim's clothes. The forensic scientists at the crime lab would most certainly be interested in comparing these fibers in as many ways as possible. It should be noted that, no matter how much testing is done, there would be no way that a conclusion could be reached that the fibers taken from the victim's clothes **definitely** came from the suspect's blue jeans. There are too many pairs of similar jeans in the marketplace, and they do not normally contain any unique characteristics. This makes it all the more important to develop as much information as possible about the known and unknown fibers so that the association between them will be as meaningful as possible.

Because these are small fibers, the forensic scientist would most likely use UV/visible microspectrophotometry to determine the exact colors of the fibers. He or she would also use infrared microspectrophotometry to determine the chemical nature of the fiber itself, what class (e.g., cotton, polyester, etc.) of fiber, and, if possible, what subclass it belongs to. If the fibers taken from the woman's clothes have the same characteristics as the known fibers taken from the suspect's jeans, then his jeans are a possible source of the unknown fibers.

CHAPTER SUMMARY

When light strikes a material, it can cause a number of effects, depending on the nature of the material and the characteristics of the light. Light can be thought of as packets of energy (photons), which can be described in terms of their wavelengths or frequencies. When the energy reaches matter, it can cause effects to the nucleus or the electrons that surround it. This is manifested by the absorption of certain wavelengths of the light. The exact number or type of wavelengths of light absorbed is characteristic of the material and the molecules from which it is made. This information helps to characterize materials.

Although many different types of light can interact with matter, forensic science is chiefly concerned with two types: ultraviolet (and visible) and infrared. Ultraviolet light causes electrons in atoms and molecules to be promoted to higher energy levels, resulting in the absorption of light. Since visible light is also in this range, accurate information about the color of an object from its absorption of visible light can be collected. If a material is not colored, it will absorb ultraviolet light only if it has certain chemical characteristics.

Infrared light is absorbed by all molecules. It causes the bonds that hold atoms together to vibrate and/or rotate. Since there are many different types of bonds within a given molecule, there are many different absorptions of infrared light for even simple molecules. Infrared spectra are so complex that each one is unique to a particular molecule.

For light absorption to be measured, an instrument must be used. All instruments for measuring light absorption are reasonably similar. They consist of a source of light, a sample holder, a way of breaking the light into individual wavelengths, and a detector to tell when light has been absorbed.

Test Your Knowledge

1. What is a wavelength? What unit of measure does it have?
2. Define frequency? What are its units? What is a wavenumber?
3. How is the relationship between wavelength and frequency expressed?
4. How is the relationship between the energy of a photon and its frequency expressed?

5. Rank the following regions of the electromagnetic spectrum in order of decreasing energy (list the highest energy one first).
   Infrared
   Radio
   Visible
   X-ray

6. What happens to the molecules of a substance when X-rays strike it? Why are X-rays called "ionizing radiation"?

7. Why is visible light spectroscopy always measured at the same time as UV light spectroscopy? What happens to molecules when light in these regions strikes them?

8. What happens to molecules when light in the infrared region strikes them?

9. What is the most common type of detector used in UV/visible spectrophotometry?

10. What is the most common type of detector used in infrared spectrophotometry?

11. What type of light interactions with molecules give rise to molecular fluorescence?

12. What does the unit "Hertz" measure?

13. The absorption of light by molecules is said to be quantized. What does this mean?

14. What is a monochromator? How is it used in spectroscopy?

15. If you had two fibers that appeared to be the same color to the naked eye, what spectroscopic technique would you use to determine if they were?

16. Briefly describe the purpose of the Michaelson Interferometer in FTIR.

17. What is diffuse reflectance? On what types of samples is it used?

18. What is a diamond cell? When is it used?

19. What spectroscopic technique would you use to identify a pure sample of an illicit drug?

20. All spectrophotometric detectors measure the amount of light that passes through a sample (or reflects off its surface). How can you determine what light is absosrbed by the sample?

Consider This . . .

1. Why is infrared spectrophotometry a good method for identifying a pure chemical substance, whereas UV/visible spectrophotometry is seldom used for this purpose?

2. How is a spectrophotometer constructed? What are the various components, and what do they do?

3. What types of samples can be analyzed by IR? How are substances that are opaque (not transparent) handled using this technique?

BIBLIOGRAPHY

Skoog, D.A., Holler, F.J., & Nieman, T.A. (1997). *Principles of instrumental analysis* (5th ed.). Florence, KY: Brooks Cole.

Humecki, H.J. (Ed.). (1995). *Practical guide to infrared microspectroscopy*. New York: Marcel Dekker.

Perkampus, H.-H., & Grinter, H.C. (1992). *Uv-Vis spectroscopy and its applications (Springer Laboratory)*. New York: Springer Verlag.

Separation Methods in Forensic Science

KEY TERMS

Adsorption

Analyte

Capillary gas chromatography

Chromatogram

Chromatography

Detectors

Diode array detector (DAD)

Electron capture detector

Electrophoresis

Flame ionization detector

Fourier Transform Infrared Spectrophotometer (FTIR)

Gradient chromatography

High performance liquid chromatography

Ionize

Isocratic

Less

Like dissolves like

Liquid phase extraction

Mass spectrometry

Mobile phase

Nitrogen-phosphorus detector

Normal phase

Partitioning

pH

Polarity

Pyrogram

Pyrolysates

Pyrolysis

Retention factor (Rf)

Retention time

Reverse phase

Separatory

Solid phase extraction

Stationary phase

Temperature program

Thermal conductivity detector

Thin layer chromatography

INTRODUCTION

Physical evidence is seldom encountered in a pure state when it is discovered and collected. Therefore, some useful separation techniques must be available when needed. There are various separation methods in analytical and forensic chemistry. They range from the basic, such as gravity filtration, to the complex, such as capillary electrophoresis. Not all separation methods are commonly used

THE CASE

Agents of a drug task force are investigating a large-scale drug operation that is suspected of importing large amounts of cocaine from South America. They focus on one particular house in a rundown neighborhood downtown. One of the undercover agents, posing as a drug buyer, purchases some white powder from the occupants of the house. The drug is rushed to the local crime laboratory, where a drug chemist will determine if a controlled substance is present. The chemist determines that cocaine is present in the powder. On the basis of this undercover buy, the task force obtains a search warrant and raids the house. They find several hundred pounds of white powder. This, too, will be sent to the forensic chemist for analysis. The task of the chemist in such cases is to separate the cocaine from the cutting agents and determine the percent composition of the cocaine. This will involve one or more separation techniques, discussed in this chapter, and analysis methods, which are discussed in Chapters 5 and 13.

in forensic science, and there is no single, universal method for separation of evidence from its surroundings. The separation method used in a case depends on what and how much evidence needs to be separated. For example, the separation of drugs from cutting agents may involve a liquid extraction if there is a large amount of material, or gas chromatography if only small amounts are involved. The separation of the components of gasoline in fire residues normally is accomplished by gas chromatography. The choice of the best extraction method depends on the nature of the analyte and whether or not eventual confirmation of a pure substance is desired. If a particular substance is to be unequivocally identified (confirmed), it must be pure. It must be separated from all impurities, substrates, cutting agents, etc. Although numerous separation methods are available to a forensic chemist, the three major techniques used in forensic science laboratories to purify evidentiary materials will be discussed in this chapter:

- Liquid phase extraction
- Solid phase extraction
- Chromatography

LIQUID PHASE EXTRACTION

INTRODUCTION

Liquid phase extraction involves the separation of two or more substances (**analytes**) through a process in which two solvents that do not dissolve in each other (immiscible) compete for the analytes. This process is called **partitioning**, which involves analytes being distributed between the two solvents according to certain chemical properties, mainly **polarity** and **pH**. A diagram of the partitioning process is shown in Figure 6.1. An example of the purification of a drug mixture by liquid phase extraction is described in the "In the Lab" box below.

POLARITY

Many chemical compounds exhibit a property known as **polarity**. This is the tendency of the compound to behave like a miniature magnet, with a positive side and a negative side. Figure 6.2 gives examples of polar and non-polar compounds.

Figure 6.1. The partitioning process. This diagram illustrates how two components of an analyte, A and B, position themselves between immiscible liquid.

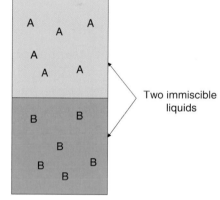

Two immiscible liquids

carbon tetrachloride

carbon dioxide

pentane

non-polar compounds

water

isopropyl alcohol

$NH_4^+ OH^-$
ammonium hydroxide

polar compounds

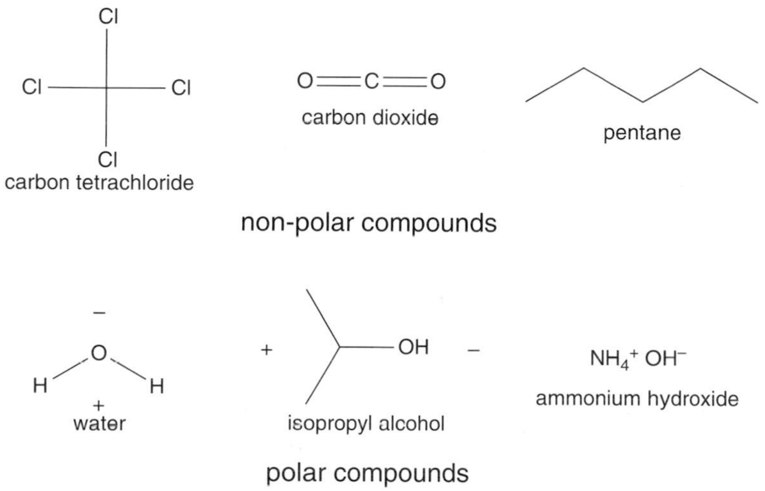

Figure 6.2. Polar and non-polar compounds. Compounds that exhibit certain types of symmetry are non-polar. The electrons pull equally in all directions in non-polar compounds.

Most organic compounds tend to be non-polar or slightly polar, whereas inorganic compounds can range from non-polar to very polar. Compounds that are symmetric tend to be non-polar. Polarity is caused by an excess of electron density on one side of a molecule and therefore a deficiency on the other side. A simple rule about solubility and polarity is **like dissolves like**. This means that polar compounds have a greater affinity for other polar compounds and, in the case of solubility, it means that polar solutes will dissolve more readily and to a greater degree in polar solvents. This advantage is exploited in liquid extractions. If there is a mixture of analytes in which some are polar and others are non-polar, they can be separated into two groups using a polar solvent such as water to remove the polar substances and a non-polar solvent such as methylene chloride to dissolve the non-polar substances.

pH

Another property of certain chemical compounds is their acidity or alkalinity relative to water (pH). An acidic substance is one that releases hydrogen ions H^+ (really hydrated so they are in the form of $H_2O \cdot H^+$ or H_3O^+) when dissolved in water. Acids have pH values between 0 and 7. An alkaline or basic substance is one that releases OH^- ions when dissolved in water. Its pH is between 7 and 14. A neutral substance is one that releases neither H_3O^+ nor OH^- ions when dissolved. Its pH is 7. Drugs, for example, can be classified as acidic, basic, or neutral. Cocaine is a basic drug. When it is dissolved in an aqueous solvent, it attracts H^+ from the solvent, leaving OH^- behind. On the other hand, barbiturates are acidic drugs, and caffeine is a neutral drug. Sugars and carbohydrates (common cutting agents in street drugs) are also neutral. Figure 6.3 shows an example of the protonation of a basic compound, in this case, cocaine.

Liquid phase extractions are ideal for separating substances of different polarity and pH; however, if two substances have similar polarity and both are acidic or basic, such as cocaine and heroin, then a liquid phase extraction may not work. A different type of separation process would be required for such mixtures.

Figure 6.3. Cocaine hydrochloride. If cocaine is dissolved in an acidic solution, an extra proton (H^+) attaches itself to the amine group (NH_2R) on the cocaine. This makes the cocaine much more polar than it is in its free form.

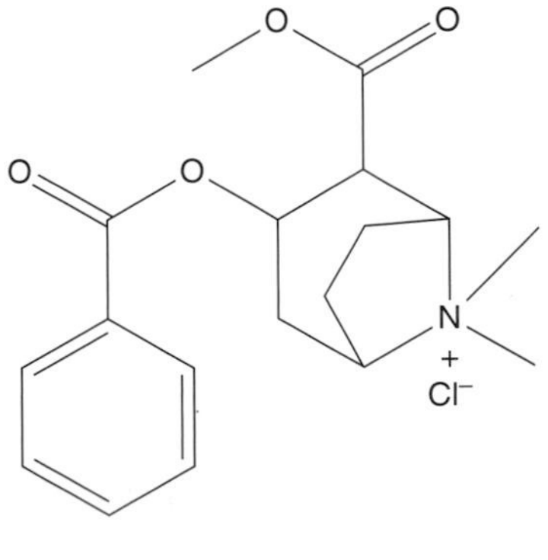

IN THE LAB:
Separation of a drug mixture by liquid phase extraction

Both the properties of polarity and pH can be used to advantage when trying to purify a drug. Suppose that a forensic scientist receives a drug sample that turns out to be 50% cocaine hydrochloride and 50% fructose, a complex sugar used as a cutting agent. Cocaine hydrochloride is a salt form of cocaine that is much more polar than cocaine free base, the naturally occurring form of cocaine. Fructose is a neutral, non-polar substance. The task is to separate the cocaine from the fructose, saving the cocaine and getting rid of the fructose. Two immiscible solvents will be employed in this process. This is an example of a liquid phase extraction.

1. The mixture is dissolved in water and filtered. All of the cocaine hydrochloride will dissolve. It is fairly polar and water is a polar solvent (remember: like dissolves like). Some of the fructose will dissolve. The filtration step removes the fructose that doesn't dissolve.
2. The filtered liquid (the filtrate) is then put into a glass **separatory** funnel, as shown in Figure 6.4. Then some weakly alkaline liquid such as ammonium hydroxide is added to the water. This gives the solution a high pH. The H^+ and Cl^- that are attached to the cocaine hydrochloride react with the ammonium hydroxide to form ammonium chloride, and the cocaine hydrochloride is converted to the free base form of cocaine, which is much less polar than cocaine hydrochloride and much less soluble in water, causing it to precipitate out.
3. Now an equal volume of a non-polar solvent such as chloroform ($CHCl_3$) is added to the separatory funnel. The non-polar cocaine free base dissolves readily in the chloroform, but the somewhat polar fructose stays in the water.
4. The chloroform and water layers are separated. The chloroform can then be evaporated, leaving the purified cocaine free base.

The foregoing is a basic extraction. The solvent containing the dissolved analyte is made basic and then extracted with a non-polar solvent. An acidic extraction can be used on mixtures containing an acidic drug.

SOLID PHASE EXTRACTION

Solid phase extraction does not involve a partitioning, or competition, mechanism. Instead, it relies on **adsorption**, a process whereby a solid, liquid, or gaseous analyte is attracted to the surface of an adsorbing material. The analyte molecules cling to the surface. If one component of a mixture is adsorbed by the adsorbing material, it can be separated using this method. A number of chemical processes affect the tendency and tenacity of the adsorption, and polarity is one of the more important ones.

One example of solid phase adsorption is the trapping of an accelerant on the surface of a plastic strip coated with finely divided charcoal, a potent adsorbing material. The strip is immersed in a can

Figure 6.4. A separatory funnel.

containing the residue from a fire suspected to contain an accelerant. As the can is heated, the accelerant molecules evaporate into the airspace above the debris and adsorb onto the surface of the coated strip. When it is removed from the can, the strip can be heated or added to a solvent to elute (remove) the adsorbed accelerant.

Another example of a solid phase extraction is the "clean-up" process of a biological sample containing a drug. A column of adsorbing material is inserted into a large tube, and a blood or urine sample is then added. It flows through the tube under a vacuum and the drug is adsorbed onto the surface of the adsorbing material, whereas the rest of the blood or urine passes through. It can then be eluted with a compatible solvent. There are a variety of solid phase adsorbing materials available for different types of drugs and other substances. Solid phase extraction can be used in many cases in which liquid phase extraction cannot be used.

SOLID PHASE MICROEXTRACTION

The solid phase microextraction technique is used for very small samples. A small wire is coated with an adsorbent such as charcoal and attached to a holder that can extend or withdraw the wire. The wire is extended into a vapor or liquid where adsorption takes place. Then the wire can be introduced into a gas chromatograph (see the next section) where the adsorbed materials are eluted and analyzed. An apparatus used for solid phase microextraction is shown in Figure 6.5.

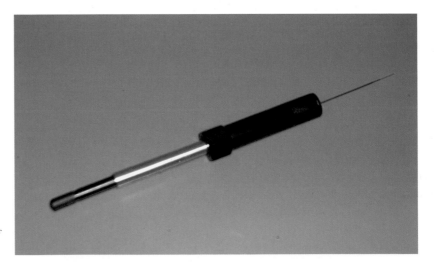

Figure 6.5. The solid phase microextraction apparatus. The coated fiber is at the very tip of the needle.

CHROMATOGRAPHY

INTRODUCTION

The term "chromatography" means literally "to write with color." This seems odd for a family of techniques that seem to have little to do with color and everything to do with separating mixtures of substances. The reason for the name is that the technique was originally developed by a Russian chemist, Tswett, who was interested in separating colored pigments in plants. He made a long, thick column of a fine, sand-like substance and poured through it a solution of dissolved plant pigments. He found that the pigments divided into colored bands at various points along the length of the column. Each band represented a component of the pigments. The bands were identifiable by their different colors. The history of chromatography is described in more detail in the box "History of Chromatography" below.

Since these original experiments, chromatography has come a long way. The term has come to represent a family of techniques that all do essentially the same thing: separate complex mixtures of substances into their individual components, based on partitioning between a stationary and a mobile phase, and then display these components so that the analyst can get information about their number and chemical nature.

There are many advantages of chromatography over solid and liquid phase extractions, and a huge variety of materials may be separated by one type of chromatography or another. Chromatography methods are also generally more sensitive than extractions. Some chromatography methods can separate millionths or billions of a gram of material.

Almost all chromatographic methods work on the same principle. They differ in how the experiment is set up and some of the analytical details. The components of the analyte will show some differences in at least one property that is exploited by the chromatography experiment. That will be the basis for the separation. The original experiment, cited earlier, where plant pigments are separated in a column of silica, illustrates the principles.

In all forms of chromatography, there are two phases present: a **stationary phase** and a **mobile phase**. The stationary phase is a finely divided solid material or a viscous liquid that is contained within a long column. The mobile phase is a liquid (or liquid solution) or a gas under pressure. The mobile phase moves through the stationary phase carrying the analyte mixture with it. Depending on their affinity for

HISTORY OF CHROMATOGRAPHY

Chromatographic separations may have been performed as far back as the fifteenth century, but there are no written records about how they were done. In 1906, Mikhail Tswett, a Russian botanist, published his first paper on his technique of liquid-solid separation of plant pigments. He was clearly the father of chromatography. It wasn't until 1941, with the development of alumina as a dependable stationary phase, that scientists finally recognized chromatography as reliable.

In 1941, Martin and Synge, working with amino acid separations, developed a technique that used two liquids rather than a liquid and a solid. This became the foundation for liquid chromatography. They also developed the first theoretical framework for describing how chromatography could be optimized by measuring its efficiency. They also speculated that a gaseous mobile phase could be paired with a liquid stationary phase to achieve effective separations. This foreshadowed the development of gas-liquid chromatography, or simply gas chromatography. For all of this development work, they won the 1952 Nobel Prize.

In 1958, new detectors were announced for gas chromatography, thus greatly extending its usefulness. These were the flame ionization and electron capture detectors. At about the same time, Golay proposed the use of narrow-bore capillaries which would have their inside walls coated with a liquid as the stationary phase. This development revolutionized gas chromatography because it greatly improved separations and efficiencies.

What is known today as high performance liquid chromatography wasn't developed until the 1970s. Reverse phase chromatography using long chain hydrocarbons coated on the surface of silica beads was developed about this time. HPLC started with long, thin columns like gas chromatography, but it was soon discovered that shorter, thicker columns containing small particles could improve resolution. This technology is widely used in forensic science.

Source: Chromatography—A Century of Discovery 1900–2000 by C.W. Gehrke, R.L. Wixom and E. Bayer, Ed., Amsterdam, Elsevier Science, 2001.

the stationary phase, the components of the analyte move quickly or slowly through the column, separating from other components along the way.

In the plant pigment separation cited earlier, the column of silica is an example of a stationary phase. The plant pigment mixture is dissolved in a solvent that is then poured through the stationary phase. The solvent is the **mobile phase**. If the stationary phase had no affinity for the analyte components, then they would all travel together right along with the mobile phase and emerge at the bottom together. The stationary phase is designed to attract or, in some cases, repel, certain members of the analyte, each one in a different way. This means that the progress of each component of the analyte will be affected by the stationary phase, one hopes in a different way from all of the others. When the mobile phase has completed its journey through the column, the various pigments have been held up by the stationary phase and form colored bands at various points in the column. To recover one or more of the pigments, they would have to be eluted off the column using a suitable solvent. Each pigment has a distinct visible color and can thus be detected visually. In modern applications of chromatography today, most analytes have no color, and there are such small quantities being separated, no color would be visible anyway. Therefore, each type of chromatography has its own types of **detectors** that use different properties of the analytes to signal their presence. A familiar analogy for the workings of chromatography is given in the "In More Detail" box below.

In chromatography, the relationship between the stationary phase and mobile phase is often described as **normal** or **reverse phase**. In normal phase chromatography, the mobile phase is less polar than the stationary phase. This is always the situation with gas chromatography, where the mobile phase is a non-polar inert gas. In reverse phase chromatography, the opposite is true. The mobile phase is more polar than the stationary phase. In liquid chromatography and **thin layer chromatography**, either normal or reverse phase can be used.

> ## IN MORE DETAIL:
> ### How does chromatography work?
>
> One way of visualizing how chromatography works to separate components of a mixture is to consider a group of tourists who are visiting a large city. They decide to go sightseeing one day. The city runs tour buses that leave a central point every few minutes. Passengers can board or alight from any bus any time. All the buses make a circuit of the city, stopping at a number of points of interest. A group of tourists all board a bus in the morning (the mixture). Along the way, some of the tourists stop and get off the bus to see an art gallery, the state capitol, or a performing arts center, etc. They get back on a tour bus after seeing the sights. Some of the passengers may stay on the bus for the whole tour as a way of getting the lay of the land. At the end of the day, all of the tourists will eventually make it to the end of the tour, but at different times. They will have been separated by their different preferences for one sight or another or for the tour bus. In this analogy, the tour buses are the mobile phase, and the various points of interest are locations on the stationary phase.

TYPES OF CHROMATOGRAPHY

Gas-Liquid Chromatography (GLC or GC)

Probably the most versatile and commonly used chromatography method is gas-liquid chromatography or, more commonly, gas chromatography. It is abbreviated "GC." In modern GC, called **capillary gas chromatography**, the stationary phase consists of a very thin (like a human hair) hollow tube whose inside walls are coated with a very viscous, high boiling liquid, with a consistency similar to molasses. Older GCs used a wider glass or metal tube (1/8–1/4 inch in diameter) that

Figure 6.6. The gas chromatograph. (1) The injector and inlet. (2) The mobile phase. (3) The stationary phase (column) in an oven. (4) The detector. (5) A computer to display and manipulate the data from the detector. (6) The chromatogram.

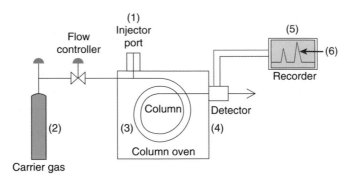

was filled with tiny, spherical, solid particles that were each coated with the same type of viscous liquid. They were called "packed columns."

The mobile phase in GC is an inert (unreactive) gas, usually helium or nitrogen. It passes through the hollow column under pressure, carrying with it the analyte components. For the analytes to be separated, they too have to be in the gas (vapor) phase. Figure 6.6 shows the parts of a gas chromatograph.

Parts of the Gas Chromatograph
THE INJECTOR

The injector is the first part of the chromatograph. It is a heated chamber where the analyte is introduced and mixed with the mobile phase. The injector can be heated from around 50–300°C. The heroin and cocaine solution is introduced into the injector with a syringe. The heat of the injector causes the chloroform, the cocaine, and the heroin to vaporize. The injector is also where the mobile phase is introduced under pressure. It carries the vaporized analyte components into the column that contains the stationary phase.

THE STATIONARY PHASE

The separations take place in the GC column. A typical capillary column is usually 12–30 meters long and is coiled up and fit into an oven. The column is kept at about the same temperature as the injector. The analyte is pushed through the column under pressure. During this journey, the heroin and the cocaine interact with the stationary phase in different ways, causing them to separate from the chloroform solvent and each other. As each component reaches the end of the column, it goes to the detector.

The nature of the stationary phase can affect the separation of components of the analyte. Changing the polarity can have a significant effect. In addition, changing the temperature of the column can have even more profound effects. Raising the temperature of the column even 10° can cut retention times of substances in half. In cases in which a mixture contains very volatile and non-volatile substances such as gasoline, using a **temperature program** can be very helpful. The temperature of the column starts out low. This will allow the very volatile substances to separate. Then the temperature is slowly raised until it peaks at a high temperature. This permits the slower, less volatile substances to elute in a reasonable time.

THE DETECTOR

Detectors in gas chromatography are designed to respond to a gas or solution of two or more gases. There is a continuous stream of the mobile phase gas streaming through the column to the detector. GC detectors are insensitive to the mobile phase and do not respond to it. When an analyte component is present, the composition of the gas changes and the detector responds to the change. GC detectors convert the signal created when an analyte component reaches it to a small electric current. This is amplified and displayed on a computer as a peak. There are a number of variations of this general principle. A few of the more commonly encountered detectors used in forensic science applications will be discussed later in this chapter.

Ideally, each substance in the mixture would take a different amount of time to traverse the column and reach the detector. A **chromatogram** is a plot of the response of the detector versus the time it takes each component to get through the instrument. The time is called the **retention time**. Every time the same substance is analyzed by GC, its retention time should be about the same, as long as the conditions of the experiment are not changed. Thus, the retention time gives a tentative identification of the substance. It is important to understand that the retention time is not a characteristic that definitely identifies a substance. For example, an exhibit may show a peak at 2.4 minutes in a gas chromatogram. A sample of known heroin also has a retention time of 2.4 minutes using the same instrument under the same conditions. The fact that both the unknown and the known have the same retention time is indicative of their being the same substance but does not prove it. There are millions of substances in the world, and several might have the same retention time. This is why chromatography is a separation technique and not an identification technique.

Figure 6.7. A gas chromatogram of partially burned gasoline. Each peak represents a component of the gasoline. Whole gasoline contains more than 300 substances.

```
File      : D:\GAS30.D
Operator  : TESSA GINGER
Acquired  :  4 Feb 103   3:16 pm using AcqMethod CARPET
Instrument :   GCD Plus
Sample Name: GASOLINE BURNED FOR A SHORT TIME ON CARPET
Misc Info :
Vial Number: 1
```

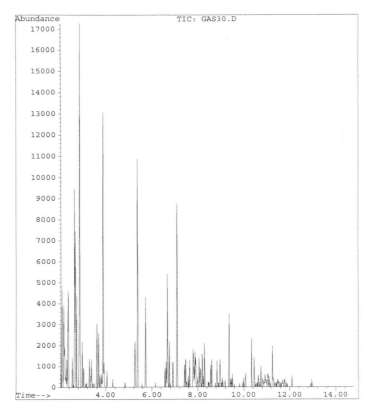

FLAME IONIZATION DETECTOR

The flame ionization (FID) detector has a small flame that causes the analyte components to **ionize**, or lose an electron. The ions cause an electric current to flow. This current can be amplified and displayed on a chart recorder or computer as a triangular peak. The more of an analyte component there is, the more current is created and the bigger the peak. Thus, the detector can be used for both qualitative and quantitative analysis. A gas chromatogram of partially burned gasoline that was obtained with a **flame ionization detector** is shown in Figure 6.7.

MASS SPECTROMETER DETECTOR

See Chapter 5 for a discussion of how the mass spectrometer works. A mass spectrometer can detect and identify each analyte component as it comes off the column. It also displays each component as a peak on a mass chromatogram. When a GC and an MS are joined together, it is possible to separate a mixture into individual components and identify each one—a very powerful analytical tool! The identification process can be very efficient if the GC-MS system contains a spectral library. This is a collection of up to thousands of mass spectra of known compounds. A reasonably powerful personal computer can take the mass spectrum of an unknown substance and use it to search the spectral library. This process may take less than one minute for 50,000 compounds. The result of the search will usually be a list of about 10 compounds whose mass spectra resemble the unknown most nearly and a number that indicates how closely each one matches. A very high number would indicate that there is a high likelihood that the known and unknown are the same substance.

OTHER GC DETECTORS

There are several other types of GC detectors that are not as widely used in forensic applications. They are listed here for completeness:

- **Nitrogen-Phosphorous**: This type of detector is a lot like the flame ionization detector except that it can detect only substances that contain nitrogen or phosphorous. This type is widely used in biological, toxicological, and environmental applications and is very sensitive.
- **Thermal Conductivity**: This type of detector relies on the change in the ability of the mobile phase gas to conduct heat as it is mixed with an analyte. It is simple to engineer and use and is very versatile.
- **Electron Capture**: This extremely sensitive detector is used on substances that have a halide such as chloride or bromide, or oxygen in the molecule.

Quantitative Analysis by GC

As previously noted, the size of the peak on a gas chromatogram is proportional to the amount of material that reached the detector. Strictly speaking, the area under the peak (geometrically speaking) is the important quantity. Suppose, in the case with the heroin and cocaine, a chemist wanted to determine the percent of cocaine in the drug exhibit. The following steps would be taken:

1. A sample of the exhibit would be weighed out and dissolved in a suitable solvent that also contained an internal standard. An internal standard is a compound that elutes near the analyte and is used to standardize the procedure.

2. A sample of known cocaine would be weighed out and dissolved in the same solvent with the same amount of internal standard.

3. Both would then be chromatographed, and the areas under the cocaine and internal standard for both the known cocaine and case exhibit would be calculated. Through simple proportions, the weight of cocaine in the exhibit could be calculated, and then the percent could be determined by knowing the weight of the powder that was dissolved.

In most jurisdictions, the percent of a drug is not a legal issue in the sense that the penalty for possession or distribution does not depend on the percent of cocaine. Why, then, would a drug chemist do this quantitative analysis? In some cases, the investigative agency asks for this data because it can help in determining how far up the distribution chain the seized drugs are. For example, if normal "street" cocaine is 40% pure and this particular exhibit is 80%, it means that the person arrested for possession of this exhibit is probably not an ultimate user, but is probably a distributor. Sometimes judges want to know the purity of a drug exhibit so that they can impose appropriate sentences when someone is found guilty of possession or distribution.

Pyrolysis-Gas Chromatography

One of the limitations of gas chromatography is temperature. The stationary phases of gas chromatographs are limited to about 300°C. There are many substances that could be otherwise analyzed by GC except that they do not vaporize to any appreciable degree at these temperatures. These substances include fibers, paints, plastics, hairs, and other polymers. A modification of gas chromatography, called **pyrolysis**, can make it possible for a gas chromatograph to handle polymers.

The term "pyrolysis" means essentially "to heat in the absence of air." If a polymer such as a fiber is heated to very high temperatures, up to 1000°C, in the absence of any oxygen, it will not burn, instead it will decompose into stable fragments, called **pyrolysates**. If this process is done under the same conditions, the number, size, and relative amounts of the pyrolysates will be the same for a particular polymer type. In pyrolysis-GC, an apparatus that can hold a small fragment of polymer is inserted directly into the injector of the gas chromatograph. Recall that there is only a mobile phase here, a gas such as helium. There is no oxygen present, so only pyrolysis—not combustion—takes place. The pyrolyzer is then heated to high temperatures, generally

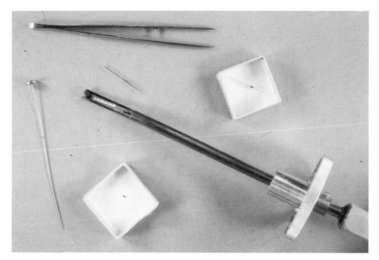

Figure 6.8. *A pyrolysis unit. The quartz tubes hold the samples and can fit inside the platinum coil on the end of the probe.*

700–1000°C and the polymer decomposes. The fragments are then separated, the same as the components of any analyte. The resulting chromatogram is called a **pyrogram**. Figure 6.8 shows a pyrolysis instrument and Figure 6.9 shows a pyrogram of a fiber. Pyrolysis gas chromatography is an example of a "hyphenated technique" where two instruments are combined to improve the analysis. Hyphenated techniques are discussed below "In More Detail: Hyphenated Analytical Techniques."

IN MORE DETAIL:
Hyphenated analytical techniques

Gas chromatography-mass spectrometry is an example of a "hyphenated" technique. This is a colloquial term used in analytical chemistry to denote two instruments that are engineered to act as a single entity. In the case of GC-MS, the individual substances separated by the GC are sent directly into the mass spectrometer, where they may be identified. The mass spectrometer can be thought of as a detector for the GC, or the GC can be thought of as an inlet device for the MS. Other hyphenated techniques are also used in analytical chemistry and forensic science. For example, there is GC-FTIR. FTIR stands for **Fourier Transform Infrared Spectrophotometer**. In this instrument, the GC separates a mixture into individual components, and then each one can be introduced into a special sample compartment in an FTIR (see Chapter 5).

High performance liquid chromatography or HPLC-MS (see the next section) can also be mated to a mass spectrometer. It works in much the same way as GC-MS.

Pyrolysis-GC, discussed earlier, is another hyphenated technique. This one is a bit different from the others in that the pyrolysis unit cannot work by itself and must have a GC, whereas the other instrument pairs can stand alone.

Figure 6.9. A pyrogram of a polyester fiber.

File: C:\CHEMPC\DATA\G3UNKF.D
Operator: Group 3
Date Acquired: 28 Jan 102   1:36 pm
Method File: FIBERS.M
Sample Name: UNKNOWN FIBER
Misc Info:
ALS vial: 1

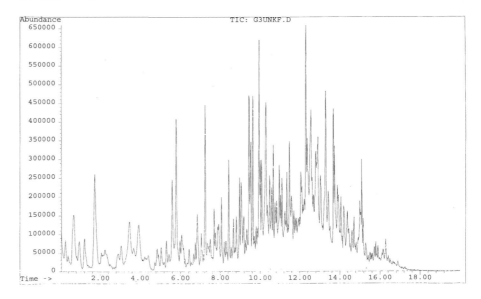

Applications of Gas Chromatography

There are numerous forensic applications of gas chromatography. The more common ones are as follows:

- **Drugs**: This is one of the most common uses of GC in forensic science today. Virtually all controlled substances can be measured both qualitatively and quantitatively using gas chromatography. This is changing somewhat today as forensic science laboratories are increasingly relying on HPLC/MS for chemical analysis. For further information, see Chapters 13 and 14.

- **Fire Residues**: Accelerants that are used to set fires for arson purposes can often be recovered from the fire scene. These are generally complex mixtures of hydrocarbons. The best example is gasoline, the most commonly used accelerant in fires. Gasoline contains more than 300 different chemicals, most of which are hydrocarbons. The gas chromatogram of gasoline shows a pattern of peaks that is unique. No other accelerant material has this same pattern. See Chapter 18 for more about accelerants.

- **Polymers**: Hairs, fibers, paints, plastics, rubbers, adhesives, and other polymers can be characterized by gas chromatography, but only if they are pyrolyzed first using a pyrolyzer that is outfitted to the injector of the gas chromatograph.

HIGH PERFORMANCE LIQUID CHROMATOGRAPHY (HPLC)

Remember the original experiments that were done on plant pigments in chromatography. These are actually examples of liquid chromatography. The analytes were dissolved in a liquid and then poured through a bed of silica. The liquid is the mobile phase, and the silica is the stationary phase. This process is pretty slow because it depends on gravity to get the mobile phase through the stationary phase. Vast improvements have been made in liquid chromatography since these first experiments. Stationary phases have become much more efficient in separating components of an analyte, and they are much more sensitive. As a result, the mobile phase can be pushed through the stationary phase using pumps. This makes the experiment go much faster while keeping the high-resolution power of the technique. This type of chromatography is called **high performance liquid chromatography** or HPLC. Some people refer to this technique as high pressure liquid chromatography, but this is technically not correct.

In HPLC, packed columns are routinely used, and the stationary phase can be similar to those in packed column GC or can be very different. In fact, one of the most popular HPLC stationary phases used in forensic science is a C_{18} hydrocarbon (octa-decane). This material has the approximate consistency of candle wax. Mobile phases are either a single liquid or a mixture of miscible liquids.

HPLC has some significant advantages over GC. In GC, for example, the stationary phase is always more polar than the mobile phase (nothing can be less polar than an inert gas such as nitrogen or helium). In HPLC, stationary phases and mobile phases can be designed so that the stationary phase is **less** polar than the mobile phase. Octa-decane is an example of a very non-polar stationary phase. In such cases, the chromatography is referred to as being **reverse phase**. This can be a great help in separating a mixture of non-polar substances that would not separate well using a polar stationary phase.

Another advantage of HPLC over GC is that the composition of the mobile phase can be altered during the run. This is called **gradient chromatography**. At times it is desirable to start with a relative nonpolar mobile phase and then gradually increase its polarity by adding more and more of a polar solvent. This can be easily accomplished using two solvents and two pumps. A computer can control the rela-

Figure 6.10. A high performance liquid chromatograph. (1) Mobile phase reservoirs. (2) Pumps to put the mobile phase under pressure so it will percolate through the stationary phase. (3) Stationary phase column. (4) Detector. The most common detector is a UV-visible diode array detector. (5) Computer for display and manipulation of data from detector.

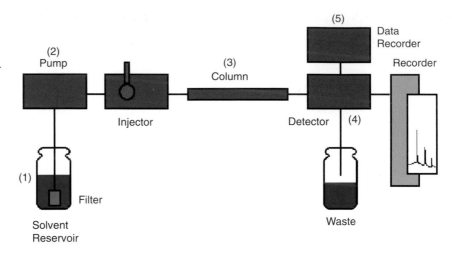

tive amounts of each solvent, thus changing the polarity of the mobile phase. This would be done in cases in which it will effect better separations among analyte components. When the mobile phase stays constant during an HPLC run, it is called **isocratic** chromatography.

Another big advantage is that HPLC can be used with compounds that are sensitive to high temperatures and with large molecules that do not have sufficient vapor pressure for GC.

Parts of an HPLC

Figure 6.10 shows a schematic of an HPLC. Each part is described in the following sections.

THE INJECTOR

A liquid chromatograph has the same parts as a gas chromatograph. They differ in how they work because the mobile phase is a liquid or liquid solution. As in gas chromatography, the analyte is dissolved in a suitable solvent, in this case one that is compatible with the liquid mobile phase. The injector usually consists of a sample loop, a small piece of hollow tubing that is isolated from the mobile phase stream until the analysis is to be done. Then the sample loop is joined to the stream, and the analyte is introduced. The mobile phase is pumped into the system via very accurate and precise liquid pumps. In the injector the analyte is pumped through the mobile phase under pressure from the pumps. An auto sampler unit can be used with HPLC, as it can with GC. The liquid samples are loaded into small vials, which, in turn, are put into a carousel. The computer controlling the process directs how the samples are analyzed.

STATIONARY PHASE

Today, most stationary phases are in packed columns. There are capillary HPLC columns, but they are not yet commonly used in forensic applications. Typically, columns are around 5 mm in diameter. The stationary phase is either a solid or a viscous liquid coated onto spherical, solid particles. The columns are generally much shorter than GC columns—usually around 25 cm. Even with this short length, efficient separations can be achieved.

DETECTORS

The detectors in an HPLC are quite different from those for a GC. Whereas GC detectors must detect solutions of gases, HPLC detectors must detect liquid solutions. When there is no analyte component present, the mobile phase flows through the detector by itself. When an analyte component is present, the properties of the liquid mobile phase change. Detectors are designed to detect changes in the concentrations of substances in the mobile phase.

The most popular detector for HPLC is a UV/visible spectrophotometer, usually a special type called a **diode array detector** (DAD). This detector measures the ultraviolet and visible spectrum of the solution as it flows through. Most mobile phases are not active in the UV or visible spectral range, so the detector does not respond to a pure mobile phase. Most of the substances that are analyzed by HPLC have a UV or visible spectrum, so the detector will detect their presence. See Chapter 5 on spectroscopy for a discussion of UV/visible spectroscopy.

A diode array detector simultaneously measures all of the wavelengths of UV and visible absorption of the analyte, so there are many ways that the data can be presented. For example, a simple chromatogram of retention volume (similar to retention time) versus absorption can be plotted. This will look the same as a gas chromatogram. In addition, however, many HPLC systems are capable of plotting absorption versus retention volume versus wavelength. This "three-dimensional" plot presents a great deal more information than a simple two-dimensional plot. An example of a 3-D plot is shown in Figure 6.11.

Other detectors also can be used for HPLC. They are briefly discussed here:

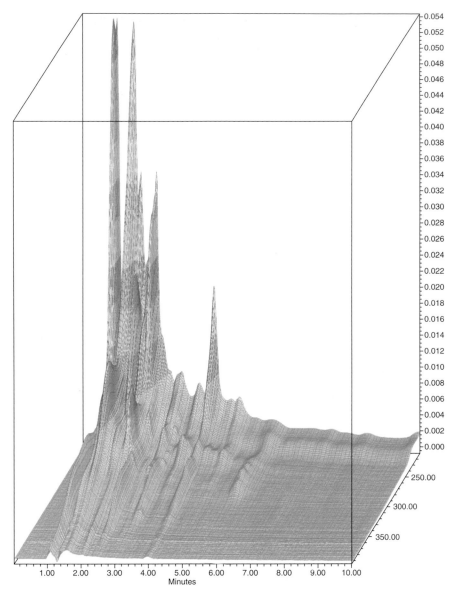

Figure 6.11. A "three-dimentional" HPLC plot. The X-axis is time, the Y-axis is wavelength from the detector, and the Z-axis is absorbance units from the detector.

- **Fluorescence**: This detector will detect only those substances that exhibit fluorescence, such as the illicit drug LSD. This limits its utility, but it is extremely sensitive.
- **Conductivity**: Most liquids will conduct electricity to a greater or lesser extent. When an analyte, especially a polar one, is introduced into the mobile phase, the conductivity changes, thus enabling detection of the analyte. This works best if the mobile phase contains at least some water.

- **Refractive Index**: The ability of liquids and other transparent materials to bend and decrease the velocity of light as it passes through is called refraction. All liquids refract light differently. A mobile phase with an analyte dissolved in it will refract light differently than the mobile phase alone. A refractive index detector can measure this change.

- **Mass Spectrometry**: LC-MS has not been around nearly as long as GC-MS, in part because of some formidable engineering hurdles that had to be overcome. When the mobile phase and analytes reach the mass spectrometer, the mobile phase is stripped away and the mass spectrum of each component of the analyte is measured. As in GC, this permits separations and identifications to take place in one step.

Applications of HPLC

HPLC has become increasingly popular in forensic science laboratories. Some of the same materials that are separated by GC methods can also be analyzed by HPLC. Also, HPLC may be the method of choice for separating analytes that are sensitive to temperature, such as explosives, which decompose, sometimes violently, when subjected to heat. Some of the more popular applications to evidence analysis are as follows:

- **Drugs**: Many controlled substances are analyzed by HPLC. In addition drugs taken from body fluids can also be analyzed. Drugs are more often identified by GC because of the ease with which a mass spectrometer can be used. Although HPLC-MS instruments are now commercially available, they are not widely used in forensic science labs.

- **Soils**: Organic extractions can be done on soils and the various substances separated. The result is a profile of the soil. The substances in the mixture are not identified, but the profile is a useful way of determining if soil found at a crime scene could have come from a particular location.

- **Explosives**: It may not be safe to run explosive extracts by GC because of the high heat, but HPLC is an ideal method for separation of explosive residues.

- **Inks and dyes**: Determination of the visible and UV spectra of inks is useful in comparing a writing instrument to writing on a document. It can also be used to follow the aging of the ink as it dries and degrades. Fiber dyes can be extracted from fibers and separated by HPLC also.

THIN LAYER CHROMATOGRAPHY (TLC)

If you have ever cleaned up an ink spill with a paper towel and noted that the ink dyes separated on the surface of the paper, you are actually performing a type of chromatography. Paper chromatography is one of the oldest methods of separation and is still used in some applications. The paper acts as a stationary phase, and the ink solvent is the mobile phase. The method is somewhat crude and in recent years has largely been replaced by more sophisticated stationary phases and complex mobile phases. The evolved technique is now called thin layer chromatography because the stationary phase consists of a thin layer of solid material that is coated onto a small glass or plastic plate.

Thin layer chromatography is very much like HPLC in its characteristics. The stationary and mobile phases are the same types. In TLC it is not possible to change the characteristics of the mobile phase on the fly, but through proper mixing, a large variety of polarities can be achieved in the mobile phase. Stationary phases of various polarities are also available commercially.

Stationary Phase

The stationary phase is a thin layer of a solid material combined with binders that is coated onto the surface of either a glass or plastic plate. These range in size from a microscope slide to more than 6 inches on a side. The coatings range from a few microns up to 1–2 mm for preparative scale work. For many samples of forensic science interest, the stationary phase coating may contain an internal fluorophore. This is a substance that will fluoresce when exposed to UV light. When a non-fluorescent sample is loaded onto the stationary phase, it will quench or blot out the plate's fluorescence and will be seen as a dark spot. This helps in determining how heavily a spot has been made and its position.

Mobile Phase

As in HPLC, the mobile phase in TLC is an organic liquid (or water) or a solution of two or more liquids. The polarity of the mobile phase ranges from the very non-polar to very polar. It can also be buffered to maintain a particular pH or ionic strength.

The TLC Experiment

Thin layer chromatography is carried out in a chamber. Its dimensions depend on the size of the coated plastic or glass plate. The plate must fit so that it does not touch the sides of the chamber. The plate must also fit entirely within the chamber. A typical arrangement may use a 5×10 cm plate and a 400 ml beaker. A piece of filter paper is put inside the chamber up against the side. This will absorb some of the mobile phase so that the entire inside of the chamber is saturated with the mobile phase. The top of the chamber is tightly covered. In many cases, Parafilm® works well.

The analyte is dissolved in a suitable solvent. The solvent must be volatile. Chloroform or methanol is often used. A well of a spot plate can be used for this procedure. A spot of the dissolved analyte is put onto the plate about 1 cm up from the bottom. This spot is kept as tiny as possible. A very narrow capillary can be used to make the spot. If necessary, the spot can be overspotted, but the original spot should be dried first by blowing on it. This will keep the spot as small as possible. One of the advantages of TLC over other forms of chromatography is that more than one sample can be run at the same time. There is generally room for several spots along the bottom of the TLC plate even with the need to keep some space between each spot. For example, in a drug case containing three exhibits where cocaine is suspected, one TLC plate can hold spots of each of the three exhibits as well as a sample of known cocaine and perhaps lidocaine, a common cutting agent for cocaine.

After the plate is loaded with samples, a small amount of the mobile phase is put in the chamber. There must be enough to travel up the plate but not enough to cover the spots. The chamber is covered and the mobile phase will travel up through the stationary phase, carrying with it the analyte spots. Using the same principles that govern all chromatographic separations, the components of the analyte will be differentially retarded by the stationary phase. When the mobile phase front has nearly reached the top of the plate, the process is stopped by removing and drying the plate. An apparatus used for performing TLC is shown in Figure 6.12.

Detection

Detection of the components of the analyte is different than with other types of chromatography. Some of the spots may show up under UV light. Some may fluoresce. Most will quench the native fluorescence of

Figure 6.12. A thin
layer chromatography
apparatus.

the plate and will show up as dark spots. In other cases there may be reagents that can be added to the spots, usually by spraying an aerosol. These reagents react with the analyte component to form a characteristic color. For example, THC, the active ingredient in marijuana, shows up as a bright, orange-red spot when sprayed with a reagent known as Fast Blue BB. Most nitrite (NO_2) containing compounds, such as most explosives, will turn red when a two-step reagent known as Greiss Reagent is sprayed on them.

Once the spots are visualized, their positions are measured as a **retention factor (Rf)**. The retention factor is the distance that the mobile phase travels up the plate divided by the distance that the component traveled. This is done in an attempt to make the results of a TLC experiment portable so that other laboratories can use the Rf data in their work. Using a ratio can eliminate many of the variables that are present in this process so that one lab can rely on the results of another lab.

Applications of TLC

TLC can be used in all of the applications that HPLC and, to some extent, GC are used for. In fact TLC is often used to "model" an HPLC experiment. If a separation needs to be optimized, it can often be done quickly and cheaply using TLC. Then the parameters such as stationary and mobile phase compositions can be transferred to HPLC.

Advantages and Disadvantages of TLC

Compared to GC and HPLC, TLC is much cheaper. No instrument is required. In addition, many samples can be run simultaneously on TLC, whereas only one sample can be run at a time on GC or HPLC. On the other hand, quantitative analysis can be easily performed using GC and HPLC, whereas this is not possible with TLC. TLC is also less sensitive than the others; more sample is required. Finally, a method of visualization is needed for analyte components separated by TLC, whereas the other methods result in visible peaks on a chart.

ELECTROPHORESIS

Electrophoresis is a type of chromatography that operates under somewhat different principles than the others previously discussed. Stationary phases are usually quite different in electrophoresis, although conventional stationary phases can be used in some examples. There are two major types of electrophoresis: gel and capillary. Gel electrophoresis is similar in some respects to TLC, whereas capillary electrophoresis is similar to HPLC. The major advantage of electrophoresis over conventional chromatography is resolution; even very similar components of an analyte can be easily separated by electrophoresis, whereas they cannot be separated at all by other forms of chromatography. The best example of this resolution power is DNA. To accomplish forensic DNA analysis, it is necessary to separate fragments of DNA and determine their size. These fragments are virtually identical in chemical composition except for their size. Capillary electrophoresis is capable of separating DNA fragments that differ by only one base pair in size. In such cases, the entire separation process is accomplished by mass action: Larger fragments of DNA travel slower.

THE STATIONARY PHASE

In gel electrophoresis, the stationary phase is a slab of a gel material such as agarose or polyacrylamide. This slab is immersed in a buffer solution to maintain a particular pH and ionic strength. Small wells are made in the gel, and the analyte, which may be extracted DNA or blood or another body fluid, is put in the well. As in TLC, many samples can be run at the same time, and in the case of DNA, several positive and negative controls are run with each case. In capillary electrophoresis, the stationary phase is a thin column, similar to a capillary used in GC. The capillary itself may be the stationary phase, or it may be filled with another material.

THE MOBILE PHASE

The mobile phase in electrophoresis is unique to chromatography: It is an electric current. When the capillary or gel slab is immersed in the buffer, a power supply is connected that will deliver hundreds or thousands of volts to the system. This will give one end of the stationary phase a strong positive charge and the other a strong negative charge. The buffer serves to put a positive or negative charge on the analyte components. During the separation, the analyte components will migrate toward the side with the charge opposite to their own. In DNA analysis, for example, the DNA fragments usually have a negative charge, and they migrate toward the positive side of the capillary or gel. DNA fragments that differ only slightly in size will still have different rates of migration and will be separated.

DETECTION

Detection in gel electrophoresis is usually accomplished by staining the analyte components so they can be seen with visible or UV light. Sometimes the fragments are made radioactive and are visualized by their ability to expose photographic film. In the case of capillary electrophoresis, the detection is usually accomplished by UV absorption of light by the analyte components, although other detection methods are available. The result is a chromatogram with a peak for each component. This is similar to the output from GC or HPLC. This type of chromatogram is called a capillary electropherogram (see Figure 6.13).

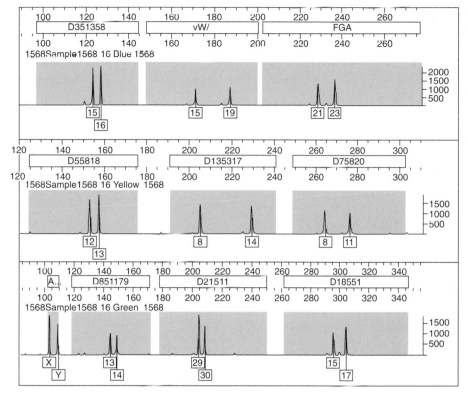

Figure 6.13.
A capillary electropherogram. This is very similar to a gas or liquid chromatogram. Each peak represents a different substance, in this case a piece of DNA.

APPLICATIONS OF ELECTROPHORESIS

- **DNA typing**: This is the only method available for separating DNA fragments suitable for forensic DNA typing. Either gel or capillary electrophoresis can be used.

- **Drugs**: Although drugs can be analyzed by a number of types of chromatography, capillary electrophoresis is the most sensitive of the methods.

- **Explosive residues**: Capillary electrophoresis has been employed because it is more sensitive and has higher resolution than HPLC. Most organic explosives can be separated by this method.

- **Gunshot residues**: Again capillary electrophoresis is very sensitive. It must be optimized for detection of inorganic substances in this application.

- **Questioned documents**: Capillary electrophoresis has just begun to be used for the separation of ink components used in pens.

BACK TO THE CASE: DRUGS

A straightforward application of gas chromatography is the separation and quantitative analysis of drugs. Unless there is only a very small quantity of the drug, it can be separated by a GC and then confirmed by mass spectrometry. Most drugs and many other substances can be separated this way. Some drugs cannot be separated because they decompose in the heated inlet and column of the GC. This is also true of some other substances such as explosives. In such cases HPLC is a good substitute because it is not a high temperature process.

CHAPTER SUMMARY

Forensic science evidence seldom appears in a pure form. Often the material of interest must be separated from other substances that are present. Such evidence commonly includes drugs, fire residues, and explosive residues. When a large amount of the material of interest is present, then a bulk extraction method can be used that takes advantage of the solubility or insolubility of materials in certain solvents. More commonly, however, the important evidence occurs in very small quantities, too small to be handled without losing it. In such cases, techniques that can separate and detect very small quantities of material are necessary. These techniques are collectively called chromatography.

Chromatography is a family of techniques that separate materials using a stationary phase and a mobile phase. The basis for the separation is that different substances will have different affinities, or attractions, for the stationary or mobile phase. In gas chromatography and liquid chromatography, the stationary phase is put into a long column. The mobile phase travels through the column under pressure. It carries with it the mixture that is to be separated. As the components of the mixture interact with the stationary and mobile phases, they are separated and emerge from the column at different times. The instrument contains a detector that reacts to the presence of the components of the mixture. In thin layer chromatography the sample is placed onto a thin plate that is coated with the stationary phase, and then the plate is dipped into a reservoir containing the mobile phase. This travels up the plate, carrying the mixture whose components then separate by interaction with the stationary phase. The type of chromatography used depends on the nature and form of the mixture.

A specialized type of chromatography, called electrophoresis, is used for the separation of substances that are so similar to each other that they cannot be separated by using conventional chromatography. In electrophoresis, a strong electric charge is used to carry the analytes through the stationary phase. Very specialized types of stationary phases are used in electrophoresis. In forensic science the major use of electrophoresis is in the separation of fragments of DNA.

Test Your Knowledge

1. What is normal phase chromatography? What types of substances are best separated this way?

2. What does the term "stationary phase" mean? What types of stationary phases are used in gas chromatography?

3. What type of substance is the mobile phase in gas chromatography?

4. List two advantages of HPLC over GC.

5. Explain how a flame ionization detector works. What other types of detectors are used in gas chromatography?

6. What is reverse phase liquid chromatography? What types of substances are best separated using this technique?

7. In HPLC, why would you want to use two pumps for the mobile phase?

8. List and describe three types of detectors used in HPLC.

9. Describe three advantages of thin layer chromatography over other types.

10. Give at least two advantages of HPLC over TLC.

11. What are the two general requirements for an analyte in order for it to be separated by GC?

12. In electrophoresis, what are the stationary phase and mobile phase?

13. What is the major advantage of electrophoresis over other types of chromatography?

14. Define pH. Why is it important in liquid extractions?

15. How do extractions differ from chromatography?

16. What is the difference between liquid phase extraction and solid phase extraction?

17. What is the difference between adsorption and partition chromatography?

18. Why can HPLC be used for the quantitation of an analyte but TLC cannot?

19. How are the spots obtained from TLC visualized?

20. How does capillary electrophoresis differ from gel electrophoresis?

Consider This . . .

1. Polymers are found in many types of evidence such as paints, fibers, plastics, etc. What type of chromatography is performed on these polymers? How is it done? What is the principle that permits comparison of knowns and unknowns this way?

2. Explosive residues are generally not analyzed by GC. Why? What method(s) can be used for separation of explosive residues? Certain drugs, like amphetamines, are difficult to separate by GC because of their interaction with the stationary phase and their thermal instability. What can be done to drugs like these to make it easier for them to be separated by GC?

3. In gas chromatography, resolution of complex mixtures of analytes can be improved by temperature programming. Explain how and why this helps. Likewise, in HPLC, mobile phase programming can be used to achieve a similar purpose. How does this work?

BIBLIOGRAPHY

Siegel, J.A., Knupfer, G., & Sauko, P. (Eds.) (2001). *The encyclopedia of forensic sciences, Vol. 1.* Boston, MA: Academic Press.

Saferstein, R. (Ed.) (1983). *Forensic science handbook, Vol. 1.* Upper Saddle River, NJ: Prentice Hall.

Saferstein, R. (Ed.) (1988). *Forensic science handbook, Vol. 2.* Upper Saddle River, NJ: Prentice Hall.

Wixom, R.L., & Bayer, E. (Eds.) (2001). Chromatography—A Century of Discovery 1900–2000. Amsterdam: Elsevier Science.

PART THREE

Biological Sciences

Pathology

KEY TERMS

Algor mortis

Asphyxia

Autolysis

Autopsy

Blunt force trauma

Carboxyhemoglobin

Cause of death

Contact gunshot wound

Contusion

Coroner

Defensive wounds

Defibrillatory

Distant gunshot wound

Embalming

Exhumation

Exsanguination

Forensic

Hematoma

Hyperthermia

Hypothermia

Histology

Incised wounds

Intermediate gunshot wound

Lacerations

Ligatures

Livor mortis

Manner of death

Mechanical trauma

Medical examiner

Medicolegal autopsy

Microtome

Pathologist

Petechiae

Postmortem clock

Post mortem lividity

Primary cause of death

Putrefaction

Rigor mortis

Secondary cause of death

Sharp force trauma

Shored exit wound

Smears

Stippling

Tardieu spots

Tetany

Toxicology screen

Trauma

Ventricular fibrillation

INTRODUCTION

A pathologist is a medical doctor who studies and diagnoses disease in humans. A *forensic* pathologist is a pathologist who has studied not only disease but trauma (wounds and damage) that leads to the death of an individual. The modern **autopsy**, from the

Greek *autopsia,* meaning seeing with one's own eyes (Oxford English Dictionary, 2005), involves the standardized dissection of a corpse to determine the cause and manner of death. Regrettably, the number of autopsies has steadily declined in the past 50 years—less than 5% of hospital deaths are routinely autopsied, compared to 50% in the years after World War II. This is a shame, really, because autopsies are a quality control tool for doctors; they provide a "reality check" on their diagnoses and give them feedback on the effectiveness of treatments. Autopsies done to help solve a murder, however, are different in many ways, such as who conducts them, when and how they are conducted, and what purpose they serve to society.

CAUSE AND MANNER OF DEATH

The **cause of death** is divided into the primary and secondary causes of death. The **primary** or **immediate cause of death** is a three-link causal chain that explains the cessation of life starting with the most recent condition and going backward in time. For example,

1. **Most recent condition** (coronary bypass surgery, for example)
   Due to, or as a consequence of:

2. **Next oldest condition** (a rupture of the heart's lining due to tissue death from lack of oxygen, for example)
   Due to, or as a consequence of:

3. **Oldest (original, initiating) condition** (coronary artery disease, for example)

Each condition can cause the one before it. At least one cause must be listed, but it is not necessary to always use all three. The **secondary cause of death**, which includes conditions that are not related to the primary cause of death but contribute substantially to the individual's demise, such as extreme heat or frigid temperatures, is typically listed.

A distinct difference exists between the standard hospital autopsy and a **medicolegal autopsy**. The hospital autopsy is conducted based on a doctor's request and the family's permission; if the family denies the request for personal or religious reasons, the autopsy is not performed. A medicolegal autopsy, however, is performed pursuant to a medical investigation of death for legal purposes. For more information on the history of the autopsy, see "History: The Autopsy."

If a person dies unexpectedly, unnaturally, or under suspicious circumstances, the coroner or medical examiner has the authority to order an examination of the body to determine the cause of death. The **manner of death** is the *way* in which the causes of death came to be. Generally, only four manners of death are acknowledged: homicide, suicide, accidental, and natural. The deceased may have met his or her end in a way that appears suspicious to the authorities, and therefore the cause and manner of death must be established. Other purposes for a medicolegal autopsy may be to identify the deceased, establish a time of death, or collect evidence surrounding the death. The cause of death is often known, but the manner and mechanism of death may not be immediately obvious and are crucial to the goals of a medicolegal autopsy.

Although a pathologist can perform a hospital autopsy, more than normal medical training is required to interpret morbid anatomy and fatal trauma. In one study by Collins and Lantz (1994), trauma surgeons misinterpreted both the number and the sites of the entrance and exit wounds in up to half of fatal gunshot wounds.

CORONERS AND MEDICAL EXAMINERS

THE ORIGINS OF THE CORONER SYSTEM

The office of **coroner** was first granted by England's *Charts of Privileges* to St. John of Beverly in A.D. 925 and before the 1194 publication of the *Articles of Eyre*, the office of coroner had become an official position throughout the country. These individuals were called "keepers of the pleas of the crown," a phrase later shortened to "crowner" and then "coroner." The position was initially that of a formidable and prestigious judicial officer in charge of collecting monies due the king, trying felony cases, and gradually narrowed to the investigation of unusual, untimely, or suspicious deaths. By the thirteenth century, coroners had to examine all bodies before burial and appraise all wounds, bruises, and other signs of possible foul play (Thorwald, 1964) (see "History: The Origins of the Coroner System").

HISTORY:
The origins of the coroner system

During the last decade of Henry II's reign, discontent had developed over the corruption and greed of the sheriffs, the law officers who represented the Crown in each English county. Sheriffs were known to extort and embezzle the populace and manipulate the legal system to their personal financial advantage: They diverted funds that should have gone to the king. A new network of law officers who would be independent of the sheriffs was established to thwart their greedy ways and return the flow of money to the king. At that time they were "reif of the shire." Later they became known as the "shire's reif," and then "sheriff."

The edict that formally established the coroners was Article 20 of the *Articles of Eyre* in September 1194. The king's judges traveled around the countryside, holding court and dispensing justice wherever they went; this was called the "General Eyre." The Eyre of September 1194 was held in the County of Kent, and Article 20 stated:

> "In Every County Of The King's Realm Shall Be Elected Three Knights And One Clerk, To Keep The Pleas Of The Crown"

And that is the only legal basis for the coroner. Coroners had to be knights and men of substance; their appointment depended on their owning property and having a sizeable income. Coroner was an unpaid position; this was intended to reduce the desire to adopt any of the sheriffs' larcenous habits.

The most important task of the Coroner was the investigation of violent or suspicious deaths; in the medieval system, this position held great potential for generating royal income. All manners of death were investigated by the coroner. Interestingly, discovering the perpetrator of a homicide was not of particular concern to the coroner—the guilty party usually confessed or ran away to avoid an almost certain hanging. The coroner was, however, concerned to record everything on his rolls, so that no witnesses, neighbors, property, or chattels escaped the eagle eyes of the Justices in Eyre. There was a rigid procedure enforced at every unexpected death, any deviation from the rules being heavily fined. The rules were so complex that probably most cases showed some slip-up, with consequent financial penalty to someone. It was common practice either to ignore a dead body or even to hide it clandestinely. Some people would even drag a corpse by night to another village so that they would not be burdened with the problem. Even where no guilt lay, to be involved in a death, even a sudden natural one, caused endless trouble and usually financial loss.

Sources: Thorwald, 1964; Wilson and Wilson, 2003.

The first American coroner was Thomas Baldridge of St. Mary's, Maryland Colony, appointed on January 29, 1637. He held his first death inquest two days later. It was not until 1890 that Baltimore appointed two physicians as the United States' first medical examiners (Thorwald, 1964).

The position of coroner can be appointed or elected, and typically no formal education or medical training is required. Today, many coroners are funeral directors, who get possession of the body after the autopsy. This can be a major source of income to such officials.

A **medical examiner**, by contrast, is typically a physician who has gone through four years of university, four years of medical school, four years of basic pathology training (residency), and an additional one to two years of special training in forensic pathology. These positions are appointed. Some states have a mixture of ME's and coroner systems, whereas others are strictly ME or coroner systems (see Table 7.1).

ON THE WEB

The website for the National Association of Medical Examiners
www.thename.org
A virtual autopsy, presenting a series of interactive cases and histories, written by Ajay Mark Verma at the University of Leicester, UK. This website won the Scientific American 2002 Sci·Tech Web Award.
www.le.ac.uk/pathology/teach/va

Table 7.1. *U.S. and Canadian Death-Investigation Systems.*

MEDICAL EXAMINER SYSTEMS STATE/PROVINCE CHIEF MEDICAL EXAMINERS		
Alaska	Manitoba	Oklahoma
Alberta	Maryland	Oregon
Arkansas[1]	Massachusetts	Puerto Rico
Connecticut	Mississippi[1]	Rhode Island
Delaware	Montana[1]	Tennessee
Dist. of Columbia	New Hampshire	Utah
Georgia[2]	New Jersey	Vermont
Guam	New Mexico	Virginia
Iowa	North Carolina[3]	Virgin Islands
Kentucky[1]	Nova Scotia	West Virginia
Maine	Labrador	Newfoundland
DISTRICT/COUNTY MEDICAL EXAMINERS		
Alabama	Florida	Michigan
Arizona		

Table 7.1. *Continued*

CORONER SYSTEMS		
American Samoa	Louisiana	Prince Edward Island
British Columbia	Nebraska	Saskatchewan
Colorado	Nevada	South Dakota
Idaho	New Brunswick	Wyoming
Indiana	North Dakota	Wyoming
Kansas	Ontario	Yukon Territory
Quebec	Northwest Territories	Nunavit
MIXED MEDICAL EXAMINER/CORONER SYSTEMS		
California	Missouri	South Carolina
Hawaii	New York	Texas
Illinois	Ohio	Washington
Minnesota	Pennsylvania	Wisconsin

[1]Also have coroners in every county.
[2]Coroners and autonomous medical examiners in some counties
[3]Coroners in some counties

THE POSTMORTEM EXAMINATION (AUTOPSY): THE EXTERNAL, OR VISUAL, EXAMINATION

The visual or external examination of a body starts with a description of the deceased's clothing, photographs (including close-ups) of the body both clothed and unclothed, and a detailed examination of the entire body. Any trauma is noted on a form, an example of which is shown in Figure 7.1, where the pathologist can make notes, sketches, or record measurements (see Figure 7.2); damage to clothing should correlate to trauma in the same area on the body. Gunshot wounds are recorded, for example, to indicate entrance and exits wounds and the

Case Number: 91-234

HISTORICAL SUMMARY

This 57-year-old black male was reportedly found in his bed with a gunshot wound of the head and a handwritten suicide note on the bedstand. He was transported to the hospital by emergency medical services but died in the emergency room at Hometown Hospital. He had a history of recent headaches. Additional details are contained in the investigator's report contained in the medical examiner case file.

EXAMINATION TYPE, DATE, TIME, PLACE, ASSISTANTS, ATTENDEES

Under the provisions of the Death Investigation Act, a complete autopsy is performed in the County Morgue on Tuesday, November 5, 1991, beginning at 1205 PM with the assistance of Angela Harden. Also in attendance is Major Gleet of the Hometown Police Bureau.

PRESENTATION, CLOTHING, PERSONAL EFFECTS, ASSOCIATED ITEMS

The body is contained in a white plastic body bag bearing a tag with the deceased's name on it and an identification number of 91-234. The hands are covered with paper bags secured at the wrists with rubber bands. A pair of white briefs are present in the pelvic area and are stained with a small amount of yellow fluid with an odor of urine. A gold-colored ring is present on the left ring finger. The briefs are discarded and the ring is removed and forwarded with the body. No other items are present with the body.

EVIDENCE OF MEDICAL INTERVENTION

An endotracheal tube exits from the right side of the mouth. Multiple perimortem needle-puncture wounds are present in each subclavian region. An intravascular cannula is inserted in the right cubital fossa. A small needle mark with underlying hematoma is present in the left radial fossa. Electrocardiographic conductor pads are located over each shoulder anteriorly and in the left lateral midthoracic area. A gauze pad is taped to the right side of the forehead and covers a wound that will be described in further detail below.

POSTMORTEM CHANGES

Rigor mortis is generalized and well developed. Livor mortis is well developed, dorsal, the usual violet color, and blanches with light pressure. The eyes show early corneal clouding. The vermilion borders of the lips are slightly dry. Other postmortem changes are absent.

POSTMORTEM IMAGING STUDIES

Postmortem radiographs of the head show a density beneath the inner table of the left parietal bone, consistent with a medium-caliber bullet.

FEATURES OF IDENTIFICATION

A hospital band on the right wrist bears the deceased's name. The body is unembalmed and that of a black male appearing slightly older than the stated age. Height measures 68 inches, and weight is 160 lb. The physique is mesomorphic. The head hair is black, coarse, measures about 1 inch in greatest length, and shows frontoparietal balding. The irides are brown. The teeth are natural with some amalgam restorations. An oblique, well-healed, 4-inch scar with crosshatched suture marks is located in the left inguinal area. The penis is uncircumcised. No tattoos are noted. The distal phalanx of the left fifth finger has been previously amputated and is well-healed. No other distinctive external markings are present.

EXTERNAL EXAMINATION

General

Body habitus and hair distribution are normal for age and gender. There is no evidence of malnutrition or dehydration. No peculiar odors or color changes are noted. There is no visible or palpable lymphadenopathy.

Head

A penetrating wound, consistent with a gunshot entry wound, is present on the right side of the head, just above the top of the right ear, 2 inches above the external auditory meatus. The wound is located 64 inches above the heel and 6 inches to the right of the anterior midline. The wound consists of a ½-inch circular hole from which extend radial tears measuring up to ½ inch in length. A ¼-inch concentric rim of purple contusion surrounds the hole. Within the superficial wound track, prominent deposits of soot and gunshot residue are visible. No soot or stippling is present on the skin surface surrounding the wound. Dry blood streaks are present posterior to the wound and within the hair. The ear canals are free of blood. The face shows no evidence of trauma. The scalp and soft tissues of the head are otherwise normal, except for palpable lump beneath the skin overlying the left midparietal skull just above the left ear. The nasal

Figure 7.1. *Although no standards exist for the format of autopsy reports, this is a suggested format from the College of American Pathologists (Hanzlick, 2000).*

Figure 7.1. *Continued*

and facial bones are without palpable fracture. The conjunctival vessels are slightly congested, and there are no ocular or facial petechiae. A small amount of blood-tinged fluid is present in each nasal vestibule. The lips, gums, teeth, tongue, and buccal mucosa are normal and free of injury. The pinnae and mastoid regions are normal.

Neck
The neck shows no indication of abrasion, contusion, swelling, asymmetry, or other abnormality.

Torso
The torso is free of injury and is symmetrical. No subcutaneous emphysema or cutaneous lesions are noted. The abdomen is moderately distended with gas. Two testes are palpable in the scrotum, which is otherwise normal. The external genitalia, perineum, and anorectal areas are normal except for a small external hemorrhoid at the 2 o'clock position. The inguinal regions and buttocks are normal.

Upper Extremities
The upper extremities are symmetrical, muscular, and well developed. No pigmented or scarred needle tracks are seen, and there are no hesitation marks or healed incised wounds. A 0.5-cm, resolving, subungual hematoma is present beneath the left thumbnail. No soot or gunshot residue is visible on the hands.

Lower Extremities
The lower extremities are well developed and symmetrical. There is slight hair loss bilaterally in a socklike distribution, and the toenails are somewhat thickened and untrimmed.

Evidence of Injury
External evidence of injury is limited to an apparent gunshot wound of the head, a resolving subungual hematoma of the left thumb, and evidence of medical intervention as described above.

Summary
External examination shows a well-developed black male with no significant findings except an apparent gunshot wound of the right side of the head.

INTERNAL EXAMINATION
Torso
Evisceration/Dissection Method. The thoracic and abdominal organs are removed using the Virchow technique (individually). Chest and Abdomen Walls and Cavities. The skin of the chest and abdomen is reflected using the usual Y-shaped incision. Subcutaneous fat and musculature are normal and free of injury. There are no abnormal fluid collections in the chest or abdomen. The ribs and sternum are intact and without fracture. No unusual odors or color changes are identified. Examination of the organs in situ shows normal organ morphology and relationships. The viscera are congested. The diaphragm is normal. The stomach is distended with air.

Organ Weights.

Heart, 485 g	Liver, 1650 g
Left lung, 450 g	Right lung, 510 g
Kidneys, 160 g each	Spleen, 140 g

Cardiovascular System. The left ventricle demonstrates concentric hypertrophy with a left ventricular wall thickness of 2.1 cm. The coronary arteries are normally distributed and are widely patent throughout their lengths, with minimal, soft, atherosclerotic plaques focally. The epicardium, valve leaflets, chordae, and endocardium appear normal. The myocardium is reddish-tan throughout, and no focal myocardial lesions are observed. The thoracoabdominal aorta and major branches show moderate, yellow, atherosclerotic streaking without ulceration. There are no vascular perforations.
The carotid arteries are pliable and patent.

Respiratory System. The trachea and bronchi are grossly normal except for focal mucosal contusion adjacent to the endotracheal tube cuff, which is positioned appropriately. The hilar nodes and structures are normal. The major pulmonary vessels are normally distributed and free of gross abnormalities. The lungs appear similar, and each lung is congested and moderately edematous, exuding a pink-white foam on manual compression. There is no aspirated blood. No consolidation is observed. There is no indication of thrombosis, embolism, infarction, or neoplasia. The visceral and parietal pleura are free of hemorrhage or perforating defects.

Digestive System. The serosa, wall, and mucosa of the esophagus, stomach, small bowel, colon, and rectum are grossly normal. The stomach is distended with air and contains approximately 1 cup of partially digested food, primarily consisting of green vegetable material. Hepatobiliary System and Pancreas. The liver shows intense congestion. There is no indication of fatty change or cirrhosis. No focal intrahepatic lesions are noted. The gallbladder contains about 15

Figure 7.1. Continued

cc of viscous green bile, no stones, and is grossly normal. The extrahepatic biliary ducts are patent. The pancreas shows the usual lobular architecture, mild autolysis, and is otherwise normal.

Reticuloendothelial System. The spleen has a tense capsule and is acutely congested. The red and white pulp are normal. Nodes of the axillary, hilar, mediastinal, abdominal, and cervical area appear normal, except to note mild anthracosis of hilar nodes. The thymus is involuted. Bone marrow of the vertebral bodies appears normal and without focal lesions or masses.

Urogenital Systems. The kidneys are symmetrical and each shows congestion of the cortex and medulla. The capsules strip easily and the cortical surfaces are smooth. The corticomedullary ratio and junction are normal, as are the pyramids, calyces, pelves, and vessels. The ureters are of normal caliber. The urinary bladder is normal and contains approximately 100 cc of amber urine. The seminal vesicles are normal, and the prostate is firm and nodular with slight enlargement.

Endocrine System. The thyroid gland is normal size, symmetrical, tan, and free of nodularity, hemorrhage, or cysts. The parathyroids are not identified grossly. The adrenals are of normal size and are free of nodularity or hemorrhage.

Head
The scalp is reflected with the standard intermastoidal incision. There is no indication of scalp trauma, except for a 4-inch circular area of full-thickness soft tissue hemorrhage around the gunshot wound in the right parietal area and a 2-inch circular area of full-thickness soft tissue hemorrhage in the left midparietal region above the left ear. The right temporoparietal bone shows a ⅜-inch circular hole with sharply defined margins on the outer table and beveling of the inner table. The outer table of skull around the hole shows black discoloration from gunshot residue, which is also visible in the diploic spaces. The left midparietal bone contains a ½-inch circular fracture, which is displaced into the overlying parietal soft tissues 2.5 inches directly above the left auditory meatus. Adjacent to the inner table of the bone fragment is a medium-caliber, slightly deformed, fully jacketed, round nose, copper-colored projectile, which is retrieved for submission to the Crime Laboratory. The bullet shows prominent lands and grooves. The dura shows ragged, roughly circular, ⅜-inch, perforating defects in both parietal areas in locations corresponding to overlying defects in the skull. Diffuse subarachnoid hemorrhage is present over the convexities. Brain weight is 1540 g. There is no evidence of significant herniation or midline shift. Coronal sections demonstrate a hemorrhagic wound track extending from the right midparietal region transversely through the brain to the left midparietal cortical surface. The wound track extends through the upper cerebral peduncles. Small cortical contusions are present on the inferior aspect of the frontal poles bilaterally. The circle of Willis contains a 0.5-cm berry aneurysm of the anterior communicating artery on the left. No focal or mass lesions are seen within the brain, and the cortex is normal to palpation. Moderate cerebral edema is noted. The basilar skull and atlanto-occipital region are intact.

Neck and Pharynx
The skin of the neck is dissected up to the angle of the mandible. There is no evidence of soft tissue trauma to the major airways or vital structures in the lateral neck compartments. The hyoid bone and thyroid cartilages are free of fracture. The carotid vessels are pliable and patent. The epiglottis is not inflamed or swollen. There is no airway mucosal edema. No foreign objects are present in the upper airway except for an endotracheal tube. The anterior cervical spine is intact. The tongue is normal.

Spinal Column and Cord
The thoracolumbar spinal column shows mild degenerative osteophytic lipping. The spinal cord is not removed or examined.

Additional Dissection None.
SUMMARY OF INJURIES
Examination shows an apparent contact gunshot wound of the right parietal area with perforating brain injury and recovery of a projectile in the left midparietal area. No other acute injuries are present.

ANCILLARY PROCEDURES, LABORATORY TESTS, AND RESULTS
     1. Vitreous for chemistries: Na, 135; K, 8.0; Cl, 120
     2. Peripheral blood for ethanol quantitation: Negative
     3. Urine for drug abuse screen: Negative
     4. Documentary photographs are prepared and filed in the case folder
     5. Retrieved bullet is forwarded to Crime Lab for firearms examination; results will be reported by the Crime
Lab

Figure 7.1. Continued

BLOCK LISTING AND HISTOLOGIC DESCRIPTION

Block 1 Heart and lungs
Block 2 Liver, spleen, pancreas, kidney
Block 3 Adrenal, thyroid, pancreas
Block 4 Routine sections of cerebrum, cerebellum, basal ganglia
Block 5 Routine sections of esophagus, stomach, small and large bowel
Block 6 Prostate
Block 7 Gunshot entry wound

The heart shows mild hypertrophic change; prostate shows benign prostatic hyperplasia. The gunshot wound shows hemorrhage, extensive gunpowder particles, and thermal changes in the collagen. Other sections are not contributory.

FINDINGS AND DIAGNOSES
1. Contact gunshot wound of right parietal area of head
    A. Perforating brain injury
    B. Fracture of left parietal bone
    C. Diffuse subarachnoid hemorrhage
    D. Cerebral edema
    E. Recovery of bullet from left parietal bone (no exit wound)
    F. Wound track right to left and slightly upward
2. Berry aneurysm, left anterior communicating cerebral artery
3. Resolving subungual hematoma, left thumb
4. External hemorrhoid
5. Concentric left ventricular hypertrophy, heart
6. Benign prostatic hyperplasia
7. Degenerative osteoarthritis, spinal column
8. Remote amputation of distal portion of left fifth finger
9. Surgical scar, left inguinal area; probable remote hernia repair

SUMMARY AND COMMENTS
Investigation and autopsy show that death resulted from a self-inflicted gunshot wound of the head. No other significant
injuries were observed. The finding of a cerebral artery berry aneurysm is a possible explanation for the history of recent headaches. Cardiac findings suggest a history of hypertension.

CAUSE-OF-DEATH STATEMENT
Perforating brain injury
due to: Contact gunshot wound of head
Based on the circumstances, the manner of death is classified as suicide.

AMENDMENTS
None as of 11/10/91.

path of the bullet through the body, as shown in Figure 7.3. **Defensive wounds**, like those shown in Figure 7.4, that are trauma caused by the victim trying to defend himself or herself against an attacker, are also noted. See "Classification of Trauma" later in this chapter for more details.

When the clothing is removed, care is taken to preserve any trace evidence that may be later submitted to a forensic science laboratory. Wet clothes are suspended to air-dry at room temperature. Folding wet clothes may obscure important evidence patterns, such as blood stains, and promote the growth of bacteria, which, besides smelling bad, can damage potential DNA evidence.

The age, sex, ancestry, height, weight, state of nourishment, and any birth-related abnormalities are noted during the external exam. The body is also checked for death-related phenomena that may

CASE NO._____   NAME _____

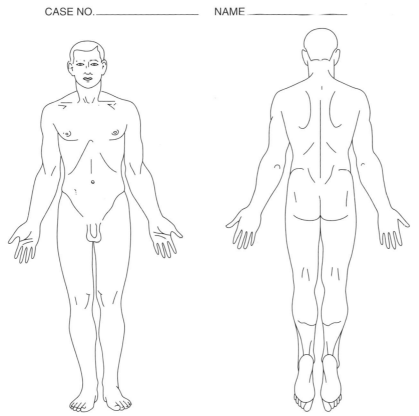

Figure 7.2. Diagrams like this are used to mark wounds, bruises, and other trauma, as well as for taking notes during the visual examination.

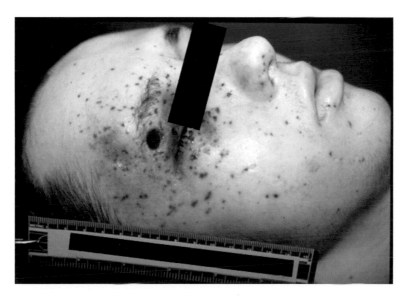

Figure 7.3. A gunshot wound to the head typically shows a small, clean entrance; note the stippling of gunpowder burns around the wound.

Figure 7.4. *In protecting themselves from attack by a sharp object, victims often have wounds indicating their attempt to ward off their attacker.*

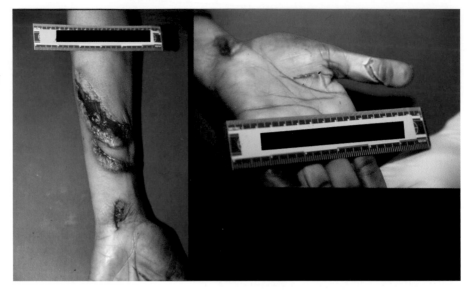

provide information to the investigation. For example, rigor mortis and livor mortis, if present, are noted. **Rigor mortis** is the stiffening of the body after death due to the membranes of muscle cells becoming more permeable to calcium ions. Living muscle cells expend energy to transport calcium ions outside the cells; calcium plays a crucial role in muscle contraction. Without this calcium transport, the muscle fibers continue to contract until they are fully contracted; the muscles release only when the tissues begin to decompose. Onset begins two to six hours after death, starting in the smaller muscles and eventually affecting even the largest ones. The stiffness remains for two to three days and then diminishes in reverse order. The rate of rigor mortis depends on activity before death and the ambient temperature; these must be taken into account by the pathologist when estimating a time since death.

Livor mortis, also known as **post mortem lividity**, is the settling of blood due to gravity after the heart no longer circulates it through the body. This results in a purplish discoloration in the skin, shown in Figure 7.5; the blood vessels also are not reaching the lungs to be oxygenated and the settled blood takes on a bluish tone. This is not true, however, of people who have died from poisons or substances that alter the color of the blood—for example, carbon monoxide, which colors the blood a bright, cherry red. Lividity begins to set in about an hour after death and peaks in about three or four hours. The blood settles in accordance with gravity and, once coagulated, does not move. The only exception to this is where pressure is applied—for example, a body lying on its back will have light patches where the blood couldn't settle, such as around the shoulder blades and the buttocks. Because of this, lividity can indicate whether a body has been moved: The pattern of lividity does not match the position of the body as it was found, as illustrated in Figure 7.6.

The eyes are also examined for a variety of indications that will provide clues to the pathologist. **Petechiae**, shown in Figure 7.7, are pinpoint hemorrhages found around the eyes, the lining of the mouth and throat, as well as other areas, often seen in hanging or strangulation victims. But petechiae are by no means conclusive evidence of strangulation or asphyxiation because other phenomena, such as heart attacks or cardiopulmonary resuscitation, can induce them. In older pathology literature, they may be

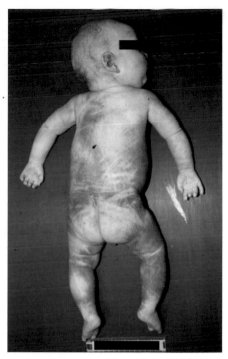

Figure 7.5. Lividity is the settling of blood cells once the heart stops pumping.

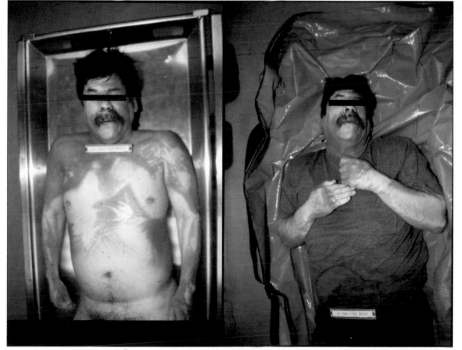

Figure 7.6. Lividity becomes fixed and, if a body is moved after this point, that fact becomes clear: The pressure of the body's weight keeps blood cells away from areas in contact with a surface. The position of the body as found (clothed) is confirmed once the clothes are removed.

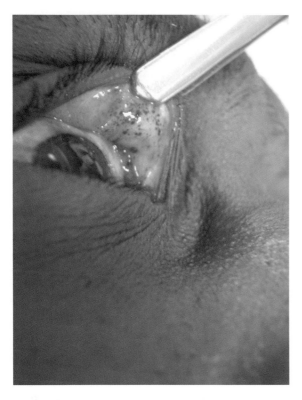

Figure 7.7. Petechiae, tiny blood vessels that burst often due to strangulation, are seen in the eyelids as well as other places.

referred to as Tardieu spots, after the doctor who first described them. The mouth area and oral cavity (the inside of the mouth) are also examined for trauma, trace evidence, and indications of disease.

CLASSIFICATION OF TRAUMA

Traumatic deaths may be classified as mechanical, chemical, thermal, or electrical. It should be noted that medical doctors and surgeons may classify wounds differently than medical examiners and forensic pathologists.

MECHANICAL TRAUMA

Mechanical trauma occurs when the force applied to a tissue, such as skin or bone, exceeds mechanical or tensile strength of that tissue. Mechanical trauma can be described as resulting from sharp or blunt force. **Sharp force** refers to injuries caused by sharp implements, such as knives, axes, or ice picks. It takes significantly less force for a sharpened object to cut or pierce tissue than what is required with a blunt object.

Blunt force trauma is caused by dull or non-sharpened objects, such as baseball bats, bricks, or lamps. Blunt objects produce **lacerations**, or tears in the tissue, typically the skin, whereas sharp objects produce **incised wounds**, a wound that has more depth than length or width. The size, shape, and kind of wound may allow the forensic pathologist to determine if a sharp or blunt object caused it. Judicious interpretations and caution are required because of the flexible nature of many of the body's tissues and the variability of the violent force. For example, a stab wound 1″ wide, 1/8″ thick, and 3″ deep could have been produced by (1) a sharp object of the same dimensions, (2) a sharp object that is 1/2″ wide, 1/8″ thick, and 2″ long that was thrust in with great force and removed at a different angle, or (3) a sharp object larger than the stated dimensions but pushed in only part of

HISTORY:
The autopsy

Physicians have been performing autopsies for thousands of years. A Chinese text, *Hsi Yuan Chi Lu*, *The Washing Away of Unjust Wrongs*, written in 1247, describes various trauma patterns, how to identify weapons from the wounds they leave, and how to tell if a victim was drowned or died in a fire.

Greek physicians, including the famous Galen who lived during the second century A.D., performed autopsies as early as the fifth century B.C. on criminals, war dead, and animals. Christian Europe discouraged and even forbade autopsies until the sudden death of Pope Alexander V in 1490. It was questioned whether his successor had poisoned him. An examination found no evidence of poisoning, however. During the reign of Pope Sixtus IV (1471–1484), the plague raged through Europe causing millions of deaths. The Pope allowed for medical students at the universities in Bologna and Padua to perform autopsies in hopes of finding a cause and cure for the savage disease.

In 1530, the Emperor Charles V issued the *Constitutio Criminalis Carolina*, which promoted the use of medical pathology by requiring medical testimony in death investigations.

Complete autopsies were not performed, however, but this did signal an advance by mandating some medical expertise to perform the inquest.

In the 1790s, the first English pathology texts were published: Baille's *Morbid Anatomy* (1793) and Hunter's *A Treatise on the Blood, Inflammation, and Gun-Shot Wounds* (1794). The next great advance came from the legendary Rudolf Virchow (1821–1902), who added microscopic examinations of diseased body tissues to the gross visual exam in his 1858 *Cellular Pathology*. Virchow's work signals the beginning of the modern autopsy process.

The first Medical Examiner's office in the United States was instituted in Baltimore in 1890. New York City abolished the coroner system in 1915 and established the Medical Examiner's office headed by Milton Helpern. Helpern added toxicological exams with the help of Alexander Gettler. In 1939, Maryland established the first statewide Medical Examiner system in the United States and, in doing so, set the position of Medical Examiner apart from the political system in the state.

Sources: Iserson, 2001; Thorwald, 1964.

its length, as represented graphically in Figure 7.8. Death from blunt and sharp trauma results from multiple processes, but sharp trauma most commonly causes death from a fatal loss of blood (**exsanguination**) when a major artery or the heart is damaged. Blunt trauma causes death most often when the brain has been severely damaged. A **contusion** is an accumulation of blood in the tissues outside the normal blood vessels and is most often the result of blunt impact. The blood pressures the tissues enough to break small blood vessels in the tissues, and they leak blood into the surrounding area. Importantly, the pattern of the object may be transferred to the skin and visualized by the blood welling up in the tissues. An extreme

Figure 7.8. *It may be difficult to determine the size and shape of the weapon by the size and shape of the wound. Here, the same size wound could be caused by a knife of the same, smaller, or larger size. Education, training, and experience are important for the forensic pathologist to make a proper interpretation.*

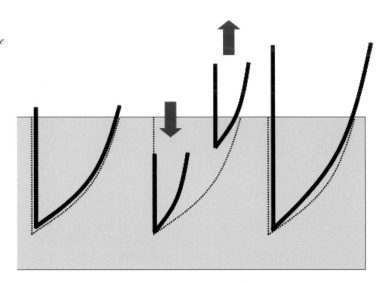

Table 7.2. *Descriptions of the major classes of gunshot wounds (GSW).*

GSW CLASS	DISTANCE	CHARACTERISTICS
Contact (entrance)	0	Blackening of the skin; lacerations from escaping muzzle gases; bright red coloration of the blood in wound from carbon monoxide gases reacting to hemoglobin in blood (**carboxyhemoglobin**)
Intermediate (entrance)	0.5 cm–1 m	Unburned gunpowder penetrates skin and burns it, causing small red dots called **stippling**; the stippling pattern enlarges as the muzzle-to-target distance increases
Distant (entrance)	>1 m	Speed of gunpowder is insufficient to cause stippling at this distance; lack blackening; no carboxyhemoglobin; circular defect with abraded rim; distance indeterminate
Shored exit	—	Skin is supported or **shored** by some material, such as tight clothing, wall board, or wood, as bullet exits; may look very similar to **entrance GSW** *except* pattern of shoring material (such as the weave of cloth) may be transferred to skin as it expands when bullet exits

contusion, a **hematoma**, is a blood tumor, or a contusion with more blood.

Gunshot Wounds

The projectile from a discharged firearm produces a special kind of blunt force trauma. Table 7.2 lists the major classes of gunshot wounds (GSW), and their characteristics.

CHEMICAL TRAUMA

Chemical trauma refers to damage and death that results from the interaction of chemicals with the human body. This is the domain of the forensic toxicologist and is discussed later in this chapter. If the damage from chemicals is external, as in the case of acid or alkaline burns, then this is still the purview of the pathologist.

THERMAL TRAUMA

Extreme heat or cold also may produce death. **Hypothermia** is too much exposure to cold, and **hyperthermia** is exposure to excessive heat. Either condition can interfere with the normal physiological mechanisms that keep body temperature at about 98°F/37°C. In both cases, the forensic pathologist may encounter few signs at autopsy that will indicate either of those mechanisms; more commonly, external or environmental factors, as well as what is *not* found, may lead to this determination. Individuals in a vulnerable state of health—typically the sick, the very elderly, or the very young—most often succumb to hypo- or hyperthermia. Other factors may contribute, such as alcohol, which reduces sensitivity to cold and dilates (opens) the blood vessels, speeding the cooling of the body. Hyperthermia deaths are common in elderly people in northern cities and infants left in automobiles during the summer. The inside temperature of a closed car in the sun can exceed 140°F/60°C and can be fatal to an infant in 10 minutes. Thermal burns tend to be localized; persons who die in a fire do so generally because of a lack of oxygen (**asphyxia**) and the inhalation of combustion products, like carbon monoxide (CO). Additionally, the level of CO in the tissues can determine whether the person was alive or dead when the fire burned him or her. A body from a burned building with 1 or 2% CO is presumed to have been dead (or at least not breathing) at the time the fire started. True deaths from thermal injuries do occur due to either massive tissue damage and/or swelling of the airway, causing suffocation.

ELECTRICAL TRAUMA

Electricity can cause death by a number of means. Circuits of alternating current (AC) at low voltages (<1,000 V) that cross the heart cause **ventricular fibrillation**, a random quivering that does not pump the blood through the body properly. A person in ventricular fibrillation for even a few minutes cannot be resuscitated. The heart fibrillates because the current is acting like a (faulty) pacemaker. AC in the United States alternates from positive to negative at 3,600 times/minute and at 2,500 times/minute in Europe; the heart can beat only about 300 times/minute at maximum. At high voltages, the amount of current causes the heart to *stop* beating (it becomes **defibrillatory**) pushing the heart into **tetany**, a sustained contraction that is broken only when the circuit is broken. Although the heart will generally start beating normally again, high voltages produce severe burns and cellular damage within a fraction of a second.

OTHER EVIDENCE COLLECTED

Other evidence is routinely collected at autopsy for submission to a forensic or toxicological laboratory. In cases in which sexual assault is known or suspected, three sets of swabs will be used to collect foreign body fluids. For females, a vaginal swab, an oral swab, and a rectal swab are collected; for males, oral and rectal swabs alone are taken. One set will be for **smears**, whereby the collected fluid on each swab is wiped across a separate clean glass microscope slide. Smears are microscopically examined for the presence of spermatozoa. The second and third sets are for serological examinations, including testing for the acid phosphatase in seminal fluid and possible blood typing. Any other stains on the decedent's clothing or body may also be swabbed for later analysis.

Known head hairs and pubic hairs are collected during the autopsy procedure. These will be forwarded to the forensic science laboratory for comparison with any questioned hairs found on the decedent's clothing or at the crime scene. A pubic hair combing is also taken to collect any foreign materials that may be associated with the perpetrator of a sexual crime.

Any **ligatures** (for binding victims), such as electrical cords, ropes, or duct tape, are extensively photographed, sketched, and then collected. The knots should be retained for later examination by the forensic science laboratory, because hairs, fibers, or other trace evidence may have been trapped in the knot when it was tied. The ligature is cut away from the knot and then labeled to distinguish that cut from any others that may have existed when the body was brought to the morgue. Be alert—not all cuts may be due to the perpetrator! Emergency medical technicians may have cut the ligatures, or clothes for that matter, in an effort to free or unbind the victim.

If the decedent's identity is unknown, a full set of fingerprints is taken to be referenced against any databases. For badly decomposed remains, the jaws may be removed to facilitate a forensic dental examination and identification.

INTERNAL EXAMINATION AND DISSECTION

After the external examination, the pathologist then removes the internal organs, either all together or individually; this latter method is called the Virchow method, after the famous pathologist Rudolph Ludwig Carl Virchow (1821–1902), known for his meticulous methodology. In the Virchow method, each organ is removed, examined, weighed, and sampled separately to isolate any pathologies or evidence of disease (Dolinak, Matshes, and Lew, 2005). The stomach contents, if any, are examined in detail because they can provide crucial clues to the decedent's last actions. The nature, amount, size, and condition of the contents are described, including the possibility of microscopic analysis to identify partially digested or difficult-to-digest materials (see "In More Detail: Cereal Killer in Spokane"). The small intestines may also be examined for undigested materials (corn kernels, tomato peels, among others) to determine the rate of digestion. Liquids digest faster than solids; 150 ml of orange juice empties from the stomach in about 1.5 hours, whereas the same amount of solid food may empty in 2 hours or more, depending on the density of the food. Finally, a toxicological exam may be requested.

Each organ is sectioned and viewed internally and externally. Samples for microscopic analysis of the cellular structure (**histology**) and for toxicology screening tests are taken. After all of the organs have been examined, they are placed in a plastic bag and returned to the body cavity.

IN MORE DETAIL:
Cereal killer in Spokane

In February 1999, the residence of James Cochran* was found engulfed in flames, and Kevin, Cochran's eleven-year-old son, was missing. Cochran claimed no knowledge of his son's location, suggesting Kevin had started the fire while playing with matches and had run off. Two days later, the fully clothed body of Kevin Cochran was found along a road north of Spokane (see Figure 7.9). Kevin's clothing, face, and mouth exhibited a large amount of creamy brown vomit. Kevin's shoes were tied but were on the wrong feet. At autopsy, the pathologist determined the cause of death to be strangulation. The boy's stomach contents, fingernail clippings, hand swabs, and clothing were collected as evidence for laboratory examination. That same week, James Cochran was arrested for embezzling funds from his employer.

James Cochran's pickup truck was seized and searched. Several droplets of light brown to pink material were observed on the driver's side wheel well hump and in various locations on the mid-portion of the bed liner. The scientist collecting these droplets noted the smell of possible vomit while scraping to recover the stains (see Figure 7.10).

Stains from the bed of the pickup truck were compared to the vomit and gastric contents of Kevin Cochran. One of Kevin's sisters stated in an interview Kevin was last seen eating cereal in the kitchen the morning of the fire. Investigators recovered known boxes of cereal from the Cochran's kitchen. Two opened and partially consumed plastic bags labeled *Apple Cinnamon Toastyo's®*, and *Marshmallow Mateys®*, among others, were submitted (see Figure 7.11). If the cereal found in the kitchen of the Cochran residence "matched" the cereal in the vomit on Kevin's clothing and was found to be similar to stains in the pickup truck, investigators may have a connection linking James Cochran to the death of his own son.

All of the cereal brands could be distinguished microscopically. The microscopical examination and comparison of stains found on the pickup truck bed liner revealed the presence of vomit with cereal ingredients similar to that found in the vomit on Kevin's clothing and gastric fluid (see Figure 7.12). The cereal ingredients were consistent with *Marshmallow Mateys®*, the final meal of Kevin Cochran. The vomit in Cochran's truck, along with other trace evidence, linked him to the death of his son, as well as the arson of his home. Investigators learned that Cochran gave a file folder containing documents, specifically the homeowner's and life insurance policies of his children, to a neighbor the night after the fire.

On Memorial Day 1999, James Cochran committed suicide in his jail cell using a coaxial cable from a television set. Investigators theorized Cochran killed his son and set fire to his house for the insurance money.

Source: Schneck, W. (2003).

*All names have been changed.

DETERMINING TIME SINCE DEATH (POSTMORTEM INTERVAL)

Following death, numerous changes occur that ultimately lead to the dissolution of all soft tissues. The importance of these changes to the forensic pathologist is that they provide a sequence of events that allow

Figure 7.9. *The victim was found along a lake road by a snow plow driver. Note that the victim's shoes are tied on the wrong feet.*

Figure 7.10. *The bedliner of the suspect's pick-up truck. Arrows point to suspected vomit stains.*

Figure 7.11.
Marshmallow Mateys® breakfast cereal. The anchor-shaped particles contain oat flour, whereas the colored particles contain processed corn starch and sugar.

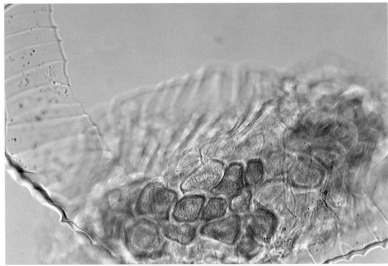

A

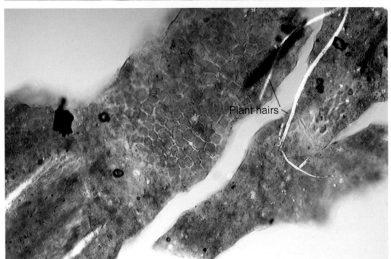

Plant hairs

B

Figure 7.12 *(A) Microscopic structures in oat flour. The lower portion of the photomicrograph shows cells in the coating of the oat bran. (B) Microscopic structures in vomit containing oat flour. The elongated fibrous structures are plant hairs common in oat flour.*

an estimate of time since death ensued. This determination is based on the principle of sequential changes called the **postmortem clock**. The evaluation may include the following phenomena:

- Changes evident upon external examination of the body, such as temperature, livor, rigor, and decomposition
- Chemical changes in body fluids or tissues
- Physiological changes with progression rates, such as digestion
- Survival after injuries, based on the nature, severity, and other factors such as blood loss

Because of the variation inherent in each of these processes, an initial time range of death is established and modified as more information becomes available. This initial time range is the interval prior to which it may be asserted with some evidence that the victim was alive, based on witness sightings, signed documents, or other established events. This initial time range is then modified by various methods of evaluating postmortem changes.

Postmortem cooling, or **algor mortis**, occurs at a rate of about 2°–2.5° per hour at first, then slows to about 1.5° during the first 12 hours, and decreases further after that. The temperature is typically taken with a rectal thermometer to capture the body's inner core temperature. Many factors, such as ambient temperature, clothing, and air currents, can affect postmortem cooling and, although it is still relied upon, it is known that the accuracy of this method is low.

The eyes are also an indicator of postmortem changes. Because the circulation of blood ceases, blood settles in the innermost corners of the eyes. If the eyes remain open, a thin film forms on the surface within minutes and clouds over in two to three hours; if they are closed, it may take longer for this film (an hour or more) and cloudiness (24 hours) to develop.

Postmortem lividity can be seen as early as 20 minutes after death or may take several hours to develop. In its early stages, lividity will blanch when pressed, but in advanced stages the eventual pressure will burst the skin capillaries, causing pinpoint hemorrhages called **Tardieu spots**. As noted earlier, rigor mortis becomes apparent within half an hour to an hour and progresses to a maximum within 12 hours, remains for about 12 hours, and then progressively disappears within the following 12 hours.

Stomach contents may be helpful in the determination of time since death. This is based on the assumption that the stomach empties at a known rate, which speeds or slows with the various types of food in it. Light meals last in the stomach for 1.5–2.0 hours, medium meals up to 3 or 4 hours, and heavy meals for 4–6 hours. Food moves from the stomach in small amounts, after having been chewed, swallowed, digested, and ground into tiny pieces. A meal eaten hurriedly or gulped will last longer because it hasn't been properly chewed. Alcoholic beverages also delay the stomach's evacuation.

Decomposition of the body begins almost immediately after death and consists of two parallel processes:

- **Autolysis**, the disintegration of the body by enzymes released by dying cells
- **Putrefaction**, the disintegration of the body by the action of bacteria and microorganisms

The body passes through four main stages of decomposition: fresh, bloated (as the gaseous by-products of bacterial action build up in the body cavity), decay (ranging from wet to mushy to liquid), and dry. These changes depend in large part on the environmental factors surrounding the decedent, such as geographical location, seasonality, clothing, sun exposure, and animals and insects in the area. Insect activity, when present, greatly assists the decomposition process.

LABORATORY ANALYSIS

HISTOLOGY

The pathologist typically requests a histology examination for evidence of cellular pathologies resulting from disease, trauma, or preexisting conditions. Small samples of the tissues of interest are taken, embedded in plastic, and sectioned using a **microtome** (a machine that makes very thin, very precise slices) to a thickness of only a few microns. A medical technologist or histologist will then examine the sections microscopically, write a report, and pass this along to the pathologist.

TOXICOLOGY

Another routine examination requested by pathologists in medicolegal autopsies is a broad-based screen test, called a **toxicology screen**, or "tox screen" for short. These tests help the forensic toxicologist determine the absence or presence of drugs and their metabolites, chemicals such as ethanol and other volatile substances, carbon monoxide and other gases, metals, and other toxic chemicals in human fluids and tissues. The results help the toxicologist and the pathologist evaluate the role of any drugs or chemicals as a determinant or contributory factor in the cause and manner of death.

AUTOPSY REPORT

The autopsy report is a crucial piece of information in a death investigation. No standard method for reporting autopsy results exists, although guidelines and headings have been suggested by the College of American Pathologists (see Figure 7.1). Because the results of an autopsy, whether hospital or medicolegal, may end up in court, it is imperative that certain basic and specific information be included in every autopsy file, such as

- Police report
- Medical investigator report
- Witness reports
- Medical history of the decedent

EXHUMATIONS

Humans have always had particular practices for dealing with the dead. Rituals, ceremonies, and wakes are all a part of how society acknowledges a person's passing life. One of the most common funereal practices in the United States is the embalming and burial of the dead. If questions about cause or manner of death arise once the deceased is buried, the decedent must be dug up or removed from his or her mausoleum; this process is called an **exhumation**. The changes wrought by death, time, and embalming practices can obliterate or obscure details that otherwise might be easily examined. **Embalming** is a process of chemically treating the dead human body to reduce the presence and growth of microorganisms, to retard organic decomposition, and to restore an acceptable physical appearance. Formaldehyde or formalin are the main chemicals used to preserve the body. These chemicals are highly reactive and can alter or mask drugs or poisons in

> **ON THE WEB:**
> Society of Forensic Toxicologists (SOFT)
> www.soft-tox.org
>
> The Society of Forensic Toxicologists, Inc. (SOFT) is an organization composed of practicing forensic toxicologists and those interested in the discipline for the purpose of promoting and developing forensic toxicology.

the body at the time of death. Toxicologists Tim Tracy of the University of Minnesota and Pete Gannett at West Virginia University have developed special methods to analyze embalmed tissues for drugs, poisons, and medications. These methods have been successfully applied in casework; more methods for other drugs of abuse in embalmed tissues are being researched (Gannet, et al., 2001; Tracy, et al. 2001).

CONSULTATIONS

The forensic pathologist, when presented with challenging cases of burned, decomposed, or dismembered bodies, may consult with any of a variety of forensic specialists. Forensic anthropologists, entomologists, and odontologists all may play a role in a death investigation. Some ME offices or forensic laboratories have one or more of these specialists on staff due to regular caseload demands. This is especially true of offices that cover a large geographical area or large metropolitan areas.

CHAPTER SUMMARY

Medical examiners study disease and trauma that lead to the death of an individual. By conducting autopsies, the dissection of a dead body to determine the cause and manner of death, they greatly assist death investigations. It is a sad fact, however, that the number of autopsies has steadily declined in the past 50 years; the medical profession loses its most valuable quality control tool when autopsies are not performed. Many times the morgue is as important as the crime scene.

Test Your Knowledge

1. What's the difference between cause and manner of death?
2. What is the primary cause of death?
3. Name the four manners of death.
4. What is the difference between a coroner and a medical examiner.
5. What is livor mortis?
6. What is another name for postmortem cooling?
7. How long does rigor mortis last?

8. What are petechiae? Where do they appear?

9. What are Tardieu spots?

10. Who was Milton Helpern?

11. What is autolysis?

12. How many stages are there to decomposition?

13. Histology is the study of what?

14. What is an exhumation?

15. What is the difference between blunt and **sharp force trauma**?

16. What is stippling?

17. Where does the term "sheriff" come from?

18. What causes rigor mortis?

19. What is putrefaction?

20. How accurate is algor mortis?

Consider This . . .

1. How does a standard autopsy differ from a medicolegal autopsy? Why?

2. What specialists might assist a medical examiner? Why? What other specialties in this textbook might aid a pathologist in his or her investigation?

3. Why do you think the number of hospital autopsies has declined? Do you think the number of medicolegal autopsies has similarly declined? Why or why not?

BIBLIOGRAPHY

Collins, K.A., & Lantz, P.E. (1994). Interpretation of fatal, multiple, and exiting gunshot wounds by trauma specialists. *Forensic Science International*, 65, 185–193.

Dolinak, D., Matshes, E., & Lew, E. (2005). *Forensic Pathology: Principles and Practice*. Amsterdam: Elsevier.

Gannett, P.M., Hailu, S., Daft, J., James, D., Rybeck, B., & Tracy, T.S. (2001). In vitro reaction of formaldehyde with fenfluramine: Conversion to N-methyl fenfluramine. *Journal of Analytical Toxicology*, 25, 88–92.

Hanzlick, R. (2000). The autopsy lexicon. *Archives of Pathology and Laboratory Medicine*, 124, 594–603.

Hutchins, G., Berman, J., Moore, W., & Hanzlick, R. (1999). Practice guidelines for autopsy pathology. *Archives of Pathology and Laboratory Medicine*, 123, 1085–1092.

Iverson, K. (2001). *Death to dust: What happens to dead bodies?* (2nd ed.) Tucson, AZ: Galen Press, Ltd.

Oxford English Dictionary (2005). Oxford, United Kingdom: Oxford University Press.

Schneck, C.O. (2003). Cereal murder in Spokane. In M.M. Houck (Ed.), *Trace Evidence Analysis: More Cases in Forensic Microscopy and Mute Witnesses.* New York: Academic Press.

Spitz, W. (Ed.). (1993). *Spitz and Fischer's medicolegal investigation of death* (3rd ed.). Springfield, IL: C.C. Thomas Publisher.

Thorwald, J. (1964) *The Century of the detective.* New York City: Harcourt, Brace & World.

Tracy, T.S., Rybeck, B.F., James, D.G., Knopp, J.B., & Gannett, P.M. (2001). Stability of benzodiazepines in formaldehyde solutions. *Journal of Analytical Toxicology,* 25, 166–173.

Wilson, C., & Wilson, D. (2003). *Written in Blood: A History of Forensic Detection.* New York City: Carroll and Graf Publishers.

Anthropology and Odontology

KEY TERMS

Acetabulum

Anterior

Appendicular skeleton

Axial skeleton

Biological profile

Buccal

Calipers: spreading, sliding

Carpals

Centers of ossification

Clavicle

Coccyx

Compact bone

Cortical bone

Cranial skeleton

Cranium

Datum

Deep

Diaphysis

Distal

Enamel

Endochondral bone

Epiphysis

Femur

Fibula

Forensic anthropology

Forensic odontologist

Frontal sinus

Humerus

Ilium

Inferior

Interstitial bone

Intervertebral disk

Intramembranous

Ischium

Lacuna(e)

Lateral

Lingual

Lumbar

Mandible

Marrow

Mastoid processes

Medial

Medullary cavity

Mesial

Metacarpals

Occlusal surface

Os coxae/innominates

Osteoblasts

Osteoclasts

Osteon

Patella

Phalanx, phalanges

Phenice method

Post-cranial skeleton

Posterior

Pre-auricular sulcus

Proximal

Pubis

Radius

Ribs

Sacrum

Scapula

Sciatic notch

Skull

Sternum

Superficial

Superior

Sutures

Symphysis

Taphonomy

Tibia

Trabecullar bone

Ulna

Vertebrae

INTRODUCTION

Anthropology is the study of humans, their cultures, and their biology. Anthropology can be divided into the study of human biology and human culture, and these areas can be further divided into the study of the past and the study of the present. This presents us with four main disciplines within anthropology, as shown in Table 8.1.

Forensic anthropology is the application of the study of humans to situations of modern legal or public concern. This typically takes the form of collecting and analyzing human skeletal remains to help identify victims and reconstruct the events surrounding their deaths. Why wouldn't a medical doctor or pathologist perform these analyses? As medical doctors, pathologists learn about the body's various organ systems; additionally, forensic pathologists learn what makes these systems stop working. Forensic anthropologists are taught about only one system in the body: the skeleton. They learn to identify minute pieces of bone, recognizing hints that might indicate what portion of what bone they are holding. Pathologists require assistance from the advanced, focused knowledge of skeletal anatomy that anthropologists have just as anthropologists require assistance from the detailed and extensive medical training that pathologists gain in medical school. Pathologists generally do not learn about the bits and pieces that are the clues forensic anthropologists use to identify human remains.

Forensic anthropology involves methods from all of the anthropological disciplines but mostly from paleoanthropology and bioanthropol-

Table 8.1. *The traditional disciplines of anthropology. Forensic anthropology applies the methods of paleoanthropology to present populations.*

	PAST	PRESENT
Biological	Paleoanthropology	Bioanthropology
Cultural	Archaeology	Ethnology

Paleoanthropology The *biological* study of *past* human populations
Bioanthropology The *biological* study of *current* human populations
Archaeology The study of *past* human *cultures*
Ethnology The study of *current* human *cultures*

ogy because of the study of the human skeleton. Archaeological methods are employed to collect the remains, and paleoanthropological techniques are used to identify and analyze the bones to determine sex, age, race, and other biological descriptors. Forensic anthropology is therefore multidisciplinary in nature and requires a professional with the proper education, training, and experience to assist investigators.

THE HUMAN SKELETON

The human skeleton consists of 206 bones, most of which are paired (left and right) or grouped by area (the skull or the spine, for example), as shown in Figure 8.1. Bone may seem like a "dead" material because it is so hard and inflexible. In reality, the skeleton is a very active organ system that can repair itself and alter its form over time. Bone, as a tissue and a structure, responds to the stresses placed on it, adding or subtracting boney material as needed. This activity that takes place throughout our lives, plus the genetic potential we inherit from our parents, results in the biological and anatomical variation we see between and within populations and individuals.

Bones perform four main functions for the body: support, motion, protection, and growth.

First, the skeleton provides the infrastructure for attachment and support of the softer tissues in our bodies. Second, these attachments allow the bones to act as levers, providing motion, powered by muscles, at the joints. The structure and arrangement of our bones sets the range of motion for our limbs and bodies. Third, the hard bones protect our soft organs from physical damage; this is especially true of the brain (encased by the skull) and the heart and lungs (enclosed within the spine and rib cage). Fourth and finally, the bones are centers of growth from infancy through to early adulthood; they also continue important physiological functions throughout our lives by housing the tissue that makes red blood cells. Bones supply us with a ready source of calcium if our dietary intake of that mineral is too low for too long.

BONE ORGANIZATION AND GROWTH

Bone growth and maintenance are complex processes that continue throughout our lives. Our skeletons must grow, mature, and repair at the macro- and microscopic levels even as we use them. An understanding of how bones grow and are organized is central to many of

The skull is the entire skeletal portion of the head, including the mandible, or lower jawbone. Without the mandible, the remainder is called the cranium. The cranium is constructed of twenty-eight separate bones in the adult. Most of these bones develop and grow as individual entities, joining at seams call sutures; all of the sutures have names but only a few of them concern us here. Many of the bones are paired and most have landmarks, either physical or determined by measurement, which are important for the analysis of the skull.

The clavicle is a short double-curved bone that is commonly known as the "collar bone". It attaches medially to the manubrium and laterally to the scapula, a flat, trapezoidal bone, also called the "shoulder blade."

The twelve ribs articulate with the second through ninth thoracic vertebrae **posteriorly.** The first seven ribs articulate with the sternum, or breastbone, anteriorly. The eighth through tenth ribs are interconnected anterior-medially by common cartilages; the last two ribs "float" anteriorly in the muscle walls of the chest.

The **superior** portion of the sternum has facets for articulating with the shoulder girdle called the clavicular notches. It is interesting to note that this joint is the only place where the arm is physically attached to the body by tendons.

The arm bone (the humerus), the bones of the forearm (the radius on the thumb side and ulna on the little finger side), the wrist bones (the carpals), and the bones of the hand (the metacarpals) and fingers (the phalanges, singular "phalanx") make up the rest of the upper limb.

Humans have twenty-four vertebrae (singular "vertebra") that constitute the spine, made up of seven cervical (neck), twelve thoracic (chest), and five lumbar (lower back). Regardless of the type, all vertebrae share some common characteristics. The vertebrae stack in a flexible integrated column, held upright by tendons and muscles.

Between each vertebra is an **intervertebral disk** made up of fibrous cartilage; these disks act as shock absorbers, cushioning the spine. The sacrum consists of, usually, five fused vertebrae that form a curved triangular bony structure that narrows inferiorly. It articulates with the last lumbar vertebra an the two bones of the pelvis. The coccyx, a vestigial tial, is variable in shape and number.

The pelvis is made up of three separate bones, one of which has already been mentioned, the sacrum. The other two are a pair of bones that form the pelvic girdle, the **os coxae or innominates.** Both os coxa have three parts: The ilium, the ischium, and the pubis. The leg allaches to the pelvis at a cup-shaped feature called the acetabulum

The femur is the sole bone of the thigh and is the largest and strongest long bone in the body. The patella, or knee cap, floats in the tendon of the largest thigh muscle, protecting the knee joint. The lower leg has two bones, the tibia (sometime called the "shin bone") and the fibula. The bones of the ankle (the carpals), foot (the metacarpals), and the toes (the phalanges) round out the bones of the lower limb.

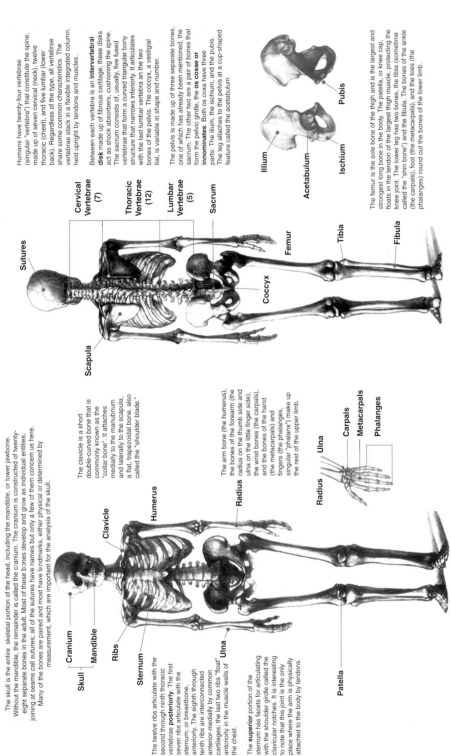

Figure 8.1. The human skeleton consists of 206 bones, which are paired or grouped by area (Source: Barcsay, 2001).

IN MORE DETAIL:
Bone growth

Two types of bone growth characterize the human skeleton: endochondral and intramembranous. **Endochondral bone** growth starts with a "model" of a bone consisting of cartilage and **centers of ossification** (see Figure 8.2). From these centers, bone is produced and infiltrates the cartilage model, which itself continues to grow. The developing shaft of the bone is called the **diaphysis**, and the ends are called **epiphyses**. The growing areas eventually meet, and the bone knits together. Not all epiphyses unite at the same time, so the sequence of union is important for estimating age at death for individuals younger than about 20 years. In **intramembranous** bone growth, instead of a cartilage model, the ossification occurs within a membrane, and this occurs in many bones of the skull. Bone differs from cartilage by having its collagenous connective tissue matrix impregnated with inorganic salts (primarily calcium phosphate and lesser amounts of calcium carbonate, calcium fluoride, magnesium phosphate, and sodium chloride). The osteoblasts, which form the osseous tissue, become encapsulated in lacunae but maintain contact with the vascular system via microscopic canaliculi. When they become encapsulated, they are referred to as osteocytes.

A characteristic feature of a cross section of the shaft (diaphysis) of a long bone is its organization in concentric rings around a central canal containing a blood vessel. This is called a Haversian system (osteon). Between neighboring Haversian systems are non-concentric lamellae, devoid of Haversian canals, termed interstitial lamellae. Vascular canals, called Volkmann's canals, traverse the long axis of the bone; they are always at right angles to Haversian canals. Their function is to link vascular canals of adjacent Haversian systems with each other and with the periosteal and endosteal blood vessels of the bone. The outer perimeter of a long bone, beneath the osteogenic connective tissue (called periosteum), is composed of circumferential lamellae, which also lack Haversian canals. This thick-walled hollow shaft of compact bone (the diaphysis) contains bone marrow. At the **distal** ends of long bones, where Haversian systems are not found, the bone appears spongy and is therefore called cancellous, or spongy, bone. The spongy appearance is misleading, because careful examination of the architecture reveals a highly organized trabecular system providing maximal structural support with minimal density of bony tissue.

The epiphyses at the ends of the diaphysis or shaft contain the spongy bone covered by a thin layer of compact bone. The cavities of the epiphyseal spongy bone are in contact with the bone marrow core of the diaphysis except during growth of long bones in young animals. Interposed between the epiphysis and the diaphysis is the cartilaginous epiphyseal plate. The epiphyseal plate is joined to the diaphysis by columns of cancellous bone; this region is known as the metaphysis.

When bone is formed in and replaces a cartilaginous "model," the process is termed endochondral ossification. Some parts of the skull develop from osteogenic mesenchymal connective tissue, however, without a cartilaginous "model" having been formed first. This is termed intramembranous ossification, and these bones are called membrane bones. In both instances, three types of cells are associated with bone formation, growth, and maintenance: osteoblasts, osteocytes, and osteoclasts. The osteoblasts produce osseous tissue (bone), become embedded in the matrix they manufacture, and are then renamed osteocytes, to reflect their change of status. They remain viable, because they have access to the vascular supply via microscopic canaliculi through which cellular processes extend to receive nutrients and oxygen. Osteoclasts actively resorb and remodel bone as required for growth; these are giant, multinuclear, phagocytic, and osteolytic cells.

Figure 8.2. *(A) Bone growth starts in centers of ossification and spreads out to meet each other. (B) The main portion of a long bone is called the diaphysis, and the ends are called epiphyses (singular: epiphysis). The pattern of epiphyseal union is important for estimating age up to 25 years (Source: Image (A) © Kentucky University Medical Center, with permission).*

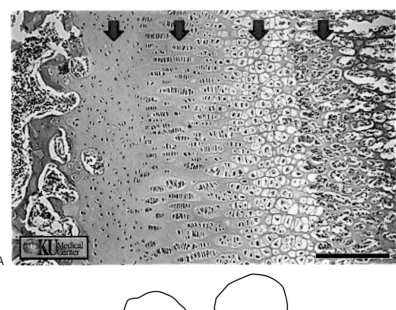

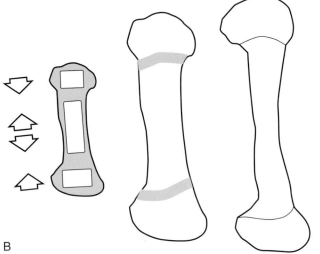

the analyses that forensic anthropologists perform (see "In More Detail: Bone Growth").

Bones consist of an outer layer of hard, smooth **compact bone**, also called **cortical bone**, pictured in Figure 8.3. The inner layer is an infrastructure of sponge-like bone called **trabecullar bone** in long bones, which increases the structural strength of the bone without additional weight. In the very center of long bones is the **medullary cavity**, which contains **marrow**, a fatty material that also houses blood-generating tissues. In life, this composite architecture creates a very strong but resilient framework for our bodies.

The microstructure of bone is quite complex and organized, as shown in Figure 8.4. Specialized growth cells (**osteoblasts**) produce bone and deposit it in layers, eventually becoming encapsulated in a self-made chamber (**lacuna**; plural **lacunae**). They maintain contact with the circulatory system and other bone cells through microscopic vascular channels through which cellular processes extend to receive nutrients and oxygen. When an osteoblast becomes fully encapsulated, it is referred to as an **osteon**.

The third main type of bone cell, **osteoclasts**, actively break down and remodel bone as required for growth. When an osteocyte reaches the end of its productivity, it dies and the bone around is reworked and made available to new osteoblasts. In response to the stresses our activities place on our skeletons, the interaction between osteoblasts, osteocytes, and osteoclasts model and shape our bones. Because new osteons are formed by remodeling existing structures, bone has a patchwork appearance at the cellular level. Bone that lies between recently reworked bone is called **interstitial bone**; the amounts of new, reworked, and old bone provide an indication of how old someone is, and we will see later how this can provide an estimate of age at death.

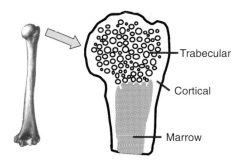

Figure 8.3. The outer portion of a bone is the compact or cortical bone and is very dense. The inner portion of a bone is trabecullar bone, which is made up of a fine web-work of thin bony spines. The center of a long bone contains marrow, where blood is made.

- Trabecular
- Cortical
- Marrow

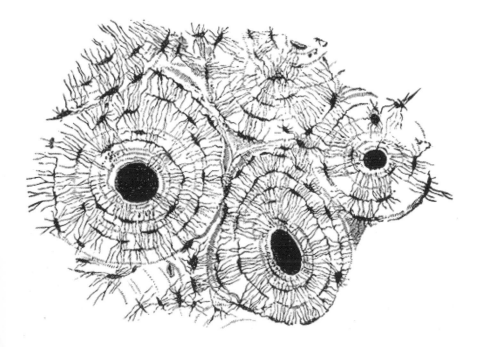

Figure 8.4. Bone grows much in the way a brick wall is made and repaired. Bone is laid down by osteoblasts (bone-generating cells) and then, in response to the stresses it undergoes, is torn down by osteoclasts (bone-destroying cells) before being reworked by the osteoblasts. Bone may seem dead, but it is a very active tissue during life.

SKELETAL ANATOMY

Before we describe the human skeletal anatomy, it is important to discuss the proper handling of human remains. Most people encounter skeletons only on Halloween or at a costume party. Given that they are potent symbols of death (which is what they represent in those contexts), it is only natural that people feel nervous or anxious when presented with the real thing. The urge to gesture, joke around, or taunt others with a bone or skull is simply a way of expressing that unease, by laughing at "the Grim Reaper." What you must keep in mind, however, is that the material you are handling was once part of a human being, like yourself, with a life, family, feelings, and dignity. Additionally, every specimen is unique and irreplaceable, so care must be taken with how it is handled. Bones should always be held over a table, preferably with a padded or protected surface. The skull is of special consideration due to its delicacy and centrality to a forensic examination. The bones of the nose and the eye orbits and the teeth are fragile. The skull should be handled by the sides and base in both hands with a firm grasp. As the noted osteologist Tim White (2000) says, "Common sense and both hands should always be used" (page 53).

Figure 8.1 shows the human skeleton in a variety of anatomical views. The **cranial skeleton** refers to the skull only; everything else is called the **post-cranial skeleton** (meaning below the cranial skeleton). The **axial skeleton** describes the spine (**vertebrae**), **ribs**, and breastbone (**sternum**). The grouping of either upper limb bones (including the shoulder) or lower limb bones (excluding the pelvis) is called the **appendicular skeleton**.

COLLECTING HUMAN REMAINS

Forensic anthropologists rarely find skeletal remains that are above ground. Often a hiker, hunter, or some other civilian in a remote or uninhabited area stumbles across the bones at a crime scene. Because the "evidence" has been found by untrained persons, securing the scene is the most effective way of initiating evidence protection. The subsequent searching of an area for bones is similar to processing other crime scenes, however, and proceeds as an orderly, careful search by trained personnel. This search may be aided by various detection

Figure 8.5. *Some crime scenes are very similar to archaeological excavations, where shovels and trowels replace magnifying lenses and fingerprint powder. Professional archaeologists or forensic anthropologists are skilled at locating and removing buried remains and should be consulted before any digging starts (Source: Ubelaker, 2000, © Taraxacum Press, with permission).*

methods, such as probes that detect the gases produced by decomposition, radar that penetrates into the ground, or even dogs trained to sniff for the smells of human decomposition, so-called cadaver dogs.

If the remains are scattered, each bone fragment should be flagged or marked. This provides a view of the pattern of dispersal and where missing bones might lie. Context is even more important with skeletal remains, so the individual bones should not be disturbed until the entire scene has been photographed and documented. All of the bones on the surface, even animal bones, should be collected.

Buried remains require more time and skill to retrieve, as pictured in Figure 8.5. Archaeological techniques are employed to excavate buried skeletal materials and should be performed only by trained personnel under the supervision of an experienced archaeological excavator. A grid is set up with one point set as a **datum**, or reference point, from which all measurements originate. Each unit in the grid is excavated separately; the units may be processed at the same time or done in series. Soil and materials are removed a thin layer at a time (usually 2–5 cm) slowly exposing the buried items. Figure 8.6 shows how each bone is carefully delineated and cleaned in place to preserve the final position of the body. Only after all the bones have been found,

Figure 8.6. To gain a clear picture of the body's last resting position, it is useful to clean down to the bottom of the remains and then clear out all the soil around the bones. This process is called "pedestalling" the body (Source: Ubelaker, 1999, © Taraxacum Press, with permission).

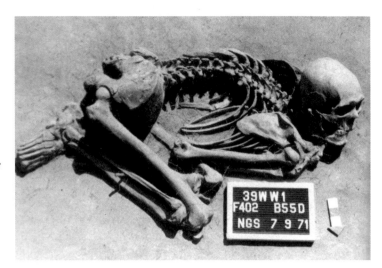

excavated, photographed, and documented will they be removed and transported for analysis.

Sometimes, humans, animals, and nature are not kind to skeletal remains. A skull or bone may not be whole when recovered, so it must be reconstructed prior to analysis. Thin wooden sticks and glue usually do the trick, although other means may need to be used depending on how damaged the bone is. Subsequent analyses need to be kept in mind (carbon 14 dating, DNA, X-rays, etc.) to minimize any obstacles to their successful completion.

ANALYSIS OF SKELETAL MATERIALS

The first question the anthropologist must ask is, "Is the submitted material really bone?" With whole bones, this is obvious. A surprising number of materials can **superficially** resemble a bone fragment, and even professionals need to be careful, especially with very small fragments. It may be necessary to take a thin section of the material and examine it microscopically for cell morphology. Elemental analysis is also very useful for small fragments, because few materials have the same elemental ratios as bone.

Once the material is determined to be bone, the second question is whether the bone is animal or human. Answering this question can present a greater challenge than it may appear at first. Pig bones, bear

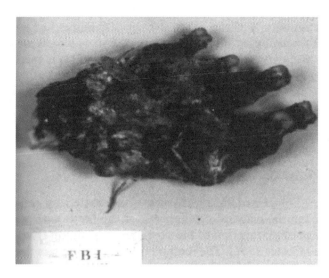

Figure 8.7. If a questioned material is bone, it still may not be human. Some animal bones look very much like human bones at first glance. A professional trained in non-human skeletal anatomy may need to be consulted (Source: Ubelaker, 1999, © Taraxacum Press, with permission).

paws, and some sheep bones can, at first, appear similar to human bones, as illustrated in Figure 8.7. A comprehensive knowledge of human anatomy and a solid grounding in animal osteology will answer most of these questions. A comparative collection of cataloged skeletal remains is crucial to an accurate taxonomical assessment: It can be as useful to know what something is, not just what it is not (see "In More Detail: When Is a Doctor Not a Doctor?").

THE BIOLOGICAL PROFILE

Once the remains are determined to be human, a **biological profile** can be developed for the individual(s) represented. The biological profile consists of assessing the sex, age at death, racial affinity, height, and any other aspects that would describe the individual class level information. The biological profile is the first step toward identifying whom the remains represent. It is a waste of time to immediately start comparing the dental X-rays or sequencing DNA samples of a 20-year-old woman when the bones recovered are from a 50-year-old man. Which bones are present and their quality will determine what methods can be applied and, in part, the accuracy of those methods.

The criteria that help determine the biological profile are either qualitative, i.e., morphological (the presence or absence of a trait, or the shape or size of a landmark), or quantitative. Physical anthropologists use many, many different measurements as a way of discrimi-

IN MORE DETAIL:
When is a doctor not a doctor?

Dr. Douglas Ubelaker (2000) writes of the following example in his book, *Human Skeletal Remains*. A bone fragment had been found in a remote part of Alaska. The bone displayed a fracture that had been repaired surgically with a metal plate (see Figure 8.8). The extensive bone growth over the surgical plate indicated the patient had received the surgery long before death. Given the nature of the surgery and the surgical efforts, the authorities began to search for the surgeon who had performed the operation. After these efforts failed, the bone was sent to Dr. Ubelaker at the Smithsonian Institution, where a microscopic section revealed the bone to have a non-human bone cell morphology, one that closely matched that of a large dog. This explains why the surgeon couldn't be found—because the doctor was a veterinarian! This is an excellent example of why you should not make assumptions and not come to a conclusion until you have all of the facts.

0   3 CM

Figure 8.8. *Bone fragment with surgical plate. The fragment was originally thought to be human because of the plate which indicated surgery.*

nating between individuals, samples, and populations. Some of this information has been cataloged (for example, at the University of Tennessee's Forensic Data Bank) and used to provide virtual "comparative collections" of measurements that can be used by anyone with a computer (FORDISC is an example of commercially available software for forensic anthropologists). As more museums and universities surrender their osteological collections for repatriation and reburial, collections of data instead of bones will become increasingly crucial to future anthropologists' research. Quantitative physical anthropology is dominated by statistical analysis, and sometimes these analyses, such as principal component analysis, are quite complex, involving many measurements, samples, and relationships.

Is This Person Male or Female?

Although in life the differences between males and females are almost always obvious, in death these differences are not always so apparent, especially when the visual cues the flesh provides are gone. Males can be up to 20% larger than females, but in some instances there is little or no difference in size. Many of the quantitative skeletal traits overlap in the middle of the distribution of their values, and statistical analysis is required to sort out equivocal examples.

Sexual differences in the human skeleton begin before birth although they are not truly diagnostic until after puberty. In general, females' post-cranial skeleton develops faster than in males, and this difference in rate can be used to infer sex in pre-pubertal individuals. Typically, however, sex should not be estimated unless the individual is of an age at which puberty has begun; above 18 years of age, sex can be determined with confidence.

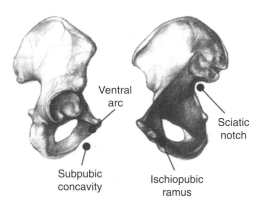

The significant differences between males and females are size and function-related morphology.

The two areas that are used most often to determine the sex of an individual in life are also the most diagnostic in death: the pelvis and the skull. Other bones can be very useful for estimating sex as well, and with only a few measurements an experienced forensic anthropologist can be accurate 70–90% of the time.

The largest number of and most accurate traits for determining sex reside in the pelvis, illustrated in Figure 8.9. See also Table 8.2. The major reason that male and female skeletal anatomy differs so much in the pelvic region is that only females carry and bear babies; human pelvic anatomy reflects this functional difference. Thus, the male pelvis tends to be larger and more robust, whereas the female pelvis is broader and can exhibit pregnancy-specific traits. A useful trait for distinguishing between the male and female pelvis is the **sciatic notch**, located on the **inferior lateral** border of the **ilium**. The sciatic notch is wide (an angle of about 60°) in females and narrow in males (about 30°).

A very reliable method for determining the sex from the pelvis is the **Phenice method**, developed by Dr. Terrell Phenice in 1969, which uses three characteristics: the ventral arc, the subpubic concavity, and the ischio-pubic ramus. The ventral arc is a ridge on the **anterior** surface of the pubic bone that is present in females but absent in males. The subpubic concavity is a depression on the **medial** border of the ischio-pubic ramus, just inferior to the pubic **symphysis**. The concav-

Figure 8.9. The main differences between males and females in the pelvis are due to females' biological ability to bear children. The most reliable method is the Phenice method, which uses three areas and the presence and absence of certain characteristics (Source: Barcsay, 2001, © heirs of Jeno Barcsay, with permission).

Table 8.2. *Traits useful for estimating sex from the pelvic bones, including those detailed by Phenice: The sub-pubic angle, the ventral arc, and the ischo-pubic ramus.*

METHOD	MALE CHARACTERISTICS	FEMALE CHARACTERISTICS
Pelvis in general	*Large, rugged*	**Smaller, gracile**
Subpubic angle	*Narrow*	**Wide**
Acetabulum (hip socket)	*Large*	**Medium to small**
Sciatic notch	*Narrow*	**Wide**
Preauricular sulcus	*Not present*	**May be present**
Ventral arc	*Not present or very small*	**Present, sometimes strongly**
Subpubic concavity	*Not present or shallow*	**Present, sometimes deep**
Ischio-pubic ramus	*Thin, narrow*	**Wide, possible ridge**

ity is wider and deeper in females and is only slight, if at all present, in males. Finally, the ischio-pubic ramus itself is flatter and thinner in males, whereas in females it is wide and may even have a ridge on it. It is possible to be accurate in sexing a pelvis with only these three traits. The Phenice method cannot be relied upon all by itself, however, because the pelvic remains may be fragmentary and the pubic bone may be absent. Numerous measurements have been used along with statistical analysis to derive more objective sexing methods than descriptive anatomy. Often, these methods are as accurate as morphological traits, but they are important for gauging slight differences between anatomically similar populations.

Sex can be estimated from the **cranium** as well as the pelvis, but the traits may not always be as obvious. As shown in Figure 8.10, males tend to be larger and have larger muscle attachments than females. The specific areas of interest are the brow ridges, **mastoid processes** (bony masses just behind the ears for attachment of neck muscles), occipital area at the rear of the **skull**, upper palate, and the general architecture of the skull.

The skull is one of the most, if not the most, studied, measured, and examined parts of the skeleton. This metric enthusiasm extends to the

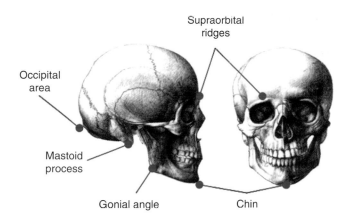

Occipital area

Supraorbital ridges

Mastoid process

Gonial angle

Chin

Figure 8.10. *The skull has many indicators of maleness or femaleness on it, but they are not as clear as those on the pelvis. It takes training and experience to become a good judge of variations in populations. A very slender male or a very robust female may have skeletal traits that fall into an overlap between the sexes.*

determination of sex. Thirty-four standard measurements are the minimum for inclusion of a skull into the National Forensic Data Base, and from these sex (and race, as we'll see later) can be estimated. These measurements are taken with specialized rulers, called **calipers**, that are either **spreading calipers** or **sliding calipers**. The measurements are taken from various landmarks around the skull. Complicated statistical techniques are used to sort out the measurements, relate them to each other, and then compare them against an appropriate reference population. Software developed at the University of Tennessee, called FORDISC, provides an easy way to analyze and compare data from skeletons, as graphically represented in Figure 8.11.

Post-cranial bones can also provide information about a person's sex, but most of these are based on size and therefore are quantitative. Many of the post-cranial bone measurements will yield an accuracy of between 58–100%. The measurement may be straightforward, but the interpretation may not be. For example, if the head of the **femur** is greater than 48 mm, then the person was most likely male; a measurement of less than 43 mm indicates a female. The area between 43 and 48 indicates that the size of the person was such that estimating sex from this measurement alone would give an inconclusive result. This is why it is very important to consider all of the recovered bones before making a judgment and, in turn, this emphasizes the need for a comprehensive search and collection of the remains at the scene.

How Old Was This Person?

As we develop in the womb, grow into adults, and age over the years, our skeletons change in known and predictable ways. For infants and

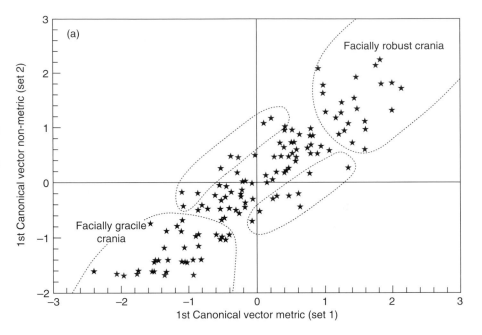

Figure 8.11. This is a plot of the calculations taken from a skull to determine its racial ancestry. Given the amount of overlap between some of the circles, you can see how even numerical data can sometimes lead to a "fuzzy" answer with race (Source: Lahr, M.M., & Wright, R.V.S. (1996)).

children, this is the appearance and development of skeletal growth areas that spread, meet, and fuse into whole bones. As adults, our skeleton's growth shifts to maintenance functions, responding to new stresses, such as exercise (or lack thereof) and job-related activities. Our later years bring with them the loss of bone mass, the slowing of our physiology, and the general degradation that accompanies our senior years. These changes are all recorded in our skeletons, and forensic anthropologists use these alterations to estimate a person's age at death.

Estimating age is conceptually different from estimating sex: There are only two sexes, but age is a continuum of 70, 80, or 90 (sometimes more) years. The age-related changes in our skeletons are predictable but not specific enough to allow for an estimate of "31 years and 8 months." The natural variation within a population and between individuals in a population prohibits a precise determination of age. Estimated age ranges, bracketed around the most likely age (25–35 years, for example) are the most acceptable way of reporting age at death. This bracketing necessarily leads to imprecision while retaining accuracy, but only up to a point. If you always estimate that an individual's age is between 1 and 95 years, you'll almost always be correct. That estimate, however, would not be very useful to investigators. By balancing the natural variation in aging and the anthropologist's skill with the methods used, an estimate that accurately reflects the precision of the sample *and* technique can be produced.

For the sake of convenience and organization, the range of human ages has been broken into various classes with associated years: fetal (before birth), infant (0–3), child (3–12), adolescent (12–20), young adult (20–35), adult (35–50), and old adult (50+). These classes represent the significant phases of growth, maturation, and decline in the skeleton and related tissues.

Bones can indicate the stage of development attained by the appearance and fusion of the various epiphyses throughout the body. Non-united epiphyses are easy to observe because the diaphyseal surface is characteristically rough and irregular in appearance. Epiphyseal appearance and union occur over the course of years and are processes, not an event; the degree of union (usually scored on a multipoint scale) must be carefully assessed because this could indicate which extreme of an age range is being observed. The three main stages of union are shown in Figure 8.12: First, the epiphysis is open; second, the epiphysis is united but the junction is still visible; and, third, the epiphysis is completely fused. Epiphyses can be small, so every effort should be made during collection to make sure none are overlooked.

Although epiphyses all over the body are uniting from infancy onward, the major epiphyses of the bones of modern populations fuse between 13 and 18 years of age. Union typically occurs in the order of elbow, hip, ankle, knee, wrist, and shoulder. Note that the beginning of epiphyseal union overlaps with the end of dental development and, therefore, these two methods complement one another. The last epiphysis to fuse is usually the medial **clavicle** (collarbone) in the early 20s. Once all of the epiphyses have fused, by about age 28 for most of the population, the growth of the skeleton stops and other age indicators must be used.

A few areas of the skeleton continue to change in subtle ways (compared with the appearance and union of epiphyses) throughout the remainder of adulthood. The main areas used for estimating adult age are found on the pelvic bones, the ribs, and the continuous remodeling of bone's cellular structure. These few, relatively small areas of the human skeleton have been intensely studied and restudied over the years by researchers trying to fine-tune the estimation of age at death for adults. Any one method alone, however, runs the risk of mislead-

Figure 8.12. *Different epiphyses unite with the main portion of a bone gradually, so the forensic anthropologist must evaluate the degree of union to correctly estimate age. Epiphyses fuse to the main portions of bone at different times, and this pattern of bone growth is an important technique for estimating age in younger individuals (Source: Buikstra and Ubelaker, 1994, with permission).*

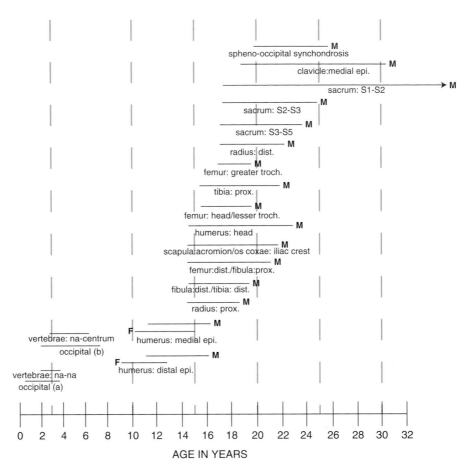

ing the investigator; therefore, all available information must be considered, including physical evidence not of an anthropological nature (clothing, personal effects, etc.).

The **pubic symphysis** (a **symphysis** is a "false" joint) is the junction of the two pubic bones lying roughly 4″–5″ below the navel. This junction is bridged by cartilage that acts as a cushion between the two bones. The symphyseal face shown in Figure 8.13 is a raised platform that slowly changes over the years from a rough, rugged surface to a smooth, well-defined area. The morphological changes of the pubic symphysis are considered by the majority of anthropologists to be among the most reliable estimators of age at death. This area was first studied in-depth by Todd (1920, 1921), who divided the changes he saw into 10 phases, each defined phase relating to an age range. Todd's work was later advanced by McKern and Stewart (1957), who broke

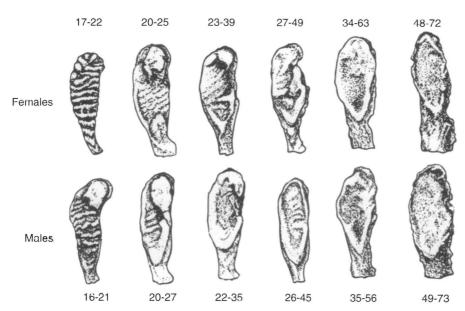

| 17-22 | 20-25 | 23-39 | 27-49 | 34-63 | 48-72 |

Females

Males

| 16-21 | 20-27 | 22-35 | 26-45 | 35-56 | 49-73 |

Figure 8.13. *The face of the pubic symphysis is an important characteristic for estimating age in adults. The surface starts out rugged and bumpy with no defined rim. Gradually, the surface flattens out, and a rim develops around the edge of the face. With advanced age, the rim begins to disintegrate although the face remains smooth.*

Todd's holistic method into a sectional evaluation to simplify the process. McKern and Stewart's work was based on young males who were killed in the Korean War, and this may have unintentionally biased their results; their work was, after all, focused on identifying soldiers of that same sex/age category. Nevertheless, the McKern and Stewart method held sway for a number of years until Judy Suchey and Allison Brooks began a large-scale collection and analysis project on the pubic symphysis by collecting samples from the Los Angeles County morgue. The intention was to collect a wide-ranging demographically accurate sample that could be assessed not only for age but also for variations due to an individual's sex. Their results are more akin to Todd's than McKern and Stewart's, although with fewer phases of development (Ubelaker, 2000).

Another area of morphological change with advancing adulthood is the sternal end of the fourth left rib. As the cartilage between the sternum and the ribs ages, it begins to ossify at a known and predictable rate. Another method of estimating age at death is the examination of the changes in the auricular surface, where the ilium attaches to the **sacrum** (the so-called "sacroilliac joint"). As age progresses, the surface of the bone becomes less bumpy and more smooth, with smallish pores opening up, creating a decrease in the organization of the surface traits.

Finally, bone never rests. It is constantly remodeling in response to the stresses placed on it. This remodeling can be seen in the microscopic structure of bone. In approximately the same way as a wall would be rebuilt, bone first needs to be torn down before it can be built up. This constant erosion and renewal leave permanent markers in bone: When we die, these changes cease. Therefore, a correlation exists between the amount of bone reworking and the amount of time the body has expended energy on this remodeling. A thin section of bone is cut, specific areas are viewed microscopically, and the various structural elements (whole osteons, fragmented osteons, **interstitial bone** fragments, etc.) are counted. Various formulae have been developed and are among the most accurate methods available for estimating age at death. A major disadvantage of this method is that some amount of bone must be removed, which may or may not be allowed because of case requirements.

Ancestry

Many of the cues we use to assess someone's ancestry in life are not well demonstrated in the skeleton. Moreover, ancestry or "race" is a difficult concept, both biologically and socially: Human physical variation is often a subtle thing, and people are sensitive to the labels other people place on them. While it is true that no pure ethnic groups exist (or have ever existed), we identify people based partly on what we perceive their "race" to be. This combination of blurred ancestral categories and popular perception, not to mention people's racial self-identity, make ancestry one of the most difficult estimations in a forensic anthropologist's examination. Nonetheless, forensic anthropologists routinely are called upon to assess skeletal remains for clues as to that person's ancestral affiliation to help lead police toward identification. The terms forensic anthropologists use to designate ancestry are typically those of the United States Census, namely, Whites, Blacks, Hispanics, Asians, Native Americans, and Other (www.census.gov).

Ancestry can be estimated by morphological or quantitative analysis, and both of these methods are centered on the skull. Features of the skull, such as the general shape of the eye orbits, nasal aperture, dentition and surrounding bone, and the face can offer indications of ancestry. Other features are more distinct, such as the scooped-out appearance of the **lingual** (tongue) side of the upper central incisors often found in individuals of Asian ancestry (so-called "shovel-shaped" incisors). But even indicators like this are not as clear as they may appear at first glance: Prehistoric Native Americans migrated into North America across the Bering Strait from Asia, and some of them show shovel-shaping on their incisors.

In hopes of rendering ancestral assessment more objective, physical anthropologists sought metric means of categorizing human populations. Currently, these means consist of numerous measurements that are then placed in formulae derived from analysis of known populations. While fairly accurate, these formulae suffer from being based on historically small samples that are not necessarily representative of modern populations. These concerns aside, given a complete skull or cranium, ancestral affiliation can be assessed with enough accuracy to make them useful for forensic investigations.

Stature

Our living stature directly relates to the length of our long bones, especially those of our lower limbs. Calculating stature from long bone lengths is relatively simple, and even partial bones can yield useful results. The only difficulty is that sex and ancestry must be known to correctly estimate height (see Table 8.4) because humans vary within and between these categories.

Table 8.4. Estimating stature is a straightforward procedure. The trick is that you must have an estimate of sex and race to determine the proper formula to use. All measurements are in cm.

White Males
Stature = 3.08 * **Humerus** + 70.45 ± 4.05
3.78 * **Radius** + 79.01 ± 4.32
3.70 * **Ulna** + 74.05 ± 4.32
2.38 * **Femur** + 61.41 ± 3.27
2.52 * **Tibia** + 78.62 ± 3.37
2.68 * **Fibula** + 71.78 ± 3.29

Black Males
3.26 * **Humerus** + 62.10 ± 4.43
3.42 * **Radius** + 81.56 ± 4.30
3.26 * **Ulna** + 79.29 ± 4.42
2.11 * **Femur** + 70.35 ± 3.94
2.19 * **Tibia** + 86.02 ± 3.78
2.19 * **Fibula** + 85.65 ± 4.08

Asian Males
2.68 * **Humerus** + 83.19 ± 4.25
3.54 * **Radius** + 82.00 ± 4.60
3.48 * **Ulna** + 77.45 ± 4.66
2.15 * **Femur** + 72.75 ± 3.80
2.40 * **Fibula** + 80.56 ± 3.24

For example, a White male with a femur length of 55.88 cm would be estimated to have been between 189 cm and 196 cm ((2.38 * 55.88) + 61.41, ±3.27, rounding up) tall during life, or about 6′1″ to 6′3″.

Facial Reproductions

To identify someone, you must have premortem records, such as X-rays. To get those, you have to have an idea of whose remains you might be studying. Sometimes, bones are found, and law enforcement investigators have no good leads as to whom they might represent. In these cases, forensic science has to turn to the world of art for assistance.

Because the shape of our faces is based on our skulls, if we were to reconstruct the soft tissues of a face on top of a skull, we could create a likeness of that individual. This is what happens in facial reproductions—an artist re-creates the likeness of a person either by sculpting the soft tissues with clay in three dimensions or by drawing, as shown in Figure 8.14. Facial reconstructions require a high degree of artistic skill, a good knowledge of human anatomy and variation, and an appreciation of the human face. These likenesses are not used for identification purposes but are meant to stir the public's recognition of otherwise unidentifiable remains. Flyers, images on television or in

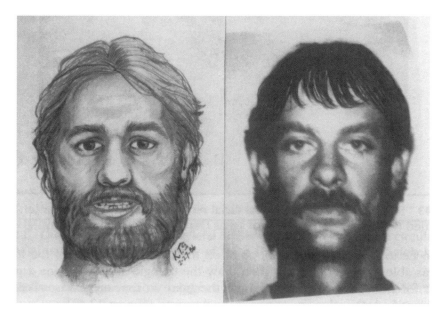

Figure 8.14. Based on science and anatomy but completed through art, facial reconstructions are helpful in approaching the public for investigative leads to a person's identity. Although they can be quite accurate, as in this example, they are not used for identification (Source: Taylor, 2001 © CRC Press, with permission).

newspapers, and police bulletins are used to distribute the likenesses in the hopes that someone will recognize them.

ODONTOLOGY

The most common role of the forensic dentist is the identification of deceased individuals. Dental identification can be conducted through comparison of dental remains to either antemortem or postmortem records. The most frequently performed examination is comparing the dentition of a deceased person to that of a person represented by antemortem to determine whether they are the same individual. The biological profile developed by the forensic anthropologist is very helpful in narrowing down the potential choices for selecting the antemortem records. If the antemortem records are available, any postmortem X-rays should replicate the view and angle in the antemortem X-rays. If antemortem records are not available, a postmortem record is created by the forensic dentist for possible future comparisons. The forensic dentist produces the postmortem record by carefully charting and writing descriptions of the dental structures and by taking radiographs.

Once the postmortem record is complete, a comparison between it and any antemortem records can be conducted. The comparison is methodical and systematic: Each tooth and structure is examined and compared. Fillings, caps, and restorations play the largest role in the identification process. Other features play a role in those individuals with good dental hygiene and few restorations. Similarities should be noted during the comparison process, as well as explainable and unexplainable discrepancies. Those differences that can be explained typically encompass dental restorations that occurred in the time elapsed between the antemortem and postmortem records. The person had a tooth pulled or a cavity filled, for example. If a discrepancy is unexplainable, such as a postmortem tooth that is not present on the antemortem record, then the odontologist will conclude that two different people are represented (an exclusion).

Dental Anatomy

The anatomy of the mouth is important to forensic science for a number of reasons. First, the teeth are made of **enamel**, the hardest substance that the body produces, and teeth can survive severe conditions and still be viable for analysis. Second, the teeth are the only part of the skeletal anatomy that directly interacts with the environment and, therefore, can reflect conditions the person experienced during life. Finally, teeth and their related structures have the potential to be used in the identification of the deceased. Because of these reasons and the complexity of fillings, braces, and other dental work, **forensic odontologists**, dental health professionals who apply their skills to legal investigations, are a specialty often relied on in cases of unidentified bodies, mass disasters, and missing person cases.

TEETH

Forensic odontologists use a variety of methods to organize and uniquely name each tooth in the mouth. The common names of teeth are also useful, but they refer to a group of teeth with the same characteristics. Typically, a numbering method is used. One of the most common is to number the teeth from the lower right molar, moving anteriorly, to the lower left molar; the next tooth would then be the upper left molar and then back around to the upper right molar (see Figure 8.15). This method sections the mouth into four quadrants: upper right, lower right, upper left, and lower left.

Each tooth has five sides: **buccal**, the side toward the cheek; **lingual**, the side toward the tongue; **mesial**, toward the midline of the body; **distal**, the side away from the midline; and the chewing surface, called the **occlusal surface**. These orientations help to describe where a cavity or filling is located. Individually, each tooth has similar structures but is shaped differently due to its functions. Every tooth has a crown, body, and root.

TOOTH DEVELOPMENT

Teeth grow from the chewing surface, or cusps, downward to the roots. This continual process is usually broken up into phases that relate to the amount of tooth development. Humans have two sets of teeth: one when we are children, called "baby" teeth but more properly termed deciduous teeth, and one when we are adults, our permanent teeth. Dentists often have a dental development chart in their offices, like the one in Figure 8.16. Different teeth develop at different rates, with incisors developing faster than molars. Teeth erupt through the gums when they are about one half to three quarters devel-

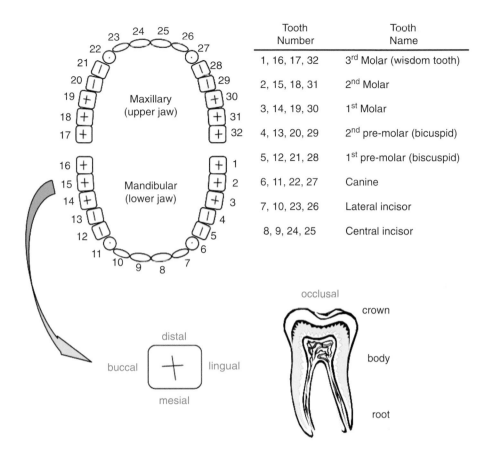

Tooth Number	Tooth Name
1, 16, 17, 32	3rd Molar (wisdom tooth)
2, 15, 18, 31	2nd Molar
3, 14, 19, 30	1st Molar
4, 13, 20, 29	2nd pre-molar (bicuspid)
5, 12, 21, 28	1st pre-molar (biscuspid)
6, 11, 22, 27	Canine
7, 10, 23, 26	Lateral incisor
8, 9, 24, 25	Central incisor

Figure 8.15. Because the terminology for teeth overlaps between top and bottom and right and left, it is important to have a unique identifier for each tooth. This aids in clear communication between forensic professionals.

oped. Notable landmarks in tooth eruption are the first deciduous incisor at about 9 months, the first permanent molar at about 6 years, the first permanent incisor at about 7 years, and the third permanent molar at sometime between 15 and 21 years; this latter tooth is notorious for irregular eruption and is not necessarily a reliable indicator of age.

IDENTIFICATION

The goal of a forensic anthropological examination is individualizing a set of human remains, often referred to as a "positive identification." This moves beyond class characteristics, no matter how narrow a classification, into the realm of uniqueness. To achieve this level of certainty, the data have to support the conclusion that the remains represent those of one, and only one, person to the exclusion of all other people.

Figure 8.16. *Teeth grow from the chewing surface, or cusps, downward to the roots. This continual process is usually broken up into phases that relate to the amount of tooth development. Humans have two sets of teeth: one when we are children, called "baby" teeth but more properly termed deciduous teeth, and one when we are adults, our permanent teeth. Dentists often have a dental development chart in their offices. Different teeth develop at different rates, with incisors developing faster than molars. Teeth erupt through the gums when they are about one half to three quarters developed (Source: © American Dental Association, with permission).*

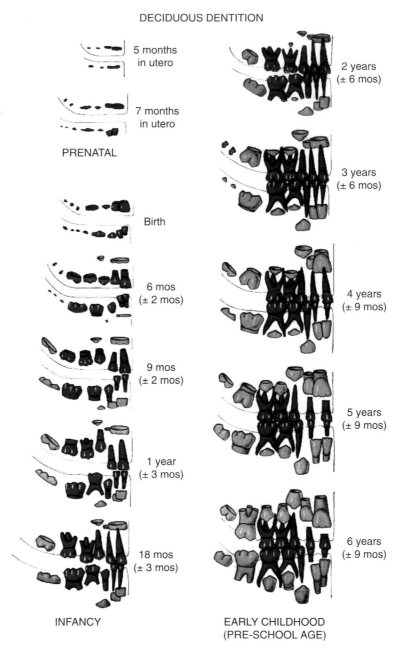

DECIDUOUS DENTITION

5 months in utero

7 months in utero

PRENATAL

Birth

6 mos (± 2 mos)

9 mos (± 2 mos)

1 year (± 3 mos)

18 mos (± 3 mos)

INFANCY

2 years (± 6 mos)

3 years (± 6 mos)

4 years (± 9 mos)

5 years (± 9 mos)

6 years (± 9 mos)

EARLY CHILDHOOD (PRE-SCHOOL AGE)

Because most people regularly visit their dentists, dental records and X-rays are the most common form of antemortem record that leads to a positive identification, as demonstrated in Figure 8.17. Because many years may have passed since the last X-ray and the forensic comparison, it may be necessary to have a skilled forensic

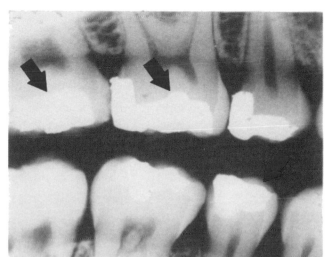

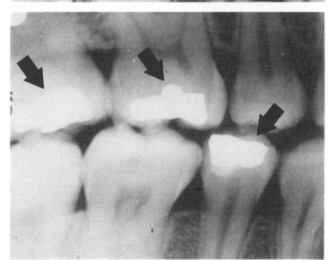

Figure 8.17. For a forensic odontologist to identify someone, he or she must have dental X-rays taken prior to the person's death. Identification is then a simple matter of comparing these with X-rays taken of the remains and looking for points of comparison, shown with arrows in the figure.

odontologist consult on the examination. Any differences between the X-rays taken before death and after death must be explainable and not be significant for the identification to be positive.

Other X-rays can lead to positive identifications as well. A structure in the frontal bone, the **frontal sinus**, is considered to be unique to a reasonable degree of scientific certainty. Likewise, the internal structure of post-cranial bones is considered to be unique as well. Surgeries, healed fractures, and disease may all be documented radiographically and also can lead to positive comparisons.

Identification through the comparison of ante- and postmortem X-rays is considered the best method for skeletal remains. People's teeth vary in size, number, and position, and the amount, size, type, location, and extent of dental work also vary enormously from person to person. Taken in combination, this natural and medical variation is such that it would be unthinkable to find two people whose teeth *and* dental work were exactly the same. X-rays can also document other individualizing traits, such as the habitual wear mentioned earlier, and some of these may be corroborated by family or friends.

INTERPRETATIONS

CAUSE VERSUS MANNER OF DEATH

The cause of death is the action that initiates the cessation of life; the manner of death is the way in which this action came about. There are, literally, thousands of causes of death, but there are only four manners in which to die: natural, accidental, suicide, and homicide. Forensic anthropologists can sometimes assist a medical examiner with assessing the manner of death (for example, see Sauer, 1984) but only rarely can one assist with the cause of death. Just because a skull exhibits entrance and exit bullet holes doesn't mean that a bullet is what caused the person to die; many people get shot each year, but only some of them die from their wounds. Likewise, a person may be strangled to death (cause: asphyxiation), but this activity may leave no markers on the skeleton. Forensic anthropologists must be very careful to stay within the bounds of their knowledge and training to provide the most useful information to medical examiners, investigators, and others who require their services.

TAPHONOMY

Taphonomy is the study of what happens to an organism from the time it dies until the time it ends up in the laboratory. In recent years, taphonomy has blossomed into a full-fledged area of study in its own right (for example, see Haglund & Sorg, 1997); this expansion has greatly assisted the various forensic sciences that relate directly to the study of the dead. This information greatly increases the ability of investigators to assess time since death, discern premortem from

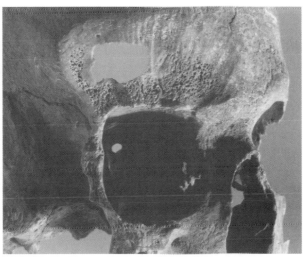

Figure 8.18.
Taphonomic marks,
such as these rodent
chewing marks, can
help the forensic
anthropologist to
determine the order of
events after a deceased
person is exposed
to the environment
(Source: White, 2000
© Academic Press,
with permission).

postmortem effects, shown in Figure 8.18, and detect subtle clues that might help lead to a killer's identity or activity.

PATHOLOGY

Forensic anthropologists work closely with forensic pathologists and may often be able to provide information beyond what a pathologist may know. Certain aspects of the pathologist's and anthropologist's work necessarily overlap, however, and these most often are in the areas of wounding and healing of bone.

The distinction of greatest importance for forensic anthropologists is the differences between premortem (before death) and postmortem (after death) injuries. Living bone has different mechanical properties than dead, dried bone, and this leads to different reactions to traumatic events. Any sign of healing in bone is definitive of a premortem injury. Wounds or breaks that occur near the time of death (called perimortem injuries) may be difficult to distinguish from trauma that occurs shortly after death because the body will not be alive long enough to begin noticeable healing. It is possible to distinguish between perimortem and long-term postmortem cuts using electron microscopy: At the edge of a fresh cut, the soft tissue will have dried and pulled back from the edge of the cut, whereas in a bone cut after the soft tissue has dried, it will be at the edge of the cut, as pictured in Figure 8.19.

Figure 8.19. It may be possible to distinguish between cuts made when the bone was fresh (perimortem) and when it was dry (postmortem). The soft tissue from a fresh cut will dry and pull back from the cut in the bone, while it will remain at the edge of the cut in the dry bone because it was already dried.

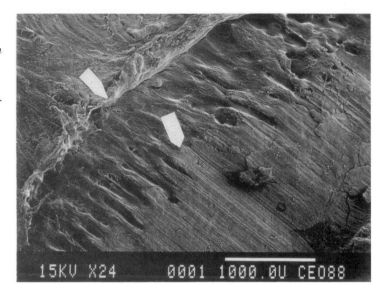

15KV X24   0001 1000.0U CE088

CHAPTER SUMMARY

Forensic anthropology plays a central role in the identification of people who are not identifiable by fingerprints or photographs: Nature has taken its course. Using their knowledge of human anatomy and variation, forensic anthropologists develop biological profiles of skeletal remains and look for individualizing traits in the hopes that the victim can be identified. They also assist other investigators, such as forensic odontologists and medical examiners, to help with the interpretation of taphonomic information and trauma.

Test Your Knowledge

1. What is forensic anthropology?
2. How is forensic anthropology different from archaeology?
3. What is a datum and how is it used?
4. How can you tell if something is bone?
5. What is a biological profile?
6. Name the two areas of the body that are the most accurate for estimating sex.
7. What is an epiphysis?
8. What is the last epiphysis to fuse?

9. Where is the pubic symphysis located?

10. What is the fourth left rib used for in forensic anthropology?

11. What needs to be known about a person before his or her stature can be calculated?

12. What is another term for a bicuspid?

13. What is forensic odontology?

14. How many teeth do humans typically have?

15. When does the first adult molar erupt?

16. What are two methods of identification for skeletal remains?

17. What is taphonomy?

18. What are some differences between the pelvises of males and females?

19. What bone is the most accurate for estimating height?

20. Name three ways to estimate age.

Consider This . . .

1. Why is ancestry such a complex concept?

2. How does forensic anthropology differ from pathology?

3. If you had only the pelvic bones of a deceased individual, what could you tell about that person?

BIBLIOGRAPHY

Barcsay, J. (2001). *Anatomy for the Artist*. London: Metrobooks.

Buikstra, J.E., & Ubelaker, D.H. (Eds.). (1994). *Standards for data collection from human skeletal remains*. Arkansas Archeological Survey Research Series Number 44. Fayetteville: Arkansas Archeological Society.

Haglund, W., & Sorg, M. (Eds.). (1997). *Forensic taphonomy*. New York: CRC Press.

Houck, M.M. (1998). Skeletal trauma and the individualization of knife marks in bone. In K.J. Reichs, (Ed.), *Forensic osteology: Advances in the identification of human remains* (2nd ed.), (pp. 410–424). Springfield, IL: C.C. Thomas Publishers.

Lahr, M.M., & Wright, R.V.S. (1996). The question of robusticity and the relationship between cranial size and shape in *Homo sapiens. Journal of Human Evolution*, 31, 157–191.

McKern, T.W., & Stewart, T.D. (1957). *Skeletal age changes in young American Males*. Technical Report EP-45. Natick, MA: U.S. Army Quartermaster Research and Development Center.

Sauer, N. (1984). Manner of death. In T. Rathbun, & J. Buikstra, (Eds.). *Human identification: Case studies in forensic anthropology* (pp. 176–184). Springfield, IL: Charles C. Thomas Publishers.

Shipman, P., Walker, A., & Bichell, D. (1985). *The human skeleton*. Cambridge: Harvard University Press.

Suchey, J.M., Wiseley, D.V., & Katz, D. (1986). Evaluation of the Todd and McKern-Stewart methods for aging the male os-pubis. In K.J. Reichs, (Ed.), *Forensic osteology: Advances in the identification of human remains* (pp. 33–67). Springfield, IL: C.C. Thomas Publishers.

Taylor, K.T. (2001). *Forensic art and illustration*. Boca Raton, FL: CRC Press.

Todd, T.W. (1920). Age changes in the pubic bone I. The male white pubis. *American Journal of Physical Anthropology*, 3, 285–334.

Todd, T.W. (1921). Age changes in the pubic bone. *American Journal of Physical Anthropology*, 4, 1–70.

Ubelaker, D. (2000). *Human skeletal remains*. Washington, D.C.: Taraxacum.

Ubelaker, D.H., & Scammell, H. (1992). *Bones: A forensic detective's casebook*. New York: Edward Burlingame Books.

United States Census (1990). www.census.gov.

White, T.D. (2000). *Human osteology*. London: Academic Press.

Entomology

KEY TERMS

Abdomen

Amctabolous

Antennae

Arthropods

Chitin

Exoskeleton

Head

Holometabolous

Incidental species

Instar

Key

Killing jar

Larva

Larvaposits

Maggot mass effect

Maggots

Mesothorax

Metamorphosis

Metathorax

Molting

Necrophagous species

Necrophilious

Nymph

Omnivorous species

Oviposits

Paurometabolous

Postmortem interval (PMI)

Predatory and parasitic species

Prothorax

Pupal stage

Puparium

Segments

Spiracles

Subspecies

Taphonomy

Taxa: kingdom, phylum, class, order, family, genus, and species

Taxonomy/taxa

Thorax

INTRODUCTION

Forensic entomology is the application of the study of arthropods (order Arthropoda), including insects, arachnids (spiders and their kin), centipedes, millipedes, and crustaceans, to criminal or legal cases. This field has been divided into three topics: urban entomology (involving insects that affect houses, buildings, and similar human environments), stored products entomology (involving insects infesting stored goods such as food or clothing), and medicolegal entomology (involving insects and their utility in solving criminal cases). Typically, the use of insects and their life cycles helps to establish a postmortem interval, or PMI, which is an estimate of how much time has passed since a person died. This estimate depends on the entomologist's knowledge of the ecology of insects and ability to accurately identify insects. Medicolegal, or forensic, entomology is what this chapter will cover, and what most professionals think of when they hear "forensic entomology." While a "specialty" science in many ways, forensic entomology is intimately linked with the disciplines of medical entomology (insects and the diseases they transmit), pathology, and **taxonomy** (the classification of living things).

Initially, insects and the law may seem an odd pairing, but wherever humans choose to live on the planet, insects are already there waiting for them. Found in nearly every habitat on land or in water, insects are the only group of animals to evolve true wings; the wings of birds and bats are modified upper limbs. This adaptation has provided insects with the means to travel far and to inhabit diverse ecologies for food and reproduction, including on and in dead animals. This may strike the average person as disgusting, but insects, particularly flies, play a vital role in the "recycling" of animal carcasses and other decomposing organic material. Often, insects are the first to find a corpse, and they colonize it in a predictable pattern. Forensically important conclusions may be drawn by analyzing the phase of insect invasion of a corpse or by identifying the life stage of **necrophagous** (dead-flesh eating) insects found in, on, or around the body. Knowledge of insect (especially fly) biology and habitats may provide information for accurate estimates of how much time a body has been exposed to insect activity. A knowledgeable entomologist will be able to tell if the insects are from the local area where the body was found; if they are not, this is a good indication that the body may have been moved. The absence of insects on a corpse is also a situation requiring the attention of a forensic entomologist.

The study of an organism from the time it dies until the time it reaches the laboratory is called **taphonomy**. The term was coined to describe the analysis of what happened to prehistoric animals, like dinosaurs, from the time they died until they became fossils sitting in a museum case. Investigating how different processes, such as wind, rain, animal or insect activity, etc., affected them, researchers could distinguish natural phenomena (animal tooth marks) from those caused by human intervention (injury) or healing. In this sense, paleontologists are a type of detective looking for clues of a prehistoric "crime." The knowledge gained from the taphonomic study of fossilized animals has been adopted by modern detectives and scientists, who have applied it to modern crimes. Forensic entomology is a good example of the application of the principles of taphonomy to legal investigations.

INSECTS AND THEIR BIOLOGY

Insects are the largest group of arthropods and are defined by having six legs and a three-segment body. Although small individually, insects are the most numerous and diverse group of organisms known, with nearly one million species described. The total number of plants and (non-insect) animals combined comprise fewer species than insects.

Unlike other animals, insects have an external skeleton, or **exoskeleton**, composed of a material called **chitin** and protein. This outer shell protects the animal's internal organs, conserves fluids, and acts as the structure for muscle attachment. Insects' bodies are divided into three **segments**, which are joined to each other by flexible joints. These segments are the **head**, **thorax**, and **abdomen**, as shown in Figure 9.1. The head contains the insect's eyes, sensory organs (including specialized **antennae**), and mouth parts. The thorax is further divided into the **prothorax**, **mesothorax**, and **metathorax**; each of these subsegments has a pair of legs. In addition, the mesothorax and metathorax are sites of wing attachment, if the insect has them. The abdomen carries much of the insect's internal organs and is segmented. Each of these segments bears a pair of holes, called **spiracles**, which the insect uses for breathing (Resh and Carde, 2003).

Figure 9.1. *Insects have three main body parts: the head, the thorax, and the abdomen. The thorax is further divided into the prothorax, the mesothorax, and the metathorax. Insects are the only animals to have evolved true wings, as opposed to specialized front limbs.*

LIFE CYCLES OF INSECTS

Distinctive of the arthropods is their variety of immature forms. As an insect grows, it passes through a series of maturation phases, and each

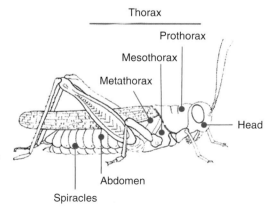

Thorax

Prothorax

Mesothorax

Metathorax

Head

Abdomen

Spiracles

219

Figure 9.2. Insects develop through various life stages, depending on the type of metamorphosis they experience. Ametabolous insects have immature forms that appear to be small adults. Paurometabolous insects emerge from hatching into a nymph form, which progresses to adult through a series of moltings. Holometabolous insects develop from eggs into larva, which then go through a separate growth stage to reach adult form. The caterpillar spinning a cocoon and emerging as a butterfly is a common example, as is the house fly shown here (Courtesy James Amrine).

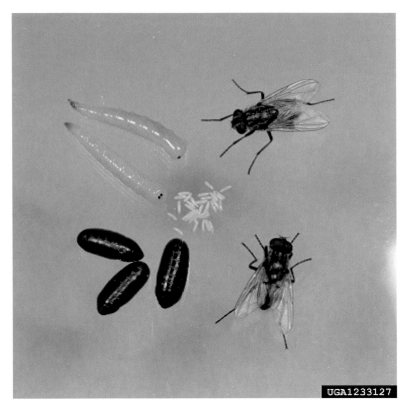

UGA1233127

phase can look quite different from the previous one or next, as demonstrated in Figure 9.2. Over millions of years of evolution, insects have developed three patterns of growth. The first and simplest is **ametabolous** ("without change") **metamorphosis**, in which the eggs yield immature forms that look like smaller forms of the adults. Eventually, these juveniles develop in size and mature sexually but otherwise undergo little structural change. This type of metamorphosis is limited to more primitive wingless insects (Apterygota).

The second type of metamorphosis is **paurometabolous**, or gradual, **metamorphosis**. The hatchlings emerge in a form called a **nymph**, which generally resembles a wingless version of the adult of the species. The nymphs and adults will occupy the same habitat and exploit the same food sources. Nymphs grow by **molting** (shedding their skin), and each successive molt produces a new **instar** or growth phase. As the nymph passes through each instar, it increasingly resembles the adult form and eventually develops wings. Different species pass through specific numbers of instars, and this can be useful in identifying immature forms. Cockroaches (Blattaria) and various predatory bugs (Hemiptera), for example, develop this way.

Holometabolous (or complete) **metamorphosis** is the third type of arthropod growth and is the most complex format of the three. The adult lays an egg (**oviposits**) or deposits a **larva** (**larvaposits**) onto a food source. The larvae (plural) start eating, or hatch from the egg and then begin eating immediately, and increase in size by molting through instars. The larval form is very different from the adult form, both in appearance and in its habitat. At the end of the instars, however, the larvae transition into an inactive phase, called the **pupal stage**. The pupa is a hardened outer shell or skin that protects the larva while it undergoes its final growth stage to the adult form. Butterflies (Lepidoptera) are a common example of holometabolous insects as they change from caterpillar into cocoon to their final, colorful adult form. There are several types of pupa in holometabolous insects, but the type most frequently encountered by forensic entomologists is that evidenced in flies (Diptera), the **puparium**. The puparium is the hardened skin of the last larval instar and tends to be darker than the normal larval skin (Resh and Carde, 2003).

Depending on the species of insect, the time it takes to go from egg to adult varies greatly: Some insects may have a few or many generations in one year. The weather, environment, season, food (abundance or lack), rainfall, humidity, and other such factors can affect the timing of insect reproduction. In the case of **necrophilious** insects ("dead loving," or those associated with decomposition), many other factors, such as location (indoors, outdoors, on land, in water, etc.), shade, slope, and where the body lies (on soil, cement, in a tree, in an attic, etc.), can have an influence on the number and timing of successive generations.

Necrophilious insects are very sensitive to chemical changes in a dead body and can detect even the slightest hint of decomposition, sometimes within minutes of death. The chemicals are by-products of the decomposition process and signal to the insect that a new food source is available. As the body decays, the signals it sends out change and communicate "food" to the different species that inhabit the body at different times and conditions. Dermestid beetles, for example, prefer dry flesh and won't colonize a body until the tissues are no longer wet or even moist; by that time, the odors and chemicals coming off the remains are very different from those emitted, say, two weeks prior. The habitat of a decomposing body is a finely tuned environment, and insects have evolved to make the most of each stage of decay.

COLLECTING INSECTS AT A CRIME SCENE

Not all of the insects mentioned here will appear on a body in equal numbers or even be present at all. The number, type, and distribution of insects drawn to a dead body will vary by the environmental conditions, time since death, location, geography, weather, and many other factors. This is why it is important for a professional forensic entomologist to collect, process, and analyze the insect data from a death scene: It is a complicated and specialized discipline. Because of this variability in the number and kind of insects, the information they provide can sometimes be quite precise because no two scenes are exactly alike in time or space.

Forensic entomologists encounter a diverse range of habitats and conditions when assisting with crime scenes. It is important that all personnel from the various agencies at a death scene cooperate with the primary investigating agency and be aware of each person's assigned responsibilities. This is especially true for the forensic entomologist because many police agencies are not familiar with this science and its requirements for evidence collection. The entomologist should discuss with the evidence technicians and the primary investigator the plan for evidence collection and the role the entomologist intends to play. The following is a suggested sequence of stages for a forensic entomological investigation (Haskell, Schoenly and Hall, 2002):

Visual observation and notes of the scene
    Recording notes
    Approximating the number and kinds of insects
    Identifying locations of major insect infestations
    Noting immature stages
    Indicating the precise location of the body
    Recording any other phenomena of note (trauma, coverings, etc.)
Collection of climatological data from the scene
    Ambient air temperature
    Ambient humidity

Ground surface temperature
Body surface temperatures
Below-body temperatures
Maggot mass temperatures
Post-body removal sub-soil temperature
Collection of specimens from:
The body before its removal from the scene
The area surrounding the body (up to 20′) before its removal
The area directly under the body after the body has been removed

Necrophilious insects, particularly flies, are attracted to dark, moist areas: On fresh bodies, this means the face (nostrils, mouth, eyes, etc.) or any open wounds, as shown in Figure 9.3. The genital or rectal areas, if exposed or traumatized, will sometimes provide shelter and moisture for ovipositing flies. The patterning and amount of ovipositing and larvae should be recorded by the entomologist with notes, drawings, and photographs.

Insects can be collected in a variety of ways, most of which will be employed at every death scene. Flying insects can be trapped in a net by sweeping it back and forth repeatedly over the body. The end of the

Figure 9.3. Insects that are attracted to dead animals as an environment for food and reproduction (necrophilious) usually inhabit dark, moist areas first, like the eyes, mouth, nose, and open wounds (Courtesy James Amrine).

net, with the insects in it, can then be placed in a wide-mouth **killing jar**, a glass jar containing cotton balls soaked in ethyl acetate. Several minutes' exposure to the ethyl acetate will kill the insects; they should then be placed in a vial of 75% ETOH to preserve them. Two labels should be prepared for each specimen (one for inside the vial and one for the outside), and they must be written in pencil; ink may dissolve in the ETOH.

Crawling insects on and around the body can be collected with forceps or fingers. The entomologist must be careful not to disturb any other potential evidence while collecting insects. If the body will be put into a body bag, it is a good idea to check for any insect activity before the body is placed inside it. Eggs and a mixture of larvae of various sizes (several hundred in total) should be collected, as well as any adults. A portion of the larvae should be preserved the same as described for flying insects; another portion should be kept alive to rear to adulthood. If the entomologist is collecting the insects in the morgue, a careful inspection of the clothing must be made in conjunction with the forensic pathologist's observations.

Once the body has been removed, the soil under the body should be sampled. An approximately $4'' \times 4'' \times 4''$ cube of soil (about the size of a one-pint container) should be taken from areas associated with the body, such as the head, torso, limbs, or wherever seems appropriate given the body's position. Additional soil samples should be taken up to $6'$ from the body in each direction. Any plant materials associated with the body or its location should also be collected for possible botanical examination.

Buried or enclosed remains present particular problems for the entomologist because insects' access to a body is limited. Some flies are barred from a body by as little as one inch of soil. Burial also slows the process of decomposition due to lower and more constant temperatures, fewer bacteria, and limited access to the body. A building can also prevent some types of insects from gaining access to a body or slow down their recognition of the chemical odors that signal decomposition. This alters the entomologist's estimation of time since death because the "clock" of insect succession rate has been altered. However, the odors may escape, but the insects may not be able to gain access: Finding a cloud of flies hovering above a car trunk may indicate that a dead body has been in there for some time. If the structure is a house or an apartment, then weather data probably will not help the entomologist devise a time since death. Rather, the thermostat settings could be substituted for "climate" data. Another example would be a car with the windows up during the summer, where the internal temperature even on a mild day can easily reach in excess of 110°F. In short, any environmental change, whether natural or artificial, needs to be measured and noted for the entomologist to make an accurate estimate.

Wide-angle and close-up photographs should also be taken, with emphasis on specific areas of insect activity. The forensic entomologist should, after returning to the laboratory from the scene, begin collecting weather data for the period of time in question.

THE POSTMORTEM INTERVAL

The forensic entomologist's main contribution to death investigation is an estimate of the **postmortem interval (PMI)**. Being able to provide a time range for when the crime occurred is of great importance in limiting the number of suspects who may or may not have an alibi. If the victim is unidentified, the PMI may also assist in narrowing the number of potential missing persons. Calculating an estimated PMI sets a minimum and maximum time since death based on the insect evidence collected and developed. The maximum limit is set by the insects present on and around the body at the time of collection; this limit is moderated by the recent weather conditions that could help or hinder those species' development. A minimal PMI is estimated by the age of developing immature insects and the time needed for them to grow to adulthood under the conditions surrounding the crime scene. A maximal PMI can be difficult to estimate because the uncertainty widens as time and decomposition continue. It is imperative that forensic entomologists conduct outdoor studies with local species in various seasonal conditions to establish a baseline reference. This data can provide invaluable information in estimating PMIs, especially in circumstances in which the environmental indicators may be vague.

THE CLASSIFICATION OF INSECTS

Estimating a PMI requires that the forensic entomologist be able to precisely identify the insects on and around a body. This classification can be properly performed only by an experienced forensic entomologist with the proper reference collections. Differentiating between closely related insects, especially certain species of flies, requires the recognition of minute anatomical details and should be attempted only by qualified professionals. The science of identifying and classifying organisms is called **taxonomy**. All organisms are categorized by their relatedness through the recognition of significant evolutionary traits. The order of relatedness is broken into these **taxa**, or related groups: *Kingdom, Phylum, Class, Order, Family, Genus*, and *Species*[1]. When referring to a specific type of organism, describing it by the genus and species (and sometime subspecies) is sufficient to set it apart from all other organisms. Thus, when talking about *Calliphora vicina, Canis familiaris*, or *Homo sapiens*, we know we're discussing the blue-bottle fly, the domesticated dog, and humans. These terms are formal, so the genus is capitalized and, because they are Latin, it is proper to format genus and species designations in italics. Some species, especially among insects, have well-defined variants or **subspecies**. A taxonomic **key** is a method for classifying organisms whereby each trait identified separates otherwise similar groups of organisms. By following a detailed key, a forensic entomologist can identify, or "key out," all of the forensically important insects that may be found on or around a body (see Table 9.1).

REARING INSECTS

A significant step in identifying immature insects, especially flies, is the rearing of larvae into adult insects. This process is not as easy as it may initially sound. Because these immature insects are the reference materials for the forensic entomologist's estimation of PMI, their growth environments (temperature, humidity, space, light, etc.) must be closely controlled and monitored. A vent hood is a necessity: As the insects' food source is rotting meat (typically beef, pork, or liver), the smell can become unpleasant. Depending on how many cases the entomologist works on at one time, the rearing laboratory must be able to keep all of the samples and insects separate. It is important to remember that fly larvae grow into adult flies, and they must be kept from zipping across the laboratory! And, finally, the laboratory must be designed so that it is easy to clean after the case work is completed.

---

[1]For example, humans have the classification Kingdom: Animalia, Phylum: Cordata, Class: Mammalia, Order: Primates, Family: Hominidae, Genus: *Homo*, and Species: *sapiens*.

Table 9.1. *Insects that may be found on or near decomposing bodies.*

ORDER	COMMON NAMES	FAMILIES	FIGURE(S)
Collembola	Springtails		See Figure 9.4a
Blattaria	Cockroaches		See Figure 9.4f
Coleoptera	Beetles	Necrophagous Silphidae (carrion beetles) Dermestidae (dermestid beetles) Cleridae (checkered beetles) Predatory Staphlynidae (rove beetles) Scarabaeidae (scarab beetles) Histeridae (clown beetles)	See Figure 9.4d See Figure 9.4b
Dermaptera	Earwigs		
Diptera	Flies	Calliphoridae (blow flies) Muscidae (house flies) Piophilidae (skipper flies) Sarcophagidae (flesh flies) Scathophagidae (dung flies)	See Figure 9.4e See Figure 9.4c
Hemiptera	True bugs	Predatory	See Figure 9.4g
Hymenoptera	Ants, bees, wasps	Predatory/necrophagous	See Figure 9.4h

Specially devised chambers for insect rearing can be purchased, but they are very expensive. Many entomology laboratories use large aquariums, disposable containers (often, pint-sized ice cream or plastic containers with holes punched in the lid for ventilation), and pie tins as inexpensive but effective larval growth chambers. Some insects require special conditions for growth that mimic the phase of decomposition that the insects recognize as appealing; for example, dermestid beetles colonize remains only after they have become dry and desiccated.

CALCULATING A PMI

The main reason for studying the presence and life stages of insects on a corpse is to establish the time since death. A method for estab-

Figure 9.4. *A list of the insects commonly found on or near dead bodies.*
a. Springtail
b. Clown beetle
c. House fly
d. Carrion beetle
e. Blow fly (blue)
f. Cockroaches
  (American)

A

B

C

D

E

F

G   H

Figure 9.4. *Continued*
g. *Assassin bug*
   (Hemiptera,
   Reduviidae)
h. *Hymenopter*
   (Hymenoptera)
(Photos a-f courtesy of
James Amrine, J.L.
Castner, University of
Florida Entomology
Department; g and h,
Wikipedia, with
permission).

lishing a PMI is based on an ecological and faunal study of the cadaver, demonstrated in Figure 9.5. Data collection must be detailed and precise if the PMI is to be accurate. The basis of the method is to study which insects, or their young, inhabit a dead body and in what sequence they do so. A body that has been dead only one or two days will have infestations primarily of blow flies, such as Cochliomyia macelleria, because these insects are attracted to a corpse almost immediately. The recognition of each species in all stages and knowledge of the time occupied in each stage allow the entomologist to estimate a time since death. Other information could be gleaned from the faunal composition, such as whether the body has been moved.

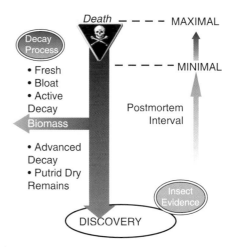

Four ecological categories exist in the cadaver community. The first are the **necrophagous species**, which feed on the carrion itself, contributing directly to the estimation of PMI. Examples of necrophagous insects are diptera, coleoptera, silphidae, and dermestidae. The second most important group, forensically speaking, is the **predatory and parasitic species**, such as certain colcoptera, silphidae, and some diptera. These insects prey on other insects, including the necrophagous ones, which inhabit the cadaver. Some of the species in this group, however, may be necrophagous while immature but become predatory in later instars. **Omnivorous species** make up the third category. Wasps, ants, and some coleoptera fit into this group because they may eat material from the body, other insects, or whatever food source

Figure 9.5. *Data collected by entomologists and other biologists, as well as weather data, all contribute to the estimation of a postmortem interval, the time since death occurred.*

presents itself. The last category of insects uses the cadaver simply as an extension of their normal habitat, such as spiders, butterflies, collembola, and others. This group can be referred to as the **incidental species.**

The faunal succession on carrion is linked to the natural changes that take place in a body following death. After death, the body temperature falls to that of its ambient environment. Cellular breakdown begins after several hours, resulting in the release of gases such as ammonia (NH_3), hydrogen sulfide (H_2S), carbon dioxide (CO_3), and nitrogen (N_2). Putrefaction follows due to the activity of microbes, especially those from the body's own intestinal flora. These chemical and microbiological consequences of death are the earliest, and tend to be the most accurate, indicators of time since death. Their accuracy and utility diminish as time moves forward, however, and other information, namely ecological, must be used.

At a minimum, the estimated age of an immature insect can provide a PMI, but this estimate does not provide a maximum limit because the amount of time between death and egg/larval deposition is unknown. Necrophagous insects appear almost immediately as the cellular breakdown begins; some species of flies are so sensitive to the chemistry of death that they appear within minutes of the cessation of life. The level of larval development can provide an estimate that is accurate from less than one day to just over one month, depending on conditions and the species reared.

A more complete, although more complicated, method of PMI estimation involves the study of the succession of insect species on and within a body. The forensic entomologist employs a model that is based on information about the ecological and environmental events between the time of death and the appearance of a particular insect species. The simplest model would be one in which the forensic entomologist estimates the age of a larva and the time between death and the insect landing on the body. The activities of the insects, especially fly larvae, accelerate the putrefaction and disintegration of the body. The number of waves of insects in the succession on a body has been interpreted to be between two and eight. Such a model provides both a minimum and maximum PMI and yields much more accurate estimates. Many environmental factors—such as whether the body is on the surface, buried, in water; the temperature; weather; humidity; amount of light/shade; season; and even manner of death—influence the number, type, appearance, and life cycles of necrophagous insects. Regardless of the complexity, the forensic entomologist must choose a model of insect development or succession, sometimes drawn from published experimental data. For an example, see "In More Detail: PMI in Hawaii."

IN MORE DETAIL:
PMI in Hawaii

One of the first cases forensic entomologist Lee Goff worked involved the death of a woman in Honolulu in 1984. She had last been seen over two weeks before her body was found near the shore by an abandoned brewery; her car, with blood on the interior, was found 30 miles away. Goff and a graduate student collected maggots, hide beetles, and scene information. "There were three species of maggots on the body, in different locations and in different stages of development," Goff (2001) notes in his book, *A Fly for the Prosecution*. "I sorted each type into two sublots. I measured the length of each of the maggots in one of the lots, and used the average of these lengths to give me some idea of their stage of development. Then I preserved them in ethyl alcohol. I put the other sublot of maggots into a rearing chamber to complete their development to the adult stage" (2001; 3)

Initially, Goff used a home-grown computer approach for calculating the PMI, the first time he had done so in a real case (remember, this was 1984). The result was disappointing: It told him that the body either didn't exist or there were two bodies! The conflicting results were caused by the computer not being able to resolve the dilemma of flesh-flies, which like soft, moist tissues to eat, being on a body at the same time as the hide beetles, whose preference is for dried flesh. To investigate a solution, Goff visited the crime scene. He found that the victim had been lying on her back, partially submerged in about 5 inches of water. This accounted for soft, moist tissues to remain on the victim's back while the Hawaii sun dried out the front.

Returning to the lab, Goff began his calculations with this new data. "For Chrysomya rufifacies, egg laying can begin quite soon after the adult females reach the body and will continue, under Hawaiian conditions, for approximately the first 6 days following death (and) completion of development ... usually requires 11 days. Since the only evidence of this species on the body was the empty pupal cases, discarded when the flies reach adulthood, I was confident that all *Chrysomya* maggots maturing on the body had completed development before it was discovered" (2001; 6). Goff noted that another fly species, the cheese skipper, which does not invade a body until days after death, were all at the same stage as samples from a test he had conducted: 19 days old.

The hide beetles were also very useful. "In lowland habitats on Oahu, (the beetles) begin to arrive between 8 and 11 days after the onset of decomposition, and during decomposition studies I have gathered larvae comparable in size to those collected from this case beginning on day 19. The remaining species ... were consistent with a postmortem interval of 19 to 20 days but did not yield more precise information" (2001; 6–7). Goff gave an official estimate of 19 days to the medical examiner.

The victim had been missing precisely that many days, since she was seen leaving the restaurant she co-owned with a tall man. This man was later identified, tried, and convicted of second-degree murder—all because of some flies and beetles.

Source: Goff, M.L. (2001). A fly for the prosecution. Cambridge, MA: Harvard University Press.

If no data are available that take into account the parameters that the forensic entomologist faces, then experimentation is required. The experimental conditions should be as close to those at the crime scene as possible; this logically means that a forensic entomologist should be collecting data on decomposition in his or her ecological zone(s) year round. The closer the experimental data are to the crime scene conditions, the lower the margin of error will be in the PMI estimate. The subjects for these decomposition studies are typically small pigs (under 50 pounds), which have been shown to be appropriate stand-ins for humans despite their smaller size. In the second part of a study funded by the National Institute of Justice, Haskell, Schoenly, and Hall (2002) showed that both pigs and humans attract a large majority of the same arthropod species, but also the most common and moderately common species. This was true regardless of the pigs' size, but small pigs are easier to physically handle than large ones, practically speaking. Importantly, their work also indicates that cross-comparison of "pig studies" from differing geographical zones may be a viable research interest.

One of the most influential factors in estimating PMI is temperature. Temperature has a direct effect on the metabolism and development of insects. This is true not only of ambient (air) temperature but also the amount of sun or shade to which a body is exposed. The larvae of necrophagous flies (**maggots**) are essentially "eating machines," and they have a metabolism and feeding rate that is much higher than other immature insect forms. When maggots are living, feeding, and moving all in approximately the same area, the temperature can soar by many degrees: This is termed the **maggot mass effect**. The temperature at the center of a maggot mass can be 100°F while the ambient temperature is in the 30°F range, and this could obviously bias a forensic entomologist's PMI estimation.

The forensic entomologist studies insect samples that were killed and preserved at the time of collection as well as those kept alive for rearing. The time when the preserved samples were collected is the starting point for the PMI, and it is from here forward that the entomologist makes his or her calculations for the maximal time since death. Because every death scene has unique circumstances and environs, no one algorithm best calculates all PMI estimates. As shown in Figure 9.6, computers are now being used to create very complex but highly realistic models that provide forensic entomologists with improved models for PMI estimates. As humans and computers

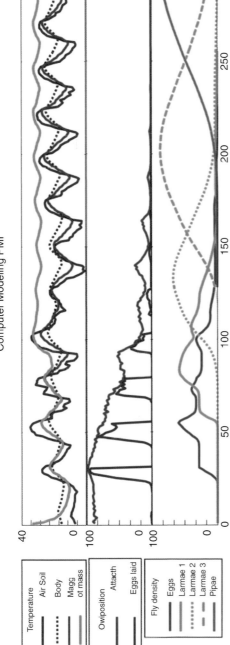

Computer Modeling PMI

The purpose of a forensic PMI model is to predict the timing of a past event. The ending point of the model is always the point where the sample was obtained: in this hypothetical case at 1:00 AM on April 39, 2001, in Farwell, Texas. Using historical hourly temperature data from a nearby weather station that was conveniently found on the internet, we run the model, here starting at a point 300 hours in the past (above).

Because the ending point of the simulation is know, but the starting point is not, it is necssary to run the model at all possible starting pints using historical environmental data to generate possible solutions from each hour in the past. Each model run assumes a time of death at the point where the model starts.

Figure 9.6. *Computers logically lend themselves to calculations involving a large amount of data, such as a PMI. This example shows the data collected, graphed, and interpreted using specific mathematical models for the particular geographical area in question. Other models would need to be employed if the victim were found in another area or at another time of year (Courtesy: D.H. Sloane).*

become more adept at handling large amounts of complex data, the estimates of PMI based on entomological information will become more realistic and accurate.

OTHER FORENSIC USES FOR INSECTS

Insects are important evidence in ways other than estimating PMI. Because they ingest portions of the bodies they inhabit, insects can ingest drugs, toxins, or other substances that are in the body at the time of death. Additionally, DNA from the victim may be obtained from the guts of insects who feed on the body. These tiny samples can offer critical evidence of poisoning or identity, for example, when no other evidence is available. Insects can also give indications of location (many species are habitat-specific), travel, or geography if associated with items, such as crates or shipments.

CHAPTER SUMMARY

Although forensic entomology may seem to be a narrow specialty, it is applicable in a wide variety of cases. Forensic entomologists identify insects associated with dead bodies and estimate time since death. They also play an important role in other areas where insects and crime intersect, such as drugs, poisons, and location of stolen goods.

Test Your Knowledge

1. What is an insect? Is it the same as an arthropod?
2. How many kinds of insect development are there and what are they?
3. What is the protective outer covering on an insect called?
4. What is an instar?
5. What is taxonomy?
6. What are the categories, in order, used to describe an organism's taxonomy?
7. In "In More Detail: PMI in Hawaii," Lee Goff used a particular fly, Chrysomya rufifacies, in his PMI estimate. What are the words "Chrysomya rufifacies"?
8. What does "necrophilious" mean?

9. What are the uses of forensic entomology?

10. What is the difference between insects that larvaposit and those that oviposit?

11. Name three factors that go into calculating a postmortem interval.

12. What are the four kinds of species that can be found on a dead body?

13. How many waves of insect invasion are there on a dead body?

14. What does "necrophagous" mean?

15. What are the larvae of necrophagous insects called?

16. What is the "maggot mass" effect?

17. What is the role of insects in decomposition?

18. What goes into the calculation of a PMI?

19. How would a forensic entomology case in Montana differ from one in Florida?

20. What is "molting"?

Consider This . . .

1. Early in this chapter it was mentioned that the complete absence of insects from a dead body is a good reason to contact a forensic entomologist. What do you think some of the reasons for this might be?

2. If a person was killed in Florida in a train's freight car and the body wasn't discovered until it reached its destination in New York days later, how might a forensic entomologist assist investigators? Would the time of year matter?

BIBLIOGRAPHY

Byrd, J.H., & Castner, J.L. (Eds.). (2001). *Forensic entomology: The utility of arthropods in legal investigations.* Boca Raton, FL: CRC Press.

Catts, E.P., & Haskell, N.H. (Eds.). (1990). *Entomology & death: A procedural guide.* Clemson, SC: Joyce's Print Shop.

Goff, M.L. (2001). A fly for the prosecution. Cambridge, MA: Harvard University Press.

Haskell, N.H., Schoenly, K.G., & Hall, R.D. (2002, February 12). Testing reliability of animal models in research and training programs in forensic entomology, Part II. NIJ Office of Science and Technology, Investigative and Forensic Sciences Division, 2002 Grantees Meeting, Atlanta, GA.

Resh, V., & Carde, R. (2003). *Encyclopedia of insects.* Amsterdam, Elsevier.

Smith, K.G.V. (1986). *A manual of forensic entomology.* Ithaca, NY: Comstock Printing Associates, Cornell University Press.

Serology and Bloodstain Pattern Analysis

KEY TERMS

ABO blood group

Acid phosphatase (AP)

Agglutinate

Angle of impact

Anti-A and anti-B

Antibodies

Antigen

Arterial spurts/gushes

Back spatter

Benzidine

Blood

Blood group

Bloodstain pattern analysis (BPA)

Brentamine Fast Blue B

Cast-off stains

Christmas tree stain

Confirmatory test

Direction angle

Directionality

Electrophoresis methods

Enzyme linked immunosorbent assay (ELISA)

Erythrocytes

Fluorescein

Fluorescence

Fly spots

Forward spatter

Genome

Hemoglobin

Human antiserum

Immune response

Leucomalachite green

Leukocytes

Luminescence

Luminol

Lymphocytes

Macrophages

Neutrophils

Ouchterlony test

Parent stain

Passive bloodstains

Phenolphthalein

Plasma

Platelets

Point of origin

Precipitin test

Presumptive test

Projected or impact bloodstains

Prostate specific antigen (p30)

Proteome

Proteomics

Saliva

Satellite droplets

Semen

Serology

Skeletonized stains

Spatter

Spermatozoa

Takayama test/ hemochromogen test

Tetramethylbenzidine (TMB)

Time since intercourse (TSI)

Transfer bloodstains

Urine

Voids

Wipe stain

INTRODUCTION

As DNA roared into the forensic laboratory in the late '80s and into the public consciousness in the early '90s, the study of bodily fluids left at crime scenes, serology, nearly became a casualty of scientific advancement. While most forensic laboratories still perform serology examinations, some have abandoned them and send potential biological stain samples directly to DNA analysis. But serology still has an important place in the modern forensic science laboratory for several reasons. First, finding someone's DNA on an item of evidence doesn't necessarily tell you the *source* of that DNA: It could be skin flakes, saliva, or semen, each of which may have different implications in the context of the case, alleged sexual contact versus mere presence, for example. Second, as a preliminary test, serology is fast, efficient, and inexpensive, thereby saving much time and effort by identifying biological fluid stains and avoiding needless DNA analysis of non-biological materials. Remember, sorting evidence by relevance is the key to interpretation, and serology does a good job of that. And, finally, many pre-DNA serology cases are being reexamined in the light of current DNA methods. The reinterpretation of historical cases through the lens of modern methods—without an appreciation or understanding of the "state of the art" at the time of the first analysis—is fraught with potential pitfalls for scientists, attorneys, and law enforcement officers alike. The work of today's forensic scientists will be judged by methods that may not even currently exist, and serology offers a good example of the dangers of incautious reanalysis. How is this so? Several hundred of these reanalyses have resulted in wrongly imprisoned people being released.

The review and reexamination of past forensic biology cases pose a danger of historical misinterpretation in regards to the specificity of serology versus DNA analysis. If a pre-DNA serology case included an individual (both the known and the questioned samples are A+, for example), it is certainly possible for DNA analysis to later exclude that individual. Does this mean the serologist was wrong, incompetent, or, worse, malicious in his or her examination? Certainly not. DNA is a far more specific comparison method than serology for many reasons. While DNA analysis is based on the groups of base units that make up our genetic code, serology identifies proteins associated with specific body fluids. Proteins are a more complicated form of biological material and are coded for by DNA. But, while proteins are specific to a body fluid type, they are generic biologically; that is, the proteins that are used to detect semen in one body are the same proteins that exist in any other semen stain. This doesn't make serology a poor test; traditional DNA methods, for example, are largely blind to the type of body fluid being tested and are not faulted for this weakness. The serology tests used today work well for their intended purpose: the quick and simple identification of body fluids. To look back and judge serology a bad test in the light of DNA analysis is to ignore the science upon which both methods are based.

Serology is a major component of crime scene processing and analysis. It also plays a large role in the processing of items of evidence in the laboratory, presumptively identifying blood, semen, saliva, and urine prior to further analysis. It is a conservative method, in that, while the stain may be identified, only a portion of that stain is further analyzed and the identified remainder is reserved for future testing. Given the intense interest in the analysis of proteins (the discipline is called **proteomics**; see "In More Detail: Proteomics—The New Serology?"), serology may be due for some exciting scientific and technological advances.

COLLECTION OF BODY FLUIDS

Serological analysis, like other forensic analyses, has two types of tests. The first is a **presumptive test**, which is highly sensitive to but not specific for a particular substance. For example, a color test may indicate that a stain may be blood but not what kind of blood (human or non-human). It may even give a false positive result; that is, the substance may not be blood but something that reacts with the chemicals used in the test. The idea of presumptive tests is that some false positive results are acceptable as long as no false negative results are obtained. False negative results are possible, of course. The presumptive positive results can be more specifically tested with the second type of test, a **confirmatory test**, which tests positive for the substance in question and only that substance. Therefore, it is known in advance that some presumptive positive results will not be confirmed. Confirmatory tests lack sensitivity, in that a relatively large amount of the substance must be available for the test to be positive, but a few false negatives are acceptable as long as no false positives are obtained. Generally, confirmatory tests are performed in a laboratory rather than in the field

IN MORE DETAIL:

Proteomics—The new serology?

The term **proteome** was first coined to describe the set of proteins coded in the genetic makeup of an individual, the **genome**. The study of the proteome, called **proteomics**, covers all the proteins in any given cell, their various forms and modifications, interactions, structure, the higher-order complexes they form, and pretty much everything that happens "post-genome." As you can imagine, this is an enormous and complicated area of study in biology and chemistry. And the goal of proteomics is no less enormous: a full description of cellular function.

The study of proteins would be impossible without the success of the study of the genome, notably the Human Genome Project (http://www.ornl.gov/sci/techresources/Human_Genome/home.shtml), which provided the "blueprint" of possible gene products (i.e., proteins). But despite having a successful template, proteomics faces several daunting challenges. While scientists studied the genome with DNA sequencing, polymerase chain reaction (PCR), and automated methods, proteomics has to work with samples that are limited and variable, degradation of proteins over time, the huge range of protein abundance (more than 10^6-fold), not to mention the modifications and changes proteins go through after they are constructed by a cell. Realizing that most of the results of proteomic research are intended for medical applications, disease and drug effects must be considered as well.

But the potential payoff is astounding. Disease-specific drugs that target only those cells or cell proteins responsible for devastating or deadly diseases, highly accurate diagnoses, therapeutic treatments that are not debilitating for the patient (as are some current treatments, like radiation or chemotherapy), and even treatment of genetic diseases by targeting the proteins involved are all realizable goals of proteomics.

What could proteomics do for forensic science? It could improve the accuracy and broaden the scope of serology, for example. Imagine a device like a personal digital assistant (PDA) with a detection and viewing screen. A biological stain is swabbed at a crime scene, and the swab is smeared across the detection screen. In a few seconds the results of a highly specific protein analysis appear— the person in question is male, late 40s, probably Hispanic, diabetic, and has a blood protein complex with a frequency of 1 in 1,340,000 in the U.S. population, for example. The next generation of forensic biologists needs to watch what happens in the rapidly advancing discipline of proteomics.

For more information about proteomics, read Sali, A., Glaeser, R., Earnest, T., & Baumeister, W. (2003). From words to literature in structural proteomics. *Nature*, *422*, 216–225.

Source: Tyers, M., & Mann, M. (2003). From genomics to proteomics. *Nature*, *422*, 193–197.

because they require more controlled conditions and additional equipment than presumptive tests.

The presumptive/confirmatory test structure allows crime scene and laboratory personnel to sort potential evidence into processing categories, such as "Test for DNA" and "No DNA Testing." Because of the demands on reconstructive sciences to sort out what is relevant from what is irrelevant, a testing structure like this greatly aids the laboratory to more efficiently process and analyze evidence.

THE MAJOR BODY FLUIDS

BLOOD

By definition, blood is a tissue, composed of several types of cells in a matrix called **plasma**. Plasma consists of about 90 percent water and 10 percent of a long list of other substances (7 percent protein, 3 percent urea, amino acids, carbohydrates, organic acids, fats, steroid hormones, and other inorganic ions). Within the plasma are three types of cells: erythrocytes (red blood cells), leukocytes (white blood cells), and platelets.

Red blood cells, **erythrocytes**, are legion in the blood: roughly 5 million per *milliliter* of blood! The purpose of these 6–8 µm diameter cells is to transport oxygen and carbon dioxide throughout the body via the circulatory system; this is accomplished by hemoglobin. **Hemoglobin**, you may recall from introductory biology, the respiratory pigment of many animals, is a conjugated protein consisting of four polypeptides, each of which contains a heme group. The heme groups contain iron and have the ability to bind with oxygen; this association is reversible, allowing for respiration. Erythrocytes are produced in the bone marrow and have about a four-month life span. They discard their nuclei as they mature and, therefore, contain no DNA.

White blood cells, or **leukocytes**, are active in the immune system but are not as numerous as erythrocytes, about 10–15,000 per milliliter of blood. The two types of leukocytes differ in their specific functions but work in a coordinated fashion to provide the body's defense against disease. The first type, **macrophages**, is the watchdog of the bloodstream, constantly on patrol for signs of incursion. Once bacteria or damaged platelets are identified, macrophages swarm the area and consume the offending materials until they literally eat themselves to death. The dead macrophages are exuded from the body as pus.

The other type, the smaller **neutrophils**, is the second line of defense and offers up a more complicated response to invaders: the **immune response**. Neutrophils work in conjunction with **lymphocytes**, which are produced in the bone marrow and the thymus gland, to engender the immune response. Lymphocytes produce **antibodies**,

241

which are protein molecules that can bind to foreign molecules. Any foreign molecule that induces antibody formation is called an **antigen**. Finally, **platelets** are only fragments of cells and contain no nuclei; they number around 15–300,000 per milliliter of blood. Platelets are involved in the clotting process.

Genetic Markers in Blood

A **blood group** is a class of antigens produced by allelic genes at a single locus and inherited independently of other genes. Genetically controlled and invariant throughout a person's life, blood groups are a robust biological marker. About 20 human blood groups are known to exist. The practical meaning of this is that a blood group is a permanent genetic trait—one that is controlled by genes and unchanging throughout a person's life. This makes blood potentially excellent evidence for classification and possible inclusion or exclusion.

Several systems are used to characterize and classify blood. The first and best known is the **ABO blood group**, discovered in 1900 by Karl Landsteiner. The letters A, B, and O refer to the antigens on the surface of the red blood cells; corresponding antibodies, **anti-A** and **anti-B**, are present in the plasma (**www.nobel.se**) (see Table 10.1).

A person with Type B blood will have anti-A antibodies in his plasma. If the plasma is mixed with Type A blood, the cells will **agglutinate**, or clump together, from the reaction of the Type A antigens and the anti-A antibodies. Very few forensic laboratories currently perform blood group testing. If a stain is tested and is presumptively positive for blood, it is sent on for DNA analysis. (See "History: Landsteiner's Breakthrough".)

Table 10.1. *The blood groups in the ABO System and their antigens, antibodies, and frequency in the population.*

BLOOD GROUP	ANTIGEN	ANTIBODY	POPULATION FREQUENCY
A	A	Anti-A	40%
B	B	Anti-B	10%
AB	A & B	None	5%
O	H	Anti-A & Anti-B	45%

HISTORY:
Landsteiner's breakthrough

Karl Landsteiner was awarded the Nobel Prize for Medicine or Physiology in 1930 for his discovery of human blood groups. A portion of the presentation speech by Professor G. Hedrén, Chairman of the Nobel Committee for Physiology or Medicine of the Royal Caroline Institute, made on December 10, 1930, follows (the full text is available at www.nobel.se):

Thirty years ago, in 1900, in the course of his serological studies Landsteiner observed that when, under normal physiological conditions, blood serum of a human was added to normal blood of another human the red corpuscles in some cases coalesced into larger or smaller clusters. This observation of Landsteiner was the starting-point of his discovery of the human blood groups. In the following year, i.e. 1901, Landsteiner published his discovery that in man, blood types could be classified into three groups according to their different agglutinating properties. These agglutinating properties were identified more closely by two specific blood-cell structures, which can occur either singly or simultaneously in the same individual. A year later von Decastello and Sturli showed that there was yet another blood group. The number of blood groups in man is therefore four.

Landsteiner's discovery of the blood groups was immediately confirmed but it was a long time before anyone began to realize the great importance of the discovery. The first incentive to pay greater attention to this discovery was provided by von Dungern and Hirszfeld when in 1910 they published their investigations into the hereditary transmission of blood groups. Thereafter the blood groups became the subject of exhaustive studies, on a scale increasing year by year, in more or less all civilized countries. In order to avoid, in the publication of research on this subject, detailed descriptions which would otherwise be necessary—of the four blood groups and their appropriate cell structures, certain short designations for the blood groups and corresponding specific cell structures have been introduced. Thus, one of the two specific cell structures, characterizing the agglutinating properties of human blood is designated by the letter A and another by B, and accordingly we speak of "blood group A" and "blood group B". These two cell structures can also occur simultaneously in the same individual, and this structure as well as the corresponding blood group is described as AB. The fourth blood-cell structure and the corresponding blood group is known as O, which is intended to indicate that people belonging to this group lack the specific blood characteristics typical of each of the other blood groups. Landsteiner had shown that under normal physiological conditions the blood serum will not agglutinate the erythrocytes of the same individual or those of other individuals with the same structure. Thus, the blood serum of people whose erythrocytes have group structure A will not agglutinate erythrocytes of this structure but it will agglutinate those of group structure B, and where the erythrocytes have group structure B the corresponding serum does not agglutinate these erythrocytes but it does agglutinate those with group structure A. Blood serum of persons whose erythrocytes have structures A as well as B, i.e. who have structure AB, does not agglutinate erythrocytes having structures A, B, or AB. Blood serum of persons belonging to blood group O agglutinates erythrocytes of persons belonging to any of the groups A, B, or AB, but erythrocytes of persons belonging to blood group O are not agglutinated by normal human blood serum. These facts constitute the actual basic principles of Landsteiner's discovery of the blood groups of mankind. However, the discovery of the blood groups has also brought with it important scientific advances in the purely practical field—first and foremost in connection with blood-transfusion therapy, identification of blood, and establishing of paternity.

The transfer of blood from one person to another for therapeutic purposes began to be practised on a considerable scale during the 17th century. It was found, however, that such blood transfusion involved serious risks and not infrequently resulted in the death of the patient. Therapeutic application of the blood transfusion had therefore been almost entirely given up by the time of Landsteiner's discovery. As a result of the discovery of the blood groups it was now possible, at least in the majority of cases, to explain the cause of the dangers linked with this therapeutic measure as previous

experience had shown, and at the same time to avoid them. A person from whom blood is taken must in fact belong to the same blood group as the patient. Thanks to Landsteiner's discovery of the blood groups, blood transfusions have come back into use and have saved a great many lives.

Already at the time of publishing his discovery of the blood groups in 1901, Landsteiner pointed out that the blood-group reaction could be used for investigating the origin of a blood sample, for instance of a blood stain. However, it is not possible to prove by determining the blood group that a blood sample comes from a particular individual, but it is possible to prove that it is not from a particular individual. If, for instance, the blood of a blood stain is from an individual belonging to blood group A, then it cannot be from an individual who is found to belong to group B, but a blood-group determination will not tell us from which person of blood group A the blood came.

Retrieved from www.nobel.se.

Presumptive Tests for Blood

Presumptive tests for blood react with the **hemoglobin** present in blood. If hemoglobin is present, one of two general results occurs depending on the test. Either a colorless reactive substance changes to a colored form (from clear to pink, for example) or light of a specific wavelength is emitted (fluorescence or chemiluminescence) in the presence of hemoglobin.

In the first type of test, the testing chemical is added to the suspected stain and then an oxidant is added, usually 3 percent hydrogen peroxide. The hydrogen peroxide reacts (oxidizes) with the hemoglobin and changes the color of the testing chemical; hemoglobin acts as the catalyst, speeding up the reaction. The most commonly used catalytic color tests are **phenolphthalein**, **benzidine**, **leucomalachite green**, and **tetramethylbenzidine (TMB)**. For sensitivity and safety reasons among these tests, the phenolphthalein test is used more often than the other tests. The sensitivity of the phenolphthalein test can detect blood diluted down to 10^{-7} (1 part in 10 million), and even decades-old bloodstains can yield positive results. Phenolphthalein is cross-reactive with other substances, such as some vegetables. The test is performed by moistening a clean cotton swab with distilled water and rubbing it on the suspected stain. A drop of phenolphthalein solution is added to the swab's tip; it should remain colorless. A drop of hydrogen peroxide is then added; if the tip turns pink, the test is presumptively positive for blood. If the swab tip remains colorless, then the result is negative for blood. The color change must be within several seconds because the tip may turn pink through normal oxidation after several minutes' exposure to air.

At times, it is not only the presence of blood that is of interest but also the pattern or distribution of the blood. The area to be tested may be large or intricate, such as floors, walls, and automobile interiors. In these instances, the testing chemical is sprayed onto the surface(s) and then observed for any emitted light (glowing). Because the light output is faint, the treated surfaces must be viewed in the dark or with an alternate light source (ALS). Specialized photographic techniques must be used to capture the images, because the effect of emitted light is temporary. These types of tests may affect subsequent tests and, therefore, caution must be employed in their use (see Table 10.2). A noted expert in serology, Robert Spaulding (2005), has suggested that

Table 10.2. *The effects of various presumptive serology tests on subsequent tests.*

REFERENCE	CHEMICAL TEST	EFFECTS ON SUBSEQUENT TESTS
Laux (1991)	Luminol	Does **not** significantly affect presumptive, confirmatory, species origin, and ABO tests. Does interfere with some enzyme and protein genetic marker systems, such as acid phosphatase, esterase D, peptidase A, and adenylate kinase.
Gross, Harris, & Kaldun (1999)	Luminol	Does **not** affect polymerase chain reaction (PCR) of DNA.
Hochmeister, Budowle, & Baechtel (1991)	Ethanolic benzidine Phenolphthalein Luminol	Does **not** affect recovery of DNA for restriction fragment length polymorphism (RFLP) but does lower the yield somewhat.
Budowle et al. (2000)	Fluorescein	Does **not** affect short tandem repeat (STR) analysis of DNA.

if the stain can be seen and collected, then this type of test should not be used.

Two chemicals, luminol and fluorescein, are predominately used for large-scale serology testing. **Luminol** (3-aminophthalhydrazide) reacts in the presence of hemoglobin, much like phenolphthalein, when an oxidizer is applied. The reaction, however, results in a blue-white to yellow-green **luminescence** (light emitted as a by-product of a chemical reaction) if blood is present. Luminol is very sensitive to hemoglobin and will detect blood in dilutions of 1 in 5,000,000. Luminol, a suitable oxidant, and water are mixed and sprayed over the area of interest. The pattern will be visible for up to 30 seconds before additional treatment is required; over-spraying, however, will result in "bleeding" of the patterns and a loss of detail.

Fluorescein is another chemical that is used to check for the presence of blood and is prepared much in the same way as luminol except that the commercial preparation contains a thickener. This makes fluorescein stay on the surface better than luminol, making it easier to use on walls and other vertical surfaces. Unlike luminol, fluorescein produces **fluorescence** (light emitted as energy loss at a longer wavelength than it is illuminated with) and must be illuminated at 450 nanometers via an alternate light source (ALS) to be seen.

Both luminol and fluorescein are characterized as irritants but are not known to be carcinogens. Nevertheless, safety precautions and protective equipment should be employed during their use.

Confirmatory Tests for Blood

Confirmatory tests for blood utilize the formation of crystals through the application of heat and testing chemicals. For example, the **Takayama test** (also known as the **hemochromogen test**) is performed by taking a small sample of the presumptive stain and placing it under a cover slip. The sample is heated briefly and, while being observed through a microscope, pyridine under alkaline conditions in the presence of a reducing sugar is added with a pipette. If blood is present, salmon-colored crystals form. The Takayama test is very sensitive, and even very old bloodstains may give a positive reaction. Heating the sample properly is key: Even when blood is present, improper heating of the sample can result in a false negative. Historically, the Teichmans test was used as a confirmatory test for blood; it is essentially the same procedure with different chemicals. A modification of the Takayama test appeared in the *Journal of Forensic Sciences* in 1993; the author, Hatch, used a reagent (dithiothreitol) to increase the rate of crystal formation (Hatch, 1993).

Practically speaking, most forensic laboratories today do not conduct confirmatory tests for blood—the sample, if presumptively positive, will go straight to the DNA Unit. If no DNA is discovered in the sample, it probably wasn't human blood.

Species Origin

Tests that determine the species from which a blood sample originated fall into two general categories: diffusion reactions and electrophoretic methods. The most common diffusion reaction test, the **Ouchterlony test**, is based on an antibody-antigen reaction between human blood and human antiserum. **Human antiserum** is typically produced by injecting rabbits with human blood. The rabbit's immune system, reacting to the foreign blood, produces antibodies to neutralize it. When the rabbit's human-sensitized blood is drawn and the serum isolated, it can be used to detect human blood because it now has antibodies that will react specifically with human blood. Other antisera, for dogs, cats, and horses, for example, can be produced in a similar way.

The Ouchterlony test is performed as follows. An agarose gel is poured into a small petri dish. A circular pattern of six wells is cut out of the gel with an additional well in the center. The antihuman serum is placed in the center well with a known human control placed in every other well. The sera and samples are allowed to diffuse. The human controls test positive (of course) with a diffusion line; if the unknown tests positive, the line extends between the adjoining known samples (see Figure 10.1). The Ouchterlony test is being supplanted by the Hematrace™ card, in which a positive test is indicated by a color change. The Hematrace card™, however, cross-reacts with ferret blood, and some question its use as a confirmatory test.

Electrophoresis methods are based on the diffusion of antibodies and antigens on an electrically charged gel-coated plate. The bloodstain extract and the human antiserum are placed in separate wells on opposite sides of the plate. When the plate is charged, a zone of precipitation forms at the juncture of the antibodies and antigens; Figure 10.2 shows an electrophoresis device.

Figure 10.1. *The most common diffusion reaction test, a precipitin test, is based on an antibody-antigen reaction between human blood and human antiserum. Human antiserum is produced by exposing animal blood to human blood, typically by injecting rabbits with it. The rabbit's immune system, reacting to the foreign blood, produces antibodies to neutralize it. When the rabbit's human-sensitized blood is drawn and the serum isolated, it can be used to detect human blood because it now has antibodies that will react specifically with human blood.*

Figure 10.2.
Electrophoresis methods are founded on the diffusion of antibodies and antigens on an electrically charged gel-coated plate. The bloodstain extract and the human antiserum are placed in separate wells on opposite sides of the plate. When the plate is charged, a zone of precipitation forms at the juncture of the antibodies and antigens.

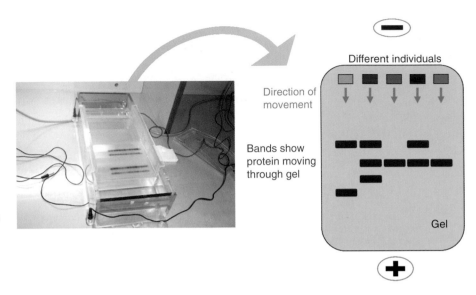

SEMEN

Semen is a complex gelatinous mixture of cells, amino acids, sugars, salts, ions, and other materials produced by post-pubescent males and is ejaculated following sexual stimulation. The volume of ejaculate varies from 2 to 6 milliliters and typically contains between 100 and 150 million **spermatozoa**, or sperm cells, per milliliter. Sperm cells, shown in Figure 10.3, are a specialized structure approximately 55 μm in length with a head containing DNA and a tail that wiggles, or flagellates, to produce movement. The presence of semen (or, by extension, intact sperm) is considered presumptive evidence of sexual contact.

Presumptive Tests for Semen

Semen contains **acid phosphatase (AP)**, a common enzyme in nature that occurs at a very high level in semen. This allows for a test, with a high acid phosphatase threshold, to presumptively identify semen. The most common test is **Brentamine Fast Blue B** applied to the sample on an alphanaphthyl phosphate substrate (see "In More Detail: Presumptive Test for Acid Phosphatase"). Brentamine Fast Blue B is a known carcinogen and must be handled accordingly. A piece of filter paper or cotton swab is moistened with sterile water and applied to the questioned stain. The reagent is added, and if an intense purple color is seen, the test is positive; if no color reaction occurs within two minutes, the test is negative. If a sample is known or suspected to be "old," a negative result should be cautiously interpreted because AP

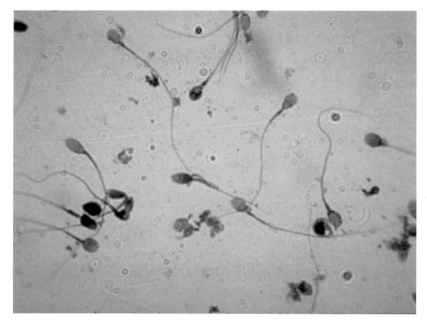

Figure 10.3. *Between 100 and 150 million spermatozoa, or sperm cells, are found per milliliter of ejaculate. Sperm cells are a specialized structure approximately 55μm in length with a head containing DNA and a tail that wiggles, or flagellates, to produce movement. The presence of semen (or, by extension, sperm) is considered presumptive evidence of sexual contact.*

IN MORE DETAIL:

Presumptive test for acid phosphatase (AP)

Reagent Preparation

Buffer: 8.21 g anhydrous sodium acetate, pH5.5

Reagent: Step 1. Dissolve 5 mg alpha-naphthyl phosphate in 5 mL buffer
Step 2. Dissolve 5 mg Fast Blue B in 5 mL buffer

Prepare a single reagent by adding equal volumes of reagents from Steps 1 and 2.

Procedure

Perform the test on a portion of the stain, an extract, or a "wipe" made from the stained area. Always test an unstained substrate control and a positive control at the same time as the unknown samples.

Two-Step Method

Apply a drop of the alpha-naphthyl solution to the unknown sample and wait 60 seconds.

Apply a drop of the Fast Blue B solution, to the unknown sample.

An immediate purple color is a positive reaction.

One-Step Method

Apply a drop of the combined reagents to the unknown sample. A purple color that develops within 60 seconds is a positive reaction.

Source: Shaler, R.C. (2002). Modern forensic biology. In R. Saferstein (Ed.), *Forensic science handbook, volume I* (p. 537). Englewood Cliffs, NJ: Prentice-Hall.

becomes less active over time. Positive results are presumptive, because many biological fluids, including vaginal secretions, contain some amount of AP. Commercial test kits are available for semen (AP).

Semen can be visualized by alternative light sources (ALS) at 450 nm and viewed with amber goggles. The method is useful because it works on light or dark surfaces, covers large areas, and is quick; the disadvantage is that some other fluids will also fluoresce, such as saliva and urine. Semen exhibits a blue-white fluorescence, and this test can help narrow down which stains to test for AP. Just because no fluorescence occurs, however, it cannot be assumed that no semen is present.

Confirmatory Tests for Semen

The presence of intact spermatozoa in a biological stain has historically been the conclusive test for semen: They are abundant in semen and should be plentiful. A lack of sperm doesn't necessarily mean the stain isn't semen, only that sperm are not present.

The traditional method for sperm identification is to use the **Christmas tree stain**, which turns the tip of the sperm's head pink, the bottom of the head dark red, the middle portion blue, and the tail yellowish-green; skin cells stain green to blue-green and are easily

---

IN MORE DETAIL:
Using the Christmas tree stain to visualize sperm

Reagents

Nuclear Fast Red
2.5 g aluminum sulfate
50 mg nuclear fast red
  Dissolve the aluminum sulfate in 100 mL warm deionized water and then add the nuclear fast red. Stir.
  After cooling, filter the solution.
Picro Indigo Carmine (PIC)
1.30 g picric acid
0.33 g indigo carmine
  Dissolve the picric acid in 100 mL of warm deionized water, yielding a saturated solution. Add the indigo carmine and stir overnight.

Procedure

Apply a portion of the extract to a microscope slide and heat fix.

Cover the dried extract with a few drops of nuclear fast red and allow to sit for at least 10 minutes. Wash the excess reagent away with deionized water.

Add a drop of PIC to the still wet slide and allow to sit for 30 seconds or less. Wash away the excess PIC with absolute ethanol.

Observe the slide microscopically.

Source: Shaler, R.C. (2002). Modern forensic biology. In R. Saferstein (Ed.), *Forensic science handbook, volume I* (p. 541). Englewood Cliffs, NJ: Prentice-Hall.

distinguished. See "In More Detail: Using the Christmas Tree Stain to Visualize Sperm" for a description of the method. An extract of the stain is dissolved in water; then a portion is applied to a microscope slide, heat-fixed, and then colored with the Christmas tree stain. The slide is then scanned at magnifications of 400 to 1000. Even if detached sperm heads are the only structure on the slide, and the Christmas tree stain assists in distinguishing them, this is still confirmative for the presence of semen. Phase contrast or dark field microscopy may also be used. The hematoxyln-eosin stain is also used but operates on the same premise (the stain colors are purple for heads and pink for the rest).

In 1978, George Sensabaugh published a paper outlining the forensic use of a **prostate specific antigen** that he named **p30** (*p* for *prostate, 30* for its molecular weight of 30,000) (Sensabaugh, 1978). A method called **enzyme linked immunosorbent assay** (**ELISA**) is typically used to detect p30 at levels as low as 0.005 ng/mL (the threshold for rectal samples is 2 ng because of reactions with other substances found there). ELISA is based on the antigen-antibody reaction: The reagent is added to the filter paper or swab, and if an intense purple color is seen, a reaction has occurred and p30 is present. The labeled p30 is attached to the antibody. In ELISA, the label is an enzyme that catalyzes a reaction where the substrate changes color—the deeper or more intense the color, the more label, and therefore the more p30, is present.

Depending on the crime, it may be useful to determine the **time since intercourse** (**TSI**) to assist in the sequence of events. Typically, this means the detection of spermatozoa, but because of natural variations the timing is rarely exact. Motile (moving) sperm can survive in the vagina for about 3 hours, ranging from 1 to 8 hours; they survive longer in the cervix, up to several days in some cases. By the time a case gets to the forensic serologist, however, the sperm are rarely motile. Time lags inevitably occur between the occurrence of the crime, collection, submission, and analysis. Forensic serologists therefore look for *intact* sperm, and these can persist in the vagina up to 26 hours and the heads alone can last up to 3 days; again, the persistence is greater in the cervix. Intact sperm can be found from 6 to 65 hours *post coitus* in the rectum (or until the next bowel movement) but rarely more than 6 hours in the mouth. Levels of p30 have been used to estimate TSI, and most p30 is eliminated within 24–27 hours after intercourse.

SALIVA

Saliva can be evidence in a number of crimes. Bite marks, licked adhesives (like envelopes and stamps), eating and drink surfaces, or even expectoration (spitting) can yield important DNA evidence. Saliva stains may be difficult to see, and detection can be tricky.

The problem is that although the enzyme amylase occurs in saliva, and tests exist for amylase, amylase also occurs in many other body fluids. It occurs in higher amounts in saliva, so the intensity of a test result could be considered presumptively positive for saliva. An old test, the radial diffusion test, used to be employed to confirm the presence of saliva, but now the sample is considered presumptively positive for saliva and simply sent on for DNA analysis. Saliva has large amounts of skin (epithelial) cells from the inner cheek walls and therefore is easy to type for DNA analysis.

URINE

Urine, the excreted fluid and waste products filtered by the kidneys, can be presumptively tested for through the presence of urea (with urease, an enzyme) or creatine (with picric acid). Also, when heated, a urine stain gives off a characteristic odor that everyone is familiar with. Urine has few skin (epithelial) cells in it, and a sample must be quite concentrated for DNA typing to be successful.

BLOODSTAIN PATTERN ANALYSIS

One of the most explicit methods of forensic science that exemplifies its reconstructive nature is the analysis and interpretation of bloodstain patterns. Bevel and Gardner, in the second edition of their *Bloodstain Pattern Analysis* (2002), define **bloodstain pattern analysis** (**BPA**) as the analysis and interpretation of the dispersion, shape characteristics, volume, pattern, number, and relationship of bloodstains at a crime scene to reconstruct a process of events. A combination of geometry, physiology, physics, and logic, bloodstain pattern analysis requires extensive training coupled with a solid scientific education to be properly applied. The International Association for Identification (IAI, www.theiai.org), for example, requires the following for its certification in bloodstain pattern analysis:

- An applicant for certification must be of good moral character, high integrity, good repute, and must possess a high ethical and professional standing.

- A minimum of forty (40) hours of education in an approved workshop providing theory, study and practice which includes oral and/or visual presentation of physical activity of blood droplets illustrating blood as fluid being acted upon by motion or force, past research, treatise or other reference materials for the student; laboratory exercises which document bloodstains and standards by previous research.

- A minimum of three (3) years of practice within the discipline of Bloodstain Pattern Identification, following the required forty (40) hour training course, must be documented.

- The applicant for certification must have a minimum of 240 hours of instruction in associated fields of study related to Bloodstain Pattern Identification in any of Photography (Evidence/Documentary) Crime Scene Investigation Technology, Evidence Recovery, Blood Detection Techniques/Presumptives, Medico Legal Death Investigation, Forensic Science and Technology.

- The course requirements must include a forty (40) hour Basic Bloodstain Evidence Course as previously outlined.

As can be seen from this brief overview of the process, a great deal of work must be accomplished to be considered certified (more details can be found at the IAI's website).

TERMINOLOGY IN BPA

Bloodstains can be grouped into three main classes: passive, transfer, and projected or impact stains. **Passive bloodstains** include clots, drops, flows, and pooling. **Transfer bloodstains** include wipes, swipes, pattern transfers, and general contact bloodstains. Finally, spatters, splashes, cast-off stains, and arterial spurts or gushes are examples of **projected or impact bloodstains**. Other patterns include fly spots, voids, and skeletonized stains.

A **wipe stain** is created when an object moves through a preexisting bloodstain. An example would be the stain resulting from a clean rag being moved through a blood pool on a floor. By contrast, a swipe stain is the transfer of blood onto a target by a moving

Figure 10.4. Forward spatter results from droplets being projected to the front of the object hitting the blood; back spatter is the opposite. Cast-off stains result from blood being flung from a bloodied object; note that cast-off stains can occur behind the item as well—for example, on the back of the assailant's shirt.

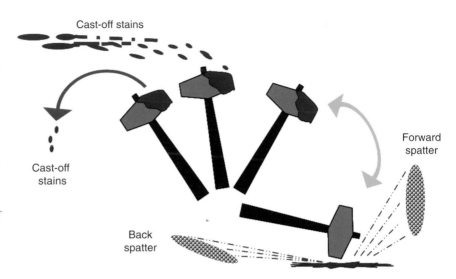

object that is itself bloodstained. Blood-soaked clothing being dragged over unstained vinyl flooring would result in a swipe, not a wipe. An easy way to remember this difference is that an object that is stained creates a swipe; an unstained object makes a wipe (no stain, no "s").

Spatter is a technical term in BPA that describes a stain that results from blood hitting a target. Two types of spatter are recognized. The first, **forward spatter**, results when blood droplets are projected away from the item creating the impact, such as a hammer. **Back spatter**, by contrast, is caused by droplets being projected toward the item; in general, back spatter will be lighter and the stains smaller than forward spatter (see Figure 10.4). Note that the word "splatter" has no technical meaning in BPA and should not be used.

Cast-off stains are the result of blood being flung or projected from a bloody object in motion or one that stops suddenly (see Figure 10.4). Cast-off stains are linear and reflect the position of the person moving the bloody object. If a criminal bludgeons a victim with a baseball bat, as his arm comes back to swing again, any blood on the end of the bat will be projected by centrifugal force in an arc. Remember that cast-off stains can arc directly behind the object and land, for example, on the back of the assailant's shirt.

The blood flowing through your arteries is under high pressure. When an artery is breached while the heart is pumping, blood will spurt or gush from the wound, as depicted in Figures 10.5A and B.

A

B

Figure 10.5. The blood flowing through your arteries is under high pressure. When an artery is breached while the heart is pumping, blood will spurt or gush from the wound Arterial spurts/gushes can vary due to the pumping action and variable pressure of the blood as it exits the wound producing a zig-zag, up-and-down pattern. The spurt pattern from the victim (A) can be seen in detail on the round ottoman (B). Courtesy of John Black, South Carolina Law Enforcement Division.

Figure 10.6. Voids are an indicator that some secondary object came between a blood spatter and the final target; this leaves an outline or "shadow" on the final target. Voids are important clues about items that may have been moved or discarded after an attack but during the criminal process.

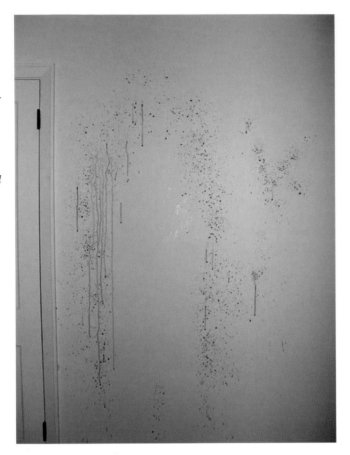

Arterial spurts/gushes can vary due to the pumping action and variable pressure of the blood as it exits the wound, producing a zig-zag, up-and-down pattern.

Fly spots are stains resulting from fly activity—and may mimic other relevant BPA patterns. Flies may regurgitate and defecate when consuming blood at a crime scene, and these spots, as well as small amounts of tracked blood, can be confusing or misleading.

Voids are an indicator that some secondary object came between a blood spatter and the final target; this leaves an outline or "shadow" on the final target, as illustrated in Figure 10.6. Voids are important clues about items that may have been moved or discarded after an attack but were present during the criminal process.

As a stain dries, the edges and borders dry first due to surface effects. If the bloodstain is wiped, these dried areas, called **skeletonized**

stains, remain behind. Skeletonized stains retain the size and shape of the original stain and indicate the passage of time.

Figure 10.7 shows some of the basic measurements for BPA. The **angle of impact** is the acute angle created by the intercept of the target with the droplet's vector. This is different from the **direction angle**, which is the angle between the long axis of the stain and a standard reference point, usually 0° vertical. The **directionality** demonstrates the vector of a droplet when it hit the target; the tail points in the direction of travel. **Satellite droplets** are small amounts of blood that detach from the **parent stain** and "splash" onto a surface.

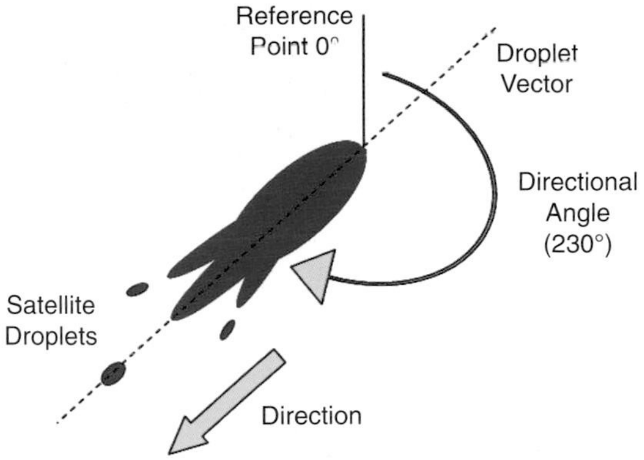

Figure 10.7. The "tail" *of the stain indicates direction, and the angle from a 0° reference point shows the directional angle. The satellite droplets are small "splashes" from the parent stain.*

DETERMINING POINT OF ORIGIN

Although it may seem a macabre game of connect-the-dots, determining the point of origin of one or more bloodstains is central to the reconstruction of a blood-related event. Whenever the direction of a bloodstain can be determined, it can be expected to have originated at a point somewhere along that line. Doing this for a number of bloodstains can demonstrate a convergence of lines (paths) indicating a possible **point of origin** for the stains, as shown in Figure 10.8. The more paths converge at this point, the more confidence the analyst has in that point being the origin of the pattern (see Figure 10.8A). The analyst must be aware that multiple paths may cross, generating a confusing or conflicting pattern (see Figure 10.8B).

A bloodstain scene may be particularly difficult to interpret. Visual aids, from simple strings and pins to advanced forensic software, are available to assist the analyst. Stringing a scene, depicted in Figure 10.9, is the easiest and cheapest method to interpret multiple bloodstain patterns. The path of a bloodstain is established, and a string is run from that point backward using pushpins or masking tape to fasten the ends; this process is repeated for numerous stains until the pattern's origin becomes clear. Rulers, protractors, and even laser pointers can be used in this method. Software, such as *Backtrack/ Images®* and *Backtrack/Win®*, available through Forensic Computing of

Figure 10.8. By finding the path for each bloodstain in a pattern, an analyst can interpret a point of origin; the more paths that converge on this point, the more likely it is the actual point of origin. The analyst must not be confused by multiple adjacent patterns, however; just because points converge doesn't mean that is the point of origin.

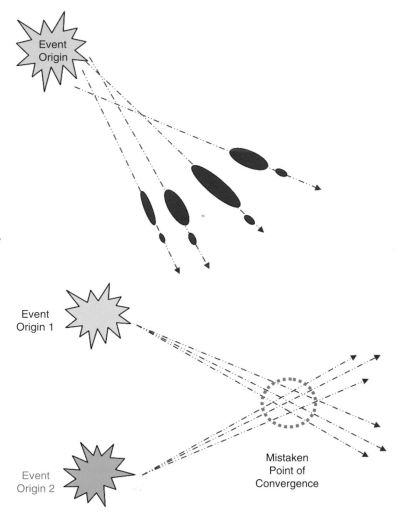

Ottawa (**www.physics.carleton.ca/~carter/index.html**), can assist through the automated calculation of vectors and angles; an example is shown in Figures 10.10A and B.

DOCUMENTING BLOODSTAINS AT THE SCENE

Presumptive serological tests can be employed to discover if the stain in question is truly blood; if it's tomato sauce, then it's probably not worth the analyst's time. Additionally, many of the enhancement tech-

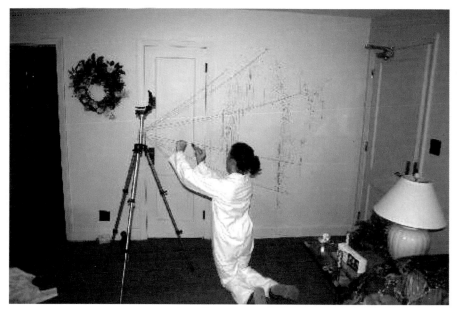

Figure 10.9. Stringing a scene is the easiest and cheapest method to interpret multiple bloodstain patterns. The path of a bloodstain is established, and a string is run from that point backward using pushpins or masking tape to fasten the ends; this process is repeated for numerous stains until the pattern's origin becomes clear. Rulers, protractors, and even laser pointers can be used in this method.

niques used for visualizing blood discussed earlier in this chapter can be used to visualize bloodstain patterns.

The documentation of bloodstains is painstaking work but crucial to a successful reconstruction. Tom Bevel and Ross Gardner (2002), two well-known bloodstain pattern analysts, recommend the following photographic guidelines:

- Document the entire scene as discovered, including "establishing" photographs.
- Photograph pattern transfers, pools, and other fragile patterns first.
- Document patterns with "establishing" photographs that show the pattern's relationship to landmarks or other items of evidence.
- Take macro and close-up photographs; *include a scale in every photograph.*
- When reconstructing point of origin, document individual stains used in the reconstruction.

Note-taking, sketching, and measurements must accompany photographic documentation to help organize and describe the stains at the scene.

Figure 10.10.
Computer software can assist in the reconstruction of bloodstain patterns by automatically calculating angles and vectors. Courtesy: Forensic Software of Ottawa.

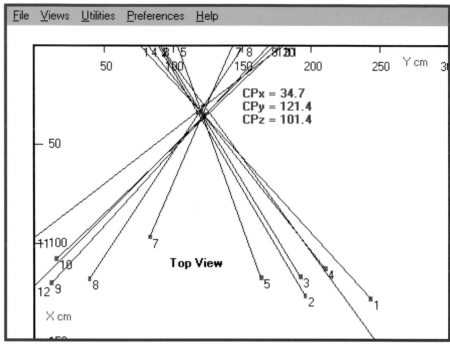

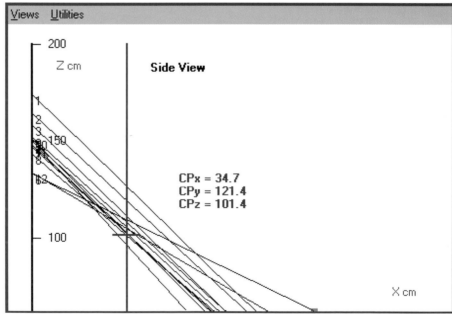

CHAPTER SUMMARY

Far from being obsolete, serology is still a significant part of the forensic biologist's toolkit. Testing for body fluids can save a great deal of time and effort later in the analytical stream of things (that is, DNA analysis). There can be no argument that serology has taken a backseat to DNA, but the argument should be made for retaining a battery of serology tests in the modern forensic laboratory. New methods being researched, such as proteomics, offer serology a renewed life and application as biologists discover more and different ways to identify the substances that make up our bodies.

Test Your Knowledge

1. What is serology?
2. What is the proteome?
3. What is the difference between a presumptive and confirmatory test?
4. Name the components of blood.
5. What are leukocytes and what do they do?
6. What is a blood group?
7. List three presumptive tests for blood.
8. What is luminol?
9. What does the precipitin test determine?
10. What is the most common presumptive test for semen?
11. What does "ELISA" stand for?
12. How long can motile sperm survive in the vagina?
13. Name the presumptive test for saliva.
14. What is the difference between a wipe and a swipe?
15. What is a cast-off stain?
16. How is point of origin determined?
17. Is there a presumptive test for urine?
18. What is angle of impact?
19. What is the difference between forward and backward spatter?
20. What is acid phosphatase?

Consider This . . .

1. Why would you perform presumptive serology tests? Why not send everything straight to the DNA Unit?

2. Why would two (or more) experts disagree about bloodstain pattern interpretations? Isn't it just geometry and physics?

BIBLIOGRAPHY

Bevel, T., & Gardner, R.M. (2002). *Bloodstain pattern analysis* (2nd ed.). Boca Raton, FL: CRC Press.

Budowle, B. et al. (2000). The presumptive reagent fluorescein for the detection of dilute bloodstains and subsequent STR typing of recovered DNA. *Journal of Forensic Science*, 45, 1090+.

Gaensslen, R.E. (1983). *Sourcebook in forensic serology, immunology, and biochemistry*. Washington, D.C.: U.S. Government Printing Office.

Grispino, R.R.J. (1990). The effect of Luminol on the serological analysis of dried human bloodstains. *Crime Lab Digest*, 17, 13+.

Gross, A.M., Harris, D.A., & Kaldun, G.L. (1999). The effect of Luminol on presumptive tests and DNA analysis using the polymerase chain reaction. *Journal of Forensic Sciences*, 44, 8837+.

Hatch, A.L. (1993). A modified reagent for the confirmation of blood. *Journal of Forensic Sciences*, 38, 1502+.

Hochmeister, M.N., Budowle, B., & Baechtel, F.S. (1991). Effects of presumptive test reagents on the ability to obtain restriction fragment length polymorphism (RFLP) patterns from human blood and semen stains. *Journal of Forensic Sciences*, 44, 597+.

International Association for Identification, **www.theiai.org**.

Laux, D.L. (1991). Effects of Luminol on subsequent analysis of bloodstains. *Journal of Forensic Sciences*, 36, 1512+.

Ponce, A.C., & Verdu Pascual, F.A. (1999). Critical revision of presumptive tests for bloodstains. *Forensic Science Communications*, 1(2). Retrieved from **www.fbi.gov**.

Sali, A., Elaeser, R., Earnest, T., & Baumeister, W. (2002). From words to literature in proteomics, *Nature*, 422, 216–225.

Sensabaug, G. (1978). Isolation and characterization of a semen-specific protein from human seminal plasma: A potential new maker for semen identification. *Journal of Forensic Sciences*, 23, 106+.

Shaler, R.C. (2002). Modern forensic biology. In R. Safertein (Ed.), *Forensic science handbook, volume I* (pp. 525–613). Englewood Cliffs, NJ. Prentice-Hall.

Spaulding, R.P. (2005). Identification and characterization of blood and bloodstains. In S.H. James & J.J. Nordby (Eds.), *Forensic science: An introduction to scientific and investigative techniques* (181–202). Boca Raton, FL: CRC Press.

Sutton, T.P. (1999). Presumptive blood testing. In S.H. James, (Ed.), *Scientific and legal applications of bloodstain pattern interpretation* (47–70). Boca Raton, FL: CRC Press.

Tyers, M. & Mann, M. (2003). From genomics to proteomics, *Nature* 422, 193–197.

DNA Analysis

KEY TERMS

Adenine (A)

Allele

Allelic ladders

Amelogenin

Autorad

Chromosome

CODIS (Combined DNA Index System)

Cytosine (C)

DNA

DNA type

Dominant

Double helix

Endonucleases

Gene

Genetic code

Genomic

Genotype

Guanine (G)

Hardy-Weinberg equilibrium

Heterozygous

Homozygous

HLA DQ alpha gene

HLA polymarker

Homozygous

Hypervariable

Ladders

Length polymorphism

Linkage equilibrium

Match

Microsatellites

Minisatellites

Mitochondrial

Monomers

Multilocus genes

Nucleotides

Nucleus

PCR (polymerase chain reaction)

Phenotype

Polyacrylamide

Polymaker

Polymer

Polymerase chain reaction (PCR)

Polymorphism

Primers

Probe hybridization

Product rule

Recombinant DNA

Recessive

RFLP (restriction fragment length polymorphism)

Restriction enzymes

Reverse dot blot

Sequence polymorphisms

STR (short tandem repeat)

Tandem repeat

Taq polymerase

Thermal cycler

Thermocycler

Thymine (T)

Variable number of tandem repeats (VNTR)

THE CASE

A woman goes to a night club with several of her friends. During the evening she has several mixed drinks. She meets a man at the club for the first time and they become friendly and move to their own table. At some point during the evening she goes to the powder room, leaving her drink on the table. Without being seen, the man slips some GHB (a powerful, colorless, odorless, tasteless, sleep-inducing drug) into her drink. When she returns she finishes her drink. A short time later she feels dizzy and sleepy. The man she met offers to take her home. Her friends at the club see her leave with the man but don't realize that she has been drugged. At her apartment, the woman passes out and is the victim of sexual assault. She wakes up the next day in bed and realizes that she has been raped. She calls a friend who takes her to the hospital where she is examined and the sexual assault confirmed. A standard rape kit is used to collect biological samples from the woman including vaginal swabs, pubic hairs, a buccal sample, and her clothing from the night before. The police interview her and she gives a description of the man who took her home.

After interviewing people at the night club, the police investigators have a good description of the man and some people at the bar recognize him and give the police his name. He is brought in for questioning and confirms that he was with the victim and, when she said she didn't feel well, he offered to take her home. He tells the police that he got her to her apartment and then left. He didn't remember if she locked the door. He said that this was the last he saw of her. Pursuant to a warrant, biological samples are taken from the man including hairs and a buccal sample. The clothes he was wearing the night before are also seized.

All of the items are sent to the crime lab for processing. One of the major foci of the evidence will be DNA taken from the woman and known samples obtained from the suspect.

INTRODUCTION

Chapter 10, "Serology," discussed non-DNA markers present in blood and other body fluids. These include red and white cell blood antigens and some enzymes. Many of these proteins are polymorphic: They exist in more than one form and everyone inherits one or more forms of each of them. This permits for subdividing a population based on the frequency of occurrence of each particular marker. Although this type of serological analysis can be quite useful, it suffers from several problems. They include a lack of stability of many of these proteins if the blood or body fluid is subjected to drying or other environmental insults. In addition there are usually only a few forms of these proteins, so the level of population discrimination is limited. The discovery of DNA, its structure, and how it carries genetic information has had profound effects on our understanding of the development of plants and animals and how some diseases are caused and perhaps cured. In addition it has caused a revolution in forensic science. Today, a successful DNA profile makes it possible to reach a conclusion that a DNA sample came from a specific individual, giving law enforcement and forensic science a new, powerful identification tool that complements fingerprints and other methods of identification.

In this chapter the development of DNA typing methods and how they help in the comparison of blood, semen, and other body fluids and tissues generated by criminal activity will be explored. The application of DNA databases to crime scene evidence will also be presented.

THE NATURE OF DNA

DNA (deoxyribonucleic acid) is a molecule that is found in nearly all cells. Notable exceptions are red blood cells and nerve cells. DNA is a special type of molecule known as a **polymer**, a molecule made up of repeating simpler units, called **monomers**. DNA is located in two regions in a cell: the nucleus and mitochondria. Both can be used in DNA typing; however, mitochondrial DNA molecules are much shorter and are used only in special circumstances. Mitochondrial DNA will be discussed later in this chapter.

NUCLEAR DNA

Nuclear DNA is described as a **double helix**. A helix is a spiral-shaped object. DNA looks like two helices that wrap around each other. Consider a ladder, made up of two poles held together by a series of rungs. Now consider taking the ladder at both ends and twisting it until it

Figure 11.1. *The DNA double helix.*

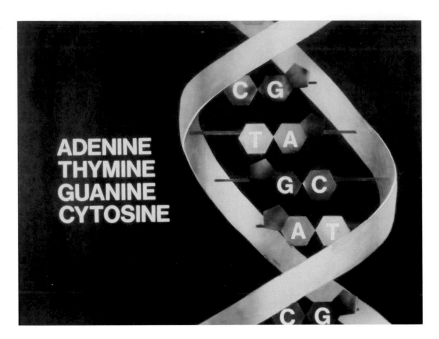

looks more like a spiral staircase—this is what the DNA molecule looks like. This is shown in Figure 11.1.

The poles of the ladder are identical in all living things. They are made up of alternating sugar molecules (deoxyribose) and phosphates. Dangling off each sugar molecule is one of four **bases** (nucleotides): **adenine (A), guanine (G), cytosine (C)**, and **thymine (T)**. When an adenine base and a thymine base come into proximity, they form a bond to each other. Likewise, when cytosine and guanine get near each other, they will bond. Neither T nor A can link with G or C. The DNA molecule, then, consists of the sugar-phosphate backbones connected by linked base pairs, and the linkages must be A-T, T-A, G-C, or C-G. Although the pairing of bases is governed by the above rule, the order is controlled by genetics. The order of the base pairs contains a sort of blueprint or **genetic code** that determines many of the characteristics of a person. An apt analogy would be a person's telephone number. Everyone has a 10-digit phone number but, in order to get the phone to ring, someone has to dial the digits in the correct sequence.

DNA in Cells

Most cells in the human body have a **nucleus**. This is where most of the cell functions are controlled. Within the nucleus, the DNA is

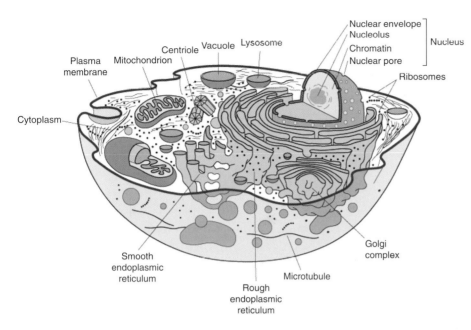

Figure 11.2. A human cell.

Labels in figure:
- Nuclear envelope
- Nucleolus
- Chromatin
- Nuclear pore
- Nucleus
- Ribosomes
- Centriole
- Vacuole
- Lysosome
- Plasma membrane
- Mitochondrion
- Cytoplasm
- Golgi complex
- Smooth endoplasmic reticulum
- Rough endoplasmic reticulum
- Microtubule

arranged into 46 structures called **chromosomes**. The chromosomes are arranged in 23 pairs. One member of each pair of chromosomes comes from the father's sperm cells, and the other member comes from the mother's egg cells. Male sperm contains 23 chromosomes, and the female ovum (egg) also contains 23. When sperm and egg unite, the 23 chromosomes from the sperm and the egg pair up, forming the 46 found in every nucleated cell in the offspring. One pair of chromosomes determines the sex of the individual. For females, both chromosomes are of the X type. In males, one of the chromosomes is X and the other is Y. A cutaway diagram of a human cell is shown in Figure 11.2.

Genes and the Genetic Code

Within the long strands of DNA are sections called **genes**. The ordering of the base pairs in genes provides the chemical instructions to manufacture particular proteins in the body. These genetic instructions are copied onto RNA (ribonucleic acid), which transmits this information to protein manufacturing sites in the cell. Each gene codes for a particular characteristic protein. For example, there are sets of genes that define hair color, eye color, skin pigmentation, and gender. However, it might be surprising to know that more than 99% of all human DNA is exactly the same. The rest of the DNA, less than 1%, contains the genetic information that differentiates one human being from another. There are estimated to be up to 100,000 genes in a human being. In recent years, a global effort has been underway to map the entire human genome. The ultimate goal of this project is to find all genes on all chromosomes and determine their function.

Variations of Genes: Alleles

Genes that determine a person's characteristics are found in particular locations on the chromosomes. Each individual has two copies of each gene. One copy comes from the person's father, and the other half comes from the mother. Some traits are determined by a single gene on one chromosome. Others, such as eye color, are determined by multiple genes at several locations. Simple observation of people's eyes indicates that there must be considerable variation in this gene. Some people have brown eyes, others have blue or green eyes, and many people have eye colors that are intermediate between two or more of these. For example, a person may inherit the eye color gene that codes for blue eyes from her mother and the same gene that codes for brown eyes from her father. The offspring's actual eye color is determined by genetic rules that may be somewhat complicated. The two different eye color genes are called **alleles**. If a person inherits the same form of a gene from the mother and the father, that person is said to be **homozygous** with respect to that gene. For example, if a person inherits a gene that codes for brown hair from the mother and from the father, then she is homozygous with respect to hair color. If, on the other hand, the person receives different forms of the same gene (blond and brown hair color), then she is said to be **heterozygous**; she has different alleles for hair color. In addition, some alleles are **dominant**, whereas others are **recessive**. For example, the allele that codes for brown hair is dominant, whereas the one that results in blond hair is recessive. If a person inherits one of each of these alleles, she will have brown hair because the brown hair allele is dominant over the blond hair allele.

Some characteristics may exist in multiple forms or alleles. Each person will inherit one of these from the mother and one from the father. If there are a large number of such alleles, then the potential exists for a great deal of variation among human beings at this location (locus). This situation provides the basis for a DNA profile wherein the variation of alleles at several loci can be combined to provide a statistical evaluation of the likelihood of a particular set of alleles in a given population.

What is meant by "different forms of an allele"? The visible manifestation of variability—eye color, for example—is called the **phenotype**. The **genotype** is the genetic description of the allele. For example, a person with the genotype XY would be a male (the phenotype).

In DNA analysis, locations (loci) that are **polymorphic** are purposely chosen. These loci exhibit variation among members of a population. The more variation there is at a locus, the more discriminating the analysis will be. For example, in the ABO blood system, type A blood is present in about 42% of the Caucasian population, type O is present in about 43%, type B about 10%, and type AB in about 5%. Thus, the locus for ABO blood type does show some variation, but by itself isn't very discriminating, since even its rarest form would still include 5% of individuals as being the source of a blood sample.

There are two types of variability in alleles. The first type is **sequence polymorphisms**. An example of this in DNA would be the following:

$$\begin{array}{cccccccc}
\text{C T C G \textbf{A} T T A A G G} & & \text{C T C G \textbf{G} T T A A G G} \\
: : : : : : : : : : \text{ and } : : : : : : : : : : \\
\text{G A G C \textbf{T} A A T T C C} & & \text{G A G C \textbf{C} A A T T C C} \\
\uparrow & & \uparrow
\end{array}$$

The two sequences of double stranded DNA are exactly the same except at the location indicated by the arrows.

The other type of variation in DNA is called **length polymorphism**. Consider the following variation in a part of Lincoln's Gettysburg Address:

Four Score and Seven Years Ago
Four Score and and Seven Years Ago
Four Score and and and and Seven Years Ago
Four Score and and and and and and Seven Years Ago

These phrases are all the same except for the "and," which repeats a different number of times in the various phrases. Now consider the length polymorphism that occurs in the following DNA sequence:

```
CATGTAC-CATGTAC
::::::: :::::::
GTACATG-GTACATG
```

```
CATGTAC-CATGTAC-CATGTAC-CATGTAC
::::::: ::::::: ::::::: :::::::
GTACATG-GTACATG-GTACATG-GTACATG
```

```
CATGTAC-CATGTAC-CATGTAC-CATGTACCATGTAC-CATGTAC
::::::: ::::::: ::::::: ::::::::::::::: :::::::
GTACATG-GTACATG-GTACATG-GTACATGGTACATG-GTACATG
```

Each of these consists of a seven base pair sequence that is repeated. In the first case it is repeated twice. In the second example it is repeated four times, and in the third, six times. Because the repeats are right next to each other, without any intervening base pairs, these are referred to as **tandem repeats**. When variation in the number of repeats occurs from one individual to the next, then this locus is described as having a **variable number of tandem repeats** (VNTR). A person's **DNA type** is a description of the types of alleles at all of the locations being analyzed on the genome.

Population Genetics

Consider the case presented at the beginning of this chapter. Suppose that forensic scientists at the lab find that the DNA type of the seminal fluid found in the victim is the same as the DNA type of the suspect. What does this mean? How likely is it that the suspect was the person who deposited that semen in the victim? These are important questions that have to be answered and explained to a jury should the case get into a court room. The science of population genetics helps answer these questions.

In forensic DNA analysis multiple loci are evaluated, giving rise to many pieces of data. Consider the situation in which there are several alleles at a particular locus. The frequency of occurrence in the population can be determined for each allele. Now consider this situation at several loci. If it can be determined which allele is present at each locus, the frequency of occurrence of all of these alleles can be determined by simply multiplying the frequency of occurrence of each one. This can be illustrated with the familiar coin toss routine. If a coin is tossed once, the probability (frequency of occurrence) for it coming up heads is $\frac{1}{2}$ since there are only two equally probabl outcomes. If the coin is tossed twice, the probability of it coming up heads both times is $\frac{1}{4}$ because there are four possible outcomes from tossing a coin twice: H-H, T-T, H-T, and T-H. Only one of these outcomes is heads twice (H-H). This probability can be determined by multiplying the probability of each toss: $\frac{1}{2} \times \frac{1}{2} = \frac{1}{4}$. Likewise the probability of getting three heads in a row is $1/8$ ($\frac{1}{2} \times \frac{1}{2} \times \frac{1}{2}$). The technique of multiplying probabilities together is known as the **product rule**.

This procedure can be invoked only when the probability of each event is independent of the other events. In genetics this means that the occurrence of each allele must be independent of all of the other ones being measured.

Linkage Equilibrium

The loci used in genotyping have been extensively tested to check for independence. Using the product rule in such cases can yield genotypes that are so rare that the chances of finding more than one person at random within a population is essentially nil. For an example of this, see the section on STRs later in this chapter.

DNA TYPING

One of the pioneers of DNA typing was Dr. Alec Jeffries of Leicester University in England. He developed the method of DNA typing that is now known as RFLP. A bit of the history of Dr. Jeffries' discoveries is given below.

RESTRICTION FRAGMENT LENGTH POLYMORPHISM (RFLP)

RFLP was the first commercial method of DNA typing. It grew out of the technology of **recombinant DNA**, a technique whereby a piece of DNA is taken from the chromosomes of one organism and is spliced into the DNA of another organism. This developing technology is being used to introduce desirable traits from one animal or plant into another using bacteria as the carriers. For example, the gene that

HISTORY OF DNA TYPING:

Dr. Alec Jeffries

In 1984, a major breakthrough in forensic science took place in England at the University of Leicester. Dr. Alec Jeffries was studying the gene for a substance called myoglobin. He noted that parts of the gene did not seem to have a role in the production of myoglobin. He noted that these parts of the gene consisted of repeating base sequences of about 10–15 units in length. These sequences are known as minisatellites. He found that the number of repeats of this core base sequence differed from person to person and there were many variations. He called these "hypervariable regions." Dr. Jeffries suspected that this finding could be of use to forensic scientists as a way of differentiating people by their DNA. He then discovered that there were many such hypervariable regions throughout the human genome. His technique of isolating and displaying these regions from DNA became known as "DNA fingerprinting."

codes for growth hormones in humans has been introduced into certain fish. This causes them to grow faster, providing a mechanism for supplying increased amounts of protein to people in need.

The pieces of DNA that contain these traits are cut out of a DNA strand by **restriction enzymes**. These enzymes, also known as **endonucleases,** are designed to cut DNA at a specific sequence of bases. In the RFLP technique, polymorphic regions of DNA are identified. These are length polymorphisms whose core sequence is between 15 and 40 base pairs in length. These regions are **hypervariable**; that is, they have a large number of alleles and are commonly known as **variable number tandem repeats (VNTRs)**. In 1980 the first DNA polymorphism was discovered that was suitable for forensic purposes. It is a gene known as (D14S1). Over the next few years, other hypervariable genetic markers were developed for RFLP. Many of them do not seem to carry genetic instructions, and their function is not well understood. They take up some of the space between genes on the chromosomes. At the same time other hypervariable regions were discovered at several loci. They were called **multilocus genes**.

How RFLP Works

Once a hypervariable region is discovered, a specific restriction enzyme is introduced that cuts the DNA on either side of the VNTR. Several different VNTRs can be analyzed at the same time, employing different specific restriction enzymes simultaneously. When DNA has been attacked by restriction enzymes, it is cut into hundreds of

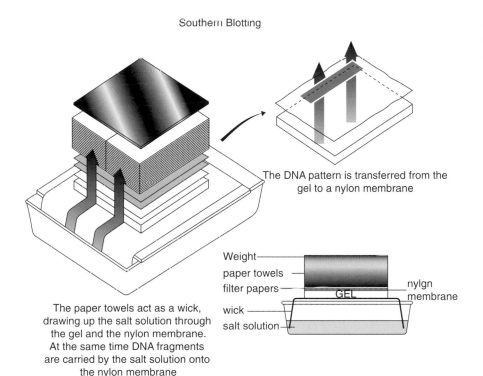

Southern Blotting

Figure 11.3. *The Southern Blotting technique.*

The DNA pattern is transferred from the gel to a nylon membrane

Weight
paper towels
filter papers
GEL
nylgn membrane
wick
salt solution

The paper towels act as a wick, drawing up the salt solution through the gel and the nylon membrane. At the same time DNA fragments are carried by the salt solution onto the nylon membrane

fragments. Mixed in among the fragments are the VNTRs of interest. They must be separated and visualized so that the DNA type can be determined.

Gel electrophoresis is used to separate the fragments of DNA. This technique is described in Chapter 5. Recall that gel electrophoresis uses a slab of agarose or polyacrylamide gel to separate the different size DNA fragments by sieving them through pores in gel. The DNA is denatured at this point so that the double strands unzip into single strands. Mixtures of DNA containing fragments of known size are also electrophoresed at the same time so that the lengths of the VNTRs can be determined when the gel is developed. These calibration standards are known as **ladders**.

This gel is quite fragile, so a technique known as Southern Blotting is used to transfer the separated DNA fragments onto a more rugged nylon membrane. This technique is shown in Figure 11.3.

Once the DNA has been transferred to the nylon membrane, a technique called **probe hybridization** is used to visualize the VNTRs. In this technique, short strands of DNA that are complementary to the

Figure 11.4. *Probe*
hybridization.

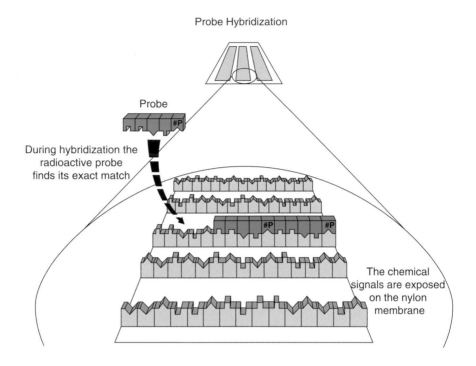

Probe Hybridization

Probe

During hybridization the
radioactive probe
finds its exact match

The chemical
signals are exposed
on the nylon
membrane

core repeating units of the VNTRs are labeled with a chemilumines-
cent substrate so that they will fluoresce. Alternatively, radioactive
labeled probes can be used. If radioactive probes are used then the
radio-labeled DNA can be used to expose a piece of X-ray film or other
large negative. Only the labeled DNA will expose the film. This film is
called an **autorad**. These probes will attach themselves only to their
complementary strands of DNA, which are the VNTRs of interest. The
excess probe is then washed off the membrane. Figure 11.4 illustrates
probe hybridization.

Special camera film can be used to photograph the membrane.
Only the hybridized strands of VNTRs will show up.

If the VNTR is homozygous, only one band will show up on the
electropherogram. If it is heterozygous, then two bands will be visible.
With the help of the calibration ladder, the lengths of each of the
VNTRs can be determined. As RFLP was developed, the number of different VNTRs increased and data was developed that determined the
population frequencies of occurrence of each allele of each VNTR.
When an RFLP was run on a case and a match occurred between a
known and unknown DNA sample, the probability that this match was
a coincidence could be determined. For example, assume that an

RFLP test was run on a blood stain found on a knife used to murder someone. A DNA sample was taken from the suspect in the case. Assume that different VNTRs were analyzed. It was found that the following frequency of occurrence existed for each VNTR:

1–1 in 20
2–1 in 50
3–1 in 25
4–1 in 75

Using the product rule, the cumlative frequency of occurrence of this DNA type would be $1/20 \times 1/50 \times 1/25 \times 1/75 = 1$ in 5×10^7. Thus, this DNA type would occur in one out of every 50 million people. The more probes used in an RFLP determination, the fewer people would have that DNA.

THE POLYMERASE CHAIN REACTION (PCR)

Dr. Kary Mullis won the Nobel Prize for developing a method for replicating DNA. A bit of the history of his work is presented in the box below. Another major advance in the mid 1980s was the development of forensic applications of the **polymerase chain reaction (PCR)**. PCR

HISTORY OF DNA TYPING:
Dr. Kary Mullis

In 1983, Dr. Kary Mullis, a biochemist, invented a way to replicate targeted parts of a strand of DNA. He coined this technique "polymerase chain reaction" or PCR. He was awarded the *Nobel Prize* for his discovery in 1993. Mullis's idea was to develop a process by which DNA could be artificially multiplied through repeated cycles of heat and cold using an enzyme called DNA polymerase.

DNA polymerase is a naturally occurring substance in all living organisms. Its function is to aid in the replication of DNA as a cell divides. During the cell division process, the double stranded DNA unzips to become single stranded. DNA polymerase acts by binding to each single strand and directs the formation of the complementary strand, thus making exact duplicates of the DNA. Originally, the DNA polymerase that Mullis used for the reaction was temperature sensitive and would decompose at the high temperatures that are needed to cause the DNA double helix to unzip. His big breakthrough came when he developed thermal cycling and employed polymerase that is found in organisms that live in or near geysers and can survive at temperatures well above those needed for PCR. Originally, he employed DNA polymerase from a bacterium known as Thermas Acquaticus (Taq). This is still widely used during the replication process.

had been used since the 1970s for making copies of (amplifying) DNA using polymerase enzymes. One of these enzymes, Taq polymerase, is stable at high temperatures, and its use in PCR permitted the process of DNA amplification to be automated using a thermal cycler. The first marker to be amplified and used forensically was DQα (now called DQA1). More recently PCR began to be used to identify short repeating sequences of DNA called **STRs (short tandem repeats)**. This has now become the most widely used new DNA typing technique in the United States.

In numerous cases the DNA recovered from the crime scene is insufficient or too badly degraded for analysis. These samples may be unsuitable for RFLP because the technique requires strands of DNA that are thousands of bases in length, making it difficult to type degraded samples. Fortunately, the markers used in PCR are much shorter, 100–400 bases long, making it possible to type degraded DNA. In addition, PCR provides a method for the replication of DNA to make millions of copies. This can provide material of sufficient quantity and quality to be used in subsequent DNA typing.

The PCR Process

PCR methods are very sensitive to contamination by foreign DNA. For this reason, DNA extractions are always done in a location physically isolated from the place where the subsequent amplifications will be performed. Once the scientist enters the amplification room with the extracted DNA, he or she will usually not leave until the extraction is done so as to minimize carrying foreign material into or out of the amplification environment.

The PCR process involves three steps. They all take place within a **thermal cycler**, which is essentially an apparatus capable of achieving and maintaining preset temperatures very precisely. The DNA samples are mixed with various solvents and reagents and then put in small vials. The samples would consist of various known and unknown samples and controls to ensure that the process is working properly. Under the control of a computer, the thermal cycler heats the vials to predetermined temperatures for each step of the process. These temperatures can be held for precise amounts of time and then quickly changed to the next level. Each set of temperatures constitutes a thermal cycle. The thermal cycler can be programmed to run through as many cycles as necessary for proper amplification.

The steps in the PCR are:

1. **Denaturation.** The DNA is added to a solvent and heated close to boiling. Under these conditions the double stranded DNA denatures. The bonds between the base pairs that hold the strands together break, resulting in single stranded DNA. As long as this temperature is maintained, the strands will remain apart. Each strand will be the template for the formation of a new piece of double stranded DNA.

2. **Annealing.** The next step in the PCR process is to attach a short strand of synthetic DNA to each of the separated strands. These are called **primers** because they will mark the starting points for addition of new bases to complete the reproduction of each strand.

3. **Extension.** Under the influence of taq polymerase, single bases (**nucleotides**) are added to the primer, one-by-one. Each base is complementary to a single nucleotide present on the strand being duplicated. In this way the entire complementary strand is built up and a new piece of double stranded DNA is produced. This process occurs at each of the single strands created by the denaturation process, so the result is that two identical pieces of double stranded DNA are produced. This occurs in one cycle of the PCR process.

After the third step, the temperature is automatically raised to effect denaturation again. The two strands of DNA denature, forming four single strands. These are in turn subjected to annealing and extension, forming four new double strands. The process continues until a sufficient amount of DNA is produced, typically 25–30 cyles. This produces millions of copies of the original DNA, enough for additional typing. The steps in PCR are shown in Figure 11.5.

DNA Typing of PCR Product

As was mentioned previously, the first DNA region that was widely subjected to amplification and typing for forensic purposes by PCR is the **HLA DQ alpha gene**. This gene exhibits sequence polymorphisms. DQ alpha and a number of other genes collectively called **polymarker** are typed using a method called **reverse dot blot**. This process involves identifying the particular alleles present by reacting them with color-forming reagents on specially treated nylon strips.

DQA1/Polymarker typing is not often used in forensic science anymore. It has been largely supplanted by STRs whose analysis is PCR

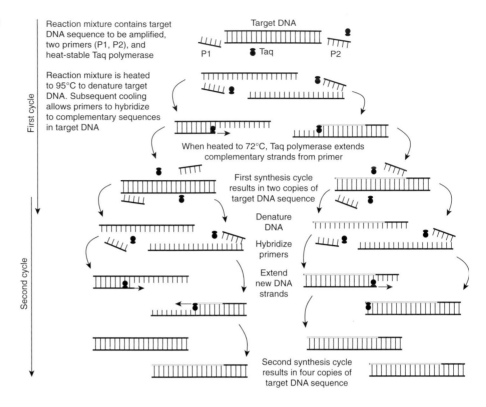

Figure 11.5. *The steps in polymerase chain reaction.*

First cycle

Reaction mixture contains target DNA sequence to be amplified, two primers (P1, P2), and heat-stable Taq polymerase

Target DNA

P1   Taq   P2

Reaction mixture is heated to 95°C to denature target DNA. Subsequent cooling allows primers to hybridize to complementary sequences in target DNA

When heated to 72°C, Taq polymerase extends complementary strands from primer

First synthesis cycle results in two copies of target DNA sequence

Denature DNA

Hybridize primers

Extend new DNA strands

Second cycle

Second synthesis cycle results in four copies of target DNA sequence

based. One reason for this is that the alleles do not vary to a great extent in the human population. More important is that it is unable to resolve separate DNA types in mixtures. As a result, it is not possible to generate a DNA type that is rare enough to associate with just one individual.

STRs

Currently, the most powerful typing method for **genomic** DNA is STR analysis. STRs are lengthy polymorphic sequences of short strands of DNA that range from three to seven base pairs long and that repeat in tandem. These types of short tandem repeats are called **microsatellites**, whereas the longer ones that are typed by RFLP are called **minisatellites**. In present methodology, STRs from up to 15 loci are typed. The DNA is amplified at these loci by the PCR. The DNA containing the repeats is targeted by the primers. They are placed in constant, non-repetitive DNA just ouside the repetitive region. The DNA is next separated. Originally, this is done by sieving the different strands of DNA through polyacrylamide gels.

Modern practice is to use capillary electrophoresis using a capillary filled with a polymer similar to polyacrylamide. The DNA is detected by means of laser-induced fluorescence. To do this, one of the primers used in the PCR process is labeled with a fluorescent dye so that the amplified fragments can be detected using a laser. They are then detected by a UV-visible detector. The result is a series of peaks known as an electropherogram. See Chapter 6 for an explanation of the workings of capillary electrophoresis. Figure 6.14 shows a capillary electropherogram of a DNA sample.

Since all 13 loci are detected during the same run, there is a large amount of data for analysis. **Allelic ladders**, which are strands of DNA made up of all common alleles present at each STR locus, are used for calibration, thus enabling the computer to estimate the size of the alleles at each locus. As in other DNA typing methods the population frequency ranges for each allele at each locus have been previously determined and, using the product rule, population frequencies for the entire genotype can be estimated. These loci are highly polymorphic (hypervariable), and thus the probability of having any one allele is usually quite low. This means that the probability of having any given DNA type from the 13 loci is extremely small; on the order of one in several billion or even trillion. When you consider that the population of the United States is about 260 million people, the chances of any two of them selected at random, having the same exact DNA type at all 13 loci is extremely remote. Around 1998, the FBI laboratory made a policy decision to count as individualized any DNA type whose odds of a chance occurrence exceeded 1,000 times the U.S. population (or about 260 billion to one). Today, virtually all DNA types exceed this threshhold. Consider the sexual assault case presented at the beginning of this chapter. The DNA would be typed using 13 STRs, and the probability of having any one DNA type will be very small. So, if the DNA from the seminal fluid taken from the victim had the same type as that of the suspect and the probability of having this DNA type is one in 500 billion, then the seminal fluid can be considered to have come from the suspect to the exclusion of all other males. Thus, the advent of STR analysis has elevated DNA typing from class to individual evidence under such circumstances.

Gender Identification

There are two approaches to gender identification using DNA typing. On one of the chromosomes there is a locus called **amelogenin**. One of the regions of this locus is six base pairs longer in males than in females. Females have two X chromosomes and will thus show only one band for amelogenin. Males have one X and one Y chromosome and will thus show two bands, one that is six base pairs longer than the other. This locus is not an STR but can be analyzed at the same time as STR.

The other approach to gender ID utilizes Y-STRs. The Y chromosome, found only in males, also contains STRs. They can be typed even on small or degraded samples or mixtures with large quantities of female DNA. Y-STRs are quite useful when typing mixed samples that have sperm and are thus guaranteed to have a male fraction. It should be noted, however, that Y-chromosome analysis produces only a haplotype and is thus not as informative as regular STRs.

MITOCHONDRIAL DNA (mtDNA)

Not all human DNA is located in cell nuclei. A small amount (1%) is located in structures within cells but outside the nucleus called **mitochondria** (singular: mitochondrion). Mitochondria have some cellular respiratory functions. The proteins that control these functions are manufactured according to a genetic code supplied by the mitochondrial DNA.

There are a number of differences between mtDNA and genomic DNA:

1. mtDNA does not exist in a twisted double helix like genomic DNA. Instead, it is circular in shape.

2. There is much less mtDNA than genomic DNA, but there are many more copies of mtDNA in mitochondria cells, compared to only a few copies of genomic DNA.

3. mtDNA contains a non-coding region of 1,100 base pairs that, in turn, contain two hypervariable regions. These regions exhibit a higher mutation rate, so that over a period of many generations, sequence variations can occur at these sites. As a result these regions can be quite useful in comparing known and questioned DNA samples.

4. All male and female mtDNA comes from the mother. There is no mtDNA from one's father. This means that, except for the mutations mentioned in the preceding paragraph, every descendent of a woman should have the same mtDNA. This makes mtDNA

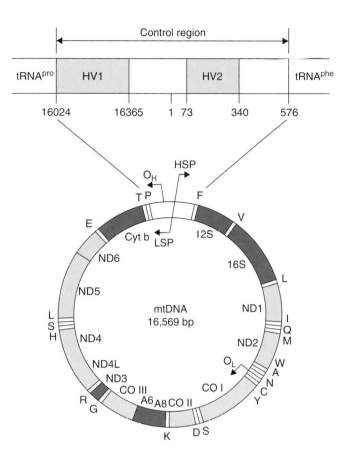

Figure 11.6.
Mitochondrial DNA.

very powerful for tracing family lines back through the maternal side.

5. mtDNA often shows a high degree of variation between unrelated people, making it a powerful tool in forensic typing.

6. Because of the large number of copies per cell, mtDNA can be useful in typing samples that have low quantities of DNA, or in exhibits that are degraded or very old.

Figure 11.6 shows the arrangement of mitochondrial DNA.

Most forensic science laboratories that perform genomic DNA analysis also do mtDNA analysis. Those that do generally use DNA sequencing; they determine the entire base pair sequence in the two hypervariable regions, rather than relying on length polymorphism. In the near future, mtDNA typing may be done using a method similar to that used in DQ alpha and polymarker analysis.

INTERPRETATION OF DNA TYPING RESULTS

PURITY ISSUES

If biological material from a crime scene is relatively clean, fresh, non-degraded, and from only one individual, then interpretation of the results of DNA typing will usually be straightforward. The vast majority of problems occur when the DNA is compromised in some way. A few of the more common situations are as follows:

- **Contamination.** If care is not used in DNA extractions and typing, then biologic material from an extraneous source, such as the evidence technician, scientist, or laboratory technician, can be introduced accidentally. This can be a serious problem, especially with PCR-based methods because the DNA from the contamination may also be amplified and can overwhelm the DNA from the sample.
- **DNA from more than one source.** This can occur in evidence from sexual assaults and crimes of violence where biological material can be easily mixed. It is sometimes but not always possible to resolve the mixture and determine the number and identity of the contributors. For example, if sperm are present in a sample, tests that look only at the male fraction of DNA such as Y-STR analysis may be useful to distinguish the male and female DNA. If one contributor dominates the other in quantity, it may not be possible to get a clear DNA type of the minor contributors.
- **Degradation.** DNA is a remarkably stable substance, but it can degrade from a number of causes including extreme heat, light, time, humidity, biological causes such as fungi and bacteria, and the presence of certain chemicals. When DNA degrades, long strands may become fragmented. This fragmentation process is exacerbated by exposure to extreme conditions for longer time periods. The RFLP technique is particularly sensitive to degradation. Successful RFLP analysis requires long intact strands of DNA. Degradation may break up the DNA into pieces that are too small for RFLP. STR analysis is better adapted to degraded DNA, and in fact, STR was developed with this problem in mind.
- **Extraneous Materials.** In addition to contaminated samples, in which some other material is introduced after the biological evidence is recovered from the crime scene, there are samples which

contain substances other than human DNA at the scene. For example, some chemicals may deactivate the enzymes that are needed for analysis. Examples include some clothing dyes, blood components, and soil. Special cleanup procedures may be needed to remove or counter the effects of these materials. There are also microorganisms such as certain bacteria that can produce substances that can destroy DNA.

COMPARISON OF DNA SAMPLES

As has been stated elsewhere in this book, the goal of all evidence analysis is to associate an item from an unknown source with one particular person or object—that is, to individualize it. For most types of evidence this involves uncovering and characterizing one or more unique characteristics. If such characteristics are found to be present only in the evidentiary material and an item taken from, for example, the suspect in the crime, then this evidence is said to be individualized to that person. With pattern evidence such as fingerprints, footwear, and handwriting, the unique characteristics are in the evidence itself, e.g., the unique patterns of the ridges in the fingerprints. In the case of DNA, the criterion for assessing the strength of the evidence lies in the genotype exhibited by the results of the DNA typing method—that is, the sum total of the alleles at all of the loci that are being tested. Statistical methods are used to determine if the genotype is rare enough to be found in just one person. Thus, if this rare DNA type were found in evidence from a crime scene and in a sample taken from the suspect in the crime, it may reasonably be concluded that the suspect deposited the DNA at the scene to the exclusion of all other people.

The term commonly used to express this comparison is **match**. The DNA in the biological evidence found at the crime scene "matched" the DNA from the suspect. But the term "match" may not be understood by a jury in the same way it is by scientists. Scientists define match as a term that means that there are no significant or unexplainable differences between the known and unknown materials. Thus, the term "match" is used to describe the relationship between two fingerprints or shoeprints. In DNA typing all of the evidence is examined. The methods used for DNA typing today look at only small, select parts of the DNA. Not all of the DNA at all the loci in all 46 chromosomes is examined. It is possible that, if an exact "match" between two samples for all of the loci tested is achieved, further examination of more loci might reveal differences. For this reason, the term "match" should not be used to describe the relationship between two samples of DNA. Scientists instead use the term **genetic concordance**.

Continuous versus Discrete Alleles

The determination of genetic concordance is approached differently when continuous alleles are measured instead of discrete alleles. RFLP is an example of a typing method using continuous alleles. The tandem repeats are fairly long, and the electrophoresis method used to separate the VNTRs does not have sufficient resolution to separate all possible alleles. For example, if two people had VNTRs that differed by only one repeat unit, electrophoresis may not be able to separate them, and the two alleles would appear to be the same; that is, only one band would be seen for the two genotypes. Another way of looking at this is to recognize that the band visualized on autorad at the end of the RFLP experiment does not represent only one possible allele. When comparisons of DNA profiles are done in RFLP, the computer that does the calculations adds in some "fudge factors" to account for the lack of resolution in the bands.

STR testing utilizes discrete alleles. At the loci that are used, there are few alleles compared to the number present at an RFLP locus. STR methods amplify and detect exact lengths of DNA. This makes interpretation easier.

Estimation of Population Frequencies

The population frequency for an allele is the number of times that it appears divided by the total number of types of that allele that exists. For example, if an allele occurs four times in a database of one million, then the population frequency for that allele would be $1/250,000$. Determining a population frequency at one locus is usually pretty straightforward. Large numbers of people are tested, and the number of times each allele appears in that population is determined. It may take a large number of people in the database to have all of the alleles represented at a locus, especially if some are very rare.

Determining a population frequency for an entire DNA profile is not so simple, especially if a large number of loci are being analyzed and the number of alleles at a given loci is large. Under these conditions a particular DNA profile may never show up in a database even if the number of people tested is large. It would then not be possible to determine the population frequency of that profile. In order to get population frequencies in such situations, other methods have to be used. These methods are not simple and a detailed description goes beyond the scope of this book. Population frequencies of genotypes made up of multiple loci can be calculated by taking the product of the population frequencies of each individual locus.

Table 11.1 shows an actual example of the calculation of population frequency for a DNA profile using 13 loci. These loci are the same ones that are used to collect data for the CODIS database (see the next section).

DNA DATABASE: THE FBI CODIS SYSTEM

The **Combined DNA Index System** (**CODIS**) is a national database of DNA profiles administered by the FBI. It began as a pilot project in 1990. Today most states regularly contribute data to CODIS, and it contains many thousands of DNA profiles. For an up-to-date list of participating states, see http://www.fbi.gov/hq/lab/codis/partstates.htm. CODIS is arranged in tiers. The first tier is the local database. All profiles originate at the local level. These profiles are then fed into the state-level database. Finally, the states input their data into the national-level database. This allows a crime lab to search the database at whatever level is necessary for that particular case.

The CODIS system consists of three databases. The first contains DNA profiles that are obtained from crime scenes (the forensic database). In most cases the source of this DNA is not known. The second database consists of profiles of criminal offenders and sometimes even those arrested for felonies and misdemeanors. Different states have different criteria for what DNA types will be contributed. If a crime occurs where DNA evidence is generated such as the sexual assault described in the case at the beginning of the chapter, CODIS may be searched to see if the offender's DNA is on file or if the DNA recovered from the scene is also found at another scene, indicating possibly that a serial criminal is at large.

The third and most recent database in CODIS is that of missing persons. Efforts are being made to make this database as inclusive as possible nationwide so as to maximize the chances of identifying a missing person who may have crossed state lines.

Table 11.1. Match Statistic for a random African-American individual typed at 13 CODIS loci.

LOCUS	GENOTYPE	ALLELE FREQUENCIES	MATCH STATISTIC
CSF1PO	10, 12	0.271; 0.300	0.163
D13S317	11, 11	0.237	0.0562
D16S539	11, 12	0.294; 0.187	0.110
D18S51	14, 18	0.0639; 0.131	0.0167
D21S11	27, 37	0.0615; 0.00559	0.000688
D3S1358	15, 17	0.290; 0.200	0.116
D5S818	8, 12	0.0500; 0.356	0.0356
D7S820	8, 10	0.174; 0.324	0.113
D8S1179	12, 12	0.108	0.0117
FGA	22, 22	0.225	0.0506
TH01	6, 9	0.110; 0.145	0.0319
TPOX	10, 11	0.0933; 0.225	0.0420
vWA	15, 16	0.236; 0.269	0.127

Random Match Statistic: 5.422×10^{-19} or 1 person in 1,837,000,000,000,000,000 chosen at random from the Black population would be expected to match by chance.

The first column shows the locus where the STR is found. The second column (genotype) shows the particular alleles that this individual possesses. Note that he is heterozygous at 10 loci and homozygous at D13S317, D8S1179, and FGA. The third column (allele frequencies) contains the allele frequencies for each allele. For example in CSF1PO, the 10 allele is found in 271 out of 1,000 people in the black population. The fourth column (match statistic) is 2 times the product of the allele frequencies when the locus is heterozygous and the square of the allele frequency in homozygous cases. To find the random match statistic, all 13 match statistics are multiplied (rule of multiplication). The final number, about 7 septillion, is astronomic. As a point of reference, it is estimated that there have been no more than 100 billion (100,000,000,000) people that have ever lived on earth. Courtesy: Orchid Genescreen, East Lansing, Michigan.

The CODIS data consists of genotypes from the 13 STR loci listed in Table 11.1.

CODIS SUCCESS STORIES

The following are some CODIS success stories taken from the FBI's CODIS website: **www.fbi.gov/hq/lab/codis/index1.htm**

Richmond, Virginia, July 1998: A rape and homicide had baffled the police since the body was discovered in 1994. Although the police had samples of blood and semen found in the victim's apartment, they were unable to develop a solid suspect in the case. A recent routine computer search on the state's DNA database identified a suspect in the case. A 20-year-old convicted offender, already serving a sentence for a different rape and murder, was arrested for the 1994 crime.

Oklahoma City, Oklahoma, February 1997: In 1992 five women were bound, gagged, and stabbed in a reported drug house in Oklahoma City. The Oklahoma State Bureau of Investigation developed a DNA profile for the killer in 1995, based on evidence at the crime scene. The California Department of Justice used CODIS to match the evidence profile against Danny Keith Hooks, who was convicted of rape, kidnapping, and assault in California in 1998.

Tallahassee, Florida, February 1995: The Florida Department of Law Enforcement linked semen found on a Jane Doe rape-homicide victim to a convicted offender's DNA profile. The suspect's DNA was collected, analyzed, and stored in the CODIS database while he was incarcerated for another rape. The match was timely; it prevented the suspect/offender's release on parole scheduled eight days later.

St. Paul, Minnesota, November 1994: A man wearing a nylon stocking over his face and armed with a knife jumped out from behind bushes and forced a woman who was walking by to perform oral sex. Semen recovered from the victim's skirt and saliva was analyzed using DNA technology. The resulting profile was searched against Minnesota's CODIS database. The search identified Terry Lee Anderson, who confessed to the crime and is now in prison.

DNA CASE BACKLOG

The success of DNA typing and the CODIS database has resulted in nearly all states passing laws that require some or all people arrested for crimes to be DNA typed and the data stored in CODIS. Most of the time, these laws do not make provision for hiring additional DNA analysts or building more facilities to handle this large caseload. In most cases, so-called CODIS samples greatly outnumber criminal DNA evidence and cause huge backlogs in many forensic science laboratory systems. At the beginning of 2004, it was estimated that the total U.S. backlog exceeded 400,000 cases. In 2004, the National Institute of Justice estimated there were 300,000 unexamined sexual assault kits in the United States. A typical laboratory may take in 4,000 cases per month and have the capability of analyzing 2,000. Each month this one lab falls 2,000 cases further behind. In some states, separate CODIS labs are being created; in others, CODIS samples are being sent to private labs at considerable expense. Since this is a nationwide criminal justice problem, there is clearly going to have to be federal as well as state government participation in the solutions.

BACK TO THE CASE

As can be seen from this chapter, genomic DNA provides a powerful method for associating biologic evidence with a particular person. The chances that two unrelated people have the same DNA type as measured forensically are so small that they are unreasonable to consider. Because of this, it is possible to absolutely exclude someone as being the source of biologic evidence even though that person would have been included as a possible source using classical blood typing methods of 20 years ago. This is how it is possible for cases to be reanalyzed by DNA typing years later, and suspects who were falsely convicted of crimes can be released. Before DNA typing the biologic evidence in the sexual assault case described at the beginning of the chapter would still have been crucial but could not have been able to positively link the suspect to the assault.

CHAPTER SUMMARY

DNA is made up of pairs of nucleotides whose arrangement determines the genetic code. There are approximately 3 billion base pairs in the human genome and less than 1% differ from person to person. DNA typing methods utilize these differences in DNA to help identify people from biological evidence. The first method that was used commercially for DNA typing was restriction fragment length polymorphism. This method isolates certain regions of DNA that are made up of core base pair sequences that are characterized by repeating a different number of times throughout the human population. The RFLP process isolates these sequences by the use of restriction enzymes and separates them by gel electrophoresis. They are visualized by radio labeling or chemiluminescence.

Modern methods of DNA typing rely on the polymerase chain reaction. Using enzymes and DNA primers, sequences of DNA can be replicated automatically using a thermal cycler. The amplified DNA is then typed using short tandem repeats. These are three to seven base pair polymorphic repeating sequences. They are separated by capillary electrophoresis. Currently, 13 loci are being typed for forensic purposes. They are sufficiently polymorphic that even the most common set of 13 alleles are essentially unique in the human population.

Mitochondrial DNA is present in all cells. There are two hypervariable regions in mtDNA. Unlike nuclear DNA, which has only one copy per cell, mtDNA contains thousands of copies in each cell. All mtDNA comes from the mother, and each person has the same DNA as his or her mother. MtDNA is useful for typing old, degraded samples.

Test Your Knowledge

1. What are the four bases that make up human DNA?

2. What is a gene?

3. What is an allele?

4. List three ways that mitochondrial DNA differs from genomic DNA.

5. What is a restriction enzyme? In what kind of DNA typing is it used?

6. What are some of the advantages of DNA typing over other methods for identifying a person, such as fingerprints?

7. In one of the early DNA typing cases in England (Colin Pitchfork), the police went to all males in a town to collect DNA samples to be tested

against crime scene evidence. What are the problems with such an approach?

8. What does PCR stand for? For what purpose was it developed?

9. What is an STR? Why has it become the method of choice for forensic DNA typing?

10. What is the importance of amelogenin?

11. What is polymarker? How is it typed?

12. In RFLP what are the two ways that the VNTRs can be visualized?

13. What is a reverse dot blot? How is it used in DNA typing?

14. Of the DNA testing methods you have learned about in this chapter, which one has the potential of generating a profile that can be considered to be unique? How is this possible?

15. What types of electrophoresis are used for the separation of DNA fragments in STR analysis?

16. What is Taq polymerase? How is it used in DNA typing?

17. What are the advantages and disadvantages of mtDNA typing compared to genomic DNA analysis?

18. What is length polymorphism? Give an example.

19. What is sequence polymorphism? Give an example.

20. What does heterozygous mean in DNA? Give an example.

Consider This . . .

1. Forensic DNA typing has evolved over time by developing analytical methods for smaller and smaller fragments that, at the same time, are increasingly variable in the human population. At this time, the standard STR method in the United States uses 13 loci for comparing DNA. Yet there are systems that use 16 or 17 loci. Since present methods permit a conclusion that a DNA exhibit can be individualized to a person, what are the advantages, if any, of going on to ever-increasing numbers of loci? Is this continued development cost effective? What is the logical end for this process?

2. What are the common objections to the use of DNA databases such as CODIS? Are these rational objections at this time? Do they have the potential of being serious problems in the future? What protections can be put into place to minimize objections?

3. Techniques such as capillary electrophoresis do not have infinite resolution, and there may arise questions about whether a peak really represents just a single base pair segment. How do we know, for example, that a peak is 29 or 30 base pairs? How is this problem, if it is one, handled in case work? Remember that any assumptions should favor the accused.

BIBLIOGRAPHY

Butler, J.M. (2001). *Forensic DNA typing: Biology and technology behind STR markers.* Boston: Academic Press.

Farley, M.A., & Harrington, J.J. (Eds.). (1991). *Forensic DNA technology,* Chelsea, MI: Lewis Publishers.

Rudin, N., & Inman, K. (2002). *An introduction to forensic DNA analysis (2nd ed).* Boca Raton, FL: CRC Press.

Waye, J.S., & Fourney, R.M. (1993). Forensic DNA typing of highly polymorphic VNTR loci. In R. Saferstein (Ed.), *Forensic science handbook, vol III.* Englewood Cliffs, NJ: Prentice-Hall.

Forensic Hair Examinations

KEY TERMS

Anagen

Buckling

Catagen

Color-banding

Cortex

Cortical fusi

Cortical/medullary distruptions

Cuticle

Epidermis

Eumelanin

Follicle

Fur hairs

Fusiform

Guard hairs

Hairs

Imbricate

Keratin/keratinization

Melanin/pigment

Melanocytes

Monilethrix

Ovoid bodies

Phaeomelanin

Pigment granules

Pili annulati

Pili arrector

Pili torti

Root

Root bulb/club root

Scale cast

Scale patterns

Scales

Sebaceous glands

Shaft

Shield

Shouldering

Subshield stricture

Telogen phase

Tip

Transitional body hairs

Vibrissa

INTRODUCTION

One of the most often recovered types of evidence is also one of the most misunderstood. Hairs make good forensic evidence because they are sturdy and can survive for many years, they carry a lot of biological information, and they are easy and cost-effective to examine. DNA can also be extracted from hairs and this adds to their forensic utility.

Figure 12.1. Hairs are fibrous growths that originate from the skin of mammals. Other animals have structures that are called hairs (like the tarantula), but only mammals have true hairs. Humans use hair as cultural and personal signs of status, gender, and art (Beach photo courtesy of philg@mit.edu).

Although a few cases of poor forensic hair examination have gathered attention by the media, especially in post-conviction reexaminations, the fault often lies more with the examiners themselves than with the method. As we will see, hairs can offer strong investigative and adjudicative information but only when examined properly, reported on conservatively, and testified to accurately.

GROWTH OF HAIRS

Hairs are a particular structure common only to mammals; they are the fibrous growths that originate from their skin. Other animals have structures that may appear to be or are even called hairs but they are not: Only mammals have hairs (see Figure 12.1). Humans use hairs as signs of culture, status, and gender, as well as for personal or artistic expression.

Hairs grow from the skin or, more precisely, **epidermis**, of the body, as shown in Figure 12.2. The **follicle** is the structure within which hairs grow; it is a roughly cylindrical tube with a larger pit at the bottom. Hairs grow from the base of the follicle upward. In the base of the fol-

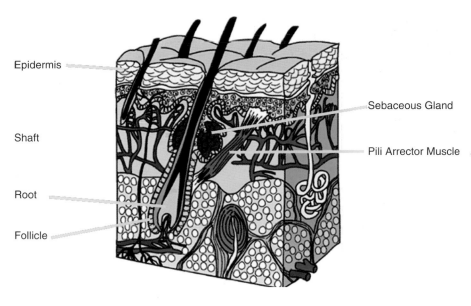

Epidermis

Shaft

Root

Follicle

Sebaceous Gland

Pili Arrector Muscle

Figure 12.2. The epidermis is a complicated structure, containing many different structures. The hair, composed of the tip, shaft, and root, develops within the follicle. As the hair grows, it slowly hardens and is fully keratinized by the time it reaches the surface of the skin. Sebaceous glands open into the follicle to secrete oils onto the hair. The pili arrector muscle controls the position of the hair, creating the "goose-bump" effect when we are chilled by contracting and pulling the hair upright.

licle, the hair is still very soft; as the hair proceeds up the follicle, it slowly begins to harden and dry out. Hair is made of **keratin**, a tough protein-based material from which hair, nails, and horns are made in animals. The hardening process of hair growth is therefore called **keratinization**. Hair is one of the most durable materials produced by nature; hairs from mummies, both natural and cultural in origin, have been found thousands of years after the person's death. Keratinization also explains why it doesn't hurt when we get our hairs cut: Hair is "dead" from the moment it peeks above the skin. The only place hair is "alive" is in the base of the follicle, which is why it *does* hurt when a hair get pulled out.

The follicle contains other structures, such as blood vessels, nerves, and **sebaceous glands**, the latter producing oils that coat hairs helping to keep them soft and pliable. Hairs even have muscles, called **pili arrector** muscles (*pilius* is the Latin word for "hair") that raise hairs when we get chilled (so-called "goose bumps").

Hairs go through three phases of growth, as depicted in Figure 12.3. In the **anagen**, or actively growing, phase, the follicle produces new cells and pushes them up the hair shaft as they become incorporated into the structure of the hair. The hair is moved up the shaft by a mechanical method. As the cells are produced, they "ratchet" up the shaft by opposing scales—much like gears in a machine! Between this mechanical method and the upward pressure from the growth of the

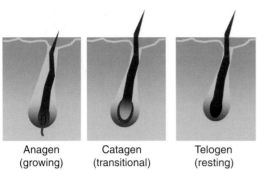

Anagen (growing) Catagen (transitional) Telogen (resting)

*Figure 12.3. Hairs grow in three phases. The **anagen**, or actively growing, phase, is when the follicle produces new cells and pushes them up the hair shaft. After two to seven years, the follicle transitions into the **catagen** phase. In this phase the follicle begins to shut down and quits producing hair in about two–three weeks. The final phase, the **telogen** or resting phase, lasts about 100 days; in this phase the follicle is shut down completely and the root is dried to a bulb. At this point, the hair is only attached mechanically and will be shed naturally.*

cells in the follicle, the hairs grow outward from the skin.

Specialized cells in the follicle produce small colored granules, called **melanin** or **pigment**, that give hairs their particular color; these cells are called **melanocytes**. Only two types of melanin are found in hairs: a dark brown pigment called **eumelanin** and a lighter pigment called **phaeomelanin**. The combination, density, and distribution of these granules produce the range of hair colors seen in humans and animals.

After the active growth phase, the hair transitions into a resting phase; this transitional phase is called the **catagen** phase. During the catagen phase, the follicle begins to shut down production of cells, the cells begin to shrink, and the root condenses into a bulb-shaped structure called, understandably, a **root bulb** or a **club root**.

Telogen phase is the resting phase for the follicle: Cell production has ceased completely, the root has condensed into a bulb, and it is held in place only by a mechanical connection at the base of the root/follicle. When this mechanical connection breaks (through combing, brushing, or normal wear), the follicle is triggered into the anagen phase again and the cycle renews. On a healthy human head of hair, about 80% to 90% of the hairs would be in the anagen phase, about 2% in the catagen phase, and about 10% to 18% in the telogen phase. When the telogen hairs are removed, new hairs begin to grow at once; clipping and shaving have no effect on growth. The time required for human follicles to regrow hairs varies from 147 days for scalp hairs to 61 days for eyebrow hairs. Humans, on average, lose about 100 scalp hairs a day; this provides for an adequate and constant source of potential evidence for transfer and collection.

Forensic hair examiners are sometimes asked if they can determine if a hair was removed forcibly, during a struggle or assault, for example, to document the severity of the assault. This is a difficult question; obviously, if the hair has a bulb root (meaning it was removed during the telogen phase), then the question can't be answered. If tissue from the follicle is attached to the root, then the hair was removed during the anagen or possibly catagen phase, that is, while the hair and the follicle were attached through active cellular growth. Because the actively growing hair is still soft and unkeratinized, the root may stretch before it is torn out of the follicle. Therefore, if the root is stretched *and* has follicular tissue attached, the examiner may state that the hair was forcibly removed, as shown in Figure 12.4. That does not, however, tell the examiner what *kind* of force was used—a violent assault, hair being caught in something, or a friendly wrestling match—and the examiner must be cautious about making unsupportable statements. Plucking hairs does not guarantee follicle tissue on the root. King, Wigmore, and Twibell showed in 1981 that only 65% of forcibly removed hairs yielded sheaths. Moreover, of the hairs removed by fast plucking, 53% had sheaths, while of those that were pulled slowly, only 11% had sheath tissue. Sheath cells were always associated with anagen and catagen hairs in this study. This study also suggests that in a bulk sample submitted as evidence, the anagen/telogen ratio may be more significant to the investigator than the presence or absence of sheathmaterial.

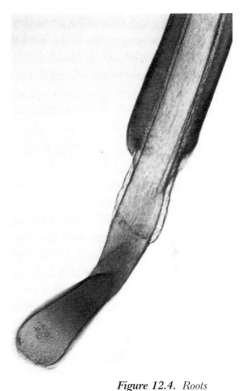

Figure 12.4. Roots that are stretched and have follicular tissue attached were probably removed by force; otherwise, this anagen phase hair would have stayed in the follicle. This does not, however, tell the examiner what kind of force was used—a violent assault, hair being caught in something, or a friendly wrestling match—and the examiner must be cautious about making unsupportable statements.

MICROANATOMY

A hair is a complicated, composite material with many intricately organized structures—only some of which are visible under the microscope. A single hair on a macro-scale has a root, a shaft, and a tip, as depicted in Figure 12.5A. The **root** is that portion that formerly was in the follicle, the proximal (the direction toward the body) most portion of the hair. The **shaft** is the main portion of the hair. The **tip** is the distal (the direction away from the body) most portion of the hair.

Internally, hairs have a variable and complex microanatomy. The three main structural elements in a hair are the cuticle, the cortex, and the medulla (see Figure 12.5B). The **cuticle** of a hair is a series of overlapping layers of **scales** that form a protective covering. Animal hairs have **scale patterns** that vary by species, and these patterns are a useful diagnostic tool for identifying animal hairs, as shown in Figure 12.6.

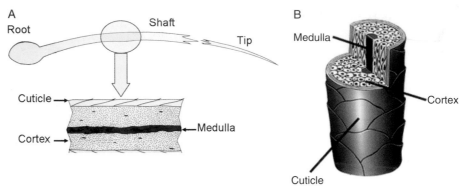

Figure 12.5. *(A) Macroscopically, a single hair has a root, a shaft, and a tip. The **root** is that portion that formerly was in the follicle, the proximal (the direction toward the body) most portion of the hair. The **shaft** is the main portion of the hair. The **tip** is the distal (the direction away from the body) most portion of the hair. Microscopically, a hair consists of three main portions. (B) The **cuticle** is an outer covering of overlapping scales, like shingles on a roof. The **cortex** is the main mass of the hair and contains numerous microanatomical features. The **medulla** is the central portion of the hair and consists of air- or fluid-filled cells.*

Figure 12.6. *The **cuticle** of a hair is a series of overlapping layers of **scales** that form a protective covering. Animal hairs have **scale patterns** that vary by species, and these patterns are a useful diagnostic tool for identifying animal hairs. Humans have a scale pattern called **imbricate** (outlined in red), but it occurs among animals as well. Despite attempts to use scales as an individualizing tool for human hairs, they are not generally useful in forensic examinations.*

Humans have a scale pattern called **imbricate**, but it is fairly common among animals and, despite attempts to use scales as an individualizing tool for human hairs, is not generally useful in forensic examinations.

The next structure is the **cortex**, which makes up the bulk of the hair. The cortex consists of spindle-shaped cells (sometimes called **fusiform**) that contain or constrain numerous other structures. **Pigment granules** are found in the cortex and are dispersed variably throughout the cortex. The granules vary in size, shape, aggregation, and distribution—all excellent characteristics for forensic comparisons. Small bubbles, called **cortical fusi**, may appear in the cortex; when they do appear, they may be sparse, aggregated, or evenly distributed throughout the cortex. Cortical fusi also vary in size and shape. Many telogen root hairs will have an aggregate of cortical fusi near the root bulb; it is thought that this is related to the shutdown of the growth activity as the follicle transitions from the catagen to telogen phase. This "burst" of fusi, then, is most likely related to physiology, so it is not necessarily useful for forensic comparisons, as pictured in Figure 12.7.

Odd structures that look like very large pigment granules, called **ovoid bodies**, may appear irregularly in the cortex. They may, in fact, *be* large, aggregated, or aberrant pigment granules, but no one knows; little, if any, research has been conducted on what ovoid bodies are. Another phenomenon that can be found in hairs is called **cortical** or **medullary disruptions**. They appear as if a small explosion occurred in the middle of the hair and may be found singly or in multiples.

Figure 12.7. Numerous microanatomical features are useful in the examination of hairs. **Pigment granules** *are found in the cortex and are dispersed variably throughout the cortex. The granules vary in size, shape, aggregation, and distribution—all excellent characteristics for forensic comparisons. Small bubbles, called* **cortical fusi**, *may appear in the cortex; when they do appear, they may be sparse, aggregated, or evenly distributed throughout the cortex. Cortical fusi also vary in size and shape. Odd structures that look like very large pigment granules, called* **ovoid bodies**, *may appear irregularly in the cortex. Another phenomenon that can be found in hairs is called* **cortical** *or* **medullary disruptions**. *They appear as if a small explosion occurred in the middle of the hair and may be found singly or in multiples.*

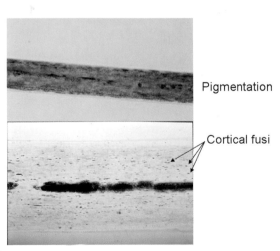

Pigmentation

Cortical fusi

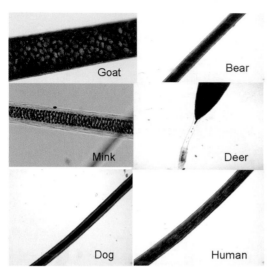

Figure 12.8. It is relatively easy to determine if a hair is human or non-human (this term is often used instead of "animal" because, technically, humans are also animals) by a simple microscopic examination. Animals have several macroscopic characteristics that distinguish their hairs from those of humans.

HUMAN VS. NON-HUMAN HAIRS

It is relatively easy to determine if a hair is human or non-human (this term is often used instead of "animal" because, technically, humans are also animals) by a simple microscopic examination, as shown in Figure 12.8. Determining *what kind* of non-human hair it is, however, may be in some circumstances quite tricky because certain animals' hairs can be similar. Animal hairs have several macroscopic characteristics that distinguish them from those of humans.

First of all, animals have three types of hairs. **Guard hairs** are large stiff hairs that make up the outer part of the animal's coat. Guard hairs should be used for microscopic identification. These hairs may have a widening in the upper half of the shaft called a **shield**. Below the shield, if it is present, may often be found a **subshield stricture**, a narrowing of the hair to slightly less than the normal, non-shield shaft diameter. A subshield stricture may be accompanied by a bend in the shaft at the stricture.

Thinner, softer **fur hairs** fill in the rest of the coat, providing warmth and bulk. Fur hairs are generic in their appearance and are typically useless for microscopic identification. The root may give an indication as to taxonomic origin, but it may also be misleading; it is best not to use fur hairs for microscopic evaluations.

Finally, animals have **vibrissa**, the technical term for whiskers, the short to long, stiff, often white hairs around the snout and muzzle. No comprehensive study has been made on the identification of taxonomic origin by vibrissa, probably because these hairs have a long life cycle and are lost comparatively less often than the myriad guard and fur hairs of a typical animal.

Some non-human hairs are **color banded**, showing abrupt color transitions along the shaft of the hair, including the tip. Raccoons, for example, have four color bands in their guard hairs; incidentally, they are the only animals known to have this many bands.

As noted earlier, scale patterns also may be useful in identifying animal hairs. The best ways to visualize scale patterns are using a scanning electron microscope or making a **scale cast** and viewing it with a light microscope. The simplest method of making a scale cast is to brush clear nail polish onto a glass microscope slide and lay the hair in the still-wet polish. Before the polish dries completely, the hair should be gently "peeled" from the polish; a cast of the exterior of the hair remains in the polish. This cast can then be examined on a light microscope.

BODY AREA DETERMINATION

Unlike other animals, humans exhibit a wide variety of hairs on their bodies; why humans have hair where they do is of interest to evolutionary biologists and one recent theory is discussed in "Issues In: Hairless and Flea-free." The characteristics of these hairs may allow for an estimation of body area origin. The typical body areas that can be determined are head (or scalp), pubic, facial, chest, axillary (armpits), eyelash/eyebrow, and limb; typically, only head and pubic hairs are suitable for microscopic comparison, as shown in Figure 12.9. Hairs that do not fit into these categories may be called **transitional body hairs**, such as those on the stomach, between the chest and the pubic region. Table 12.1 lists the characteristics generally associated with the different body hair types.

Buckling is an abrupt change in direction of the hair shaft with or without a slight twist. **Shouldering** is an asymmetrical cross-section of hairs.

In some instances, it may be difficult or impossible for the forensic scientist to make a clear decision as to whether a hair is "chest" or "axillary" in origin; it may also not matter to the circumstances of the crime. Labeling the hair as a "body hair" is sufficient and may be the most accurate conclusion given the quality and nature of the hair.

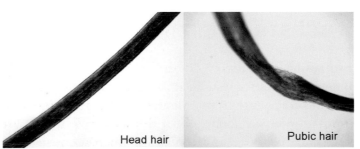

Head hair   Pubic hair

Figure 12.9. Unlike other animals, humans exhibit a wide variety of hairs on their bodies. The characteristics of these hairs may allow for an estimation of body area origin. The typical body areas that can be determined are head (or scalp), pubic, facial, chest, axillary (armpits), eyelash/eyebrow, and limb. Hairs that do not fit into these categories may be called **transitional body hairs,** *such as those on the stomach, between the chest and the pubic region.*

Table 12.1. *General descriptions of human body area hair traits. Compare with the photographs in Figure 12.10.*

AREA	DIAMETER	SHAFT	TIP
Head	Even	Straight or curly; some waviness; may be very long	Usually cut
Pubic	Varies	**Buckling**; sometimes extreme waviness or curl	Usually pointed; may be razor cut
Facial	Wide; even	Triangular in cross-section; some **shouldering**	Usually cut; may be scissors or razor
Chest	Even; some variation	Wavy to curly; some more straight	Usually pointed
Axillary	Even; some variation	Less wavy/curlier than chest	Usually pointed; may be colorless
Limb	Fine; tapering	Slight arc	Usually pointed
Eyebrow/Eyelash	Tapering	Arc; short	Pointed

This determination may have important consequences for a case: One of the authors (MMH) worked a case involving the identification of an adult pubic-area hair on a preadolescent victim. A girl of that age could not have produced a pubic-area hair: Those hairs are generated by the hormones associated with puberty. DNA from the hair was the same as that from the suspect; this, in addition to overwhelming trace evidence associating the suspect with the crime, led to a guilty plea (see Ryland and Houck, 2000, in *Mute Witnesses* for more information).

ANCESTRAL ESTIMATION

Estimating the ethnicity or ancestry of an individual from his or her hairs is just that: an estimate. A study of forensic hair examiner trainees conducted by one of us (MMH) showed that their accuracy for racial estimation on a standard set of tests was 85%, not bad for trainees, considering that this was based on a microscopical examination alone.

ISSUES IN . . .
Hairless and flea-free

Did humans lose their thick fur to cut down on parasites or to lose heat more efficiently? A new theory comes down on the side of being bug-free, not staying cool. "The nakedness of humans is a glaring difference between humans and other mammals," says evolutionary biologist Mark Pagel of the University of Reading, U.K. And Pagel and Walter Bodmer, a geneticist at the University of Oxford, believe hairlessness is tied to humans' uniquely civilized behavior. When early humans began to don clothing and build shelters, they no longer needed protective fur, the researchers say. And those with less hair may have been healthier because it was easier to keep free of parasites, which thrive where animals make permanent homes.

Sexual selection might have speeded up the evolution of hairlessness, as exposed skin signaled a healthier prospective mate, Pagel and Bodmer argued in a paper published online 9 June [2003] in *Biology Letters*.

Evolutionary biologist Robin Dunbar of the University of Liverpool notes that the theory needs testing—for example, by seeing if people in high-parasite areas have less hair. He adds that it would radically change our image of early humans. The cooling-off theory suggests that we lost most of our hair more than 2 million years ago, after taking to two legs; if the parasite idea is correct, nakedness would likely have evolved 1.5 million years later.

Source: Holden, C. (Ed.) (2003). Random Samples. *Science, 300*: 2028–2029.

Anthropologists can be more accurate using skeletal measurements, but they use several measurements on different bones and then compare them to a large population of similar measurements. This makes the anthropologists' estimate more accurate, but regrettably these are not options for microscopical hair examinations.

The general morphology and color of a hair can give an indication of a person's ancestry. Humans are more variable from one to another in their hair morphology than any other primate. This variation tends to correlate with a person's ancestry (see Table 12.2) although it is not an exact correlation. For simplicity and accuracy, three main ancestral groups are used: Europeans, Africans, and Asians. In the older anthropological and forensic literature, these groups were referred to as, respectively, Caucasoids, Negroids, and Mongoloids; these terms are archaic now and should probably not be used. They are no better at describing the intended populations than the geographic terms listed previously—Caucasoid/European hair descriptions include some Hispanics and peoples of the Middle East, for example—but the geographic terms are as accurate and less offensive.

Typically, head and pubic hairs provide the clearest evidence for ancestral estimates. It may be possible with certain other hairs, espe-

Table 12.2. *Various characteristics of hair by ancestry.*

ANCESTRY	DIAMETER	CROSS SECTION	PIGMENT DISTRIBUTION	CUTICLE	UNDULATION
African	60–90 µm	Flat	Dense; Clumped	Thin	Prevalent
European	70–100 µm	Oval	Even	Medium	Uncommon
Asian	90–120 µm	Round	Dense to very dense	Thick	Never

cially facial hairs, but body hairs should be viewed with a cautious eye. Asians, for example, have less body hair than other populations and, in some areas, may have none.

Some examiners include a fourth category: mixed race. Technically, everyone is "mixed race," so this is a misnomer; "other" might be more accurate. In one study, two researchers, one experienced in hair examinations (>14 years) and one not as experienced (<1 year), did a blind study of hair from children of known "mixed" marriages. Both researchers showed positive correlation between non-Black ancestral assessment and increasing European ancestry—the less experienced examiner had a correlation of 0.23 (1.0 being a perfect 1:1 correlation), while the experienced examiner had a stronger correlation of 0.61. This is important to remember: Just because an examiner estimates a hair to be from a person of a certain ancestry doesn't mean that is how that person identifies himself or herself racially.

DAMAGE, DISEASE, AND TREATMENTS

Humans do many different things to their hair depending on their culture: cutting, dyeing, braiding, even shaving—and this isn't limited to just the scalp. Some diseases affect the hairs or the follicles and are distinctive but rare.

The tips of hairs can provide good information about how the hair has been treated. Scissor-cut hair has a clean, straight border, whereas razor-cut hair is angled. In a hit-and-run incident or an explosion, flying glass cuts hair in a unique way, leaving a long curved "tail." Burnt hair is blackened and may appear bubbled or expanded, crushed or split hair is also easy to recognize, as depicted in Figure 12.10.

Bleaching of the hair oxidizes the pigmentation and removes its color. The treatment may stop at this point, or a new color may be added to the hair. Coloring hair is much like dyeing wool fibers (both are hair) or other textile fibers. As the hair continues to grow, the point where the bleaching/coloration was applied is visible as an abrupt color change, as shown in Figure 12.11. If the length of the natural hair color portion is measured, the examiner can make an estimate of the time interval between the cosmetic treatment and the time the hair was lost. Head hairs grow an average of $\frac{1}{2}''$ (1.3 cm) per month, so the natural portion length in inches would be multiplied by 0.5 to yield the approximate number of months.

Figure 12.10. Singed hair is blackened and may appear bubbled or expanded. Split tips may be due to cosmetic treatment or weakened hair.

The diseases that affect the hair morphology are rare, but this makes them excellent evidence for identifying a source. **Pili annulati** refers to hairs with colored rings. In pili annulati the hairs have alternating light and dark bands along its length, like tiger or zebra stripes. People with dark hair may have pili annulati but not know it because their hair color masks the condition. **Monilethrix** makes hairs look like a string of beads (the name comes from the Greek words for "bead" and "hair"). Along the length of the hair are nodes and constrictions making the hair vary in diameter. This hair beading weakens the hair and people suffering from monilethrix have patchy hair loss. **Pili torti** is, as the name suggests, a twisting of the hair along its length, creating a spiral morphology. There

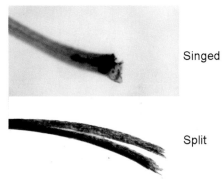

Singed

Split

Figure 12.11. Coloring hair is much like dyeing wool fibers (both are hair) or other textile fibers. As the hair continues to grow, the point where the bleaching/coloration was applied is visible as an abrupt color change. This hair has been bleached.

may be several twists in one hair. The cuticle is present, but the twisting creates stress that leads to fractures in the cuticle and cortex.

Vermin (such as lice), dandruff, or fungus may also be present on hairs, and this should be noted. They add to the classification of the hairs as coming from individuals with these traits.

Misconceptions abound about hairs and what can be derived from their examination. Age and sex cannot be determined from looking at hairs; gray hairs may occur from a person's 20s onward, and long hair doesn't mean "female" just as short hair doesn't mean "male." Hairs do not grow after you die (how could they?)—the skin shrinks from loss of water, making the hairs more prominent (likewise with nails). And, some studies to the contrary, shaving does not stimulate hair growth.

COMPARISON OF HUMAN HAIRS

The goal of most forensic hair examinations is the comparison of a questioned hair or hairs from a crime scene to a known hair sample. A known hair sample consists of anywhere between 50 and 100 hairs from all portions of the area of interest, typically the head/scalp or pubic area. The hairs must be combed and pulled to collect both telogen and anagen hairs. A known sample must be representative of the collection area to be suitable for comparison purposes. Braids, artificial treatment, graying—all must be noted and collected for a suitable known sample.

A comparison microscope is used for the examination. A comparison microscope, as shown in Figure 12.12, is composed of two transmitted light microscopes joined by an optical bridge to produce a split image. The sample on the right appears in the right-hand field of view, and the sample on the left appears in the left-hand field of view. This side-by-side, point-by-point comparison is central to the effectiveness and accuracy of a forensic hair comparison. Hairs cannot be compared properly without one.

The hairs are examined from root to tip, at magnifications of 40× to 250×. Hairs are mounted on glass microscope slides with a mounting medium of an appropriate refractive index for hairs, about 1.5. All of the characteristics present are used. The known sample is characterized and described to capture its variety. The questioned hairs are then described individually. These descriptions cover the root, the microanatomy of the shaft, and the tip (see Chart 12.1).

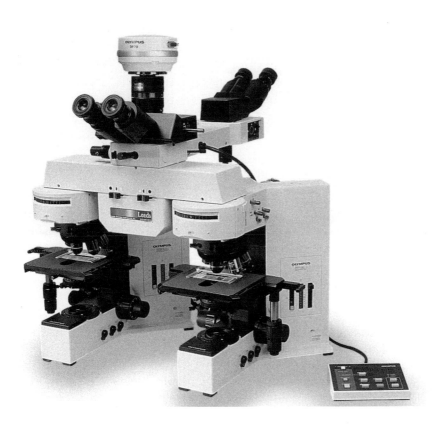

Figure 12.12. A comparison microscope is used for the examination. A comparison microscope is composed of two transmitted light microscopes joined by an optical bridge to produce a split image. The sample on the right appears in the right-hand field of view, and the sample on the left appears in the left-hand field of view. This side-by-side, point-by-point comparison is central to the effectiveness and accuracy of a forensic hair comparison. Hairs cannot be compared properly without one (Courtesy: Olympus USA).

Three basic conclusions can be drawn from a forensic hair comparison. First, if the questioned hair exhibits the same microscopic characteristics as the known hair sample, then it could have come from the same person who provided the known sample. Hair comparisons are not a form of positive identification, however. Second, if the questioned hair exhibits similarities but slight differences to the known hair sample, then no conclusion can be drawn as to whether the questioned hair could have come from the known source. Finally, if the questioned hair exhibits different microscopic characteristics from the known hair sample, then it can be concluded that the questioned hair did not come from the known source. Dick Bisbing (2001), a noted forensic hair expert, sums up the decision-making process of a forensic hair comparison this way:

> [I]f two samples have originated from one individual, there must always be sufficient agreement on several other characteristics that have no fundamental dissimilarities. An association does not rest, therefore, merely on a similar combination of identifying traits

Chart 12.1. Traits. *A sample list of hair characteristics used to describe known and questioned hairs. Some of the traits can be further defined, such as "scissor cut tip," "razor cut tip," "glass cut tip," etc. This list was produced by the Forensic Resource Network Hair Project (a National Institute of Justice-funded program).*

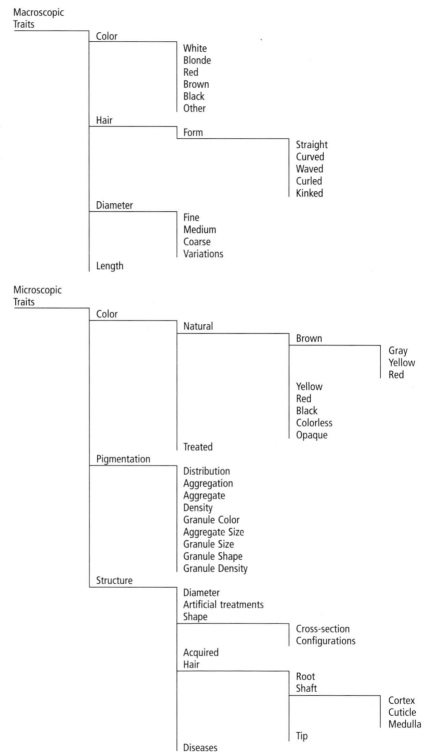

Macroscopic Traits
- Color
  - White
  - Blonde
  - Red
  - Brown
  - Black
  - Other
- Hair
  - Form
    - Straight
    - Curved
    - Waved
    - Curled
    - Kinked
- Diameter
  - Fine
  - Medium
  - Coarse
  - Variations
- Length

Microscopic Traits
- Color
  - Natural
    - Brown
      - Gray
      - Yellow
      - Red
    - Yellow
    - Red
    - Black
    - Colorless
    - Opaque
  - Treated
- Pigmentation
  - Distribution
  - Aggregation
  - Aggregate
  - Density
  - Granule Color
  - Aggregate Size
  - Granule Size
  - Granule Shape
  - Granule Density
- Structure
  - Diameter
  - Artificial treatments
  - Shape
    - Cross-section
    - Configurations
  - Acquired
  - Hair
    - Root
    - Shaft
      - Cortex
      - Cuticle
      - Medulla
    - Tip
  - Diseases

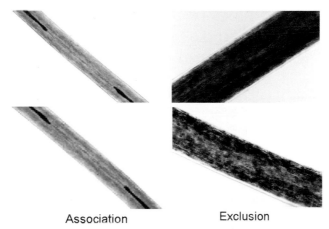

Association   Exclusion

Figure 12.13. A hair comparison is a good method of demonstrating possible association between questioned hairs and individuals. A suitable known sample of hairs from the same body area is necessary to conduct a comparison. Hairs are not a means of positive identification, however; statistics or frequency estimates cannot currently be applied to microscopic hair comparisons.

(though this condition must always be fulfilled) but also on a coexistent lack of basic divergences between the questioned and standard hair. . . . It is necessary to demonstrate *not only* that the unknown hair has the traits of the known hair *but also* that the variations that occur in the unknown hair are similar to the variations in the (known sample) (p. 408).

This evaluation and balancing of microscopic traits within and between samples is key to the comparison process; Figure 12.13 shows hairs that are positively and negatively associated. But how to interpret the results of a forensic hair comparison quantitatively? It's not as simple as might be imagined (see "In More Detail: Statistics and Hair Examinations").

DNA AND HAIRS

The advent of forensic mitochondrial DNA in the mid 1990s heralded a new era of biological analysis in law enforcement. This was especially true for hairs because it offered a way to add information to microscopic hair examinations. The microscopic comparison of human hairs has been accepted scientifically and legally for decades. Mitochondrial DNA (mtDNA) sequencing added another test for assessing the significance of attributing a hair to an individual. Neither the microscopic nor molecular analysis alone, or together, provides positive identification. The two methods complement each other in the information they provide. For example, mtDNA typing can often distinguish between hairs from different sources although they have similar, or insufficient,

IN MORE DETAIL:
Statistics and hair examinations

Given the list of traits in Chart 12.1, you might think that hairs could be coded, entered into a database, and eventually frequency information could be derived. This would be of immense help in determining the significance of hairs as evidence. A hair's traits could be entered as a query, and at the push of a button, a frequency of occurrence for a population could be calculated. But it's not that easy.

Barry Gaudette, a former hair examiner with the Royal Canadian Mounted Police and one of the premier forensic hair examiners, did a clinical study to attempt to determine the specificity of microscopic hair examinations (1974). Gaudette's work involved brown head hairs of European ancestry, coded and inter-compared. The study determined that only 9 pairs of hairs were indistinguishable, resulting in a frequency of 1 in 4,500. He did further work with pubic hairs that resulted in a frequency of 1 in 1,600 (1976).

Although critics complained that the study was flawed and the frequencies are not valid for any other sample, it was the first clinical study of its kind. Some examiners quoted these frequencies in their testimony to quantify the significance of their findings—a completely unjustified and erroneous application of the study.

A later paper by Gaudette's colleagues (Wickenheiser and Hepworth, 1990) elaborated on his study and refined the frequencies. Other smaller studies provided additional insights into what the potential specificity of microscopic hair examinations might be but, to date, no universal approach for calculating significance has been published.

And probably none will be (Gaudette 1978; 1982). Hairs are a very complicated composite biological material, and the expression of hair traits across the population is highly variable. Because hairs are three-dimensional makes quantifying the traits that much more difficult. While a computer could be used to analyze digital images and categorize the hairs, a human could do it much faster and just as accurately. And now that DNA analysis is more accessible, this approach is hardly justified.

microscopic hair characteristics. Hair comparisons with a microscope, however, can often distinguish between samples from maternally related individuals where mtDNA analysis is "blind."

In a recent study (Houck and Budowle, 2003), the results of microscopic and mitochondrial examinations of human hairs submitted to the FBI Laboratory for analysis were reviewed. Of 170 hair examinations, there were 80 microscopic associations; importantly, only nine were excluded by mtDNA. Also, 66 hairs that were considered either unsuitable for microscopic examinations or yielded inconclusive microscopic associations were able to be analyzed with mtDNA. Only six of these hairs did not provide enough mtDNA, and another three yielded inconclusive results. This study demonstrates the strength of combining the two techniques.

It is important to realize that microscopy is not a "screening test," and mtDNA analysis is not a "confirmatory test." Both methods, or either, can provide important information to an investigation. One test is not better than the other because they both analyze different characteristics. The only question left, then, as posed by James Robertson of the Australian Federal Police, is

> to what extent preliminary microscopic examinations should be conducted prior to DNA analysis . . . it may well be the case that there will *be little if any reduction in the level of microscopic examination as it will be both necessary and desirable to eliminate as many questioned hairs as possible* and concentrate mtDNA analysis on only key hairs. (emphasis added) (p. 127)

The data in the FBI study support the usefulness of both methods—and this is echoed in the expanding use of both microscopical and mitochondrial DNA examinations of hairs in forensic cases.

CHAPTER SUMMARY

Hairs are among the most often collected and potentially useful types of trace evidence. Information about people and animals is readily apparent from a simple microscopical examination. Microscopical comparisons can provide additional information by including or excluding individuals from consideration. Mitochondrial DNA enhances this information by adding genetic information to the morphological observations made by the forensic hair microscopist.

Test Your Knowledge

1. What types of hairs do animals have?
2. What are the growth stages of hairs?
3. How can you tell if a hair may have been forcibly removed?
4. What are some of the differences between human and animal hairs?
5. What characteristics are used to determine body area?
6. Why is estimating ancestry from hairs difficult?
7. Name three ways in which a hair can be cosmetically treated.
8. What constitutes a suitable known hair sample?

9. What is a comparison microscope?

10. What are the three conclusions that can be drawn from a microscopical examination of hairs?

11. How does the use of mtDNA assist in hair comparisons?

12. How would you distinguish between males and females using hair?

13. Can you tell how old someone is by looking at his or her hair?

14. What are cortical fusi?

15. What are the three macroscopic parts of a hair?

16. What are three main microscopic parts of a hair?

17. How would you distinguish between a head hair and a pubic hair?

18. Can you use statistics to describe the significance of hair comparisons?

19. Refer to Chapter 3, "The Nature of Evidence." How can microscopy of hair help a laboratory be more efficient?

20. How could you tell if a "hair" is really a synthetic fiber from a wig? What would you look for?

Consider This . . .

1. Why are hairs *not* a form of positive identification? What prevents them from being so?

2. If hairs are not a form of positive identification, why do hairs make good evidence?

BIBLIOGRAPHY

Bisbing, R. (2001). The forensic identification and association of human hair. In R. Saferstein (Ed.), *Forensic science handbook* (2nd ed.), Englewood Cliffs, NJ: Prentice-Hall.

Gaudette, B.D., & Keeping, E.S. (1974). An attempt at determining probabilities in human scalp hair comparison. *Journal of Forensic Sciences, 19,* 599–606.

Gaudette, B.D. (1976). Probabilities and human pubic hair comparisons. *Journal of Forensic Sciences, 21,* 514–517.

Gaudette, B.D. (1978). Some further thoughts on probabilities and human hair comparisons. *Journal of Forensic Sciences, 23,* 758–763.

Gaudette, B.D. (1982). A supplementary discussion of probabilities and human hair comparisons. *Journal of Forensic Sciences, 27,* 279–289.

Holden, C. (Ed.) (2003). Random Samples. *Science, 300,* 2028–2029.

Houck, M.M., & Budowle, B. (2002). Correlation of microscopic and mitochondrial DNA analysis of hairs. *Journal of Forensic Sciences, 45*(5), 1–4.

Houck, M.M. (2001). A bibliography of hair references for the forensic scientist. *Forensic Science Communications, 4*(1). Retrieved from www.fbi.gov.

Houck, M.M., & Koff, C.M. (2000, July 24–28). Racial assessment in hair examinations. 9th Biennial Scientific Meeting of the International Association for Craniofacial Identification, Washington, DC.

King, L.A., Wigmore, R., & Twibell, J.M. (1981). The morphology and occurrence of human hair sheath cells. *Journal of the Forensic Science Society, 22*, 267–269.

Robartson, J. (1999). Forensic and Microscopic Examination of Hair, in J. Robartson (Ed.) *Forensic Examination of Hair*, London: Taylor and Francis.

Ryland, S., & Houck, M. (2000). Only circumstantial evidence. In *Mute Witnesses*. San Diego, CA: Academic Press.

Wickenheiser, R.A., & Hepworth, D.G. (1990). Further evaluation of probabilities in human scalp hair comparisons. *Journal of Forensic Sciences, 35*, 1323–1329.

Wilson, M.R., Polanskey, D., Butler, DiZinno, J.A., Replogle J., & Budowle, B. (1995). Extraction, PCR amplication and sequencing of mitochondrial DNA from human hair shafts. *Biotechniques, 18*, 662–668.

Wilson, M.R., DiZinno, J.A., Polanskey, D., Replogle, J., & Budowle, B. (1995). Validation of mitochondrial DNA sequencing for forensic casework analysis. *International Journal of Legal Medicine, 108*, 68–72.

PART FOUR

Chemical Sciences

Illicit Drugs

KEY TERMS

Addiction

Amphetamine

Barbiturates

Benzodiazepines

Bureau of Narcotics and Dangerous Drugs

Cocaine

Codeine

Controlled substances

Controlled Substances Act

Confirmation

Crack

Cystolithic hairs

Dependency

Depressant

Designer drug

Diluent

Duquenois-Levine test

Drug abuse

Drug Enforcement Agency (DEA)

Ecstasy (MDMA)

Ephedrine

Ergotism (St. Anthony's Fire)

Erlich's test

Excipient

Flashbacks

Hallucinogen

Harrison Narcotics Tax Act

Hash oil

Hashish

Illicit drugs

Marijuana (*Cannabis*)

Meprobamate

Methamphetamine

Methaqualone

Microcrystal test

Narcotic

Naroctic Drug Control Act

Naturally occurring

Opiate

Opium poppy

Peyote cactus

Phencyclidine (PCP)

Plant extract

Pure Food and Drugs Act

Ruybal (or Scott) test

Semisynthetic

Sinsemilla

Stimulant

Synthetic

Uniform Controlled Substances Act

Usable quantity

THE CASE

An illicit drug task force, jointly operated by the Federal Drug Enforcement Administration and the State Police, has been investigating clandestine laboratory activity in and around a major city. They have been following suspicious characters, who have been purchasing what appear to be chemicals that can be used to manufacture methamphetamine. When the suspects have purchased everything they need, the members of the task force begin surveillance on the home where the lab is suspected to be located.

The next day, a neighbor reports smelling chemicals emitting from the house. A chemist is sent to sniff the smell. She identifies it as a chemical that is used in the final step of the production of methamphetamine. On that basis, the investigators get an undercover "snitch" to go to the house and buy some of the methamphetamine. The police take the powder to the local forensic science lab, where a drug chemist identifies it as containing methamphetamine.

On the basis of this analysis, the police and DEA obtain a search warrant, enter the house, and seize all of the paraphernalia and chemicals and drugs. This is all sent to the crime lab for analysis. The occupants will be charged with manufacture of methamphetamine.

INTRODUCTION

A drug is a natural or synthetic substance that is designed to produce a specific set of psychological or physiological effects on the human body or, in some cases, other animals. Most drugs are produced legitimately by drug manufacturers and are prescribed for particular illnesses, injuries, or other medical problems. These drugs are most often taken and used for the intended purpose. Sometimes, however, they have effects that people find pleasurable and thus are taken for other than their intended purposes. **Drug abuse** occurs when people take drugs for purposes other than for which they are intended, usually for their psychoactive effects.

In addition to legally produced pharmaceutical drugs, there are also chemical or naturally occurring substances that have no legitimate, recognized medicinal purpose, but are produced and ingested entirely for their psychoactive effects. Many of these drugs are naturally occurring substances or are extracted or derived from natural substances, usually plants. Legally produced drugs that are abused and drugs produced for no reason other than abuse are called abused drugs, drugs of abuse, or illicit drugs. In the United States many of them are **controlled substances**, a term which refers to their inclusion in a part of the Federal Code called the *Controlled Substances Act* (Act 21 U.S.C. 812). Throughout the world, many of these terms are used to describe abused drugs. The most common one is **illicit drugs**, and that is the term that will be used in this chapter.

ORIGINS OF DRUGS

THE CONTROL OF ILLICIT DRUGS IN THE UNITED STATES

Why are some drug substances prohibited or controlled in the United States, whereas others are taken freely? The reasons are complex and have to do with how people perceive the notion of what is good for the public. In addition, questions of morality, personal choice, social order, and health are part of the debate. Over time the issue of drug control has been complicated by the emergence of facts and fallacies about certain drugs. What is clear is that historically our drug control laws and regulations have been somewhat disjointed and uncoordinated and have resulted from society's responses to various social crises throughout our short history.

Prior to the beginning of the twentieth century, there was little in the way of drug control in the United States. This changed with the passage of two federal laws, one in 1906 and the other in 1914. In part, these laws were passed due to public reaction to opium smoking among Chinese immigrants, the rise of cocaine use, and increased activity by purveyors of patent medicines. The result was, in 1906, the passage of the *Pure Food and Drugs Act*, which prohibited interstate commerce in mislabeled or adulterated food or drugs. Among the substances targeted by the law were marijuana, cocaine, heroin, and opium. This act was administered by the Department of Agriculture.

In 1914, Congress passed the *Harrison Narcotics Tax Act*, which is properly known as "An act to provide for the registration of, with collectors of internal revenue, and to impose a special tax upon all persons who produce, import, manufacture, compound, deal in, dispense or give away opium or coca leaves, their salts, derivatives, or preparations, and for other purposes." This law was enforced and administered by the Bureau of Internal Revenue in the Treasury Department. It gave the federal government broad control over cocaine and narcotics traffic in the United States.

At the time the Harrison Act was passed, the climate in the country seemed to favor continuing to supply addicts with their needed drugs while simultaneously closing down dealers and purveyors of the illegal drugs. This was the way that the Act was enforced early on. Later, in the late 1920s the mood shifted, and it was felt that drug addicts could be easily cured if their drugs were taken away. This resulted in a crackdown on physicians who had been heretofore legally supplying addicts with drugs. Slowly, the view was changing from drug abuse being a medical problem to a law enforcement problem.

In 1930, Congress formed the Bureau of Narcotics within the Treasury Department. The Bureau stepped up law enforcement against illicit drugs, particularly narcotics and cocaine and marijuana. At this time, anyone who wanted to buy or import or sell any of these drugs had to register and pay a tax. Because marijuana was included, it was labeled a narcotic in all relevant federal laws, a label that stuck until the early 1970s.

After World War II, testimony before Congress indicated that half of all crime in cities in the United States was related to illegal drug use. This led, in 1956, to the *Narcotic Drug Control Act*, which called for increased penalties for illicit use of these drugs. Stiff jail sentences went to all but first-time offenders, and anyone who sold drugs to a minor faced the death penalty. This law also had another important feature. If a new drug that had a potential for abuse came into the marketplace, a recommendation to control it could be made by the Food and Drug Administration to the Secretary of Health, Education and Welfare. Drugs such as amphetamines, barbiturates, and LSD were brought under control during this time. Rather than labeling them narcotics, they were referred to in the law as "dangerous drugs." The Bureau of Narcotics was changed to the **Bureau of Narcotics and Dangerous Drugs**, and its officers became the chief enforcers of the new laws.

In 1970, the Congress passed the *Comprehensive Controlled Substances Act*. This comprehensive law repealed or updated all previous laws that controlled both narcotics and dangerous drugs. This law put all controlled substances in the federal realm. This meant that the federal government could prosecute anyone for a drug offense regardless of whether interstate trafficking was involved and irrespective of state laws.

This new law resulted in a number of major changes in drug enforcement in the United States:

1. Control of drugs became a direct law enforcement activity, rather than through registration and taxation.

2. Enforcement was moved from the Treasury Department to the Justice Department and the Bureau of Narcotics and Dangerous Drugs became the **Drug Enforcement Administration (DEA)**.

3. The decision on which drugs should be controlled rests with the Secretary of Health and Human Services, which delegates to the FDA the determination of which drugs should be controlled. In making decisions about whether a drug should be controlled, the FDA evaluates such factors as pharmacological effects, ability to induce psychological dependence or physical addiction, and whether there is any legitimate medical use for the substance (as defined and recognized by the FDA).

Under this law, tobacco and alcohol products are excluded. Controlled drugs are put into five schedules. See Table 13.1 for a summary of the schedules and the drugs that are found in each one. More comprehensive information about the federal schedules can be found on the DEA website at **http://www.dea.gov/concern/abuse/chap1/contents.htm.**

Table 13.1. *Federal Schedules of controlled substances.*

SCHEDULE I
• The drug or other substance has a high potential for abuse.
• The drug causes physical addiction or psychological dependence.
• The drug or other substance has no currently accepted medical use in treatment in the United States.
• There is a lack of accepted safety for use of the drug or other substance under medical supervision.
• Some Schedule I substances are heroin, LSD, marijuana, PCP, and methaqualone.

Table 13.1. *Continued*

SCHEDULE II
• The drug or other substance has a high potential for abuse.
• The drug or other substance has a currently accepted medical use in treatment in the United States or a currently accepted medical use with severe restrictions.
• Abuse of the drug or other substance may lead to severe psychological or physical dependence.
• Schedule II substances include morphine, PCP, cocaine, methadone, and methamphetamine.

SCHEDULE III
• The drug or other substance has a potential for abuse less than the drugs or other substances in Schedules I and II.
• The drug or other substance has a currently accepted medical use in treatment in the United States.
• Abuse of the drug or other substance may lead to moderate or low physical dependence or high psychological dependence.
• Anabolic steroids, codeine and hydrocodone with aspirin or Tylenol, and some barbiturates are Schedule III substances.

SCHEDULE IV
• The drug or other substance has a low potential for abuse relative to the drugs or other substances in Schedule III.
• The drug or other substance has a currently accepted medical use in treatment in the United States.
• Abuse of the drug or other substance may lead to limited physical dependence or psychological dependence relative to the drugs or other substances in Schedule III.
• Included in Schedule IV are Darvon, Talwin, Equanil, Valium, and Xanax.

SCHEDULE V
• The drug or other substance has a low potential for abuse relative to the drugs or other substances in Schedule IV.
• The drug or other substance has a currently accepted medical use in treatment in the United States.
• Abuse of the drug or other substance may lead to limited physical dependence or psychological dependence relative to the drugs or other substances in Schedule IV.
• Over-the-counter cough medicines with codeine are classified in Schedule V.
Controlled Substances Act (Act 21 U.S.C. 812)

Table 13.2 does not include marijuana, which is penalized according to a different method. This is shown in Table 13.3.

The Comprehensive Controlled Substances Act of 1970 remains the law of the land today. The only major changes in the law have been an increase in the number of controlled substances and changes in penalties associated with possession and distribution of the drugs. In addition, the DEA developed and recommends a model state law titled the *Uniform Controlled Substances Act*. Most states have adopted this as a framework to replace their existing drug laws. Under this Act, states use the same scheduling system for controlling illicit drugs. Some states have added schedules, changed the specific drugs within a schedule, or have changed penalties for possession or distribution of drugs, but the basic framework remains the same as for the federal laws.

WHY ARE CERTAIN DRUGS REGULATED?

As mentioned previously, there are a number of reasons why governments seek to control the use of illicit drugs—why some substances are labeled "illicit," whereas others are not. The major reason seems to be our notion of the "public good." There is strong sentiment that everyone should be a productive member of society so that the society will prosper and grow. If people spend their otherwise productive time in pursuit of the hedonistic pleasures derived from the abuse of substances such as illicit drugs, then they are not acting in the interest of the public good. Of course, rationales for penalizing drug abuse go beyond the nature of the public good. There is also a widespread belief that drug abuse is an immoral activity.

Of course, there are those people who believe that a person should be free to engage in such "victimless" pursuits as recreational drug use in their own homes, that no one is being hurt by this practice, and that government is making an unwarranted intrusion into people's lives when it punishes casual drug users. But then, on the other side of this coin are those who espouse economic, social, and health arguments against drug abuse. Many people become "hooked on" (addicted to) certain illicit drugs and spend much of their lives in pursuit of the drugs they need. This leads to a marked increase in crime as people steal, burgle, and rob to support their habit. In many prisons and jails, more than half of the inmates are there because of some drug offense.

Table 13.2. *Current penalties for offenses for the various schedules of controlled substances.*

CSA	2ND OFFENSE	1ST OFFENSE	QUANTITY	DRUG	QUANTITY	1ST OFFENSE	2ND OFFENSE
I and II	—Not less than 10 years, Not more than life.

—If death or serious injury, not less than life.

—Fine of not more than $4 million individual, $10 million other than individual. | —Not less than 5 years, Not more than 40 years.

—If death or serious injury, not less than 20 years, or more than life.

—Fine of not more than $2 million individual, $5 million other than individual. | 10–99 gm pure or 100–999 gm mixture

100–999 gm mixture

500–4,999 gm mixture

5–49 gm mixture

10–99 gm pure or 100–999 gm mixture

1–9 gm mixture 40–399 gm mixture

10–99 gm mixture | Methamphetamine

Heroin

Cocaine

Cocaine Base

PCP

LSD

Fentanyl

Fentaryl Analogue | 100 gm or more pure or 1 kg or more mixture.

1 kg or more mixture

5 kg or more mixture

50 gm or more mixture

100 gm or more pure or 1 kg or more mixture

10 gm or more mixture

400 gm or more mixture

100 gm or more mixture | —Not less than 10 years, Not more than life.

—If death or serious injury, not less than 20 years, or more than life.

—Fine of not more than $4 million individual, $10 million other than individual. | —Not less than 20 years, Not more than life.

—If death or serious injury, not less than life.

—Fine of not more than $8 million individual, $20 million other than individual. |

CSA	DRUG	QUANTITY	1ST OFFENSE	2ND OFFENSE
	Others (Law does not include marijuana hashish, or hash oil.)	Any	—Not more than 20 years. —If death or serious injury, not less than 20 years, not more than life. —Fine $1 million individual, $5 million not individual.	—Not more than 30 years. —If death or serious injury, life. —Fine $2 million individual, $10 million not individual.
III	All (Includes anabolic steroids as of 2–27–91.)	Any	—Not more than 5 years. —Fine not more than $250,000 individual, $1 million not individual.	—Not more than 30 years. —If death or serious injury, life. —Fine $2 million individual, $10 million not individual.
IV	All	Any	—Not more than 3 years. —Fine not more than $250,000 individual, $1 million not individual.	—Not more than 30 years. —If death or serious injury, life. —Fine $2 million individual, $10 million not individual.
V	All	Any	—Not more than 1 year. —Fine not more than $100,000 individual, $250,000 not individual.	—Not more than 30 years. —If death or serious injury, life. —Fine $2 million individual, $10 million not individual.

Source: www.USDOJ.GOV/DEA/PUBS/ABUSE/DOA-P.PDF

Table 13.3. *Current penalties for marijuana abuse.*

DESCRIPTION	QUANTITY	1ST OFFENSE	2ND OFFENSE
Marijuana	1,000 kg or more mixture; or 1,000 or more plants.	—Not less than 10 years, Not more than life. —If death or serious injury, Not less than 20 years, Not more than life. —Fine not more than $4 million individual, $10 million other than individual.	—Not less than 20 years, Not more than life. —If death or serious injury, Not more than life. —Fine not more than $8 million individual, $20 million other than individual.
Marijuana	100 kg to 999 kg mixture; or 100–999 plants.	—Not less than 5 years, Not more than 40 years. —If death or serious injury, Not less than 20 years, Not more than life. —Fine not more than $2 million individual, $5 million other than individual.	—Not less than 10 years, Not more than life. —If death or serious injury, Not more than life. —Fine not more than $4 million individual, $10 million other than individual.
Marijuana	50 to 99 kg mixture; or 50 to 99 plants.	—Not more than 20 years. —If death or serous injury, Not less than 20 years, Not more than life. —Fine $1 million individual, $5 million other than individual.	—Not more than 30 years. —If death or serious injury, Not more than life. —Fine $2 million individual, $10 million other than individual.
Marijuana Hashish Hashish Oil	Less than 50 kg mixture 10 kg or more 1 kg or more	—Not more than 5 years. —Fine not more than $250,000, $1 million other than individual.	—Not more than 10 years. —Fine $500,000 individual, $2 million other than individual.

In addition to this social cost of drug abuse, there are the health arguments. Drug abuse, especially serious situations in which someone is addicted, can cause great harm to physical and mental health. If such people cannot afford health insurance, then they become a public burden on the health system, to the detriment of everyone in society.

From all of these considerations arises a dilemma that has been here as long as our society has been regulating drugs; namely, should our resources be put into stopping the flow of drugs into the hands of users, or should society concentrate on prevention of drug abuse through education and treatment? Should we treat drug abuse as a crime or as a medical condition? The United States has vacillated back and forth among these strategies, sometimes preferring one to the other and then, with a new administration, changing tactics. These are fundamental public policy questions that are not easily answered.

One of the major concerns that clouds these arguments is the extent to which illicit drugs are **addictive** as opposed to "merely" causing (psychological) **dependency**. An addiction occurs when the body makes profound physiological and biochemical changes to accommodate a drug. When the addict stops taking the drug, especially suddenly or "cold turkey," a set of physical symptoms (withdrawal syndrome) occurs. This syndrome can be intensely uncomfortable and painful, and in some cases, can even cause death. An addicted drug user will do almost anything to avoid withdrawal, and his or her life may be consumed by a constant search for a reliable source of the drug. On the other hand, when a drug user does not become physically addicted to the drug, it may still exert powerful psychological effects (dependence). The user finds the effects of the drug so pleasurable or satisfying that he or she becomes dependent on the drug. This craving is not physical, but psychological. Such users may exhibit similar behaviors to addicts; they constantly seek the drug, but there are no physical symptoms of withdrawal when the drug is stopped. In either case, drug users in the grip of an addictive or dependence-causing drug may exhibit the same behaviors that are considered antisocial by governments. The fact is that, for more than a century, the focus has been on regulation and distribution of illicit drugs. In the forensic science context, however, the task is not to settle the issue of whether or how drugs should be controlled, but rather to describe how drugs are classified and how forensic scientists identify and characterize them.

CLASSIFICATION OF ILLICIT DRUGS

There are a number of ways of classifying illicit drugs. As mentioned previously, the federal laws put them in one of five schedules based on their abuse potential and pharmacology and the existence of a legitimate medical use. Another convenient way of classifying drugs is by origin. Under this scheme, drugs are put in one of four classes:

1. **Naturally occurring.** These substances are found in nature in plants. Part of the plant is ingested, and the drug is extracted and used by the person. Examples include
   A. **Marijuana**: The leaves are dried and smoked.
   B. **Psylocybin mushrooms**: These are eaten. They contain psylocybin and psyclocin, which cause hallucinations (hallucinogen).
   C. **Peyote cactus**: The cactus buttons are eaten. They contain mescaline, a hallucinogen.
2. **Plant extracts.** These are naturally occurring substances that are extracted from plants and then ingested. Examples include
   A. **Cocaine**: Extracted from the coca plant.
   B. **Morphine and codeine**: Extracted from the opium poppy.
3. **Semisynthetic.** These substances are derived chemically from a naturally occurring substance. Examples include
   A. **Heroin**: Manufactured from morphine.
   B. **LSD**: Manufactured from lysergic acid.
4. **Synthetic.** These substances are totally man-made. Examples include
   A. **Amphetamines**
   B. **Barbiturates**
   C. **PCP**
   D. **Oxycodone**

The major classification system for drugs in use today is by predominant pharmacological effect. Under this scheme, illicit drugs are put into one of four classes: stimulant, depressant, narcotic, and hallucinogen. Each of these classes will be discussed individually along with prominent members of the class.

STIMULANTS

Stimulants are drugs that elevate mood. They help people who are sad or depressed to feel better. They give people extra energy. Other claims are also made for stimulants: They make people stronger, faster, have better sexual experiences; they make people smarter.

Stimulants can range from the mild, such as caffeine, which is not an illicit drug, to strong. The most common examples of the latter are **amphetamine** and **cocaine**. Both of these drugs have been used for many years and are still quite popular.

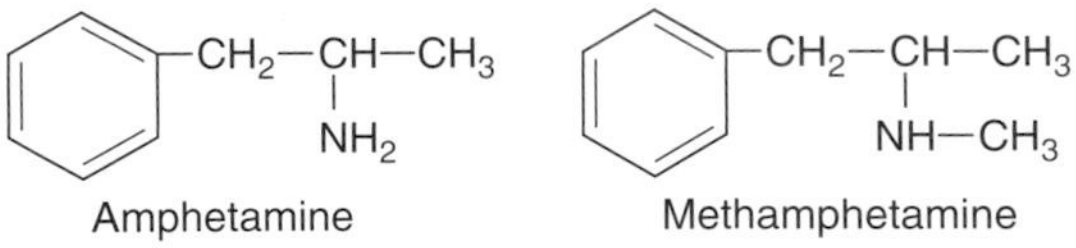

Amphetamine

Methamphetamine

Figure 13.1. *The chemical structure of amphetamine and methamphetamine.*

Amphetamine

Many drugs are derived from the basic amphetamine backbone. Two are common, illicit derivatives worldwide: amphetamine itself and **methamphetamine**, which is far more popular as an illicit drug. Amphetamine arose from the desire of pharmacologists to find a substitute for **ephedrine**, the active ingredient in a group of herbs that have been used for thousands of years. Ephedrine is used today to dilate bronchial passages in treatment of asthma, and this was one of the early uses of amphetamine. Amphetamine has also been used in the treatment of narcolepsy, a disease that causes its victims to fall asleep suddenly many times a day. It has also been used as an appetite suppressant and in the treatment of hyperkinesias, a disease that causes hyperactivity, mainly in children. Figure 13.1 shows the chemical structures of amphetamine and methamphetamine.

In the 1960s amphetamine and methamphetamine began to be widely abused for their stimulant properties. Instead of taking the drugs orally, which was the preferred route of administration for pharmaceutical uses, abusers started taking them intravenously, which magnified their effects. Even today, clandestine methamphetamine labs flourish in many parts of the country. There are many routes to the preparation of methamphetamine, and they can be found in underground publications and on the web. Because of its potent stimulant effects, methamphetamine has been called "speed" since the 1960s, and the stimulants in general are called "uppers."

In recent years, the widespread abuse of amphetamines has led to increasingly tighter control of legitimately manufactured doses, so more of the abused drugs have been made clandestinely. In an attempt to overcome the unpleasant habit of intravenous injection of these drugs, some users developed a smokeable form of methamphetamine called "ice." Ice is made by slowly evaporating a solution of methamphetamine, forming large crystals. One of the forms of clandestine amphetamine tablets is shown in Figure 13.2.

There is some disagreement as to whether amphetamine and methamphetamine are addictive. It is now generally believed that when taken intravenously, these substances become addicting, whereas when taken orally or smoked, they cause strong psychological dependence, but not addiction.

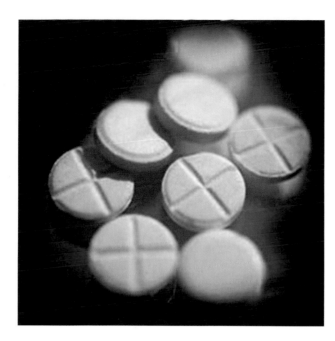

Figure 13.2.
Clandestine
amphetamine tablets.

Cocaine

The other major stimulant that is abused heavily in the United States is cocaine. Cocaine is a naturally occurring substance that is found only in the *Erythoxylon coca* plant that grows almost exclusively on the Amazon slopes of the Andes Mountains in South America. Figure 13.3 shows cocaine hydrochloride.

Evidence from as far back as 500 A.D. indicates that coca leaves were being used by indigenous people. Five hundred years later, coca plants were being extensively cultivated in Peru. Today, it is believed that more than 50 million kilograms (about 110 million pounds) are produced annually in South America, chiefly in Bolivia, Peru, and Colombia, and only a small fraction of that is used domestically or legally exported. Cocaine is extracted from the coca leaves domestically and then exported to the United States and other countries. It takes about 500 kg of coca leaves to produce 1 kg of cocaine powder. Domestically, people have used cocaine for hundreds of years by chewing the leaves of the coca plant, rather than by extracting the cocaine chemically.

One of the early proponents of cocaine for its medicinal effects was Sigmund Freud, who experimented with it extensively and praised

Figure 13.3. *(top) Cocaine hydrochloride. This is sometimes called "snow" or "flake" or "blow." (bottom) Coca leaves.*

its stimulant effects and the feelings of well-being that it caused. He especially recommended it as a method for curing morphine addiction. Conan Doyle wrote about cocaine as a habit-forming drug, describing Sherlock Holmes' experimentation with it. A motion picture of the 1980s, *The Seven Percent Solution*, is a fictional account

Figure 13.4. *Crack cocaine. This is a form of cocaine free base. It is smoked rather than snorted.*

of how Sigmund Freud cured Sherlock Holmes' addiction to cocaine.

By the turn of the twentieth century, cocaine was extremely popular in the United States as a pleasure drug. A large number of products containing cocaine were available. They could be drunk, snorted (inhaled), or injected. In 1906 with the Pure Food and Drugs Act and the Harrison Act of 1914, cocaine use was severely curtailed and the drug became heavily controlled. As amphetamines became easy to obtain in the 1930s, cocaine use declined and then rose when amphetamines became strictly controlled in the 1970s.

For many years it was believed that cocaine was not a physically addictive drug. This opinion has changed, however, with the advent of **crack**, a form of crystalline cocaine free base. When smoked, crack cocaine can cause physical addiction. Crack cocaine is shown in Figure 13.4.

Cocaine provides an excellent case study of the huge profits that are to be made in illegal drug trafficking. A major drug trafficker can obtain 500 pounds of coca leaves from a farmer for about $250. This, in turn, will produce about 1 pound of pure cocaine. This is sold to a refiner/exporter for about $1,000. Broken down into smaller packages and diluted, the final product, up to 5 pounds of diluted cocaine, can sell for as much as $25,000 on the streets of the United States, a hundred-fold increase in value.

DEPRESSANTS

When one thinks of depressant substances that are abused, the first one that comes to mind is ethyl alcohol, used to make beer, wine, and spirits. It is important to note, however, that alcohol is not a listed controlled substance in the United States; that is, it is not covered under the laws that control drugs. Alcohol is covered under a separate set of regulations. This topic will be covered in Chapter 14 on forensic toxicology.

The Barbiturates

Depressant drugs are known by a variety of names including sedatives, hypnotics, and the street drug term "downers." They all have the common effect, to one degree or another, of decreasing brain activity. In small doses these drugs may be taken to reduce anxiety and are termed "sedatives." In larger doses the same drugs are taken as sleeping pills and are called "hypnotics." The major class of illicit depressants in the United States is **barbiturates**. This is a group of chemical substances based on the compounds barbituric acid and thiobarbituric acid. Barbituric acid itself is not a central nervous system depressant, but over 2,500 derivatives have since been produced. Barbiturates are grouped pharmacologically into three groups: short, intermediate, and long acting.

As a group, the barbiturates can be highly addictive and withdrawal can be difficult and dangerous. Sudden ("cold turkey") withdrawal of some of the more powerful barbiturates can be fatal. There is also a great danger of the interaction of alcohol and barbiturates. Several people, including some celebrities, have died from an accidental (or deliberate) overdose of alcohol and barbs. Other people become addicted to barbiturates and amphetamines, taking both in great quantities. Detoxification from this potentially lethal combination of drugs can take many months or even years.

Other Depressants

Besides barbiturates, other depressants have been abused over the years. One of the more popular of these was **meprobamate** (Miltown), which was introduced in the 1950s. This never became a widely used or abused drug except for a short time in the 1960s and 1970s.

Since barbiturates became popular, scientists have searched for alternative drugs that would be less toxic and addicting. They thought they had one with **methaqualone**, marketed as "Quaalude" and as

"Sopor," thus earning the drug the nicknames of "ludes" and "sopers." College students heavily abused the drug in the 1960s and 1970s, and it didn't take long to prove that it was just as dangerous as barbiturates. It was finally taken off the market in 1985, and abuse today is almost nonexistent.

Another attempt to develop a more suitable drug to replace barbiturates was the **benzodiazepines**. The first one was called "Librium" (chlordiazepoxide). Librium was so called because it liberated people from their anxieties. It rose to become the most often prescribed drug in the United States in the late 1960s. It was later supplanted by "Valium" (diazepam), which is more potent. Valium soon became the country's most popular prescribed drug. It also became a very popular illicit drug, and before long, doctors discovered that withdrawal was causing addiction and dependence-like symptoms. Overdose deaths rose markedly, and Valium soon became more tightly controlled by the Federal Government and enforced by the DEA. Today, Valium is still prescribed for anxiety but at greatly reduced rates. Valium injections are used today as an anesthetic in some medical procedures and minor surgeries.

NARCOTICS

Do you ever wonder why Dorothy fell asleep in the field of beautiful red flowers on her way to the castle of Oz? Those were the flowers of the **Opium poppy**, *Papaver somniferum*, a plant with more than 5,000 years' history as a medicinal. Opium, the resin from the poppy plant, has been, at least until the past 100 years or so, the one reliable, naturally occurring substance that physicians could use for pain and suffering, and for diarrhea and dehydration from dysentery. It has also given users pleasure, peace, rest, and relief from anxiety. These latter effects have caused opium and its derivatives to be abused throughout its history. Its highly addictive nature has also placed many of these people in its grip for much of their lives. The opium poppy pod is shown in Figure 13.5. The opium resin is contained within the pod.

Narcotics Derived from Opium

Because of their ability to relieve pain and cause sleep, opium and its derivatives became known as **narcotics**. This term is derived from the Greek word *narkotikos*, which means "sleep." During the early part of the twentieth century, the term "narcotic" became synonymous with all drugs that were considered dangerous and in need of control. Because

Figure 13.5. A pod from the opium poppy plant. At harvest time, the pods are slit open, allowing the resin to ooze out. This is gathered and dried. It contains about 10% morphine and a lesser amount of codeine. The resin itself has been smoked for thousands of years.

of this, "narcotic" became a pejorative label, and any drug that was classified as a narcotic was painted with this negative image. Many states labeled marijuana and cocaine as narcotics, and this usage became incorporated into some state laws until as recently as the 1970s.

Today, some narcotics are still used as legitimate pharmaceuticals. Morphine, which is extracted from opium, is a powerful analgesic (pain killer), used in surgical procedures. It is interesting to note that **heroin**, a drug that is easily made from morphine and is approximately 10 times more potent, is used legitimately as an analgesic in some other countries. But heroin has such a bad name in the United States that it is a schedule I controlled substance and has no legitimate medical use here.

Besides morphine, **codeine** is the other popular derivative from opium. Codeine is a popular cough suppressant in liquid preparations and is also mixed with mild analgesics such as aspirin or Tylenol® to boost their analgesic (pain management) effects. Figure 13.6 shows some elixirs containing schedule I and II controlled substances. These examples were stolen from a drug store during an evening robbery.

Synthetic Narcotics

Besides the naturally occurring narcotics, a number of synthetic narcotics are also still prescribed for pain. They include Demerol (meperidine), Oxycodone, Hydrocodone, Fentanyl, and many others. Of the members of this group, a few bear special mentioning. Methadone is a synthetic narcotic that is used in the United States as a heroin substitute to get addicts off heroin under close medical supervision. Fentanyl (China white) is the backbone of a series of **designer drugs**, illegal substances that are synthesized with very particular pharmacological characteristics designed for abuse purposes. Oxycodone (Percodan) is,

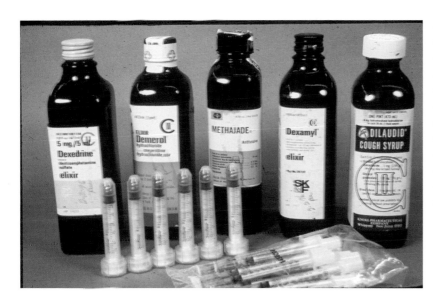

Figure 13.6. *Some elixirs containing controlled substances. Demerol, Methajade, and Dilauded are all synthetic narcotics. Dexedrine is an amphetamine, and Dexamyl is a combination of an amphetamine and a barbiturate.*

at the time this is being written, making a large comeback in the United States as an abused drug and is known as "Oxycontin." Fatal overdoses among young users are skyrocketing.

The preferred illicit narcotic for many years in the United States has been heroin. The preferred mode of ingestion of heroin has been by intravenous injection. Because addicts often have difficulty getting sterile needles, they often contract blood-borne diseases such as hepatitis or AIDS from sharing contaminated syringes. Another problem with heroin use over the years has been purity. Over the years purity has varied from less than 3% to more than 10%. This has led to many problems with overdosing, where the user may take a more potent dose than he or she is used to. Along with this is the often bizarre variety of cutting agents used in heroin, some of which can cause illness or even death. Heroin addiction is also expensive and because many addicts cannot hold down a job or earn money, they must steal to support their habit. This causes inflated crime rates traceable to the drug abuse and fills U.S. prisons with addicts.

HALLUCINOGENS

The types of drugs that have been discussed thus far—stimulants, depressants, and narcotics—all have at least some members that have accepted, legitimate medical uses. This is not the case with the hallucinogens. Most of the important members of this class are naturally occurring substances, although a few, such as PCP, are synthetic. None of the prominent members of this class have any legitimate medical use in the United States.

Many plants contain substances that can cause pain, suffering, sickness, and even death if ingested in sufficient quantity. This is, no doubt, part of their evolution that gives them an edge for survival. People and animals are not likely to try these plants twice if doing so results in unpleasant experiences. It is also true, however, that man has learned to ingest some plants or plant substances in amounts and ways that cause pleasant effects without the dangerous or uncomfortable side effects. These make up the illicit hallucinogens, virtually all of which are in federal Schedule I. Hallucinogens fall into a number of classes based on the chemical structure of the psychoactive substance. A few of the more common ones will be discussed in this chapter.

LSD (d-Lysergic Acid Diethyl Amide)

LSD is not a naturally occurring substance but is derived from ergot alkaloids derived from a grain fungus. This mold, if incorporated into bread products made from infected grain, can result in a disorder called **ergotism** or **St. Anthony's Fire**. (See "In More Detail: St. Anthony's Fire".) LSD is the most potent and certainly the most famous or infamous of the hallucinogens and the one most responsible for popularizing this genre of drugs. As little as 50 μg (a droplet smaller than a period on this page) of LSD can cause auditory and visual hallucinogens that can last up to 12 hours. LSD is not addictive, and psychological dependence has been rarely recorded, but it can cause psychosis in borderline cases and often gives rise to **flashbacks**, episodes of hallucinations months or years after a dose of the drug was taken. One of the most common forms of LSD is "blotter acid," whereby LSD is poured onto blotter paper and then is cut up into small squares. This form is shown in Figure 13.7.

Psilocybin

Psilocybin is similar to LSD in that its chemical structural backbone is the same (a substance called an "indole"). It is found in a variety of mushrooms with the most potent one being *Psilocybe mexicana*. Psilocybin's effects are similar to both LSD and mescaline, and there appears to be some cross tolerance among these three hallucinogens. Dried mushrooms contain approximately 0.5% psilocybin. The most

IN MORE DETAIL:
St. Anthony's fire

If not properly stored, rye grain is susceptible to attack by a fungus known as *Claviceps purpurea*. One of the major alkaloids in this fungus is lysergic acid, a parent compound of LSD. If lysergic acid is ingested in large quantities, **ergotism**, or St. Anthony's fire, results. As far back as the ninth century, there have been reports of people suffering a plague of blisters and a sensation that their arms and legs were falling off (they weren't). Ergot poisoning is characterized by gangrene and extreme pain and burning sensation. Convulsions, hallucinations, and even death can result from acute ergot poisoning.

St. Anthony is the patron saint of victims of ergotism, and the Order of St. Anthony treated many of the sufferers. There were sporadic reports of outbreaks of ergotism throughout the middle ages, some of them called "dancing mania." Some historians claim that women accused of witchcraft had St. Anthony's fire and that explained their behavior. This claim has never been proven, however.

In the seventeenth century, the ergot fungus was identified as the cause of St. Anthony's fire. Damp growing conditions in Europe contributed to outbreaks of the disease. The most recent reported outbreak took place in Pont St. Esprit in France in 1951. Four people died and several hundred exhibited symptoms.

LSD is normally taken in unusual dosage forms. It is a colorless, odorless, tasteless liquid. It may be dissolved in a volatile solvent and then impregnated onto absorbent paper, which is often decorated with colorful cartoon characters. Once dried, the paper is cut up into small squares and ingested. This form is called "blotter acid." Other forms of LSD include tiny colorful tablets (microdots) or small pieces of impregnated, dried gelatin (window panes). In some cases a dose of LSD was put behind a postage stamp on a letter and mailed to another user.

Source: www.angelfire.com

Figure 13.7. Blotter acid. LSD is dissolved in a solvent and poured onto absorbent paper. The solvent is allowed to evaporate, leaving the LSD. The paper is cut into small squares and eaten.

Figure 13.8.
Marijuana leaves.

effective dose seems to be about 5 mg, with higher doses causing some unpleasant effects.

Marijuana

Marijuana is unique among the controlled substances. It has some analgesic properties, albeit mild, and was, at one time, classified as a narcotic for pharmacologic and political reasons. It also has some sedative properties, also mild. It is most famous for its mild hallucinogenic effects, and here is where it seems best classified. Marijuana leaves are shown in Figure 13.8.

Unlike the other hallucinogens discussed in this section, marijuana has been purported to have beneficial medical effects. They include antiemetic (relieves nausea and upset stomach) effects for symptoms caused by anticancer drugs. It has also been linked to the reduction of eye pressure caused by glaucoma. It is important to note that, as of now, the Food and Drug Administration does not recognize any of these or other medical effects of marijuana; therefore, the federal government maintains the position that marijuana has no accepted medical use.

Marijuana is a plant that grows worldwide. It belongs to the genus *Cannabis*. In the 1960s there was controversy over how many species there are in this genus. Some botanists felt that there was evidence for four species; *sativa, indicus, gigantias,* and *ruderalis*. Others believed that there is only one species, *sativa,* and the others are regional variants of this plant. When the federal government first controlled marijuana, they referred to it as *Cannabis sativa,* as did many state legislatures. In the 1960s and 1970s there were many legal problems with prosecutions of marijuana cases in state and federal courts because it is virtually impossible to determine which species is present if one is presented with dried, chopped-up leaves. Forensic scientists were being challenged to prove that what was seized met the requirements of the law, namely that it was *Cannabis sativa* and not a different species. To respond to these challenges, Congress and state legislatures took note that all marijuana, no matter what species it might be, contains the active ingredients that give the plant its hallucinogenic properties; the cannabinoid alkaloids, chiefly Δ^9-tetrahydrocannabinol (THC). They changed the legal description of marijuana to include all members of the genus *Cannabis,* and that ended the legal skirmishing. Marijuana is currently controlled as a federal Schedule I substance.

Marijuana has been used for thousands of years. The bark of the plant is especially useful for making a type of rope called hemp, and marijuana is sometimes called the "hemp plant". In some countries, (e.g., Japan), marijuana is still grown for its hemp. In U.S. colonial times, George Washington was known to cultivate marijuana for hemp.

The THC content of marijuana varies regionally. In some low grade plants it is less than 1%. With careful breeding and cultivation and removal of male plants, a form of female, seedless marijuana called **sinsemilla**, can be grown with THC content as high as 10%. There is also a federal government–supported experimental farm in Mississippi, where marijuana with THC content over 30% has been cultivated.

The THC in marijuana is concentrated in the resin, which is most abundant in the flowering tops and leaves of the plant. The stalks, seeds, and roots contain almost no THC and are generally removed. The leaves and flowers are dried and then smoked. In many jurisdictions, marijuana stems and stalks are not controlled, and seeds are controlled only if they are viable (they can be germinated).

Concentrated forms of marijuana are prepared in various ways and known by different names all over the world. In the United States, the plant material can be boiled in methyl alcohol, filtered, and then evaporated down to a thick, gooey liquid known as **hash oil**. **Hashish** is prepared by collecting the resin from the live plants with leather straps and cloths. It consists of resin and hairs from the plant. Hash oil and hash are generally smoked in small pipes that are

Figure 13.9. *Hashish (in dish on the right), marijuana leaves (left), and some of the different types of pipes that are used to smoke it.*

designed for this purpose. High-quality hash oil may contain more than 50% THC. The dish on the right of Figure 13.9 shows some pieces of hashish.

Marijuana has some interesting effects on some people in that it appears that smoking and enjoying the drug involve a learning process. This consists of learning just how long to keep the smoke in the lungs before exhaling and learning how to recognize and cultivate the mild hallucinogenic effects. These behavioral characteristics may lead to a type of "reverse tolerance" whereby the effects seem to be stronger as the person gets more experienced in smoking. It also seems to have an appetite stimulating effect on many people, especially inexperienced users who report that they get the "munchies" and want to eat great quantities of food after smoking marijuana. Some of this reverse tolerance may be due to a very long half-life for THC, which has been found in some tissues months after the last ingestion.

Mescaline

Mescaline is one of a group of substances (see following sections for more examples) whose chemical structure is similar to amphetamine but, because of substitutions on the benzene ring, have hallucinogenic rather than stimulant properties. Mescaline is found in the upper crown of the **peyote cactus** that grows extensively in Mexico and the

Figure 13.10. *Peyote cactus buttons.*

southwest part of the United States. The crown is sliced into wafers called mescal buttons. They are then softened in the mouth and then rolled up into balls and swallowed. Intense hallucinations follow which can last for many hours. Mescaline use has been part of the religious ceremonies of Native Americans for many hundreds of years and remains so today. Figure 13.10 shows peyote cactus buttons obtained during a law enforcement raid on a house in Virginia.

MDMA (Ecstasy)

The substance known as **MDMA** or "**Ecstasy**" has been around for many years and is considered by some psychiatrists to be a true hallucinogen. It has even been used in psychotherapy, at least until the mid 1980s when Congress moved to control it. Since then its use has exploded in the United States. It has become the drug of choice in the popular drug and alcohol parties known as "raves." It is a Schedule I controlled substance.

Phencyclidine (PCP)

PCP was first marketed by Parke, Davis & Company as one of a new class of intravenous anesthetics. It was tried in surgical procedures on both large mammals and on humans. It turned out to be an excellent anesthetic for animals, principally monkeys, but did not work well on humans. It also appeared to cause hallucinations. It also caused some bizarre side effects in humans that include a feeling of no pain sensations, superhuman strength, rage, loss of memory, and paranoia. In the 1970s it was taken off the market even as an anesthetic for animals. It is now considered to have no legitimate medical use and is in federal Schedule I.

PCP is abused in a variety of forms. It showed up in the 1960s as a small, white tablet and was called the "peace pill." At about the same time, it was sprinkled or recrystallized onto marijuana, which was then sold as high grade marijuana ("Acapulco Gold" or "Colombian") and was also known as "wobble weed" or "sherms" (the drug hit like a Sherman tank). In the 1970s, someone decided that contaminating marijuana with PCP was a waste of good marijuana and started impregnating otherwise non-hallucinogenic plant material such as oregano or parsley with PCP. In this context, the drug is called "angel dust" or the aforementioned "wobble weed." On the East coast in the 1960s and 1970s, clandestine PCP lab activity was intense. PCP is among the easiest of the illicit drugs to manufacture in a home laboratory. A typical laboratory raid netted many pounds of the drug and great quantities of dried parsley.

DRUG ANALYSIS

There are a number of important considerations in designing methods for drug analysis. Chief among these is the desired information that is to be gained from the analysis. A number of questions arise that must be answered before embarking on a scheme of analysis:

1. How are the controlled drugs defined and described in this jurisdiction?
2. Are the weight of the drug and/or the aggregate weight of the exhibit important?
3. Must the identity of the drug be established and then confirmed?
4. Is it necessary to determine the purity of the drug exhibit (quantitative analysis)?
5. Is it necessary to identify any of the cutting agents present in the exhibit?

Each of these questions will be taken up in turn. From the answers, it can be shown how an acceptable scheme of analysis can be developed that will stand up to the scrutiny of a courtroom.

HOW ARE DRUGS DESCRIBED LEGALLY?

Even though most states subscribe to some form of the model controlled substance legislation propounded by the federal government, there are still some differences in how drugs are defined by a legislature. For example, there were many legal problems for drug chemists when marijuana was narrowly defined as the specific species *Cannabis sativa*. Likewise, cocaine was at one time subject to legal challenges over its description as having to be derived from coca leaves. Cocaine can occur as two mirror image isomers, only one of which is derived from coca leaves. It is then left to the drug chemist to prove that an exhibit is not the other isomer. These examples reinforce the necessity that the drug chemist knows how the law defines a particular controlled substance so that the chosen scheme of analysis takes the definition into account.

WEIGHT AND SAMPLING

In some jurisdictions, not only is the identity of the illicit drug important, but also the quantity. This can be understood in three contexts. First, there is a desire among prosecutors and police to concentrate their law enforcement efforts on major drug dealers rather than the low-level user. If the possession of large quantities of drugs is punished more harshly than small amounts, then this might discourage large-scale drug dealers from plying their trade. The important question here is: what constitutes the weight of the drug? Illicit drugs are seldom sold in a pure form. They are almost always adulterated, or cut with other substances. In most states that have weight laws, it is the aggregate weight that counts. This is the weight of the drug and any cutting agents present. So, for example, many states have laws that penalize with life imprisonment someone who possesses a substance containing cocaine or heroin that is over a certain aggregate weight. The cocaine or heroin doesn't have to weigh that much; it is the total weight that counts. So an exhibit that contains 1 g of cocaine cut with 650 g of sugar will still qualify under this law. Of course, it goes without saying that there must be proof that the balance that is used to determine these aggregate weights is accurate.

Another context where weight is important is where a government wishes to punish the possession of one form of a drug more harshly than another. This is the way that the federal government sanctions cocaine. Under federal law, the possession of equal weights of cocaine flake (or salt) and crack carry vastly different penalties, with the crack form carrying a much higher penalty.

Weight considerations also come into play at the opposite end of the spectrum, where there is very little material present. In some states, there must be a **usable quantity** of a drug present in order for a law to be broken. A usable quantity is defined as an amount of a drug that is likely to have a demonstrable psychoactive effect on an average person. Mere traces are not enough.

There is also another weight-related consideration and that is sampling. This comes into play when there are very large exhibits or when there are a large number of exhibits and the question arises as to how much of the material must actually be analyzed.

In the case of large exhibits, it is usually not necessary to identify every particle in the exhibit. However, if the entire mass of substance is to be characterized as being or containing a controlled substance, the samples taken for analysis must be representative of the whole exhibit. For example, the exhibit depicted in Figure 13.11 is a brick of marijuana that was compressed in a trash compactor. Several samples were taken from the exterior and interior of the brick and independently tested in order to show that the entire 38 pounds was marijuana.

Another major sampling issue arises with cases that have a large number of exhibits. Many times a drug chemist will receive a case that consists of hundreds or thousands of exhibits. It would be unduly consumptive of time and effort to fully analyze every exhibit. However, if one is to report that all of the exhibits contain a particular illicit drug, then it is necessary to show that the samples that were taken for analysis are representative of the whole. Let's look at one possible approach. Suppose that a case contains 1,000 exhibits. Five hundred consist of white powders, each wrapped in foil. Each appears to be approximately the same size and weight. The other 500 exhibits are small, plastic baggies, each containing approximately the same amount of white powder. At the start, these would be treated as two types of exhibits. A random sample of the foil-wrapped drugs would be opened, examined, and weighed. If they all appeared the same and weighed approximately

Figure 13.11. *A 38.5 lb brick of marijuana. This solid mass of plant material was formed in a trash compactor.*

the same, then one might assume for now that all 500 contain the same thing. The same would then be done on a random sample of the plastic-wrapped exhibits. Then a random sample of each of the two types would be tested and the drugs identified. The total weight of each set of 500 exhibits would then be estimated, and the qualitative and quantitative results would be reported for the whole case. If, at any time, one or more of the randomly selected exhibits appeared to be different in appearance or chemical properties than the rest, then the analysis scheme would have to be changed.

Finally, there is the opposite situation, when very little of a drug is present in an exhibit. Examples are "roaches" (marijuana cigarette butts), "cookers" (usually bottle caps containing the residue of injected heroin), or even bloody syringes that were used to inject heroin, which may contain traces of the drug. States that have usable quantity laws require that a certain minimum amount of a drug be present in order to prosecute someone for possession of an illicit drug. It is unlikely that any of the three examples given here would qualify as usable quantities of a drug under these statutes. In states where there are no usable quantity laws, then these exhibits could be analyzed and reported out as controlled substances. In some cases, there may not be enough material to do a complete analysis, and a confirmatory test may not be done. This will be discussed later.

DRUG PURITY

As mentioned previously, drugs are nearly always contaminated by impurities. These fall into two categories: excipients and diluents. **Excipients** are substances that may mimic the activity of the main illicit drug present in order to make it more difficult for the user to know just how much of the drug there really is in the exhibit. For example, a common excipient in cocaine exhibits is lidocaine. Lidocaine, like cocaine, is a topical anesthetic. When cocaine is snorted, it numbs the nasal membranes as does lidocaine. Thus, if lidocaine is present, the user cannot determine, at least by the numbing action, how much cocaine is present. **Diluents** are chemicals that are used to dilute an illicit drug and to give it more bulk. This is done to cut the purity of the drug and increase the amount and thus the profits.

In most states there is no requirement that cutting agents be identified in drug exhibits. In some areas including in some federal cases, it is necessary to identify and quantify all cutting agents present with an illicit drug. This is done chiefly for intelligence purposes so that the law enforcement agents can track a case up through the distribution chain and identify the possible origin of the drugs.

DEVELOPING AN ANALYTICAL SCHEME

Once the weight and sampling issues have been settled, then it is necessary to develop a scheme of analysis for the exhibits. This scheme will ordinarily proceed from general types of tests toward specific tests. For most drug analyses, the goal is to unequivocally identify all controlled substances in an exhibit. In certain cases this **confirmation** may not be done. These situations will be discussed later.

Each exhibit should be treated as an unknown substance. Police officers or drug enforcement agents may already have a good idea what is contained in a drug exhibit and will usually relay this information to the drug chemist. Sometimes this information is based in part on some field testing that the officer may have performed. Several companies sell field test kits that contain the chemicals necessary to perform a preliminary or presumptive test on a suspected controlled substance. These tests may be necessary for the law enforcement agency to establish probable cause to obtain a search warrant or for other purposes. Even if this test is performed the results are only preliminary and cannot be relied on for identification purposes. Therefore, although information about a drug case may be interesting and perhaps useful, it should not dissuade the drug chemist from treating each case as an unknown situation.

Preliminary Tests

The first test done on a drug exhibit is visual. The package should be opened (protective gloves should be worn; LSD can be absorbed through the skin, causing significant hallucinations). The appearance of the exhibit should be noted. Plant materials may be treated differently than white powders or marked tablets. The exhibit is then weighed.

Over the past 50 years or so, a number of presumptive tests have been developed that react with various common controlled drugs, usually resulting in a visual color change to the reagent when the drug is added. These are valuable tests in that they can give the examiner a direction for continuing the analysis. For example, the **Ruybal (or Scott) test** consists of three reagents that are added in turn to an exhibit suspected to be or contain cocaine. The final result of the test is a turquoise color in a chloroform layer at the bottom of the test tube where the test is run. This test is said to be presumptive for cocaine. It is not dispositive; there are other substances that give similar results to this test, but it does give the analyst a direction to proceed, namely to continue to test for the presence of cocaine. Some of the more common presumptive tests and the drugs they are used on are given in Table 13.4.

Table 13.4. *Presumptive tests for drugs.*

PRESUMPTIVE TEST	DRUG(S)	RESULTS
Duquenois-Levine	Marijuana	Purple bottom layer
Ruybal (Scott)	Cocaine	Turquoise bottom layer
Marquis	Opium derivatives	Purple
Marquis (+ water)	Amphetamine	Bright orange fluorescence
Dillie-Koppanyi	Barbiturates	Purple
Erlich's	LSD	Purple

Figure 13.12. A field
test kit.

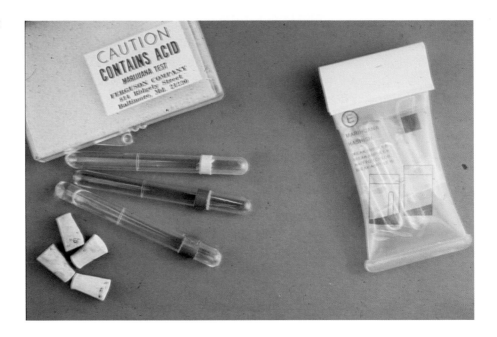

Figure 13.12 shows a field test kit commonly sold to narcotics agents.

Microcrystal Tests

In a microcrystal test, a reagent or series of reagents are added to a suspected drug under carefully controlled conditions so that the drug forms a complex with the reagent(s) and is allowed to slowly recrystallize. The appearance, color, shape, and speed of deposition are characteristic of particular types of drugs. Not all drug laboratories use microcrystal tests; their use has been declining over the past 25 years. In some parts of the country, especially in certain labs in California, microcrystal tests are used as confirmatory tests for a few selected drugs. In these situations, at least two different microcrystal tests are run along with other presumptive tests before a confirmation is declared. A certain percentage of these cases are also checked by another confirmatory test to make sure that the microcrystal tests are working properly. In other labs where microcrystal tests are used, they are considered to be presumptive tests that require further testing for confirmation.

Chromatography

As discussed in Chapter 6, chromatography tests are utilized for separation purposes. They are not used for confirmation of a drug. Once the preliminary tests have been run on an exhibit, the drug chemist may run a thin layer chromatography test to determine how many substances are present in the exhibit and what they might be. TLC will usually show a spot for each drug and most excipients, but will usually not indicate any diluents. Some labs will utilize gas chromatography or liquid chromatography for these purposes. Chromatography may be used as an essential part of confirmation by methods such as GC/MS where compounds that have identical mass spectra may have different retention times.

Quantitative Analysis

When quantitative analysis of a drug exhibit is called for, gas chromatography (or GC/MS) is almost always used. Liquid chromatography (HPLC) will work just as well on many drugs. To get accurate results by chromatography, an internal standard must always be used.

Confirmation

Most drug exhibits must be confirmed. Two tests are generally recognized as confirmatory for controlled substances. It is only necessary to run one of these tests. The most common one in modern drug labs is GC/MS. If an analyst has a white powder that is suspected of containing cocaine, then a small portion of the powder is dissolved in a suitable solvent such as chloroform and then filtered. The cocaine will be separated by the GC and then identified by the mass spectrometer. Many commercial GC/MS instruments have computer-based libraries of mass spectra of illicit drugs that can be compared to the spectrum of the drug exhibit. The retention time of the drug peak and its mass spectrum may give an absolute identification of the drug.

The other confirmatory test for drugs is infrared spectrophotometry (usually FTIR). An infrared spectrum is unique to each individual chemical compound (except certain isomers and homologs). Infrared spectrophotometers often come with computerized, searchable spectral libraries that aid in identification of the spectrum. One drawback to FTIR relative to GC/MS is that the drug must be separated from the cutting agents prior to obtaining the IR spectrum. In the case of small or dilute samples, this can pose a problem.

Some Unusual Situations

In some situations the usual schemes of analysis may not be used. These occur where there is visual or other evidence that the controlled substance present is most probably of a certain type. The most common situation that illustrates this is marijuana. As was previously mentioned, marijuana is most often used by smoking the dried, crushed leaves and flowers of the cannabis plant. This plant material has some distinctive features that can be used to help in identification. These features include the presence of **cystolithic hairs** on one side of the leaf surface and numerous, small white hairs on the other side. The seeds of the cannabis plant are also quite distinctive in appearance. Once the drug analyst recognizes these characteristic features, then it is only necessary to show that the cannabinoid alkaloids that are associated with cannabis are present. They include THC, cannabinol, and cannabidol as well as others. This is accomplished by using the presumptive color test for marijuana, the **Duquenois-Levine test**, followed by some form of chromatography, usually thin layer chromatography. The vast majority of crime laboratories throughout the world consider this scheme to be sufficient to declare that such plant material is marijuana. If the exhibit occurs as suspected hash oil, where there is virtually no plant material to identify, then this is usually treated as more of an unknown and a confirmatory test must be done.

Another situation that is often treated in a different manner than the usual drug is LSD cases. Most often, LSD occurs as blotter acid—impregnated into absorbent paper, which is then eaten. There is no other controlled substance that looks like this. LSD is intensely fluorescent and measurement of the fluorescence spectrum is highly indicative of LSD or one of its known isomers such as LAMPA (lysergic acid methylpropyl amide). There is also a good presumptive color test for LSD: **Erlich's test**. The Erlich's reagent is also a good spray reagent for thin layer plates and colors LSD light purple. A combination of Erlich's test, TLC, and fluorescence is considered a valid scheme of analysis for LSD by many labs.

The last important exception to the confirmation rule for drugs is marked tablets and capsules. Legitimately manufactured tablets and capsules are normally marked with a logo and some identifying numbers and/or letters. The size, shape, and color of the tablet or capsule can provide valuable and helpful information about the identity of that drug. There are compendia such as the *Physician's Desk Reference* (PDR) that have pictures of many common tablets and capsules and which are helpful in identification. In many laboratories, such an identification coupled with one test such as thin layer chromatography is all that is needed to identify the tablet or capsule. It must be recognized, however, that there are some very good counterfeit tablets and that many capsules can be easily opened and adulterated, so the physical identification is not enough for forensic confirmation.

CLANDESTINE LABORATORIES

Illicit drugs that are not naturally occurring must be manufactured somewhere. If they are manufactured by legitimate drug companies, then they can be obtained for abuse by either stealing them from the manufacturing plant or from a doctor or pharmacy or by forging a prescription. Drugs that are manufactured in other countries can be smuggled into this country. For many illicit drugs, these methods are virtually the only ones available to users.

For some drugs there are no legitimate manufacturing plants or, if there are, it may still be difficult to obtain these drugs by the aforementioned route. The alternative is then the clandestine laboratory. For LSD, PCP, ecstasy, and other popular abused drugs, clandestine manufacture is the only way to get them. For amphetamine and methamphetamine, which are legally manufactured, the clandestine lab is still the most popular source.

Numerous books and internet sites have recipes for manufacturing various drugs. Using these recipes and some fairly rudimentary chemicals, glassware, containers, and appliances, a person can manufacture huge quantities of illicit drugs—if he or she has a source for the chemicals. It used to be fairly easy to obtain precursor chemicals from chemical supply houses or chemistry laboratories, but in recent years, this has become much more difficult. Chemical suppliers are more particular about whom they will furnish drugs to and will report large sales of chemicals that are suspected to be precursors of popular illicit drugs to the DEA or the police. In addition, the federal government and many state governments now control the distribution of precursor chemicals by putting them in a Federal Schedule, making their distribution much more difficult. In spite of all of these efforts, illegal manufacture of illicit drugs goes on and, in some areas of the country, is on the rise.

As you might expect, the sophistication of clandestine laboratories varies widely. Some labs are extremely modern and would be the envy of a legal manufacturer. Most, however, are rudimentary, dangerous, filthy places with virtually no health and safety precautions. In fact, this characteristic is one of the more prominent reasons why clandestine labs are

Figure 13.13. A *portion of a clandestine methamphetamine laboratory. Note the messy, dangerous conditions, conducive to a destructive fire.*

discovered. Huge piles of trash containing many chemical containers or strange smells emanating from the drug lab can tip off neighbors, who then call the police. Fires and explosions also occur occasionally in clandestine labs. A clandestine drug lab is shown in Figure 13.13.

When drug enforcement agents are alerted to the possible presence of a clandestine laboratory by neighbors or by reports of the purchase of precursor chemicals, they will generally set up surveillance or use undercover agents to discover what is being produced and at what stage the production is. The best strategy is to wait until the final product has been produced. If the arrest takes place at this point, the doers can be charged with manufacture of an illicit drug and possession with intent to distribute that drug. These offenses carry the stiffest penalties. If, however, the lab is raided before the final product has been made, then the government can only charge the suspects with attempted manufacture. This crime generally carries lesser charges than does manufacture, and the government must prove that the final product could have been produced with the chemicals, apparatus, and recipes that were found at the lab. Such prosecutions are much more difficult than those for manufacture. It is worth noting that people who grow marijuana plants, peyote cactus, or psilocybin mushrooms in their home or on their property are also charged with manufacture if they are caught.

BACK TO THE CASE

This case is typical of those encountered in crime labs. All of the chemicals, apparatus, and any instructions and manuals will be seized and taken to the crime lab. There the chemists will perform complete analyses on all of the chemicals. They will show that, using the chemicals and apparatus available and the instructions, the clandestine lab operators would have been involved in the manufacture of methamphetamine. If the actual product is found, then the government can charge the suspects with manufacture of methamphetamine. If no product is found, then the charge would be attempted manufacture.

CHAPTER SUMMARY

Illicit drugs represent the largest volume of criminal cases that are examined by forensic science laboratories. They can occur naturally as with marijuana or cocaine, or they can be prepared from naturally occurring substances such as the case with heroin, or they can be totally synthetic as is the case with amphetamines and most other prescription drugs.

Illicit drugs can also be classified by major effects. There are four major types: stimulants, depressants, narcotics, and hallucinogens. Illicit drugs in the United States are controlled both by the federal and all 50 state governments. The model laws adopted by the federal government are embodied in the Uniform Controlled Substances Act, which puts drugs in one of five Schedules, according to potential for abuse and approved medical use.

Illicit drugs seldom occur as pure substances and therefore must be separated from the cutting agents. This can be accomplished by liquid extraction or by some form of chromatography. Virtually all illicit drug cases must have a confirmatory test, such as gas chromatography-mass spectrometry in order to be presented in court.

Test Your Knowledge

1. What are the four major classes of illicit drugs? Give an example of each.

2. What is a semisynthetic drug? Give an example of one.

3. What are the two major criteria for deciding if a drug shall be put in a Federal Schedule?

4. When will a drug be put in Schedule 1?

5. What Schedule would drugs that require a doctor's prescription and which are not listed in another Schedule be put in?

6. What is "speed"? What family of substances does it belong to? Where is it scheduled?

7. What was the first act passed by Congress to control drugs? What were its main provisions?

8. What was the Harrison Act? What was its purpose? What drugs was it aimed at principally?

9. What is the significance of a "usable quantity" in drug control? Give an example.

10. What is an excipient? What is its purpose?

11. What is a diluent? Why are they used? Give an example.

12. What is meant by the term "aggregate weight" as it applies to drug control? Give an example.

13. Give an example of a spot or field test and the drug or class of drugs it is used on. When are these tests used?

14. What type of test is used to determine the percentage or quantity of a drug in a mixture?

15. What tests are used for the confirmation of drugs?

16. The analysis of marijuana is considered an exception to the general scheme of analysis of drugs, especially those in powdered mixtures. Why is this so? What are the differences?

17. Both FTIR and mass spectrometry are used for confirmation. When would it be appropriate to use one or the other?

18. LSD is considered unique in the family of illicit drugs. Why? How does its occurrence differ from other drugs?

19. When a clandestine laboratory is raided and no final product is found, what charge is usually levied against the perpetrators? What must be proven in such cases?

20. Under what conditions may it not be possible to perform a confirmatory test upon a drug exhibit? How might an analyst's conclusion about the exhibit be altered in such cases?

Consider This . . .

1. The federal government and many states treat "crack" differently than cocaine. What is crack? How is it made? Why do law enforcement agencies and courts treat these substances differently? How do the penalties for possession and distribution differ federally?

2. Some forensic chemists maintain that tablets and capsules that contain stamped or printed manufacturer's markings can be treated differently than powdered mixtures of drugs. Why would this be the case? How would you take these markings into account in an analytical scheme? Why are marked capsules (two piece) treated more like unknown powders than are marked tablets?

3. What are some of the important considerations that come into play when confronted with large exhibits, e.g., a bale of 25 pounds of suspected cocaine? How would this be sampled? What if an analyst receives 25 pounds of suspected cocaine in the form of 400 1-ounce packages?

BIBLIOGRAPHY

Blachford, S.L., & Krapp, K. (Eds.). (2002). *Drugs and controlled substances information for students.* Farmington Hills, MI: Gale Group.

Controlled Substances Act, Title II of the Comprehensive Drug Abuse Prevention and Control Act of 1970 (Act 21 U.S.C. 812).

Harrison Narcotic Tax Act—1914

Ray, O., & Ksir, C. (1987). *Drugs, society and human behavior* (4th ed.), St. Louis, MO: Times Mirror/Mosby.

Physicians Desk Reference 2005, Medical Economics Co. Montrale, NJ 2005.

Siegel, J. (1988). Forensic identification of controlled substances. In R. Saferstein (Ed.), *Forensic Science Handbook, Vol. 2.* Upper Saddle River, NJ. Prentice-Hall.

United States Drug Enforcement Administration, Drugs of Abuse, US Government Printing Office, 2005. www.USDOJ.GOV/DEA/PUBS/ABUSE/DOA-P.PDF.

www.angelfire.com/wizard/kimbrough/Textbook/ergotism_blue.htm.

Forensic Toxicology

CONTENTS

KEY TERMS

Absorption

Agonist

Alcohol dehyrodgenase (ADH)

Alveoli

Antagonist

Blood alcohol concentration (BAC)

Breath alcohol concentration (BrAC)

Clinical pharmacology

Cut-off level

Dependence

Distribution

Drug

Drug recognition experts (DREs)

Elimination

Enzyme multiplied immunoassay test (EMIT)

Field sobriety test

Forensic toxicologist

Henry's Law

Horizontal gaze nystagmus

Immunoassay

Metabolism

Metabolite

Neurotoxin

Pharmacodynamics

Pharmacokinetics

Pharmacology

Preliminary breath testing instruments (PBTs)

Radioimmunoassay (RIA)

Receptor

Screening tests

Tolerance

Toxicology

Widmark curve

INTRODUCTION

Many people think of poisons when they hear the term toxicology. But actually, toxicology involves virtually any non-food substance taken by a living organism. It is concerned not only with how much of a substance was taken, but also what are the physiological and psychological effects of these substances. Since this book is about forensic science, this chapter will be concerned with **forensic toxicology**, which is mostly about humans and how they are affected

THE CASE

A police officer, patrolling in an urban area, spots a man staggering down the street in obvious distress. He pulls over and gets out of his cruiser to question the man. Before the officer can get to him, the man falls to the pavement, unconscious. The officer calls for an ambulance. The man is taken to the emergency room of the city hospital, where triage and treatment are administered, but the man dies. An autopsy is performed, but the forensic pathologist can find no obvious cause of death. Samples of blood and other body fluids and tissues are sent to the **forensic toxicologist** to see if there are any drugs or poisons present and, if so, if they could have contributed to the cause of death.

by drugs and poisons. When someone ingests a drug or poison, it is absorbed into the bloodstream, circulates throughout the body where it has its intended and/or unintended effects, and then is eliminated from the body by a variety of processes. In most cases forensic toxicologists become involved when a person has died under suspicious circumstances. The toxicologist works with the forensic pathologist to help determine the cause and manner of death. This can be a complex problem. The toxicologist must learn the person's medical history, drug use pattern, physical condition at the time of ingestion, the amount and duration of ingestion, and the identities and amounts of other drugs that are in the body at the same time.

Forensic toxicologists also work on certain cases in which people have taken drugs or poisons but don't die. These cases mainly involve ethyl alcohol and driving. The job of the forensic toxicologist in these cases is to determine the amount of alcohol in the body and its likely effects on the person while operating a motor vehicle.

This chapter describes how drugs get into the body and how they are distributed and eliminated. It does not discuss the effects of drugs on humans. That part of **pharmacology** is beyond the scope of this introductory forensic science book. Because it is so prevalent in our society and takes up so much of the forensic toxicologist's time and effort, alcohol in drunk driving cases will be used to illustrate the principles of forensic toxicology in this chapter.

FORENSIC PHARMACOLOGY AND FORENSIC TOXICOLOGY

In the broadest sense, pharmacology is the science that studies the relationships between drugs and living things. There are many branches of pharmacology, but the one that is applicable to forensic science is **clinical pharmacology**, which is concerned with the effects of drugs on humans. In forensic cases, the pharmacologist is concerned with the harmful effects of drugs on humans. This branch of pharmacology is called **toxicology**. A forensic toxicologist is a scientist who determines what chemicals, principally drugs, are present in a human being and what effects they have on that person. In the case described at the beginning of this chapter, the forensic toxicologist will determine what, if any, drugs or poisons were present that could have caused or contributed to death. Forensic toxicologists often work closely with forensic pathologists to help in the determination of the cause and manner of death in cases of suspicious death.

DRUGS AND POISONS

A **drug** is a chemical or chemical mixture that is designed to have an effect or series of physiological and/or psychological effects upon an animal (or plant). The difference between drugs and other substances is that the drug is manufactured or designed to cause a particular response in the organism. A poison, on the other hand, is a substance that has a toxic (life-threatening) effect on a living organism. Drugs can also be poisons if too much is ingested or if two or more drugs are taken such that their cumulative effects cause a toxic response. Actually, nearly any substance can be a poison if taken in excessive quantities that can cause harm to living beings.

Direct effects of drugs are those that are intended by the drug's maker. These are distinguished from side effects. Side effects are unintended. They may be mild or severe and do not occur in everyone who takes the drug. For example, some drugs meant to treat some of the symptoms of allergies may have the side effect of causing drowsiness. Some side effects may even cause death. From a toxicology standpoint then, side effects may be as important as direct or intended effects from some drugs. For the purposes of forensic toxicology, ethyl alcohol is included even though it is not considered to be a drug.

THE FORENSIC TOXICOLOGIST

The forensic toxicologist must accomplish a number of tasks in order to reach conclusions about the role of a drug in causing death:

- Determine the identity of all drugs and poisons present in the body
- Determine the quantities of all drugs and poisons present at the time of death
- Determine what **metabolites** (secondary products of drugs as they are acted on by the liver) of these drugs are present
- Determine what interactions (e.g., synergisms) may exist among the particular combination of drugs that are present
- Help determine the history and patterns of drug use by the person involved and the role that **drug dependence** may play in this case
- Help determine the role that **tolerance** may play in this case

Of all of these tasks, the most critical ones are the first two—the identification of the drugs and determination of their quantities. Drug identification will be covered later in this chapter. For now, the factors that affect the ultimate concentration of drug in the human body will be discussed.

PHARMACOKINETICS

The science of **pharmacokinetics** studies how drugs move into and out of the body. Four processes define pharmacokinetics. At times, only one or two of these processes are taking place; at other times all four are going on simultaneously. The four processes are **absorption, distribution, metabolism**, and **elimination**. Once the drug is ingested and all four processes are operating, a dynamic equilibrium is set up within the body. This means that the drug concentration at any given time in any given part of the body is dependent on all of the processes acting simultaneously. It is dynamic because the concentration at any time is determined by which of the processes is dominant at that time. Depending on the amount of drug taken and the time since ingestion, any of the processes may predominate and the concentration changes with time, sometimes rapidly.

ABSORPTION

Drugs may be introduced into a body by a number of means. They include oral, intramuscular, intravenous, rectal, topical, subcutaneous, and inhalation. Different methods are used for different drugs, and there is usually a preferred method for a given drug. All methods involve the passage of the drug through a tissue barrier such as stomach or intestine, nasal passages, skin, etc. The chemical nature of the drug dictates how easily the drug can cross the barrier. Once the barrier has been breached, the drug will enter the bloodstream. Although a drug may be distributed locally by diffusion through tissues, global distribution through the body is accomplished through the bloodstream. When a drug is taken orally, it is generally absorbed through the stomach and/or small intestine. The rate of absorption will depend in part on what else is already in the stomach at the time of ingestion.

DISTRIBUTION

The bloodstream reaches virtually every cell in the body, and drugs are mainly distributed this way. Some portion of a given drug may be bound to blood proteins and thus unavailable for interacting with the brain or other organs, and this must be taken into account when determining the effective concentration of a drug. Although a drug in the bloodstream will reach all tissues in the body, this doesn't mean that the concentration of the drug is the same everywhere. Some organs, such as the brain, heart, and liver, receive more of a blood supply than less vital organs; therefore, they would be exposed to more of a given drug. Also many drugs have a chemical structure that causes them to collect in particular types of tissue preferentially. For example, some pesticides tend to be attracted to adipose (fatty) tissue. Once these drugs get absorbed by this tissue, they are very hard to remove. Over time, large concentrations build up. This can cause great harm or even death. If pesticides are dumped into lakes and rivers, and fish ingest them, the pesticides will collect in the fatty tissues of the fish. When humans eat the fish, large doses of the pesticides can be transferred to the humans.

METABOLISM

Metabolism is a process whereby a drug or other substance is chemically changed to a different form, called a **metabolite**. Metabolism serves at least three purposes:

1. It may deactivate the drug so that it has fewer or milder effects on the body.

2. The metabolite is generally more water soluble than the parent drug. This makes it easier to eliminate through urination.

3. It may convert the drug into a substance that can be used by the body's cells for energy. This also aids in elimination of the drug.

Most metabolism takes place in the liver where enzymes cause chemical changes. A drug may undergo a series of metabolic reactions whereby a first metabolite undergoes further metabolism to form additional metabolites.

An example of metabolism that renders a substance less harmful than the primary drug or substance can be found in alcohol. Ethyl alcohol, the substance that is found in beer, wine, and spirits, is a **neurotoxin** (kills nerve cells). The liver metabolizes alcohol to acetaldehyde, which is then metabolized to acetic acid (vinegar is a dilute form of acetic acid). Neither acetaldehyde nor acetic acid is a neurotoxin. Acetic acid is very water soluble and can be easily removed from the body in urine. It is also easily used by the cells in oxidation to produce energy.

ELIMINATION

There are a number of ways that drugs can be removed by the body. The predominant mechanism is by excretion in urine. Since urine is mainly water, the drug must be water soluble before it can be effectively eliminated this way. Metabolism by the liver often accomplishes the task of rendering a drug more water soluble. If a drug is volatile (easily converted to a vapor), it can also undergo elimination by respiration; it can be exhaled from the lungs. Some volatile substances can also be partially eliminated by perspiration. Again, an example of a substance that is somewhat eliminated by perspiration and respiration is alcohol. Even so, the vast majority of alcohol is eliminated by metabolism followed by urination.

DRUG ACTIONS: PHARMACODYNAMICS

The study of how drugs act in the body is called "**pharmacodynamics**." Certain organs in the body, such as the brain, contain cells that have active sites called **receptors**. Drugs are designed to bind to a particular type of receptor. When the drug finds and binds to its receptor, the receptor causes the cell to fulfill a particular function or process. For example, when a drug binds to a certain receptor on the pancreas, it causes the pancreas to secrete insulin into the bloodstream. A drug that binds to a receptor and causes it to exert its function on the cell is called an **agonist**.

Some drugs may bind to a receptor but not cause it to exert the action of the cell to which the receptor is attached. These substances are called **antagonists**. Antagonists can serve to block or reverse the actions of agonists. An example of this agonist/antagonist relationship can be found in heroin or other narcotics. Heroin is a powerful central nervous system (CNS) depressant. An overdose of heroin can cause death by depressing the CNS so much that respiration ceases and the person dies. A person who has such an overdose can be given Nalaxone, a heroin antagonist. Nalaxone will compete with heroin for the same receptor sites in the brain, but as an antagonist will not cause CNS depression.

Determining the effective dose of a therapeutic drug can be a very complex process. It depends on the person and his or her physical condition and the severity of the illness being treated. The whole process is based on knowledge of pharmacodynamics and the characteristics of the receptors for the drug being tried.

DEPENDENCE, TOLERANCE, AND SYNERGISM
Addiction and Withdrawal
When a person becomes addicted to a drug, he or she may have a potent craving for it. The person's whole life becomes a constant search for the money to buy the drug or for the drug itself. This dependence may be more than psychological. There may be an actual physical dependence on the drug. How can you tell if a person is psychologically dependent on a drug or physically addicted? Outward actions and reactions may not reveal dependence or addiction since they have many behaviors in common. The way to find out is to suddenly stop taking the drug. If there is a physical dependence on the drug, then the person will undergo a withdrawal syndrome. This is a well-defined set of physical symptoms including high temperature, physical discomfort, pain, etc. In certain cases, such as severe addiction to barbiturates, sudden or "cold turkey" withdrawal can be fatal.

Tolerance

Chemical **tolerance** is a phenomenon whereby the body's organ systems adapt to the drug. Then it takes ever-increasing doses to achieve an equivalent psychoactive effect. Most drugs will eventually show a tolerance, but it is more pronounced in some types including opium-based narcotics such as morphine and heroin, and also cocaine and barbiturates.

The cause of tolerance appears to be a decrease in sensitivity or number of receptors for the drug. It usually takes many repeated doses of the drug over an extended period of time for tolerance to be manifested. If the drug is stopped, it may take a long time period for the body to recover. In some cases of extreme tolerance buildup, sudden (cold turkey) withdrawal of the drug can be fatal. This has been demonstrated with barbiturate addiction.

Synergism

You may have heard the expression "The whole is greater than the sum of the parts." In pharmacology, this means that the total effect on the body of two or more drugs taken together is greater than the effects would be if the drugs were taken separately. The drugs work together to magnify effects or create effects that would not have occurred otherwise. This is called "synergism." A forensic toxicologist must be aware of synergism when making conclusions about the role of a drug in the cause and manner of death. The toxicologist must know what drugs were taken and what drugs were already present when another one is taken.

One of the most well-known synergisms in toxicology is that of alcohol and barbiturates. Barbiturates are central nervous system (CNS) depressants. They slow down many functions of the body and may induce drowsiness or even sleep. Alcohol is also a CNS depressant, but its mechanism of action is different than that of barbiturates. When alcohol and barbs are taken together, the depressant effects are greatly magnified over what they would be if the alcohol and drugs were taken separately. The effects can be magnified to the point where a person may die even from sublethal quantities of both the alcohol and barbiturates. This combination of substances was apparently the cause of death of rock stars Janice Joplin and Jimi Hendrix. There are many reports of both accidental and deliberate cases of overdose by alcohol and barbs in the United States each year.

Sometimes it is difficult to determine if two drugs are acting synergistically. For example, if someone takes two central nervous system depressant drugs and has a significant reaction, it may be that the drugs' effects were merely additive, rather than enhanced. Synergism is easiest to detect if someone takes two drugs and has effects that would not be expected from one of the drugs.

IDENTIFICATION OF DRUGS IN THE BODY

Drug identification can be a difficult process. There may be little or no information about what drug or drugs a person may have taken or when or how much was taken. Many drugs require only small doses to be effective, and they may be in very dilute concentrations in the body. Some drugs have a preference for certain tissues or organs and may be hard to find. Other drugs present problems because they have a short lifetime in the body or because they break down into naturally occurring substances. GHB is an example of this. See "In the Lab: GHB." The process of drug identification involves the following very important steps:

- Sampling
- Extraction
- Screening
- Confirmation

SAMPLING

The types of samples taken from the body for drug identification are dictated by the condition of the body. If the person is alive, then blood, urine and, increasingly, hair are the preferred samples. If the person is dead, then all of the above may be available. In addition, other tissues such as brain or liver can be used. Urine concentrations of drugs and metabolites may be much greater than blood, so urine makes an ideal medium for screening for most drugs. On the other hand, the concentration of the drug in urine may not correlate well with blood concentrations, and using urine concentrations may lead to misleading conclusions about the magnitude of the drug's effects. Therefore, it is usually good scientific practice to use urine samples to screen for the possible presence of a drug, but to use blood for determination of the concentration of the drug and for confirmation of its presence.

IN THE LAB:
GHB

Gamma-hydroxy butyrate (GHB) has been around since the 1960s when it was used as an anesthetic. It has also been used by body-builders as a supplement. In recent years, however, it has become known and occasion-ally used as a so-called "date rape drug." GHB is very soluble in water, and there have been cases when unsuspecting victims, usually women, have had GHB slipped into their drink at a bar or party. This drug causes the victim to become drowsy and then fall asleep. She wakes up hours later, many times in a strange place, having been sexually assaulted.

From a toxicology standpoint, GHB presents a number of challenges. First, it is a naturally occurring substance, a by-product of amino acid degradation, and is present in many foods as well as in humans in small quantities.

Most drugs metabolize to form stable, well-characterized metabolites that persist in a human body for many hours or days. Cocaine, for example, metabolizes into benzoylecgo-nine. The **half-life** of cocaine is about 72 hours. A half-life is the time it takes for a drug to reduce its concentration by half through metabolism and/or elimination. This means that cocaine and/or benzoylecgonine would be found in a blood sample taken from a user for several days after ingestion. Unlike most drugs, GHB metabolizes very rapidly to succinic acid, another naturally occurring substance. This, in turn, is rapidly consumed to make energy for cells. Thus, GHB and its metabolites are virtually undetectable in the body in just a few hours after ingestion. This means, in many cases, by the time the victim awakes and real-izes what has happened and has gotten to a hospital, it is too late to collect a blood sample and find the drug. GHB is quite stable in aqueous solutions and, if it is administered through an alcoholic beverage, it can be iso-lated from the drink. This requires that someone preserve any of the drink that might be left over after the victim has drunk it. Unfor-tunately, this seldom occurs. When GHB is used as a date rape drug, and the victim is unaware, it is very difficult for a forensic toxicologist to provide chemical proof that GHB was used.

In the past few years, hair has become an increasingly important specimen for drug analysis. Unlike blood or urine, hair can trap and hold drugs for many months, thus yielding information about drug use patterns and frequency of ingestion. Caution must be used in hair analysis for drugs owing to the possibilities for direct uptake by hair from the outside and to the effects on hair treatments such as bleach-ing on drug concentrations. The subject of using hair for the analysis of certain drugs will be discussed further in the section "Drug Testing in the Workplace."

EXTRACTION

Liquid Phase

Once the decision has been made as to which tissue or body fluid will be used to find and identify drugs, it will be necessary to extract and concentrate the drug. This is especially important where more than one drug may be present and it is necessary or desirable to isolate the drugs from each other. In order to accomplish an efficient separation, the pH range of the drug must be known. Most drugs are either acidic or basic in aqueous solutions. A few drugs, such as caffeine, are neutral. An acidic drug is one whose pH is between 1 and 7, and a basic drug has a pH between 7 and 14. The pH dictates the method of extraction. As an example, consider a urine sample containing cocaine. Urine is very acidic and cocaine is a basic drug. In an acidic medium, then, cocaine would exist in what is known as the salt (ionic) form. The cocaine has an H^+ attached to it, and there would be an anion such as Cl^- present. This form of cocaine is very soluble in urine. If the solution is made basic, the cocaine reverts to its free base form as plain cocaine. This form is not soluble in urine or other aqueous solutions, and the cocaine would precipitate out. An organic solvent is then added in which the cocaine is very soluble. This solvent doesn't mix with the urine; thus, the cocaine can be efficiently separated from the urine this way. Acidic drugs such as barbiturates can be extracted from urine using an acid extraction rather than a basic extraction. Liquid phase extraction is not a suitable technique when two or more acidic or basic drugs are known to be present. The reason is that all of the acidic or basic drugs will be extracted together, and they cannot be separated by this method. More on liquid phase extractions can be found in Chapter 6, "Separation Methods in Forensic Science."

Solid Phase

To overcome the limitations of liquid phase extraction and increase sensitivity, solid phase extraction of drugs has become popular in recent years. In solid phase extraction a small column or coated wire is used to extract the drug out of the urine or blood. The blood or urine is poured through a special solid bed that selectively removes the drug. The bed is then washed or eluted to strip the drug off for further analysis.

A recent refinement of solid phase extraction is solid phase microextraction (SPME). Here, a specially coated wire or a synthetic fiber is inserted into the fluid that contains the drug or drugs. The drugs are selectively adsorbed onto the wire, which is then removed. The drug can then be eluted for further analysis. SPME is capable of capturing and measuring very small amounts of drug, so it is ideal for situations in which limited sample is available. Figure 14.1 shows an apparatus used for solid phase microextraction.

Figure 14.1. Solid phase microextraction. The coated wire on the end is immersed into the blood or urine sample that contains the drugs of interest. The drugs adsorb onto the surface of the wire. The wire can then be eluted with a solvent or introduced directly into the inlet of a gas chromatograph where the heat will remove the drugs and they will then be swept into the gas chromatograph.

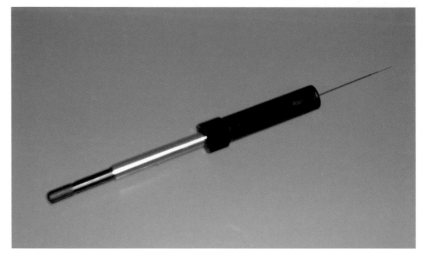

SCREENING

Screening tests are a type of preliminary test for drugs in body fluids. They are designed to give a preliminary result that indicates a drug may be present, but they do not absolutely confirm the presence of a drug. Screening tests fall into two general categories: chromatographic and immunoassay. Chromatographic tests are usually thin layer chromatography or gas chromatography (GC). More information on these tests is given in Chapter 6, "Separation Methods in Forensic Science," and Chapter 13, "Illicit Drugs."

Immunoassay Techniques

Immunoassay techniques take advantage of the reaction that takes place between antigens and antibodies in human blood. This technique is described in more detail in Chapter 10, "Serology and Bloodstain Pattern Analysis." An antigen is a substance that is resident on red blood cells and in the blood plasma itself. If a substance such as a virus or bacterium enters the body, it acts like a foreign antigen and causes the immune system to produce antibodies that can attack and immobilize the antigen before it can cause harm. The antibodies produced are specific for the type of antigen that is introduced. In forensic toxicology, this antigen/antibody reaction is exploited to detect very small quantities of drugs. A major advantage of immunoassay tests for drug screening is that they do not require prior extraction of the drugs from the urine.

One of the most popular immunoassay tests is called the **Enzyme Multiplied Immunoassay Test**, or EMIT. A rabbit is injected with a form of the drug for which detection is sought—for example, methamphetamine. The rabbit's immune system produces an antibody to the methamphetamine. The rabbit's serum is then removed. Next, a known amount of the rabbit antibodies is added to the urine of the human subject being tested for drugs. If there is any methamphetamine in that person's urine, it will immediately bind to the antibody. Next, a known amount of methamphetamine that has been labeled with an enzyme is added to the urine. If there are any antibodies from the rabbit that did not react with the methamphetamine in the subject's urine, they will now react with the enzyme-labeled methamphetamine, thus decreasing its concentration. The amount of the remaining enzyme-labeled methamphetamine is now measured. This will correlate to the concentration of the antibodies that reacted and thus to the concentration of the antibodies that reacted with the methamphetamine that was in the urine. From this, the concentration of the methamphetamine can be determined.

Another related immunoassay technique is called **Radioimmunoassay** (RIA). RIA works in a similar manner to EMIT, but the antigens are radioactively labeled. The major problem with RIA and EMIT is that the antibodies produced in the rabbit from injecting a drug are not always entirely specific for that drug. For example, antibodies produced as a result of the injection of methamphetamine into a rabbit may also bind to pseudoephedrine, a proprietary antihistamine sold in drug stores for allergy relief. Relying solely on an immunoassay test for identification of methamphetamine would mean that a false positive result would occur. Thus, all immunoassay tests as well as chromatographic screening tests must be confirmed with a suitable second test.

CONFIRMATION

Once a screening test has been completed, any drugs indicated must be confirmed. The only accepted method for drug confirmation in forensic toxicology is mass spectrometry. This technique is explained in Chapter 5, "Spectroscopic Techniques." Some laboratories take the position that performing two independent screening tests based on different chemical principles is an acceptable method for confirmation of a drug. An example of this would be to perform an EMIT test followed by thin layer chromatography. In the practice of forensic toxicology this is not acceptable. No amount of presumptive, preliminary testing will add up to confirmation. Only mass spectrometry will provide acceptable confirmation of a drug in the body.

CUT-OFF LEVELS

Every analytical technique, be it mass spectrometry or a screening test, has a limit of detection—that is, a level below which a reliable result cannot be determined. The reason is that every instrument creates some electronic noise that shows up in the chart or graph of the analysis. If the signal that indicates the presence of a drug is too weak, it will not be seen above the noise. To avoid the possibility that noise will be mistaken for a signal, each laboratory sets a **cut-off level** for each drug. This is always used in cases in which a living person is being tested for drugs in a preemployment or incarceration situation. If a drug is found at a level at or below this cut-off, the result will be reported as "drug not detected." For example, the cut-off for cocaine using EMIT may be set at 50 ng/ml (nanograms of drug per milliliter of urine). This means that a result of "drug not detected" for cocaine in this test means only that the drug was either not present or was present at a level below 50 ng/ml. If a different, more sensitive test were used with a lower cut-off, this amount of cocaine may be reported. In postmortem cases or cases in which it is necessary to determine if someone has taken any amount of a drug, then cut-offs are not used. The Department of Health and Human Services has established initial and confirmatory cut-off levels for common drugs. They are shown in Figure 14.2.

DRUG TESTING IN THE WORKPLACE

In recent years there has been increased emphasis on testing employees to make sure that they are not using drugs while on the job. This practice started with workers in sensitive situations or who worked in dangerous environments, such as police officers, locomotive engineers, pilots, etc., but has since spread to many other occupations. There have been a number of problems uncovered with workplace drug testing that are not found in other areas of forensic toxicology.

The following cutoff concentrations are used by certified laboratories to test urine specimens collected by Federal agencies and by employers regulated by the Department of Transportation:

Initial Test Cutoff Concentration
(nanograms/milliliter)

Marijuana metabolites	50
Cocaine metabolites	300
Opiate metabolites	2000
Phencyclidine	25
Amphetamines	1000

Confirmatory Test Cutoff Concentration
(nanograms/milliliter)

Marijuana metabolite (1)	15
Cocaine metabolite (2)	150
Opiates:	
Morphine	2000
Codeine	2000
6-Acetylmorphine (4)	10
Phencyclidine	25
Amphetamines:	
Amphetamine	500
Methamphetamine (3)	500

Footnotes:

(1) Delta-9-tetrahydrocannabinol-9-carboxylic acid
(2) Benzoylecgonine
(3) Specimen must also contain amphetamine at a concentration >= 200 nanograms/milliliter
(4) Test for 6-AM when morphine concentration exceeds 2000 nanograms/milliliter

Figure 14.2. Initial and confirmatory levels of common drugs. Data supplied by the Department of Health and Human Services.

SAMPLING

In a typical workplace situation, the subject is asked to supply a urine sample. This is generally done in an unsupervised fashion. It is possible for the subject to dilute the urine sample or even smuggle in drug-free urine. Many people consider it an invasion of privacy to have to urinate in front of someone else.

MIXING UP SAMPLES

Mixing up samples can be a major problem in situations in which forensic science labs are not used. If the urine sample is taken at a clinic and proper labeling and chain of custody procedures are not followed, then it is possible for samples to get mixed up. Likewise, if the testing laboratory and/or the courier don't use proper chain of custody techniques, the samples could once again be easily mixed up.

IMPROPER ANALYSIS

Many clinical laboratories that test body fluid samples for drugs do not use proper confirmation techniques. They may use two or three screening tests and report the drug as being definitely present, when no mass spectrometry confirmation test had been done. As was mentioned above, this is not acceptable forensic science procedure and runs the risk of false positive results.

FORENSIC TOXICOLOGY OF ETHYL ALCOHOL

The analysis of ethyl alcohol provides an excellent illustration of the basic principles of forensic toxicology. Two independent measuring systems (blood [BAC] and breath [BrAC]) for alcohol correlate reasonably well with each other. On top of the scientific issues with the measurement of alcohol, there are a number of social issues that provide another layer of concern for forensic scientists. Operating an automobile (or other conveyance) while under the influence of alcohol has been a serious problem in the United States for many years. An entire body of laws and regulations has been created by the Congress and the 50 state legislatures to deal with this problem. Central to all of these is the notion, unique to alcohol in the world of drugs and poisons, that the degree of sanction or punishment for operating a motor vehicle under the influence of alcohol is tied directly to the concentration of alcohol in the body. This puts an extra burden on the forensic toxicologist who is faced not only with accurately determining the amount of alcohol in a person, but also being called on to opine what the concentration may have been a few hours earlier and, sometimes, what effects a given level of alcohol may have on a person. These considerations require a thorough knowledge of all stages of the pharmacokinetics of alcohol: ingestion, absorption, distribution, and elimination. Alcohol is very similar to water in its chemical structure and is soluble in water in all proportions. Because blood and body tissues are largely made up of water, alcohol is easily absorbed, distributes itself rapidly to all parts of the body, and is eliminated easily in urine.

For forensic purposes, blood is taken as the reference for alcohol concentrations, and measurements of alcohol in other tissues or body fluids refer back to **blood alcohol concentration (BAC)**. In blood, alcohol is measured in weight/volume units, specifically the number of grams of alcohol in 100 ml (1 deciliter) of blood. In breath, a weight/volume measure is used. **It is the number of grams of alcohol in 210 liters of deep, alveolar (lung) air (BrAC).** There is an approximate correlation between BAC and BrAC of 2100:1 ± 300. Some breath testing instruments use 2100:1 conversion to blood alcohol, but some experts believe that the variation in this factor is so large that separate legal limits should be established for BAC and BrAC. In some states there are two parallel sets of regulations that sanction drunk driving behavior. One refers to BAC concentrations and the other to BrAC.

PHARMACOKINETICS OF ALCOHOL

Once alcohol is ingested, it is absorbed quite rapidly into the bloodstream. At this point a dynamic equilibrium is reached that is affected by the rate of absorption, distribution, metabolism, and elimination. For a given person, the rates of distribution, metabolism, and elimination are fairly constant. This means that the ultimate level of alcohol in the blood depends on the rate that it is absorbed from the gastrointestinal tract. The faster it is absorbed, the higher the ultimate BAC will be.

Absorption

Alcohol is absorbed mostly in the upper part of the small intestine. About a quarter of the alcohol is absorbed through the stomach lining and a few percent directly from the mouth into the bloodstream. A number of factors determine the rate at which alcohol is absorbed from the gastrointestinal tract into the blood:

- **The nature of the drink**: More concentrated mixed drinks cause more rapid absorption of alcohol than do dilute drinks such as beer or wine. Alcohol from beer is absorbed more slowly than an equivalent amount of alcohol mixed with water. This is due to the carbohydrates and other additives in the beer.
- **The rate and speed of drinking**: The faster a person consumes drinks, the more rapidly the alcohol is absorbed owing to the higher concentration in the stomach and intestine at a given time.
- **The contents of the stomach at the time of drinking**: This is a major factor in affecting the rate of alcohol absorption. The pyloric valve

connects the stomach to the upper part of the small intestine. When the stomach is empty, the pyloric valve is open, and the alcohol passes directly from the stomach to the small intestine, where it is rapidly absorbed. If there is food in the stomach when the person drinks alcohol, the pyloric valve remains closed until digestion is complete. This means that alcohol cannot pass into the intestine. It must then be absorbed through the stomach lining. This process is slower than in the intestine, and the alcohol must compete with the food for absorption into the blood. The alcohol is also diluted by the food. The result is much slower absorption. It is also subject to alcohol dehydrogenase metabolism in the stomach, thus reducing the amount of alcohol reaching the blood-stream and ultimately the maximum BAC. Figure 14.3 shows a diagram of the human digestive system.

Distribution

Once alcohol gets into the blood, it circulates rapidly through the body; it will be distributed to all parts of the body in approximate proportion to the water content of each part. Hair and bones have little water content and will not trap much alcohol. The other parts of the body have fairly consistent water content and, since equilibration occurs rapidly, good estimates can be made of the relationship between BAC and alcohol content in other parts of the body. For example, brain tissue has about 85% of the amount of alcohol that would be in whole blood.

Elimination

There are two routes to elimination of alcohol by the body. Metabolism accounts for more than 90% of elimination. It takes place mainly in the liver, which has an enzyme, **alcohol dehydrogenase (ADH)**, that first converts ethyl alcohol to acetaldehyde and then to acetic acid. The acetic acid is used by cells for energy and forms carbon dioxide and water.

Metabolism takes place at somewhat different rates in different people. The average rate is about $0.015 \pm 0.003\%$/hour. This rate of elimination is approximately equivalent to less than one drink containing one ounce of 100 proof (50%) alcohol per hour. If an average man takes in enough alcohol to reach a maximum BAC of 0.22%, it would take more than eight hours to reduce the level below 0.1%, which is the legal limit in some states to be charged with operating under the influence of alcohol (OUIL); see equation 14.1.

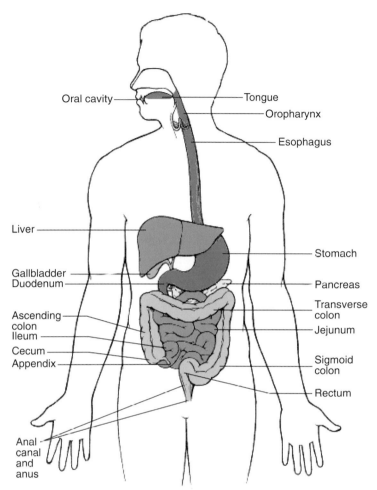

Figure 14.3. The human digestive system. Alcohol is absorbed into the blood stream chiefly through the small intestine. When there is food in the stomach, alcohol will be absorbed through the stomach lining.

Oral cavity

Tongue

Oropharynx

Esophagus

Liver

Stomach

Gallbladder
Duodenum

Pancreas

Transverse colon

Ascending colon
Ileum

Jejunum

Cecum
Appendix

Sigmoid colon

Rectum

Anal canal and anus

Equation 14.1 $\dfrac{(0.22\% - 0.1\%)}{0.015\%/\text{hour}} = 8 \text{ hours}$

The remaining few percent of alcohol is eliminated from the body as ethyl alcohol by excretion. This includes urine, breath, and perspiration. From this information, it can be easily seen that exertion, showering, and other activities will have little or no effect on BAC. Removal of alcohol from the body is largely under the control of the liver, which doesn't react to contrition, anger, or pleading. It should also be noted that ingesting large amounts of caffeine in coffee, for instance, will not help much. Caffeine is a central nervous system stimulant and alcohol is a depressant, but they work on different brain receptors and do not cancel each other out. Taking caffeine while being drunk will leave you wide awake . . . and drunk.

BAC v. Time: The Widmark Curve

When the blood alcohol concentration is plotted against elapsed time, a curve results that shows how the BAC charges over time. This is called the **Widmark curve**. Assume that a 150 lb male imbibes the equivalent of 5 oz. of 80 proof liquor all at once on an empty stomach. Absorption of alcohol would be quite rapid, and he would reach a maximum of around 0.1% within about an hour. During that steep rise in the BAC, elimination would also start to take place after a while, and the rate of increase in the BAC would start to slow. Then the BAC would start to drop, slowly at first, because absorption is still taking place to a lesser degree. Finally, after absorption is completed, the decrease in BAC would take place at a constant rate equal to the rate of metabolism in the liver until the BAC is zero. A plot of this BAC versus time is shown in the thinner line part of Figure 14.4.

If the same man were to drink the same amount of alcohol in the same amount of time, but this time after eating a meal of, say, potatoes, the rate of absorption of alcohol would be much lower, owing to the competition for absorption into the blood with the food and the fact that most of the absorption would be from the stomach. The time it would take to reach the maximum BAC would be longer, and the

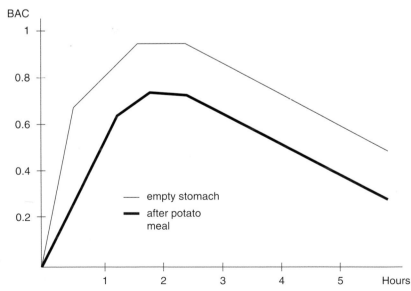

Figure 14.4. The Widmark curve. The thinner line graph represents the BAC level after drinking on an empty stomach, whereas the thicker line represents the BAC level after a meal is eaten.

IN MORE DETAIL:
How much do you have to drink to be drunk?

In many states you are considered to be under the influence of alcohol when your BAC reaches 0.1% and the trend is now to lower that to 0.08%. How much do you have to drink to reach that level? You may be surprised to see how little alcohol you need to take to be drunk.

By way of an illustration, let's assume that you are drinking on an empty stomach. Further, let's assume that you are imbibing 100 proof (50%) alcohol. Some hard liquors such as bourbon or scotch may reach this concentration. The amount of alcohol of this type you would need to reach a given level of BAC depends on your weight (in pounds) and your gender. The following formulas can be used to approximate the alcohol needed to reach a certain level of BAC:

For a male, use the following formula:

$$Vol = \frac{Wgt \times BAC}{3.78}$$

For a female, use the following formula:

$$Vol = \frac{Wgt \times BAC}{4.67}$$

where Vol = # of ounces of 100 proof alcohol, Wgt is in pounds, and BAC is in weight/volume percent. As an example, consider a 125-pound female. What volume of alcohol would be required for her to reach a level of 0.1% BAC?

$$Vol = \frac{125 \, lbs \times 0.1\%}{4.67} \qquad Vol = 2.67 \, oz$$

The average shot of a mixed drink is about 1.2 oz. This means that a woman could drink about 2.5 mixed drinks, and her BAC would be over the limit. Wine is about 12% alcohol. This would mean that the same woman could drink about 11 oz of wine, or about two, 6 oz glasses. Beer generally runs about 4% alcohol, so this woman could drink about three, 12 oz beers.

For a man of the same weight, the numbers would be about 3.3 oz of 100 proof alcohol, 13 oz of wine, or 42 oz of beer.

---

maximum BAC would be lower because metabolism would be able to better keep up with absorption. The latter part of the elimination curve in this case would have the same shape and slope as the curve generated by the drinking on an empty stomach because it also reflects metabolism by the liver. This can be seen by the thicker line plot in Figure 14.4.

The approximate amount of alcohol that it takes to reach a particular BAC level can be calculated. See "In More Detail: How Much Do You Have to Drink to Be Drunk?"

Alcohol in the Breath

The blood circulatory system is closed. Oxygen gets into the blood, and carbon dioxide and other volatile waste products are eliminated from the blood by diffusion from and to air that is inhaled into the lungs via the brachial tube, which is, in turn, connected to the mouth and nose. The pulmonary artery branches out into millions of capillaries,

Figure 14.5. *Diagram of human respiratory system.*

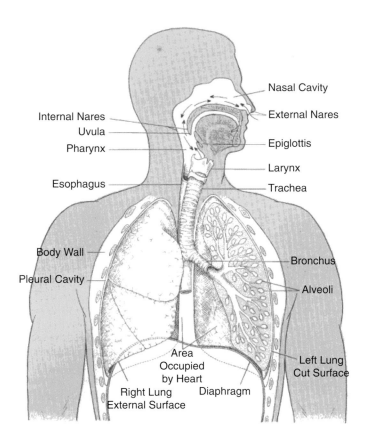

which end in small sacs called **alveoli**. Diffusion of oxygen and carbon dioxide are exchanged between the blood and air through these alveoli. If there is alcohol in the blood, then it too is eliminated through the alveoli and is exhaled from the mouth and nose. Figure 14.5 shows the human respiratory system.

The amount of alcohol that gets into the breath from the alveoli is proportional to the amount of alcohol in the blood and is governed by **Henry's Law**. This law states that when a volatile substance, such as alcohol, is dissolved in a liquid, such as blood, and then that liquid is brought in contact with a closed air space, such as alveolar breath, the ratio of the concentrations of alcohol in the blood and breath is a constant at a given temperature. If the temperature is raised, the equilibrium shifts toward the concentration in the air.

In the United States most states have adopted 2100:1 (1 ml of blood contains as much alcohol as 2.1 liters of air) at 34°C for the ratio between blood alcohol and breath alcohol, although research indicates that the actual ratio at this temperature is closer to 2300:1. This ratio is somewhat person-dependent, and the lower 2100:1 ratio favors the subject. There is a lot of variation among people.

MEASUREMENT OF ALCOHOL IN THE BODY

Breath and blood are the most commonly used specimens for alcohol analysis. In the past, urine has been used, but today it is seldom used, owing to the wide variation in measurements and lack of stable equilibrium between blood and urine concentrations.

Blood

Blood is the preferred medium for blood alcohol measurement because it provides the best surrogate for brain alcohol levels, and because of this, most states have statutes that relate sanctions for drunk driving to blood alcohol levels. The ideal blood sample would be arterial blood because it most closely tracks brain alcohol content, whereas venous BAC tends to lag behind. Nonetheless, whole venous blood is most commonly used in drunk driving cases.

The most widely used method for the analysis of BAC for forensic purposes is gas chromatography. This has several advantages including high specificity, accurate quantitative analysis, and ease of automation. Most forensic toxicology laboratories today use headspace alcohol, which is injected into the GC. In fact, most laboratories employ auto-samplers for their headspace analysis and will run these at night. The next morning, the results can be easily interpreted and reported. A typical chromatogram of alcohol is shown in Figure 14.6.

Besides GC there are also enzymatic methods for the analysis of alcohol. For example, the coenzyme nicotinamide adenine dinucleotide (NAD) will react with alcohol with alcohol dehydrogenase as a catalyst. The alcohol is converted to acetaldehyde and the NAD is reduced to NADH. The reaction is monitored by UV/visible spectrophotometry at 340 nm, where NADH absorbs, to measure the concentration of alcohol.

Breath Alcohol Testing

Breath testing or alcohol is, by far, the most widely used method for alcohol testing in use today, especially given its near universal use in drunk driving cases. Originally, breath testing instruments converted the BrAC to BAC using a constant ratio (usually 2300:1). As breath testing instruments evolved and research indicated that this ratio was

Figure 14.6. *Gas chromatogram of ethyl alcohol.*

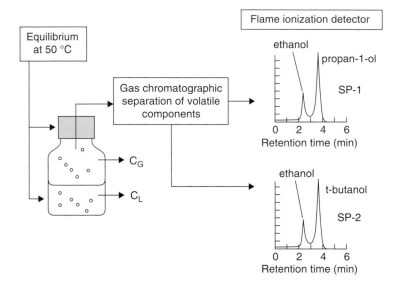

too imprecise, many states added laws that link a sanction directly to a breath alcohol concentration.

Breath testers can be divided into two general groups: those that are used primarily as **preliminary breath testing instruments (PBTs)** and those that are accepted for evidentiary purposes. There is some overlap, as some PBTs are usable as evidentiary instruments in some jurisdictions. Recent surveys indicate that the Alco-Sensors® by the Intoximeter Company are the most popular PBTs, and the Intoxilyzers® by CMI, Inc., are the most popular evidentiary instruments.

The first commercially popular and successful BrAC instrument was the Breathalyzer®. This instrument worked on the principle that acidified potassium dichromate will oxidize alcohol to acetic acid, while at the same time being reduced to chromium sulfate. For the complete equation, see equation 14.2:

Equation 14.2

$$2 K_2Cr_2O_7 \quad + \quad 3C_2H_5OH \quad + \quad 8H_2SO_4$$

potassium dichromate   ethyl alcohol   sulfuric $\rightarrow$

(yellow)   acid

$$2Cr_2(SO_4)_3 \quad + \quad 3CH_3COOH \quad + \quad 11H_2O$$

chromic sulfate   acetic acid   water

(green)

PBTs

Most of the PBTs work by either chemical oxidation or by fuel cell technology. Those that work by chemical oxidation operate using similar principles to the Breathalyzer® although potassium permanganate may be employed in place of potassium dichromate.

EVIDENTIARY BREATH TESTING INSTRUMENTS

The Intoxilyzer® is an example of an instrument that uses infrared spectroscopy for the measurement of alcohol. The alcohol is trapped in a chamber, and infrared light is passed through the sample to a detector. The more alcohol there is in the chamber, the less light gets through to the detector. The intensity of light that reaches the detector (or is absorbed by the analyte) is directly proportional to the amount of substance absorbing the light. Early instruments measured absorbance of alcohol at one wavelength of light only. There were concerns that other substances that may be present in the blood such as ketones or other alcohols could interfere with the measurement of the quantity of ethyl alcohol, so modern instruments measure the amount of alcohol at two wavelengths, with the added one chosen such that it is relatively insensitive to ketones and other alcohols. Figure 14.7 shows a diagram of the Intoxilyzer®. Other breath testing instruments in current use include fuel cell–based instruments such as the "intoximeter" and dual IR-fuel cell instruments such as the Draeger 7110. The latter instrument has the advantage of giving two independent readings using two different technologies on the same sample.

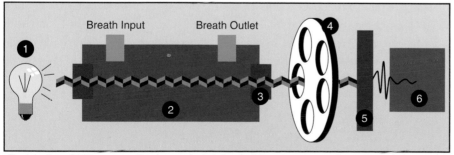

The Intoxilyzer 5000 measures the degree alcohol absorbs infrared anergy . . . the more alcohol present the greater the absorption. As shown, a quartz lamp (1) generates IR energy, which travels through a sample chamber (2) containing the subject's breath. Upon leaving the chamber, a lens (3) focuses the energy onto the chopper wheel (4) containing three or five narrowband IR filters. The IR energy passed by the filters is focused on a highly sensitive photo detector (5), which converts the IR pulses into electrical pulses. The microprocessor (6) interprets the pulses and calculates the Blood Alcohol Concentration, which is then displayed.

Figure 14.7. The *Intoxilyzer. Courtesy of CMI, Inc.*

FIELD SOBRIETY TESTING

Often, when a driver is stopped by a police officer for driving under the influence of alcohol, the subject is requested to attempt some simple tests that might shed light on his or her fitness to operate a motor vehicle. At one time, these tests were a necessary component of proof of operating under the influence (OUIL) or driving while impaired. Although these tests are no longer required as a component of proof in most states, they still may have an important function. In some states now, specially trained police officers, called **drug recognition experts** (DREs), can administer a battery of **field sobriety tests** to a driver. These tests, taken as a whole, can provide strong evidence that the driver is impaired or OUIL. They can provide probable cause to require further, quantitative alcohol testing.

First developed in California, this battery of tests has spread widely across the United States. A typical protocol calls for three tests. The first is the **horizontal gaze nystagmus**. In this test, the subject is asked to follow, with his or her eyes only, a pencil or other object as the officer moves it slowly back and forth across the subject's field of vision. If a person is sober, he or she will be able to follow the pencil easily and his or her eyes will move smoothly. If, however, the person is under the influence of alcohol (or some other drugs), the eyeballs will jerk as they move.

The other two tests measure dexterity, which would be expected to deteriorate as BAC increases. Common ones are the "walk and turn," whereby a person must walk in a straight line putting one foot directly in front of the other and then turn around and come back. The other test is to have the person close his or her eyes and touch his or her nose with the tip of a finger.

When all three tests are used as a set, there is a very high correlation between drunk or drugged behavior and the results of these tests.

BACK TO THE CASE

In the case that was presented at the outset of this chapter, there is no obvious cause of death that can be determined solely by the autopsy. Tissues and body fluids submitted to the forensic toxicologist will be screened for panels of commonly abused drugs and any other drugs that the victim may have taken. If any of these drugs are found, they will be confirmed by mass spectrometry. If there are significant metabolites present, they will be noted. It may be necessary to determine the quantity of any drug that is present to determine how much of a role it may have played in the death.

CHAPTER SUMMARY

Forensic toxicology is a part of the science of pharmacology, which is concerned with the quantities and effects of various drugs and poisons on human beings. In forensic toxicology, the chief interest is in the extent to which drugs and poisons may have contributed to impairment or death.

More than half of the cases received by forensic toxicologists involve drinking alcohol and driving. Every state and the federal government has laws that prohibit drinking and driving and set levels above which a person is either impaired or operating under the influence of alcohol. Forensic toxicologists are called upon to determine the level of alcohol present in the body and, sometimes, the level at a previous time and the effects on the person.

In cases involving drugs and poisons, forensic toxicologists usually get involved only when death has occurred. The toxicologist works with the medical examiner or coroner to help determine the cause and manner of death. The toxicologist will use data about what drugs are present and at what levels at the time of death, along with drug usage history and general health, to determine the role that drugs or poisons played in death.

Test Your Knowledge

1. Define pharmacology. How does this differ from toxicology?

2. How do forensic toxicologists work as a team with forensic pathologists? What role does forensic toxicology have in determining the cause and manner of death?

3. In the typical forensic toxicology laboratory, what is the major type of case that is handled by the toxicologists? What makes this type of case unique?

4. What is metabolism as it applies to toxicology? What role does the liver play in metabolism?

5. What is the science that describes the fate of drugs in the body from the time they are taken until eliminated?

6. What factors can affect the rate at which alcohol is absorbed from the stomach into the blood stream?

7. What are the major routes of elimination of drugs from the body?

8. For alcohol, what is the most important route of elimination from the body?

9. What is dependence? Tolerance? How do they differ?

10. When a pharmacologist has to determine the role that a given blood level of a drug may have had in the cause of death, why is it important that the drug history of the victim be known?

11. What test(s) is (are) commonly used for confirmation of a drug in the body?

12. What are immunoassay tests? When are they used? What are the two major types?

13. What is a cut-off level? Why does a toxicology laboratory use cut-off levels?

14. Why is blood generally the best sample for drugs for toxicology? Under what circumstances might blood not be the best source?

15. What is the Widmark curve? What does it measure?

16. The last portion of a typical Widmark curve has a straight-line slope? What does this measure?

17. What is the principle of operation of the Breathalyzer?

18. Most states use a conversion of blood to breath alcohol in the Breathalyzer of 2300:1. Why do other states reject this and have separate levels for blood and breath alcohol?

19. What types of instruments are used for measuring blood (not breath) alcohol?

20. Describe how the horizontal gaze nystagmus test works? When is it used?

Consider This . . .

1. There is a lot of urban folklore surrounding the ingestion of alcohol and its effects and after-effects. Some of the cautions and remedies include:

- Never mix your drinks. You won't get as drunk if you stick to one type of drink.
- You can sober up faster by drinking coffee.
- You can sober up faster by exercising heavily.
- You can counteract the effects of a hangover by having a drink the next day (hair of the dog that bit you).

Explain why all of the above are not true or don't work in light of the principles of forensic toxicology.

2. In the case cited at the beginning of the chapter, what information would be useful to the forensic toxicologist in making a determination about the role of any drugs that the person might have been taking in the cause of death.

3. One of the long-term effects of chronic alcohol abuse is cirrhosis of the liver. This is a progressive disease that gradually reduces the ability of the liver to carry out its toxicological functions, among others. If the man who died in the case cited above was found to have cirrhosis and was found to have ingested a drug, how would the pharmacologist take the liver condition into account when determining the cause of death? Would it be possible in such a case for a person to take a dose of a drug that would be sublethal for most other people and still die from it? Why?

BIBLIOGRAPHY

Baselt, R.H., & Cravey, R.H. (1996). *Disposition of toxic drugs and chemicals in man* (4th ed.). New York: Year Book.

Garriott, J.C. (1996). *Medicolegal aspects of alcohol.* Tucson: Lawyers and Judges.

Jones, A.W. (1996). Measuring alcohol in blood and breath for forensic purposes —a historical review. *Forensic Science Review*, 8, 13–14.

Moffat, A.C. (Ed.). (1986). *Clarke's isolation and identification of drugs.* London: Pharmaceutical Press.

Textile Fibers

KEY TERMS

Anisotropic

Becke line

Birefringence

Block polymer

Color

Copolymer

Courses

Crimp

Cross-sectional shape

Delustrants

Denier

Dry spinning

Dry twist test

Dye

Fabric

Fiber

Filaments

Fluorescence

Homopolymer

Interference colors

Isotropic

Knit fabric

Lumen

Manufactured fiber

Melt spinning

Metameric colors

Microfiber

Microspectrophotometer (MSP)

Natural fiber

Non-woven fabrics

Pigment

Plied yarn

Polymers

Refractive index

Simultaneous contrast

Spinneret

Spinning

Spinning dope

Spiral elements

Staple fiber

S-twist

Synthetic fiber

Technical fiber

Tex

Voids

Wales

Warp yarn

Weft yarn

Wet spinning

Woven fabric

Yarn

Z-twist

INTRODUCTION

Textile fibers, as a class, are a ubiquitous type of evidence. They are "common" in the sense that textiles surround us in our homes, offices, and vehicles. We are in constant contact with textiles on a daily basis. We all move through a personal textile environment of clothing, cars, upholstery, things we touch, and people we encounter. Fibers from textiles are constantly being shed and transferred to people, places, and things—some are better "shedders," like fuzzy sweaters, than others, a tightly woven dress shirt, for example. Certain textiles also retain fibers better than others, depending on their construction, purpose, use, and other factors—like how often they are cleaned.

Textile fibers are also one of the most neglected and undervalued kinds of forensic evidence. Fibers provide many qualitative and quantitative traits for comparison. Textile fibers are often produced with specific end-use products in mind (underwear made from carpet fibers would be very uncomfortable) and these end-uses lead to a variety of discrete traits designed into the fibers.

Color is another powerful discriminating characteristic. About 7,000 commercial dyes and pigments are used to color textiles, and no one dye is used to create any particular color and millions of shades of colors are possible in textiles (Apsell, 1981). It is rare to find two fibers at random that exhibit the same microscopic characteristics and optical properties, especially color.

Applying statistical methods to trace evidence is difficult, however, because of a lack of frequency data. Very often, even the company that made a particular fiber will not know how many products those fibers went into. Attempts have been made to estimate the frequency of garments in populations; for example, based on databases from Germany and England, the chance of finding a woman's blouse made of turquoise acetate fibers among a random population of garments was calculated to be nearly 4 in 1,000,000 garments. Cases such as the Wayne Williams case in Atlanta, Georgia, or the O.J. Simpson case in Los Angeles, California, also demonstrate the usefulness of forensic textile fiber analysis in demonstrating probative associations in criminal investigations.

TEXTILE FIBERS

A textile **fiber** is a unit of matter, either natural or manufactured, that forms the basic element of fabrics and other textiles and has a length at least 100 times its diameter. Fibers differ from each other in their chemical nature, cross-sectional shape, surface contour, color, as well as length and diameter.

Fibers are classified as either natural or manufactured. A **natural fiber** is any fiber that exists as a fiber in its natural state. A **manufactured fiber** is any fiber derived by a process of manufacture from any substance which, at any point in the manufacturing process, is not a fiber. Fibers can also be designated by their chemical make-up as either protein, cellulosic, mineral, or synthetic:

- Protein fibers are composed of polymers of amino acids.
- Cellulosic fibers are made of polymers formed from carbohydrates.
- Mineral (inorganic) fibers may be composed of silica obtained from rocks or sand.
- **Synthetic fibers** are made of polymers that originate from small organic molecules that combine with water and air.

The generic names for manufactured and synthetic fibers were established as part of the Textile Fiber Products Identification Act enacted by Congress in 1954 (see Table 15.1). In 1996, lyocel was named as a new, subgeneric class of rayon.

The diameter of textile fibers is relatively small, generally 0.0004 to 0.002 inch, or 11 to 50 μm. Their length can vary from about 7/8 inch (2.2 cm) to, literally, miles. Based on length, fibers are classified as either filament or staple fiber. **Filaments** are a type of fiber having indefinite or extreme length, such as silk or a manufactured fiber. **Staple fibers** are natural fibers (except silk) or cut lengths of

Table 15.1. *Federal Trade Commission Textile Products Identification Act, 1954, Definitions.*

FIBER	DESCRIPTION
acetate	A manufactured fiber in which the fiber-forming substance is cellulose acetate. Where not less than 92% of the hydroxyl groups are acetylated, the term "triacetate" may be used as a generic description of the fiber.
acrylic	A manufactured fiber in which the fiber-forming substance is any long-chain synthetic polymer composed of at least 85% by weight of acrylonitrile units.
anidex	A manufactured fiber in which the fiber-forming substance is any long-chain synthetic polymer composed of at least 50% by weight of one or more esters of a monohydric alcohol and acrylic acid.
aramid	A manufactured fiber in which the fiber-forming substance is any long-chain synthetic polyamide in which at least 85% of the amide linkages are attached directly to two aromatic rings.

Table 15.1. *Continued*

FIBER	DESCRIPTION
glass	A manufactured fiber in which the fiber-forming substance is glass.
nylon	A manufactured fiber in which the fiber-forming substance is any long-chain synthetic polyamide in which less than 85% of the amide linkages are attached directly to two aromatic rings.
metallic	A manufactured fiber composed of metal, plastic-coated metal, metal-coated plastic, or a core completely covered by metal.
modacrylic	A manufactured fiber in which the fiber-forming substance is any long-chain synthetic polymer composed of less than 85% but at least 35% by weight of acrylonitrile units.
novoloid	A manufactured fiber in which the fiber-forming substance is any long-chain synthetic polymer composed of at least 85% of a long chain polymer of vinylidene dinitrile where the vinylidene dinitrile content is no less than every other unit in the polymer chain.
olefin	A manufactured fiber in which the fiber-forming substance is any long-chain synthetic polymer composed of at least 85% by weight of ethylene, propylene, or other olefin units.
polyester	A manufactured fiber in which the fiber-forming substance is any long-chain synthetic polymer composed of at least 85% by weight of an ester or a substituted aromatic carboxylic acid, including but not restricted to substituted terephthalate units and parasubstituted hydroxybenzoate units.
rayon	A manufactured fiber composed of regenerated cellulose, as well as manufactured fibers composed of regenerated cellulose in which substituents have replaced not more than 15% of the hydrogens of the hydroxyl groups.
lyocel	A manufactured fiber composed of precipitated cellulose and produced by a solvent extrusion process where no chemical intermediates are formed.
saran	A manufactured fiber in which the fiber-forming substance is any long-chain synthetic polymer composed of at least 80% by weight of vinylidene chloride units.
spandex	A manufactured fiber in which the fiber-forming substance is any long-chain synthetic polymer composed of at least 85% of a segmented polyurethane.
vinal	A manufactured fiber in which the fiber-forming substance is any long-chain synthetic polymer composed of at least 50% by weight of vinyl alcohol units and in which the total of the vinyl alcohol units and any one or more of the various acetal units is at least 85% by weight of the fiber.
vinyon	A manufactured fiber in which the fiber-forming substance is any long-chain synthetic polymer composed of at least 85% by weight of vinyl chloride units.

Source: Textile Fiber Products Identification Act, 1954.

filament, typically being 7/8 inch to 8 inches (2.2 to 28.5 cm) in length.

The size of natural fibers is usually given as a diameter measurement in micrometers. The size of silk and manufactured fibers is usually given in denier (in the United States) or tex (in other countries). Denier and tex are linear measurements based on weight per unit length. The **denier** is the weight in grams of 9,000 meters of the material fibrous. Denier is a direct numbering system in which the lower numbers represent the finer sizes; and the higher numbers, the larger sizes; glass fibers are the only manufactured fibers that are not measured by denier. A one-denier nylon fiber is not equal in size to a one-denier rayon fiber, however, because the fibers differ in density. **Tex** is equal to the weight in grams of 1,000 meters (one kilometer) of the material. To convert from tex to denier, divide the tex value by 0.1111; to convert from denier to tex, multiply the denier value by 0.1111.

YARNS

Yarn is a term for continuous strands of textile fibers, filaments, or material in a form suitable for weaving, knitting, or otherwise entangling to form a textile fabric; a yarn is diagramed in Figure 15.1. Yarns may be constructed to have an **S-** or **Z-twist** or no twist at all. A yarn may be constructed as a number of smaller single yarns twisted together to form a **plied yarn**; each ply will have its own twist as well

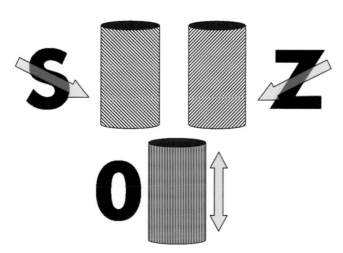

Figure 15.1. Yarns are continuous strands of textile fibers, filaments, or material in a form suitable for weaving, knitting, or otherwise entangling to form a textile fabric. Yarns can have an S-twist, Z-twist, or 0 (zero) twist.

as the overall twist of the plied yarn. Do not confuse the words "yarn" and "thread": Thread refers to the product used to join pieces of fabric together, typically by sewing, while: yarn is the product used to make fabric.

FABRIC CONSTRUCTION

Fabric is a textile structure produced by interlacing yarns, fibers, or filaments with a substantial surface area in relation to its thickness. Fabrics are defined by their method of assembly. The three major types of fabrics are woven, knitted, and non-woven.

WOVEN FABRICS

Fabrics have been woven since the dawn of civilization. **Woven fabrics** are those fabrics composed of two sets of yarns, called warp and weft, and are formed by the interlacing of these sets of yarns. The way these sets of yarns are interlaced determines the weave. **Warp yarns** run lengthwise to the fabric and **weft yarns** run crosswise; weft may also be referred to as filling, woof, or picks, as shown in Figure 15.2. An almost unlimited variety of constructions can be fashioned by weaving.

Figure 15.2. Woven fabrics are composed of two sets of yarns, called warp and weft, formed by the interlacing of these yarns. The way these sets of yarns are interlaced determines the weave. Warp yarns run lengthwise to the fabric, and weft yarns run crosswise.

KNITTED FABRICS

Knit fabrics are constructed of interlocking series of loops of one or more yarns and fall into two major categories: warp knitting and weft knitting. In warp knits the yarns generally run lengthwise in the fabric, whereas in weft knits the yarns generally run crosswise to the fabric. The basic components of a knit fabric are **courses**, which are rows of loops across the width of the fabric, and **wales**, which are rows of loops along the length of the fabric. Unlike woven fabrics, in which warp and weft are made up of different sets of yarns, courses and wales are formed by a single yarn.

NON-WOVEN FABRICS

Non-woven fabrics are an assembly of textile fibers held together by mechanical interlocking

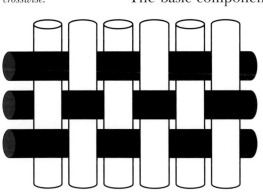

in a random web or mat, by fusing of the fibers, or by bonding with a cementing medium. Felt is a good example, but a wide variety of non-woven construction methods are currently in use, and other examples are bandage pads, automotive textiles, and medical fabrics.

FIBER CHARACTERISTICS

The shape of fibers relates to their identification. Natural fibers are used for certain products, such as cordage and rugs, more than others. Characteristics are imparted to manufactured fibers with particular end-uses in mind. Beyond fiber size and type, many other traits serve to differentiate textile fibers.

Crimp is the waviness of a fiber expressed as crimps per unit length. Crimp may be two-dimensional or three-dimensional in nature. Some fibers are naturally crimped, like wool, whereas others are more linear, such as silk. Crimp must be imparted to manufactured fibers.

Color is introduced to manufactured fibers with dyes or pigments, while natural fibers may be originally white, off-white, or a shade of brown. Natural fibers may be bleached to remove any natural color so they may be dyed more easily. The color may vary along a fiber due to differential dye uptake or because the color has been printed onto the fabric, rather than dyed. All of these traits should be noted.

Cross-sectional shape, the shape of an individual filament when cut at a right angle to its long axis, is a critical characteristic of fiber analysis. Shapes for manufactured fibers vary by design; there are about 500 different cross-sections used for synthetic fibers. The cross-section of plant or animal fibers may assist the examiner in identifying the source.

A fiber's length may be an indication of its intended end-use. All natural fibers are staple fibers except silk; manufactured fibers originate as filaments but may be cut to staple form. All fibers, natural and manufactured, are chain-like macromolecules called **polymers**, which are hundreds or thousands of repeating chemical units called monomers linked together.

NATURAL FIBERS

The first textiles were made of natural fibers. Currently, over half of the fibers produced each year are natural fibers, and the majority of these are cotton. In fact, about half of all fibers produced annually are cotton. Natural fibers come from animals, plants, or minerals. Because they are used in many products, it is important for the forensic fiber examiner to have a thorough knowledge of natural fibers and their significance in casework.

Animal Fibers

Animal fibers come either from mammals (hairs) or from certain invertebrates, such as the silkworm. Animal fibers in textiles are most often from wool-bearing animals, such as sheep and goats, or from fur-bearing animals, like rabbits, mink, and fox. A comprehensive reference collection is critical to animal hair identifications and comparisons. The microscopic anatomical structures of animal hairs are important to their identification.

Plant Fibers

The three major sources for fibers derived from plants are the seed (bast fibers), the stem, and the leaf, depending on which source works best for a particular plant. Plant fibers are found in two principle forms: the **technical fiber**, used in cordage, sacks, mats, etc., or individual cells, as in fabrics or paper. The examination of technical fibers should include a search for internal structures, such as the **lumen**, **spiral elements** or crystals, and the preparation of a cross section. Technical fibers should be mashed, fabrics teased apart, and paper repulped for the examination of individual cells. The relative thickness of the cell walls, the size, shape, and thickness of the lumen; cell length; and the presence, type, and distribution of dislocations should be noted. The most common plant fibers encountered in case work are cotton, flax, jute, hemp, ramie, sisal, abaca, coir, and kapok. See Table 15.2.

MANUFACTURED FIBERS

Manufactured fibers are the various families of fibers produced from fiber-forming substances, which may be synthesized polymers, modified or transformed natural polymers, or glass. **Synthetic fibers** are those manufactured fibers that are synthesized from chemical compounds (e.g., nylon, polyester). Therefore, all synthetic fibers are manufactured, but not all manufactured fibers are synthetic. The microscopic characteristics of manufactured fibers are the basic features used to distinguish them. Manufactured fibers differ physically in their shape, size, internal properties, and appearance.

Table 15.2. *Various natural fibers and their microscopic characteristics.*

KIND	PLANT	*GENUS SPECIES*	CHARACTERISTICS
Bast (stem) Fibers	Flax (linen)	Linum usitatissimum	The ultimates are polygonal in cross-section, with thick walls and small lumina. Microscopically, the fibers have dark dislocations that are roughly perpendicular to the long axis of the fiber.
	Jute	Corchorus capsularis	Appears bundled microscopically and may have a yellowish cast. The ultimates are polygonal but angular with medium-sized lumina. It can be distinguished easily from flax by its counterclockwise twist. The dislocations appear as angular x's or v's and may be numerous.
	Ramie	Boehmeria nivea	Ramie has very long and very wide ultimates. The walls are thick and, in cross section, appear flattened. Ramie has frequent, short dislocations and longer transverse striations. In cross section, radial cracks may be present.
	Hemp	Cannabis sativa	With the ultimates more bundled, a wider lumen and fewer nodes, hemp is easy to distinguish from flax. Cross-sectioning hemp helps in distinguishing it from jute because hemp's lumina are rounder and more flattened than jute's. Hemp may also have a brownish cast to it.
Leaf Fibers	Sisal	Agave sisilana	Sisal is relatively easy to identify due to its irregular lumen size, crystals, spiral elements, and annular vessels. In cross section, sisal looks somewhat like cut celery.
	Abaca	Musa textilis	Although potentially difficult to distinguish from sisal, its ultimates have a uniform diameter and a waxy appearance; often it is darker than sisal. Its ultimates are polygonal in cross section and vary in size. Abaca may present spiral elements but often will have small crown-like structures.
Seed Fibers	Cotton	Genus Gossypium	Mature cotton has a flat twisted ribbon-like appearance that is easy to identify. Cotton fibers are made up of several spiraling layers around a central lumen.
	Kapok	Ceiba pentandra	Kapok fiber is used primarily for life preservers and upholstery padding because the fibers are hollow, producing very buoyant products. But they are brittle, which prevents spinning or weaving.
	Coir	Coco nucifera	Coir comes from the husk of the coconut and, accordingly, is a very dense, stiff fiber easily identified microscopically. On a slide mount, coir appears very dark brown or opaque with very large, coarse ultimates.

FIBER MANUFACTURE

Synthetic fibers are formed by extruding a fiber-forming substance, called **spinning dope**, through a hole or holes in a shower head-like device called a **spinneret**, as shown in Figure 15.3; this process is called **spinning**. The spinning dope is created by rendering solid monomeric material into a liquid or semi-liquid form with a solvent or heat.

Optical properties, such as **refractive index**, birefringence, and color, are those traits that relate to a fiber's structure or treatment revealed through observation. Some of these characteristics aid in the identification of the generic polymer class of manufactured fibers. Others, such as color, are critical discriminators of fibers that have

Figure 15.3. Synthetic fibers are formed by extruding a fiber-forming substance, called spinning dope, through a hole or holes in a shower head-like device called a spinneret. This photograph appeared in the photographic essay, "The Reign of Chemistry," by W. Eugene Smith in Life Magazine, January 5, 1954.

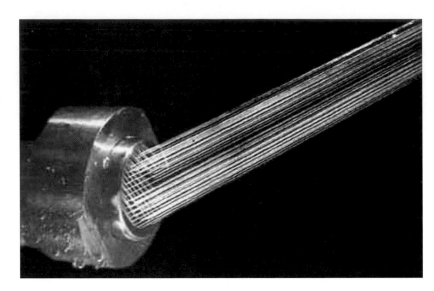

been dyed or chemically finished. A visual and analytical assessment of fiber color must be part of every fiber comparison.

The fluorescence of fibers and their dyes is another useful point of comparison. Thermal properties relate to the softening and melting temperatures for manufactured fibers and the changes the fiber exhibits when heated (see Table 15.3).

Based on a fiber's polymer composition, it will react differently to various instrumental methods, such as Fourier Transform Infrared Spectroscopy (FTIR) or pyrolysis-gas chromatography (P-GC), and chemicals, such as acids or bases. These reactions yield information about the fiber's molecular structure and composition.

MICROSCOPIC CHARACTERISTICS

A polarized light microscope is the primary tool for the identification and analysis of manufactured fibers. Many characteristics of manufactured fibers can be viewed in non-polarized light, however, and they provide a fast, direct, and accurate method for the discrimination of similar fibers. A comparison light microscope is required to confirm whether the known and the questioned fibers truly present the same microscopic characteristics.

The cross section is the shape of an individual fiber when cut at a right angle to its long axis. Shapes for manufactured fibers vary with

Table 15.3. *Melting temperatures for some fiber types.*

FIBER TYPE	TEMPERATURE (°C)
Acetate	224–280
Acrylic	Does not melt
Aramid	Does not melt
Modacrylic*	204–225
Nylon 6 6,12 6,6	 213 217–227 254–267
Olefin Polyethylene Polypropylene	 122–135 152–173
Polyester [PET]	256–268
Rayon	Does not melt
Saran	167–184
Spandex	231
Triacetate	260
Vinal	200–260

*Some members of this class do not melt.
From Carroll, G.R. (1992). Forensic fibre microscopy. In J. Robertson (Ed.), *Forensic examination of fibres*. New York City, NY: Ellis Horwood.

the desired end result, such as the fiber's soil-hiding ability or a silky or coarse feel to the final fabric. Figure 15.4 shows some variations in fiber cross sections. The particular cross section also may be indicative of a fiber's intended end-use: Many carpet fibers have a lobed shape to help hide dirt and create a specific visual texture to the carpet.

The way a fiber's diameter is measured depends on its cross-sectional shape; there is more than one way to measure the diameter of a non-round fiber. Manufactured fibers can be made in diameters from about 6 μm (so-called **microfibers**) up to a size limited only by

Figure 15.4. The cross section of a fiber may relate to its end-use. Carpet fibers, for example, often have cross sections that are meant to keep the fiber upright (making the carpeting feel plush) and to hide dirt. About 500 cross-sectional shapes are used in the manufacture of fibers.

Figure 15.5. Tiny grains of material, usually titanium dioxide, are incorporated into a fiber as it is spun; these are called delustrants. Delustrants break up the light entering the fiber and make it appear dull or give a matte finish. A fiber with no delustrant is described as being "bright."

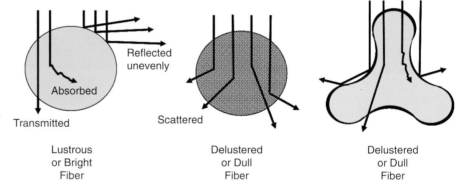

the width of the spinneret holes. By comparison, natural fibers vary in diameter from cultivated silk (10–13 μm) to U.S. sheep's wool (up to 40 μm or more) and human head hairs range from 50–100 μm.

Delustrants are finely ground particles of materials, such as titanium dioxide, that are introduced into the spinning dope. These particles help to diffract light passing through the fibers and reduce their luster, as illustrated in Figure 15.5. The size, shape, distribution, and concentration of delustrants should be noted.

OPTICAL PROPERTIES OF MANUFACTURED FIBERS

The examination of the optical properties of manufactured fibers can yield a tremendous amount of information about their chemistry, production, end-use, and environment. A crucial step in the identification and later comparison of textile fibers is the careful measurement and analysis of these properties.

POLARIZED LIGHT MICROSCOPY

Polarized light microscopy is an easy and quick non-destructive way to determine the generic polymer class of manufactured and synthetic textile fibers. Beyond the immediate characteristics used to discriminate between polymer types, the examination of fibers in polarized light provides valuable information about the production and finishing of the fiber after spinning.

REFRACTIVE INDEX

Fibers vary in shape but are almost always thicker in the center than near the edges. Thus, they act as crude lenses, either concentrating or dispersing the light that passes through them. If a fiber has a higher refractive index than the medium in which it is mounted, it acts as a converging lens, concentrating light within the fiber. If the fiber has a lower refractive index than the mounting medium, it acts as a "diverging" lens, and the light rays diverge from the fiber, as illustrated in Figure 15.6 with a fragment of glass. In most fibers, the light rays only slightly converge or diverge and thus appear as a thin bright line, called the **Becke line**, after the Austrian mineralogist Fredrich Becke who first described the phenomenon, at the interface between the fiber and the mounting medium (Good and Rothenberg, 1998). While the analyst observes the fiber, the working distance on the microscope is increased (the stage is moved down); if the fiber has the higher refractive index, the Becke line moves toward the fiber as the working distance is increased. If the mounting medium has the higher index, the Becke line moves toward the medium (away from the fiber) as the working distance is increased (see Figure 15.6). If fibers are mounted in a medium that has a refractive index of 1.52, such as Permount®, then a fiber can be described as being greater than, equal to, or less than 1.52. The refractive indices of a fiber can be measured directly by placing the fiber in a series of liquids of specific refractive indices until the refractive indices of the fiber and liquid are the same. At this point, the fiber "disappears" because it and the liquid are now **isotropic**, meaning that light is traveling at the same speed through both the fiber and the liquid (Good and Rothenberg, 1988).

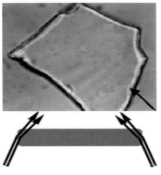

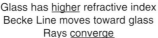

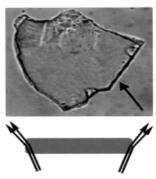

Glass has <u>higher</u> refractive index
Becke Line moves toward glass
Rays <u>converge</u>

Glass has <u>lower</u> refractive index
Becke line moves toward medium
Rays <u>diverge</u>

Figure 15.6. Fibers act as crude lenses, either focusing light into it or away from it. This has to do with the refractive index of the fiber—that is, the ratio of the speed of light in the fiber material over the speed of light in a vacuum. Appearing as a line of bright light, the Becke line can be used to determine the relative refractive index of a material, in this case, a fiber. When the distance between the microscope lens and the sample is increased, the Becke line moves toward the material with the higher refractive index. Photographs courtesy of Sarah Jones.

Fiber under
polarized light
(one filter)

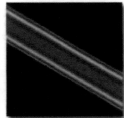

Fiber under cross-
polar light (both filters)
producing <u>interference
colors</u>

Figure 15.7. Indicative of the fiber's polymer and molecular organization, interference colors can be used to help determine what kind of fiber is being examined. Courtesy of Sarah Jones.

BIREFRINGENCE

One of the more distinctive traits of a fiber is its **birefringence**. The **interference colors** seen after crossing the polarizing filters relate to a fiber's material nature, orientation, and crystallinity (see Figure 15.7). For the sake of comparison, most natural and synthetic fibers have birefringence from 0.001 to 1.8 but birefringence as high as 2.0 or more has been reported for specialty fibers (see Table 15.4).

FLUORESCENCE MICROSCOPY

Many dyes used to color textiles have fluorescent components and their response to certain wavelengths of light can be useful in comparing textile fibers. Not all textile dyes fluoresce, but fluorescence comparisons should be performed regardless: If the questioned and known fibers both fail to fluoresce, that is another point of meaningful comparison.

Table 15.4. *Table of refractive indicies and birefringences of various fiber types.*

FIBER TYPE	$N_{\parallel}$	$N_{\perp}$	$N_{\parallel}-N_{\perp}$
Acetate			
Dicel	1.478	1.473	0.005
	1.476	1.473	0.003
Triacetate			
Tricel	1.469	1.469	0
Arnel	1.469	1.468	0.001
Acrylic			
Acrilan 36	1.511	1.514	−0.003
Orlon	1.51	1.512	−0.002
Acrilan	1.52	1.525	−0.005
Modacrylic			
Dynel	1.535	1.533	0.002
Teklan	1.52	1.516	0.004
SEF	>1.52	>1.52	−[low]
Verel	1.535	1.539	−0.004
Vinyon			
Fibravyl	1.54	1.53	0.01
Rhovyl	1.541	1.536	0.005
Vinyon HH	1.528	1.524	0.004
Rayon			
Viscose (regular)	1.542	1.52	0.022
Viscose (regular)	1.545	1.525	0.02
Viscose (high tenacity)	1.544	1.505	0.039
Vincel (high wet modulus rayon)	1.551	1.513	0.038
Fortisan	1.547	1.523	0.024
Fortisan 36	1.551	1.52	0.031
Cuprammonium	1.553	1.519	0.034
Tencel	1.57	1.52	0.05
Olefin			
Courlene (PP)	1.53	1.496	0.034
Polypropylene	1.52	1.492	0.028
SWP (PE)	1.544	1.514	0.03
Courlene X3 (PE)	1.574	1.522	0.052
Polyethylene	1.556	1.512	0.044
Nylon			
Enkalon (6)	1.575	1.526	0.049
ICI nylon (6,6)	1.578	1.522	0.056
Qiana	1.546	1.511	0.035
Rilsan (11)	1.553	1.507	0.046
nylon 6	1.568	1.515	0.053
nylon 6,6	1.582	1.519	0.063
nylon 11	1.55	1.51	0.04
Silk (degummed)	1.57	1.52	0.05

Table 15.4. *Continued*

FIBER TYPE	$N_\parallel$	$N_\perp$	$N_\parallel - N_\perp$
Aramid			
Nomex	1.8	1.664	0.136
Kevlar	2.35	1.641	0.709
Polyester			
Vycron	1.713	1.53	0.183
Terylene	1.706	1.546	0.16
Fortrel/Dacron	1.72	1.535	0.185
Dacron	1.7	1.535	0.165
Kodel	1.632	1.534	0.098
Kodel II	1.642	1.54	0.102
Spandex			
Lycra/Vyrene	1.561	1.56	0.001
Others			
Vicara (Azlon)	1.538	1.536	0.002
Teflon	1.38	1.34	0.04
Calcium alginate	1.524	1.52	0.004
Saran	1.61	1.61	0
Novoloid	1.5–1.7	1.5–1.7	0
Kynol (drawn)	1.658	1.636	0.022
Kynol (undrawn)	1.649+	1.649	<0.001
Polyacrylostyrene	1.56	1.572	−0.012
Darvan (Nytril)	1.464	1.464+	0
Polycarbonate	1.626	1.566	0.06

Sources: AATCC, 1996; ASTM, D 276–87, 1996; McCrone, Delly, Palenik, 1979; Rouen and Reeve (1970); The Textile Institute (1985).

Fluorescence occurs when a substance is excited by specific wavelengths of light. A light of relatively short wavelength illuminates a substance, and the substance absorbs and/or converts (into heat, for example) a certain, small part of the light. Most of the light that is not absorbed by the substance is reemitted, which is called fluorescence. The fluorescent light has lost some of its energy, and its wavelength will be longer than that of the source light.

Certain dye combinations may produce fluorescence of a particular intensity and color, both of which should be noted during the examination. Fibers dyed with similar dyes should exhibit the same fluorescence characteristics, unless the fiber and/or dye(s) have been degraded by UV exposure, bleaching, or some other similar means. It is important to consider these factors when collecting known samples.

COLOR IN TEXTILES

Color is one of the most critical characteristics in a fiber comparison. Almost all manufacturing industries are concerned with product appearance. Everything that is manufactured has a color to it, and often these colors are imparted to the end product. Particular colors are chosen for some products rather than others (it's difficult to find "safety orange" carpeting, for example), and these colors may indicate the end product. The number of producible colors is nearly infinite, and color is an easy discriminator.

COLOR PERCEPTION

The perception of color by a human observer is subject to a variety of factors, such as genetics, age, and environment. The human visual system is complex and adaptive. The phenomenon called **simultaneous contrast**, the perception of color based on context, is shown in Figure 15.8; the gray arrow on the left should look to be lighter or darker than the arrow on the right. In fact, they are the same gray. Your perception of the gray is affected by the background colors of yellow and blue. Another example of contextual color perception is known as the chameleon effect, illustrated in Figure 15.9, in which colors change based on the surrounding colors. Because of the factors influencing human color perception, any visual comparison, while effective as a first approximation of color, must be checked by an objective method of color measurement.

Figure 15.8. *The gray arrow on the left should look to be lighter or darker than the arrow on the right; in fact, they are the same gray. Your perception of the gray is affected by the background colors of yellow and blue. This phenomenon is called simultaneous contrast.*

Figure 15.9. *This phenomenon is called the "chameleon effect," where colors change based on the surrounding colors. The red and the green are the same colors in all of the graphics.*

DYES AND PIGMENTS

Over 80 dyers worldwide are registered with the American Association of Textile Chemists and Colorists (AATCC), and almost 350 trademarked dyes are registered with this association. Some trademarked dyes have as many as 40 variants. Over 7,000 dyes and pigments are currently produced worldwide. Natural dyes, such as indigo, have been used throughout history, while synthetic dyes have gained prominence largely since World War I (Apsell, 1981).

A **dye** is an organic chemical that is able to absorb and reflect certain wavelengths of visible light. **Pigments** are microscopic, water-insoluble particles that are either incorporated into the fiber at the time of production or are bonded to the surface of the fiber by a resin. Some fiber types, such as olefins, are not easily dyed and therefore are often pigmented.

Based on the desired end-product effects, the fiber substrate, and the type of dye used, there are more than 12 different application categories for textile dyes. Very few textiles are colored with only one dye, and even a simple dye may be put through 8–10 processing steps to achieve a final dye form, shade, and strength. When all of these factors are considered, it becomes apparent that it is virtually impossible to dye textiles in a continuous method; that is, dyeing separate batches of fibers or textiles is the rule rather than the exception. This color variability has the potential to be very significant in forensic fiber comparisons.

COLOR ASSESSMENT

The three main methods of analyzing the color in fibers are visual examination, chemical analysis, and instrumental analysis. Each of these methods has strengths and weaknesses that must be considered by the fiber examiner.

The most basic method is simple visual examination of single fibers with the aid of a comparison microscope. Visual examination is quick, and comparison is an excellent screening technique. However, it is a subjective method, and because of day-to-day and observer-to-observer variations, it is not always a repeatable method. Additionally, the dilemma of metameric colors exists. **Metameric colors** are those that appear to match in one set of lighting conditions, but not in another.

By their nature, metamers are difficult to sort out visually. Visual examination must be used in conjunction with an objective method.

Chemical analysis involves extracting the dye and characterizing or identifying its chemistry. Typically, thin-layer chromatography (TLC) is the method of choice although others may be employed. Chemical analysis addresses the type of dye(s) used to color the fiber and may help to sort out metameric colors. It can be difficult to extract the dye from the fiber, however, because forensic samples typically are small and textile dyers take great pains to ensure that the dye stays in the fiber. Dye analysis is also a destructive method, rendering the fiber useless for further color analysis. Because very light or very small fibers have little dye in them, weak or equivocal responses may result.

Instrumental analysis offers the best combination of strengths and the fewest weaknesses of the three methods outlined. Instrumental readings are objective and repeatable, the results are quantitative, and the methods can be standardized. Importantly, this type of analysis is not destructive to the fiber, and the analysis may be repeated. Again, very light fibers may present a problem with weak results, and natural fibers may exhibit high variations due to uneven dye uptake.

The **microspectrophotometer** (MSP) is an instrument that allows for the color measurement of individual fibers. The MSP is essentially a standard spectrophotometer with a microscope attached to focus on the sample. A spectrophotometer compares the amount of light passing through air with the amount of light transmitted through or reflected off a sample. The ratio of these measurements indicates the percentage of light reflected or transmitted. At each wavelength of the visible spectrum, this ratio is calculated, stored, and recorded. The light is broken into smaller regions of the visible spectrum by a monochrometer, which acts like a prism dividing the light into its spectral components, as shown in Figure 15.10.

Color is a major factor in comparing textile fibers. If an analyst is searching tape lifts, it is the predominant factor in selecting fibers for further comparison. Very fine gradations of color difference can be seen once fibers have been mounted; it is necessary, however, to train the observer's eye to make these distinctions in a uniform manner. The microspectrophotometer is crucial to the comparison process because it can segregate colored fibers that appear visually the same but are subtly different. Objectively distinguishing between otherwise identical fibers is necessary to ensure a reliable comparison method.

Figure 15.10.
Schematic of a
microspectrophotometer
(MSP), what the
instrument looks like,
and an absorbance
spectrum from a red
fiber (fiber photo inset).
Courtesy of Paul
Martin, CRAIC
Technologies.

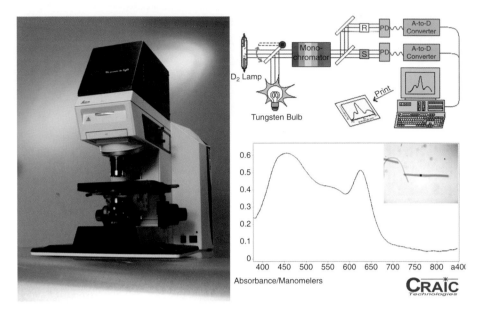

CHEMICAL PROPERTIES

While microscopy offers an accurate method of fiber examination, it is often necessary to confirm these observations. Analyzing the fibers chemically not only offers a confirmation of the microscopic work, but also may provide additional information about the specific polymer type or types that make up the fiber. For most of the generic polymer classes, various subclasses can assist in discriminating between optically similar fibers. Fourier transform infrared spectroscopy (FTIR) and pyrolysis-gas chromatography (P-GC) are both methods of assessing the chemical structure of polymers. FTIR is the preferred method because it is not destructive of the fibers.

Manufactured fibers also can be characterized by their reaction to certain chemicals, and this was a popular method prior to the introduction of instrumentation in crime laboratories. Solubility schemes tend to lack the specificity of instrumental methods and are destructive but can still be an effective means to confirm a manufactured fiber's generic class.

Solubility tests should be performed on both the known and questioned fibers side-by-side either on spot plate or on a microscope slide with a coverslip. A hot-stage microscope may be required for some methods. Numerous solubility schemes exist, and one should be

Table II—Solubilities of Fibers in Reagents Used in the Chemical Methods
Chemical Method

	NO. 1 100% CH₃COCH₃	NO. 2 20% HCl	NO. 3 59.5% H₂SO₄	NO. 4 70% H₂SO₄	NO. 5 NaOCl	NO. 6 90% HCOOH
ACETATE	S	I	S	S	I	S
ACRYLIC	I	I	I	I*	I	I
COTTON	I	I	SS	S	I	I
HAIR	I	I	I	I	S	I
HEMP	I	I	SS	S	I	I
LINEN	I	I	SS	S	I	I
MODACRYLIC	S or I*	I	I	I	I	I
NYLON	I	S	S	S	I	S
OLEFIN	I	I	I	I	I	I
POLYESTER	I	I	I	I	I	I
RAMIE	I	I	SS	S	I	I
RAYON	I	I	S	S	I	I
SILK	I	PS	S	S	S	PS
WOOL	I	I	I	I	S	I

*Depending on type

KEY TO SYMBOLS: S = SOLUBLE
PS = PARTIALLY SOLUBLE (Method not applicable)
SS = SLIGHTLY SOLUBLE (Useable but correction factor required)
I = INSOLUBLE

Figure 15.11. A solubility scheme for fibers. From American Association of Textile Chemists and Colorists Technical Manual, 1997.

chosen with available chemicals, equipment, and safety in mind; one scheme is shown in Figure 15.11.

INTERPRETATIONS

What does a positive fiber association mean? Numerous studies have shown that, other than white cotton, indigo-dyed cotton (denim), and certain types of black cotton, no fiber should be considered as being "common." These studies include looking for specific fibers on a wide variety of clothing, cross-checking fibers in particular locations (movie theater seats, for example), and frequency studies. One study cross checked fibers from 20 unrelated cases, looking for incidental positive associations; in over 2 million comparisons, no incidental positive associations were found. This makes fiber evidence very powerful in demonstrating associations.

CHAPTER SUMMARY

Fibers make good evidence for a number of reasons: They vary greatly, are easy to analyze, and are everywhere there are textiles. Fibers have figured prominently in many high-profile cases and are researched extensively by forensic and textile scientists alike. Textile fibers are one of the most frequently encountered types of physical evidence. Color is one of the most underutilized traits of a textile fiber; the color of fibers should be analyzed spectrally or chemically in any positive association. The combinations of characteristics make fibers very specific evidence: It is rare to find two fibers at random that exhibit the same characteristics.

Test Your Knowledge

1. What is a fiber?
2. What is a yarn? How is it different from a thread?
3. How are woven and knitted fabrics different?
4. What is the difference between a manufactured fiber and a synthetic fiber?
5. What is the **dry twist test** used for?
6. What is a spinneret?
7. What is the fiber-forming substance called before it is spun into fibers?
8. What is denier?
9. What is a microfiber?
10. What is refractive index? How is the Becke line used to determine it?
11. Why are fibers birefringent?
12. What is a metameric pair?
13. Why is cross-sectional shape of a fiber important?
14. What is a delustrant? How is it used in fibers?
15. How many cross-section shapes are used in making manufactured fibers?
16. How many commercial dyes are available?
17. How many colors can be produced in textiles?
18. What is a microspectrophotometer? What is it used for?

Consider This . . .

1. Why is it important to use some other means of assessing a fiber's color than just visual examination? Why isn't comparison microscopy alone

sufficient? If you compared two fibers' color and they looked the same, what else could another method tell you?

2. Numerous transfer and target fiber studies have shown that, other than white cotton, indigo-dyed cotton, and some black-dyed cotton, it is exceedingly rare to find two fibers at random that are analytically indistinguishable (for example, see Roux and Margot, 1997, or Houck, 1999). Why is this?

3. Inventory a portion of your closet by color and fiber type (look at the labels). How many different combinations are there? Do you think someone else's closet would be exactly the same?

BIBLIOGRAPHY

American Association of Textile Chemists and Colorists (AATCC) Technical Manual, Research Triangle Park, NC: American Association of Textile Chemists and Colorists, 1997.

Apsell, P. (1981). What are dyes? What is dyeing? In *Dyeing primer*. Research Triangle Park, NC: American Association of Textile Chemists and Colorists.

ASTM, D 276–87 Standard Test Methods for Identification of Fibers in Textiles, Philadelphia, PA: ASTM. 1996.

Carroll, G.R. (1992). Forensic fibre microscopy. In Robertson, J. (Ed), Forensic Examination of Fibres. New York City, NY: Ellis Horwood.

Delly, J. (2003, July). The Michel-Levy interference color chart— microscopy's magical color key. *Modern Microscopy*. Retrieved from **www.modernmicroscopy.com**.

Eyring, M. (2002). Visible microscopical spectrophotometry in the forensic sciences. In R. Saferstein (Ed.), *Forensic science handbook, volume I* (2nd ed.) Englewood Cliffs, NJ: Prentice-Hall.

Fenichell, S. (1996). *Plastic: The making of a synthetic century*. New York: Harper Collins.

Gaudette, B. (1988). Forensic fiber examinations. In R. Saferstein (Ed.), *Forensic science handbook, volume II*. Englewood Cliffs, NJ: Prentice-Hall.

Good, G.A., & Rothenberg, M. (Eds.). (1998). *Sciences of the Earth: An Encyclopedia of Events, People, and Phenomena*. London, Routledge.

Hatch, K. (1993). *Textile science*. St. Paul, MN: West Publishing.

Houck, M. (1999). Statistics and trace evidence: The tyranny of numbers. *Forensic Science Communications*, 1(3). Retrieved from **www.fbi.gov**.

McCrone, W., Delly, J.G., & Palenik, S.J., *The Particle Atlas*. Chicago, IL: McCrone Associates, 1979.

Robertson, J., & Grieve, M. (Eds.). (1999). *Forensic fibre examinations.* London: Taylor and Francis.

Rouen, R.A., & Reeve, V.C. (1970). Journal of Forensic Sciences 15:410–432.

Roux, C., & Margot, P. (1997). An attempt to assess the relevance of textile fibres recovered from car seats. *Science and Justice*, 37, 225.

Scientific Working Group for Materials (SWGMAT). (1999). Forensic fiber examinations. *Forensic Science Communications*, 1(1). Retrieved from **www.fbi.gov**.

Siegel, J., & Houck, M. (1999). Forensic fiber examinations. In C. Wecht (Ed.), *Forensic sciences.* St. Paul, MN.: Westlaw Publishing.

Smith, W.E. (1954). "The Reign of Chemistry," *Life Magazine*, January 5.

The Textile Institute (1985). *Identification of Textile Materials*, Manchester, United Kingdom: The Textile Institute.

Paint Analysis

KEY TERMS

Architectural paints

Art paints

Backscattered electrons

Batch lot

Binder

Clearcoat

Coatings

Dye

Enamel

Lacquer

Latex

Metamerism

Microtome

Paint

Paint Data Query (PDQ)

Pigment

Polarized light microscopy (PLM)

Pretreatment

Primer

Product coatings

Secondary electrons

Shellac

Smear

Solvent

Special-purpose coatings

Stain

Topcoat

Varnish

Vehicle

INTRODUCTION

The forensic analysis of paints, more properly called coatings to encompass any surface coating intended to protect, aesthetically improve, or provide some special quality, is one of the most complex areas of the forensic laboratory. The reason is that the manufacture and application of paints and coatings is one of the most complex and complicated areas in all of industrial chemistry. A forensic paint examiner, even if he or she specialized in that one material, could never grasp the entire range of coatings, paints, and materials used throughout the world. As John Thornton (2002), at Forensic Analytical Specialties, Inc., in Hayward, California, has noted,

The paint industry . . . [utilizes] more than a thousand kinds of raw materials and intermediates—more than virtually any other manufacturing enterprise. A thorough understanding of the use, properties, and identification of only the most commonly used materials may represent the entire professional career of a paint chemist. It is unrealistic to expect the same comprehension of the subject by the forensic scientist, but it is entirely reasonable to expect a basic familiarity with those aspects of paint chemistry that may affect either the analysis or the interpretation of paint evidence (2002; page 430).

This complexity is in the forensic scientist's favor, however, because variety and variation make for a more specific categorization of classes. More specificity presents the potential for a tighter interpretation and greater evidentiary significance in court.

WHAT IS PAINT?

A **paint** is a suspension of pigments and additives intended to color or protect a surface. A **pigment** is fine powder that is insoluble in the medium in which it is dispersed; that is, the granules do not dissolve and remain intact and are dispersed evenly across the surface, as shown in Figure 16.1. Pigments are intended to color and/or cover a surface; they may be organic, inorganic, or a mixture. The additives in paint

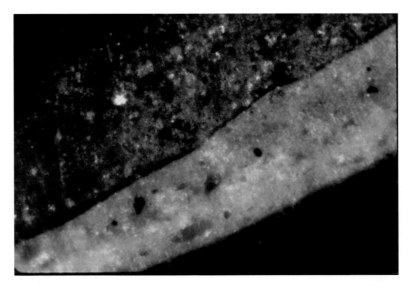

Figure 16.1. Paints contain pigments, fine powders that do not dissolve but are dispersed evenly across the surface. Pigments are intended to color and/or cover a surface; they may be organic, inorganic, or a mixture. Courtesy of Paul Martin, CRAIC, Inc.

come in a dizzying variety but have some constants. The **binder** is that portion of the coating, other than the pigment, which allows the pigment to be distributed across the surface. The term **vehicle** typically refers to the solvents, resins, and other additives that form a continuous film, binding the pigment to the surface. If the binder and vehicle sound similar, they are: The terms are sometimes used interchangeably in the coatings industry and the forensic laboratory. **Solvents** dissolve the binder and give the paint a suitable consistency for application (brushing, spraying, etc.). Once the paint has been applied, the solvent and many of the additives evaporate; a hard polymer film (the binder) containing the dispersed pigment remains to cover and seal the surface (see Table 16.1).

Table 16.1. *The components of a hypothetical gloss enamel architectural paint (from Thornton, 2002, p. 435).*

INGREDIENT	FUNCTION	POUNDS/GALLON
Ultramarine blue	Pigment (coloring agent)	Trace
Thymol	Anti-mildew agent	0.01
Cobalt naphthenate	Drier	0.02
Soya oil	Oil	0.03
Calcium naphthenate	Drier	0.03
Zirconium naphthenate	Drier	0.06
Zinc oxide	Pigment	0.2
Calcium carbonate	Extender	0.5
Mineral spirits	Solvent	1.05
Titanium dioxide	Pigment	2.8
Soya alkyd resin	Binder	5.7
Total		10.4

Paints can be divided into four major categories. The first is **architectural paints**, sometimes called household paints, and are those coatings most often found in residences and businesses. **Product coatings**, those applied in the process of manufacturing products including automobiles, are the second major category. Because automobiles play a central role in society and, therefore, in crime, much of this chapter will focus on the paints and coatings from the automotive manufacturing industry. The third kind, **special-purpose coatings**, fulfill some specific need beyond protection or aesthetic improvement, such as skid-resistance, water-proofing, or luminescence (as on the dials of wristwatches). Finally, **art paints** are occasionally encountered in forgery cases. Modern art paints are similar in many respects to architectural paints, but many artists formulate their own paints, leading to potentially unique sources; see Figure 16.2. See "In More Detail: Coating Definitions" for additional information regarding coatings.

IN MORE DETAIL:
Coatings definitions

Many of the words we use to describe **coatings**, such as paint, varnish, and lacquer, in reality have very specific technical definitions used by the coatings industry. To avoid confusion between the casual and professional meanings, we have listed some of these definitions:

Architectural Paint: Coatings encountered around a typical household.

Dye: A coloring agent that is soluble in the medium in which it is dispersed.

Enamel: A pigmented coating that has a high gloss (luminous reflectivity) when it dries.

Lacquer: Clear or pigmented coatings that dry quickly through evaporation of the solvent.

Latex: A suspension of a pigment in a water-based emulsion of any of several resins.

Paint: A suspension of a pigment in an oil vehicle; more broadly, any surface coating designed for protection and/or decoration of a surface.

Pigment: A fine powder that is insoluble in the medium in which it is dispersed.

Shellac: A solution of melted lac, a resinous excretion of the Lac insect (*Coccus* or *Carteria lacca*) dissolved in alcohol used as a sealant, adhesive, or insulating varnish.

Stain: A solution of dye or a suspension of a pigment designed to color, but not protect, a wood surface. Technically speaking, a stain colors the wood but does not coat it.

Varnish: A clear solution of oils and organic or synthetic resins in an organic solvent.

Sources: Thornton, 2002; Wikipedia, **www.wikipedia.com**

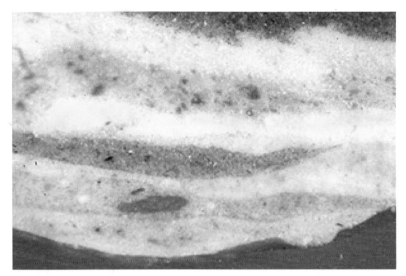

Figure 16.2. Modern art paints are mass-produced, but many artists formulate their own paints. The application process for art paints is obviously more varied and unstructured than for product coatings. Courtesy of Paul Martin, CRAIC, Inc.

PAINT MANUFACTURING

AUTOMOTIVE FINISHES

One of the most commonly encountered kinds of paint evidence is automotive paint. Cars, trucks, and similar vehicles are so deeply integrated into our daily lives, it's easy to see why this is so. Automotive paints are also a good example of how manufacturing styles and variation contribute to the significance of forensic evidence.

The automotive finishing process for vehicles consists of at least four separate coatings. The first is a **pretreatment**, typically zinc electroplating, applied to the steel body of the vehicle to inhibit rust. The steel is then washed with a detergent, rinsed, treated a second time, and then washed again. The significance of this coating is for the forensic paint analyst to be aware that any zinc found during elemental analysis may come from this coating and not necessarily the paint itself.

The second coating is a **primer**, usually an epoxy resin with corrosion-resistant pigments; the color of the primer is coordinated with the final vehicle color to minimize contrast and "bleed through." The steel body of the vehicle is dipped in a large bath of the liquid primer and plated on by electrical conduction. The primer coating is finished with a powder "primer surfacer" that smoothes the surface of the metal and provides better adhesion for the next coating.

Figure 16.3. A green vehicle paint chip, showing the layer structure common to most automotive paints. Courtesy of Mark Sandercock, Royal Canadian Mounted Police.

The **topcoat** is the third coating applied to the vehicle and may be in the form of a single-color layer coat, a multilayer coat, or a metallic color coat; this is the layer that most people think of when they think of a vehicle's color (see Figure 16.3). Topcoat chemistry is moving toward water-based chemistries to provide a healthier atmosphere for factory workers and the environment. For example, heavy metals, such as lead or chrome, are no longer used in the formulation of topcoats. Metallic or pearlescent coatings, growing in preference for new model vehicles, have small metal or mica flakes incorporated to provide a shimmering, changing color effect. Metallic pigments, including zinc, nickel, steel, and gold-bronze, give a glittering finish to a vehicle's color, while pearlescent pigments, mica chips coated with titanium dioxide and ferric oxide, try to replicate the glowing luster of pearls. The topcoat is often applied and flashed, or partially cured, and then finished with the next and final coating, the clearcoat.

Clearcoats are unpigmented coatings applied to improve gloss and durability of a vehicle's coating. Historically, clearcoats were acrylic-based in their chemistry, but nearly half of the automotive manufacturers have moved to two-component urethanes.

It is important for forensic paint analysts to keep up to date with the latest trends and techniques in the paint industry. It is equally important, however, for them to be aware of the previously used formulations and manufacturing techniques because they constitute the bulk of vehicles currently on the road. A three- to five-year-old pickup truck is far more likely to be encountered in a forensic case than the newest model sports car, so the forensic analyst must not be surprised by "history." Repaired and repainted vehicles are an additional consideration because they may have been coated with virtually anything, including spray paint.

Another issue related to vehicle coloration is that of the newer plastic substrates. Vehicle bodies are no longer made exclusively of steel; various plastics are now commonly used. For example, fenders may be nylon, polymer blends, or polyurethane resins; door panels and hoods may be of thermosetting polymers; front grills and bumper strips have long been plastic or polymer but now may be colored to match the vehicle. Braking systems, chassis, and even entire cars (BASF unveiled an entirely plastic car in 1999, as an extreme example) are now constructed from plastics. It wouldn't be unusual for the forensic paint examiner to encounter steel, aluminum, and polymer parts on the same vehicle, each colored by a very different coating system.

COLLECTION

Because it is possible a physical match exists between known and questioned paint samples, as demonstrated in Figure 16.4, the collection of paint samples should proceed with caution. This type of evidence carries great significance, and care should be taken to preserve any potential physical matches.

Samples from the crime scene (questioned) should include all loose or transferred paint materials. Nearly any object or surface may retain a paint transfer and things as varied as tools, architectural structures and elements (floors, wainscoting), glass fragments, fabrics, hairs, fingernails, roadways and signs, and, of course, vehicles. Evidentiary items with paint transfers should be packaged and submitted to the laboratory in their entirety, if possible. Depending on the size of the object, packaging could prove problematic, so often sampling of paint transfers must take place in the field. It is also important to remember that cross-transfer could have occurred. Known and questioned samples should be collected from both surfaces.

Paint evidence should be first photographed and then removed manually with non-metallic tools, such as small wooden sticks, tooth-

Figure 16.4. The strongest evidence of an association between a paint sample and a source is a physical match, considered unique and individualizing. These are somewhat uncommon and therefore carry strong probative value. Here, two paint chips are aligned to show the common border demonstrating that they were at one time one continuous coating. Courtesy of Mark Sandercock, Royal Canadian Mounted Police.

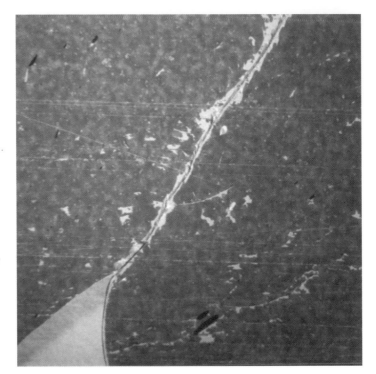

picks, or plastic forceps. If tape lifts are to be used, the paint evidence should be collected first. Because of their structure, fragility, and size, if paint is collected with tape lifts, it could be very difficult to impossible to easily manipulate paint samples that are sticky from the tape's adhesive. Additionally, the adhesive's components, or material stuck to the adhesive, could contaminate the paint sample and change its apparent chemistry (see Figure 16.5).

Flakes of paint can be removed from a surface in a variety of ways. Lifting or prying out loose flakes, cutting samples of the paint with a clean knife or blade, and dislodging them by gently bumping the opposite side of the painted surface are all examples of appropriate collection methods. If the samples are cut, the blade should go all the way through the paint layers to the subcoating surface. The sampling method will vary considerably given the circumstances of the crime, the evidence items, their location, and environmental factors; no single method will work all of the time.

When a painted object strikes a glancing blow to another object, it can transfer paint in the form of a **smear**. A smeared transfer can be very confusing and difficult to work with because components from

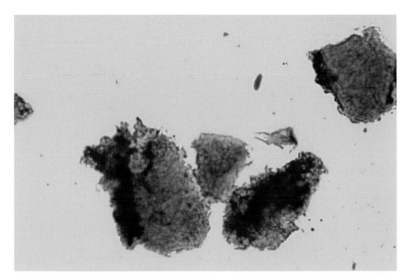

Figure 16.5. *Paint fragments must be handled with care; being too aggressive in their collection can damage or contaminate them.*

several layers of coatings can be commingled; this can reduce the forensic scientist's ability to accurately analyze the smeared paint. Even the best collection efforts can confuse the issue even more. When the forensic scientist is dealing with a smeared paint transfer, it is best to submit the entire object to the laboratory, if possible.

Paint, like any mass-produced material, varies. It is important when collecting known paint samples, therefore, that they be collected from areas as close as possible to, but not within, the point(s) of damage or transfer. This is important for two reasons. First, the damaged area itself is usually not suitable for providing a known sample: Subcoating and other incidental materials may lie within the damage and confuse or confound analysis. Second, because of manufacturing variation, detectable differences may exist between parts of an object. On an automobile, for example, the paint on the right rear quarter panel may be analytically different from the hood. Or, as another example, the hood may have been repainted because of previous damage—it could even be a new hood. In the laboratory, the analyst may not be aware of the sampling that took place at the crime scene and, because of sample source variation, find that the paint samples are analytically distinguishable when the proper samples would have been the same in all tested respects. All paint samples should be clearly labeled as to origin, with drawings or photographs as documentation.

Because paint is a multilayer composite material, the known samples should contain all layers of the undamaged paint. Differences in the thickness and sequence of layers can be significant over even short distances on a painted surface. This concept is very important with architectural paints, where substantial reworking of the surface (sanding, damage, over-painting) may have occurred. It could be important to collect known paint samples from several areas of an object if variation is noticeable or suspected; these samples from different areas should be packaged separately and labeled appropriately.

Depending on what it is made of, the subcoating surface under the suspected transfer area should be included for analysis when possible. This may extend to portions or objects near the evidentiary location, such as portions of walls, doors, window frames, handles, fenders, and decorations. These additional samples may be useful to assess any difference that may exist between the known and questioned samples.

ANALYSIS OF PAINT SAMPLES

As with any other examination, the initial step in forensic paint flake analysis is to simply look at the sample; a stereomicroscope is an invaluable aid in this process. Often, the first step may be the last: If significant differences are apparent in the known and questioned samples, the analysis is completed and the paints are excluded. The visual evaluation begins with the packaging and paperwork, looking for signs of potential cross-contamination between the submitted samples. If none is detected, then the paint samples are described, noting their condition, weathering characteristics, size, shape, exterior colors, and major layers present in each sample. The examiner's notes should include written descriptions, photographs, and drawings, as necessary. Because significant changes can be made to a portion of a sample in the process of preparation and examination, it is crucial to document how that sample was received.

PHYSICAL AND MICROSCOPIC EXAMINATIONS

A combination of microscopes (both stereo, transmitted light, and polarized light) at magnifications of 2× to 100× are used to examine the layers in a paint. Many layers will be visible without preparing the sample but definitive paint layer identification often requires some sample preparation techniques. The paint layer structure can be seen by cutting through the sample with a scalpel blade at an angle; this

Figure 16.6. A *microtome is a mini-vice that holds a sample in place while a heavy and very sharp glass or diamond-edged knife slices off sections a few tens of microns thick. Microtomes are useful for paint analysis by conserving sample consumption and preserving the samples for subsequent analyses. Courtesy of Olympus, Inc.*

increases the visible area of the sample. The structure of the layers, as well as any irregularities and inhomogeneities are typically easier to see after this sectioning.

Very thin sections of the paint can be accomplished with a steady hand and a fresh scalpel blade; a device called a **microtome** can also be used. A microtome is a mini-vice that holds a sample in place while a heavy and very sharp glass or diamond-edged knife slices off sections a few tens of microns thick, as shown in Figure 16.6. Cross sections of the paint sample, either embedded in a material for support or thin-section preparations, provide information about the layers, their thicknesses, and colors, in additional to the size and distribution of pigments. Embedded preparations, as shown in Figure 16.7, work well because the sample is easily handled and can be subjected to many analytical techniques with a minimum of additional preparation.

Microscopical comparisons of paint layers can reveal slight variations between samples in color, pigment appearance, flake size and distribution, surface details, inclusions, and layer defects. Any visual comparisons must be done with the samples side by side in the same field of view (or with a comparison microscope), typically at the same magnification. Visual memory is quirky, and samples must be seen next to each other at the same time so that subtle details are not overlooked.

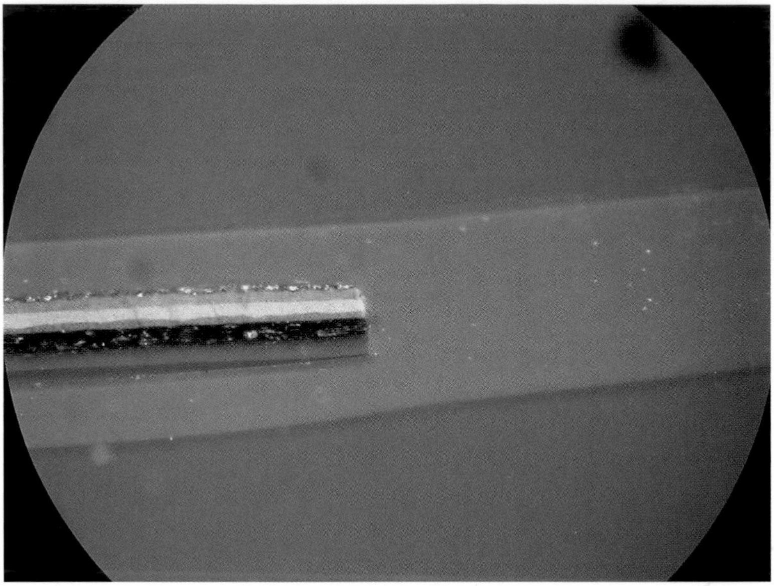

Figure 16.7. *Embedding paint chips in an epoxy resin supports and preserves the sample, allows for thin-sectioning, and makes a fragile specimen safer to handle and store. Courtesy of Paul Martin, CRAIC, Inc.*

Polarized light microscopy (PLM) is appropriate for the examination of layer structure as well as the comparison and/or identification of particles present in a paint film including, but not limited to, pigments, extenders, additives, and contaminants. The use of the PLM in the identification of pigments is detailed in "Application of Particle Study in Art and Archeology Conservation and Authentication," written by Dr. Walter McCrone (1979), in *The Particle Atlas.* Other components of paint, including extenders, are large enough that they can be identified by PLM, although some pigment particles may be too small to be identified this way. The use of PLM for the identification of paint components requires a good deal of intensive training and experience; many training courses are available in PLM for a fee, and similar analytical courses may be offered by materials science or geology departments at universities.

SOLVENT AND MICROCHEMICAL TESTS

Solvent and microchemical tests (hereafter referred to as "microchemical tests") have been used to discriminate between paint layers of different pigment and binder composition that are otherwise visually similar. The basis for these tests is that the different layers of paint have a different chemical composition and will, therefore, react differently with oxidizing, dehydrating, or reducing agents. Microchemical tests are destructive; therefore, they should be applied to known samples first to evaluate their utility to specific samples. They should also be used only when sufficient questioned samples are available to avoid consuming the entire sample.

Microchemical tests should be performed on the questioned and known materials at the same time and their effects recorded immediately; the effects may develop over time, so effects should also be recorded at intervals for the duration of each test. Additionally, microchemical tests can be applied to peeled individual layers of paint, to avoid interactivity with adjacent layers, as well as intact paint chips. Doing so may make the specific reactions clearer than subjecting an intact chip to the tests.

When a chemical is applied, the paint chip or layers may soften or wrinkle, swell, or curl. Entire layers may dissolve or disaggregate. Pigment fillers may bubble or "fizz" or flake apart (called "flocculation"). Apparent color change may be seen in some layers. As you can tell, these are not clearcut results that are easily quantified: They are mainly descriptive in nature but provide good discrimination at the early stages of an investigation and may help to initially classify a paint.

INSTRUMENTAL METHODS

Given the complex chemistry of paints and related coatings, it is not surprising that many instrumental methods are available for their analysis. Rarely will all of the instruments listed in this section appear in one laboratory—even if they did, the laboratory's analytical scheme would probably not include all of them—and the order of examination will be keyed to the instrumentation on hand.

Infrared spectroscopy (IR) can identify binders, pigments, and additives used in paints and coatings. Most IRs used in forensic science laboratories employ a microscopical bench, as shown in Figure 16.8, to magnify the image of the sample and focus the beam on the sample. The bench is a microscope stage attached to the instrument chassis with optics to route the beam through the microscope and back to the detector. Most modern IRs will also be Fourier transform infrared spectrometers (FTIRs), which employ a mathematical transformation (the fast Fourier transform) that translates the spectral frequency into wavelength. The analysis of paints by FTIR can be done in transmittance

Figure 16.8. An infrared spectrometer with a microscope attachment or "bench." Normal IRs require a sample to be pressed into a pellet or placed on a special specimen card to which the instrument is "blind." The microscope attachment allows for the handling and analysis of microscopic samples too small for either pellets or cards. This also provides for positional information about the sample to be analyzed; in the case of paints, individual layers or particles can be analyzed in place with no additional preparation.

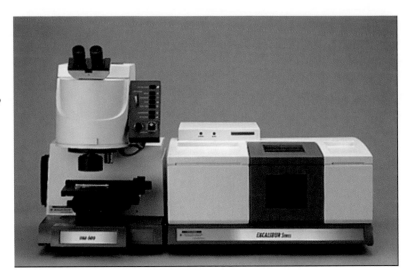

(where the beam passes through a very thin sample and then on to the detector) and reflectance (where the beam is bounced off the sample and then to the detector), but transmittance is preferred because it equalizes the signal as well as the sample geometry; also, and probably more importantly, most of the reference information available from publications and instrument vendors is in transmittance.

An IR-related technique that is gaining application in forensic science is Raman spectroscopy, which is based on light *scattering* rather than absorption. Because of this, Raman spectra provide complementary information to that obtained from IR spectroscopy. Raman spectroscopy shows great promise for a number of evidence types but, for budget and training reasons, it will be some time before Raman spectroscopy becomes a standard method in forensic science laboratories.

Pyrolysis-gas chromatography (P-GC or PyGC) disassembles molecules through heat (pyroloysis). This destructive technique uses the breakdown products for comparison of paints and identification of the binder type. P-GC is influenced by the size and shape of the samples and instrument parameters, such as rate of heating, the final temperature, the type of column, and gas flow rates. This can make P-GC vary from day to day and sample to sample—this has several methodological implications. The conditions from one analysis to the next should be the same and should be run very close in time to each other. It is important to select the known samples as carefully as possible because of the influence of size and shape on the final chromatographic results.

As little as 5 to 10 micrograms of sample are required for PGC. The patterns of peaks in the known and questioned sample chromatograms (also called "pyrograms") are compared, and the peaks must coincide for the identification to be determined.

If the instrumentation is available, pyrolysis products may be identified by pyrolysis gas chromatography-mass spectrometry (PGC-MS). The resulting reconstructed total ion chromatogram may help to identify additives, organic pigments, and impurities in addition to binder components.

Because one of the major purposes of paints and coatings is to impart color to an object, the analysis of color has been integral to the coatings industry nearly since its inception. The gross visual color of paints can be categorized systematically by one of many color systems currently in use. Two of the main systems traditionally used are the Munsell system (developed in 1915 by Alfred Munsell, an artist) and the Commission International de l'Eclairage (CIE) system, which is described in the ASTM International Standard Method D 1535 and Test Method E 308. Color systems are used to classify colors for description and communication of color information and for databases only; absorption spectra of any known and questioned paint samples are compared in forensic paint comparisons.

Absorption spectroscopy, using a microspectrophotometer (or MSP for short), has been used to categorize and discriminate between otherwise visually similar paints. MSP can also differentiate between metameric samples. **Metamerism** is the condition in which two colors appear similar under one set of conditions but different under others. One of the benefits of employing MSP is that it adds an objective method to the analysis of color. The instrumental parameters can be easily reproduced between instruments or laboratories, and this provides a basis for inter-laboratory testing and quality control. Careful reference sampling is essential to the success of color comparisons of such surfaces.

Comparison of paint layers by transmission MSP of paint thin sections is a more definite method of color analysis than reflectance techniques, but transmission MSP demands more careful preparation. The sample thickness and measurement location, for example, are critical for significant analytical comparisons, as illustrated by Figure 16.9.

One of the most generally useful instruments in forensic paint analysis is the scanning electron microscope outfitted with an energy dispersive X-ray spectrometer (SEM/EDS). SEM/EDS can be used to

Figure 16.9.
Microspectrophotometry (MSP) of paint layers by transmission of paint thin sections is an excellent method of discriminating between paint colors but demands more careful preparation. The sample thickness and measurement location, for example, are critical for significant analytical comparisons. In this figure, the small black square is the sampling area for the spectrum; it is located off the sample to collect a background spectrum. Courtesy of Paul Martin, CRAIC, Inc.

characterize the structure and elemental composition of paint layers. The SEM uses an electron beam rather than a light beam and changes the nature of the information received from the paint. The electron beam rasters over the area of interest; the electrons interact with the sample and generate a variety of signals, including surface information (**secondary electrons**), atomic number (**backscattered electrons**), and elemental information (X-rays). Secondary electrons impact the surface of the sample and are reflected to the detector, providing a visual representation of that surface, pictured in Figure 16.10. Backscattered electrons penetrate the surface of the sample and are kicked back out of the sample, with more being kicked out from the atomically denser regions. Therefore, backscattered electrons create an image in which brightness is proportional to atomic number. These types of imaging can be of great assistance in distinguishing paint layers and structures within the layers.

The primary reason for analyzing paint samples with an SEM/EDS system is to determine the elemental composition of the paint and its layers. When the electrons impact the surface, X-rays are produced as a result of high energy electrons creating inner shell ionizations in sample atoms, with subsequent emission of X-rays unique to those

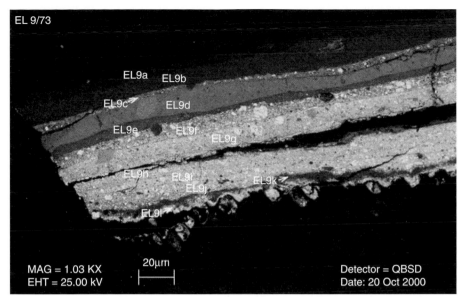

Figure 16.10. *Secondary electrons impact the surface of a paint sample and are reflected to the detector, providing a visual representation of that surface. Electrons carry no color information; their resolution, however, is very good and can provide images magnified thousands of times. Samples can also be imaged by backscattered electrons (BSE). Here, brightness is proportional to atomic number with larger numbered elements (iron, titanium) being brighter than smaller ones (silicon, oxygen). BSE images are useful for delineating paint layer structures and pigments. This is a sample of paint from a room in Osborne House on the Isle of Wight, bought by Queen Victoria in 1845. The paint samples were mounted in resin and polished to show the successive layers of paint in cross section. Since the raw materials used to produce paints have varied with time, the analytical information can be used to work out which paint schemes are contemporary. The SEM image shows a sample taken from the entrance lobby. The image was taken in BSE mode. Courtesy of English Heritage.*

atoms. The minimum detection limit under many conditions is 0.1%. Elements with atomic numbers ≥ 11 are customarily detectable. Detection of elements with atomic numbers ≥ 4 is possible using a detector with an organic film window or a windowless detector. Analysis can be performed in a rastered beam mode for bulk layer analysis, or static beam (spot) mode for individual particle analysis. Goldstein et al. (2003) present a general treatment of all aspects of SEM and X-ray microanalysis.

Mapping of elements across the cross section of a multilayer paint can be useful for explaining or demonstrating elemental distributions and elemental associations. Another technique that provides good visualization of elemental differences is atomic number contrast with backscattered electrons. These images can be used to characterize and compare the structure of paints, including layer number, layer thickness, distribution and size of pigment particles, and the presence of contaminants.

INTERPRETATIONS

Statistically evaluating trace evidence, including paint, is difficult. Most of the statements that can be made relate to samples tested in clinical or research trials although some are based on actual casework samples. A consensus of forensic paint examiners agrees that the following factors strengthen an association between two analytically indistinguishable paint samples:

- The number of layers
- The sequence of layers
- The color of each layer
- Cross-transfer of paint between items

One of the best known forensic paint examiners in the United States, Scott Ryland (1995) of the Florida Department of Law Enforcement forensic laboratory in Orlando, and his colleagues have stated that an association between two paint samples with six or more correlating layers indicates the chance that the samples originated from two different sources is "extremely remote." In cases with evidence this strong, merely stating that the two samples "could have had a common origin" is not enough—that level of statement undermines the strength of a six-layer-plus association. Although it is not a statistical or mathematical answer doesn't mean the statement isn't accurate, valid, and sound.

The significance of architectural paints varies and is in general not as well documented in the literature. This is most likely due to the enormous variability in colors, application styles, and the application of the paint itself (not all brushstrokes are equal, resulting in highly variable layers *between* samples). The situation is similar with spray paints, about which even less is known.

Instances of generating statistics to assess the evidentiary value of paint have been attempted in both the clinical literature and in casework. They are based, as are most manufacturing inquiries, on the concept of a **batch lot**, a unit of production and sampling that contains a set of analytically indistinguishable products. For example, a batch tank of automotive paint of a given color may hold 500 to 10,000 gallons, which would color between 170 to 1,600 vehicles. This would then be the unit of comparison for the significance of an automotive paint comparison—the manufacturing batch lot. If analytically identifiable differences can be determined between batch lots, then the base population is set for any other analytically indistinguishable paint samples. The final significance will be determined by the number of vehicles in the area at the time of the crime and other characteristics that set that sample apart (very rare or very common makes or models). By comparison, a batch lot of architectural paint may be from 100 to 4,500 gallons.

As with other forensic sciences, a reference collection is essential for training and casework, and paint is no exception. For paint, these take two forms: Documented samples and data. From 1974 until 1989, the National Bureau of Standards (now the National Institute of Standards and Technology) published a reference collection of automotive paint colors derived from actual production samples. Later in the collection's history, chemically accurate samples were added to it. Although limited in time frame and representation (it contained only U.S.-manufactured colors), this type of collection is invaluable for training and make and model searches. The FBI Laboratory Division maintains a color and chemical reference collection that is housed at the new laboratory in Quantico, Virginia; it is not a lending library, however, and the samples must remain at the laboratory.

A different approach overcomes these limitations. The **Paint Data Query (PDQ)** project is run by the Royal Canadian Mounted Police (RCMP) and stores the data from many databases, such as the FBI, the German Federal Police (BKA), the European Forensic Institute, and the Japanese National Police. Forensic laboratories can obtain the data by participating in PDQ by submitting 50 automotive paint samples per year, with some rare entries earning "double points." The database contains information on layer structures, primer colors, binders, pigment chemistry, and topcoat chemistry, in both visual and spectrometric formats. The technical liaison between participating agencies in PDQ is the Scientific Working Group for Materials Analysis Paint Subgroup, an FBI-sponsored group of subject matter experts.

CHAPTER SUMMARY

The forensic analysis of coatings is an important part of many investigations. From art forgeries to hit-and-run accidents to kidnappings, coatings can be a powerful form of physical evidence. Coatings, especially paint, provide samples that are chemically complex yet relatively straightforward to analyze and yield distinctive results. Automotive paints, in particular, are an important forensic analysis conducted in the modern forensic laboratory. The database PDQ provides important information about automotive manufacturers and styles that provides for accurate sourcing of known and questioned paint samples.

Test Your Knowledge

1. What is the difference between a paint and a coating?
2. What is a binder?
3. What are the categories of coatings?
4. What is a varnish?
5. What is a primer?
6. What is a clearcoat?
7. How should paint be collected if it is fragile or fragmentary?
8. What does a microtome do?
9. What is a problem with solubility testing of paints?
10. What types of instrumentation are routinely used to analyze paints?
11. What is a color system? Name two.
12. What is a metameric pair?
13. What factors strengthen an association between two indistinguishable paints?
14. What is PDQ?
15. What does a solvent do in regards to paint application?
16. Are physical matches possible with paint chips?
17. What is a batch lot?

Consider This . . .

1. How does the size of a batch lot of paint relate to the significance of a paint "match"? How would you find out how large the batch lot was?
2. How does the age of a vehicle affect the significance of a paint "match"? Why?

3. How many 1987 cars do you think are registered to drivers? How would you go about finding out?

BIBLIOGRAPHY

ASTM E-1610-02. (2002). Standard guide for forensic paint examination. West Conshohocken, PA: ASTM International.

Buckle, J., MacDougal, D., & Grant R. (1997). PDQ—Paint data queries: The history and technology behind the development of the Royal Canadian Mounted Police Laboratory. *The Canadian Society of Forensic Science Journal*, 30, 199.

Goldstein, J., Newbury, D.E., Joy, D.C., Lyman, C.E., Echlin, P., Lifshin, E., Sawyer, L.C., & Michael, J.R. (2003). *Scanning Electron Microscopy and X-Ray Microanalysis*, New York, Plenum.

Houck, M. (1999). The tyranny of numbers: Statistics and trace evidence. *Forensic Science Communications*, 1(3). Retrieved from **www.fbi.gov**.

Maehly, A., & Stromberg, L. (1981). *Chemical criminalistics*. New York: Springer-Verlag.

McCrone, W.C. (1979). Application of particle study in art and archeology conservation and authentication. In *The Particle Atlas*, volume V (pp. 1402–1413). Ann Arbor, MI: Ann Arbor Science.

Ryland, S.G. (1995). Infrared microspectroscopy of forensic paint evidence. In H.J. Humecki (Ed.), *Practical guide of infrared microspectroscopy*. New York: Marcel Dekker.

Ryland, S., & Kopec, J. (1979). The evidential value of automobile paint chips. *Journal of Forensic Science*, 24, 140.

Ryland, S., Kopec, J., & Somerville, P. (1981). The evidential value of automobile paint, part II. frequency of occurrence of topcoat colors. *Journal of Forensic Sciences*, 26, 64.

Scientific Working Group for Materials Analysis (SWGMAT). (1999). Forensic paint analysis and comparison. *Forensic Science Communications*, 1(2). Retrieved from **www.fbi.gov**.

Scientific Working Group for Materials Analysis (SWGMAT). (2002). Standard guide for using scanning electron microscopy/X-ray microspectrometry in forensic paint examination. *Forensic Science Communications*, 4(4). Retrieved from **www.fbi.gov**.

Thornton, J.I. (2002). Forensic paint examination. In R. Saferstein (Ed.), *Forensic science handbook* (2nd ed. vol. 1). NJ: Prentice-Hall.

Wikipedia, **www.wikipedia.com**

Soil and Glass

KEY TERMS

Amorphous solid

Becke line method

Borosilicate glass

Concentric cracks

Crystal lattice

Density

Density gradient

Dispersion plot

Energy dispersive X-ray analysis (EDX)

Forensic geologist

Fluids

Float glass

Fracture match

Glass

Hot stage

Humus

Immersion method

Mechanical fit

Minerals

Radial cracks

Refraction

Refractive index (**n**)

Soda-lime glass

Soil

Tempering

Thermocouple

INTRODUCTION

In this chapter we present two types of trace evidence: soil and glass. They are in this chapter together because they represent evidence that is most often characterized by optical or physical properties or by elemental content. Soil and glass are primarily mineral in content, and the methods used to analyze them relate to their compositional nature. In addition, only under unusual circumstances is it feasible to individualize glass or soil to a specific object or an exact location. Very few forensic laboratories have geologists on staff; most soil or glass examinations are conducted by forensic chemists. Soil and glass are often undervalued as evidence; they are dismissed as "only 'could have' evidence." In some cases, however, soil or glass can be extremely important, and this evidence should not be overlooked.

THE CASE: SOILS

Many years ago, Alfred Hitchcock made a movie titled *The Trouble with Harry*. In this movie, Harry is dead and the protagonists have to keep digging him up and reburying him to stay ahead of the cops. There have actually been a number of real cases in which someone or something is exposed to different types of soil. In a "Harry" case, a body may be buried in some location without the benefit of a proper funeral or casket. In many such cases, the body is clothed at the time of burial. At that time, some of the soil in the burial area will become transferred to the body. If it becomes necessary to disinter the body and then rebury it in another spot, it will then pick up soil from this second location. If this happens again, the body will collect more soil. When finally recovered, the body will have soil samples from each of the burial locations. This evidence can be compared to known samples of soil taken from each of the burial locations. Although this is not individual evidence, associating several soils with disparate locations can provide powerful circumstantial evidence.

Soil represents one of the paradoxes in forensic science: It is very commonly encountered in the environment and is found at many crime scenes, but it is seldom recognized as valuable evidence. In fact, soil is most often used as evidence when a shoeprint or tire tread impression is found in it! The soil is thus the carrier of the evidence rather than the evidence itself. One of the problems with the interpretation of soil evidence is that people easily transport small to large amounts of it over great distances. This can make it difficult to draw meaningful conclusions about the composition of soil at a given place where a lot of soil movement has taken place. On the other hand, overlooking soil evidence can mean that a crucial part of the puzzle of reconstructing a crime is not being collected and analyzed. Soil can make important contributions to the reconstruction of a crime and sometimes is the only reliable way to associate a suspect with a crime. Several soil cases will illustrate how this can be done.

Glass is also very common in our environment. There are estimated to be more than 700 different types of glass used in commercial products in the United States alone. Glass has an unusual physical structure. School teaches us that all matter is solid, liquid, or vapor, yet glass does not fall into any of these categories. It has properties of both solids and liquids. Glass is also an excellent example of trace evidence. Most often it occurs as tiny particles, although in some cases large enough pieces are found that can be fit together like pieces of a jigsaw puzzle. In this chapter, the formation of glass and how it occurs in crimes and how it is analyzed in the laboratory will be covered.

WHAT IS SOIL?

The definition of **soil** depends on who is working with it. Farmers are concerned with the topsoil in which they plant crops, and they define soil in terms of the top few inches on the ground. Engineers, on the other hand, view soil as a component of construction material and are more concerned with its physical properties than its chemical makeup. Soil scientists study changes in soil chemistry and composition and are concerned with minute alterations and variations in soil.

Forensic geologists look at soil a different way. Much of the soil that they study comes from areas that have been filled in, such as garbage dumps, gardens, or soil patches around homes or businesses, grave sites, etc. They are concerned with the transfer of soil particles from such locations to objects such as cars or clothing, either accidentally or purposefully. The case described at the beginning of this chapter is a good example. The various soils found on the body came from the locations where it was repeatedly buried. The people who buried Harry didn't realize that these soils would provide evidence of the locations of the body.

A forensic soil scientist might define soil as earth material, either natural or manmade (concrete, gravel, other building materials), that is transferred from a crime scene to a person or object, or vice versa. This soil may then be recovered from that person or object or may have been shed and found at a different location. The objective of forensic soil analysis is thus to associate soil found at a crime scene to its source. This requires, of course, that a source has been identified and known soil samples have been collected.

Natural soil contains both organic and inorganic materials. The organic materials are essentially decayed and decaying vegetative and animal matter such as grass and other plants, and insects, animal droppings, animal parts, etc. Sometimes this collective organic fraction of soil is called **humus**. The inorganic part of soil is generally crushed rock and clay materials. These are made up of **minerals**. Minerals are generally combinations of metal and non-metal ions. For example, iron and oxygen combine to form various minerals including ferric oxide (Fe_2O_3). Minerals are all crystalline solids with regular arrangements of atoms. Many have distinct colors that would be imparted to a soil sample. There are more than 2,000 minerals, but only about 50 occur commonly in U.S. soils. Rocks are generally made up of several mineral types with definite percent compositions, and the soil near the rocks will reflect this diversity.

The emphasis on the analysis of soil from a forensic science standpoint is to compare soil from the original location to the person or object to which it was transferred, in order to show that the person or object was, at one time, in the location where the soil originated. In this comparative analysis, the forensic geologist seeks to measure those

ISSUES IN SOIL:
The Adolph Coors Kidnap Murder

Adolph Coors owned a ranch in the foothills of the Rocky Mountains near Denver. While driving on his land, he was kidnapped by an employee of his company, who had been planning the crime for many months, with the intent of gaining a large ransom for Coors' return. During the kidnapping, a fight ensued and Coors was killed by the would-be kidnapper. He put the body in the trunk of his car and drove to a remote area of the mountains where he buried Coors in a shallow grave. He then took off across the country to New Jersey, where he abandoned the car in a dump and set it on fire to destroy any evidence that might link it to the crime.

The FBI investigated the case. FBI officers located the area where the kidnap/murder took place when other inhabitants of the ranch reported having heard gunshots on the property. They then located the body in Colorado when witnesses reported the presence of a strange car and observed someone digging a hole in the ground. A thorough search of personnel records of the victim's company turned up a suspect, and the FBI discovered that he had been using an alias. They determined his real name and were able to trace him to New Jersey, where they were able to locate the remains of the car used in the kidnapping. Even though the car was badly burned, the investigators were able to recover layers of soil that had been deposited in the wheel wells. They collected this soil as well as soil samples from the location of the kidnap/murder, the burial site, and other areas in the mountains where the suspect had lived and had been known to travel.

Analysis of these soil samples showed that all were represented within the layers of soil taken from the suspect's car. Thus, the investigators were able to show that the car had been in all of the locales that were relevant to the case. This was crucial evidence in the conviction of the suspect for kidnapping and murder.

physical and chemical properties that are most effective in determining if two soil samples could have come from the same location, given the time, material, and instrumental constraints that are always a part of forensic science work. An example of the use of soil analysis is shown in "Issues in Soil: The Adolph Coors Kidnap Murder".

COLLECTION OF SOIL EVIDENCE

As with all types of forensic evidence, two types of soil evidence are of concern to the forensic geologist: the questioned or unknown samples and the known or control samples. In the Coors kidnapping case described earlier, the questioned samples were those that were found on the suspect's car. The investigators wanted to establish that the kidnapper used that car to transport the body from the kidnapping site to the burial site and then to his "hideout" in New Jersey and finally to the dump. It was therefore necessary to collect soil samples from the car and determine their points of origin. In cases such as this, where several soil types are found, care must be taken to collect all of the layers of soil from the undercarriage at once and intact, because the order of the layering can help establish the order in which the car came in contact with the soil present at the various important locations. The reason is that the layer of soil that adheres to the body of the car will be from the first place that the car came in contact with soil. Succeeding layers of soil will be deposited on top of the original in the order that the car came in contact with them. A problem with these questioned samples, and for that matter in all cases, is that the forensic geologist has no control over the nature and amount of the evidence. There could be problems with limited size and amount, contamination, fragility, and stability of evidence. Thus, it is important to collect as much questioned evidence as possible and to preserve it so that it remains in substantially the same condition throughout. Soil evidence often occurs in layers, as shown in Figure 17.1. These layers must be collected very carefully because their order can be important.

Control or known soil samples present different problems for the forensic geologist. If the scene covers a large area, care must be taken to get the control samples as close to the evidentiary soil location as possible. If there is some doubt, then samples should be collected from several areas to account for natural variation of soil. In cases in which

Figure 17.1. *Layers of soil.*

the known sample is to be taken from a hole, such as a grave, the vertical layering is as important as the horizontal, so control samples should be taken from the same depth as the questioned sample came from. Time can also be an important consideration. If a long time has passed between the crime and the time that the controls are collected, there may have been disruption to the area where the incident took place. For example, suppose a rape occurred on someone's back lawn and the victim got soil stains on her clothes. A few weeks later, the back yard was resodded, and then the control samples were obtained. The resodding may mean that the controls really aren't controls; the type of soil in the new sod may be very different from the soil that was there previously.

Dry soil samples can be put in plastic bags or film canisters or other airtight containers. Wet samples, however, should be put in paper or cloth sacks so the water can evaporate; otherwise, there can be irreversible chemical changes to some of the minerals. The location and circumstances concerning soil evidence can be important as shown in "In More Detail: Soil Evidence."

IN MORE DETAIL:
Soil evidence

Thom Holpen, one of the country's leading trace evidence analysts, describes the value of soil evidence in the book *Trace Evidence Analysis: More Cases in Mute Witnesses*. One of the cases he describes shows that the location and circumstances surrounding the deposition of evidence such as soil can be very important. The summary of this case is reprinted here:

> On a warm summer evening, a burglary was attempted on a business with the perpetrator trying to gain entrance through a foundation crawlspace having a dirt floor. A man meeting the description of the suspect was stopped a block from the business by police and taken in for questioning. The suspect wore no shirt and his jeans were soiled. When questioned as to how his jeans became soiled, he said he had been playing baseball at a local park several blocks away. His clothes were submitted to the laboratory along with soil samples from underneath and around the perimeter of the business as well as from the park. Examination of the jeans revealed several clumps of soil deposited on the inside of the front waistband. Analysis and comparison of the soil from the jeans with the reference samples revealed it to be consistent only with the soil from underneath the business. The suspect pled guilty when confronted with the evidence. How did the soil get inside the waistband? It is believed that when the suspect was crawling on his belly, and with no shirt to block it, the waistband of the jeans scooped the soil up. His perspiration moistened the soil and helped tack it to the inside material of the jeans. (p. 98)

Source: Houck, M.M. (Ed.). (2004). Trace evidence analysis: More cases in mute witnesses. Amsterdam: Elsevier.

ANALYSIS OF SOILS

Soil possesses many physical and chemical properties that can be exploited in comparison of known and unknown samples. The properties that are chosen for measurement depend on the nature of the case. Most crime laboratories do not have personnel who possess the knowledge and skills to exploit soil comparisons fully. Instead, they rely on expertise that may reside at a local university, such as a mineralogist, geologist, or soil scientist. The majority of forensic soil cases consist of footwear where someone has left a shoeprint in soil. This usually doesn't involve analysis of the soil per se, although it probably should, but is limited to comparison of the shoeprint with the shoe.

REPRESENTATIVE SAMPLES OF SOIL

Soil, like some other types of evidence, may occur in large quantities in a given case. When this happens, it may be necessary to make some decisions about how much of the exhibits will actually be analyzed. It is usually not necessary to examine all of the soil in an exhibit as long as the samples taken from the bulk exhibit are representative of the whole.

There are a number of ways of determining how many samples to take, their size, etc., but the first consideration is to decide if and how an exhibit of soil is to be homogenized so the samples taken will be representative. This presents some unique situations with soil evidence. One of the most important tests done on soils is to determine the particle size distribution. Some soils are coarse and contain large chunks of material, whereas others are finely divided. If a known and questioned sample of soil came from the same location, their particle size distributions are usually similar. This means that the determination of particle size distribution must be made before the soil is pulverized in an attempt to homogenize it. In this context, particle size distribution is an inventory of the various sizes of the particles present in a sample of soil.

Once the particle size distribution is determined (see the next section) for a discussion about how this is done), then the soil can be crushed and pulverized to make it consistent throughout so that samples taken for analysis represent the bulk exhibit and the physical and chemical analyses on the samples can be extrapolated to the bulk sample.

PHYSICAL PROPERTIES

In general, physical properties are fairly easy to measure and are inexpensive and not too consumptive of material. Standard methods of analysis are available from the American Society of Testing and Materials and the U.S. Geological Survey.

The most common physical tests are color and particle size distribution. The color of soil is affected by moisture content, mineral distribution, and location. Dusty, dry soils tend to be light tan or white, owing to lack of moisture. Agricultural or tropical soils tend to be dark brown, owing to the high humic content. The naked eye is a very good discriminator of color when comparing soil samples, and standard

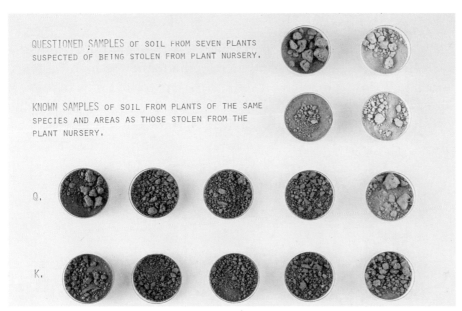

QUESTIONED SAMPLES OF SOIL FROM SEVEN PLANTS
SUSPECTED OF BEING STOLEN FROM PLANT NURSERY.

KNOWN SAMPLES OF SOIL FROM PLANTS OF THE SAME
SPECIES AND AREAS AS THOSE STOLEN FROM THE
PLANT NURSERY.

Q.

K.

Figure 17.2. Different color soils. This photo was taken from a case in which several plants were uprooted and stolen from a plant nursery. Samples of soil taken from the plants were compared with samples taken from the pots the plants were in at the nursery.

color charts can help with exact color characterization. Figure 17.2 shows a case involving several different soil colors.

The best way to analyze mineral content is to look at the minerals under a polarizing microscope and identify the actual minerals present and their relative amounts. Many, if not most, soil comparisons begin and end here. Particle analysis is key to understanding the nature of the samples under examination. Interested students should seek out the references and resources from the McCrone Institute on polarized light microscopy of particles and minerals.

An easy method of ascertaining particle size in soil samples is by sieving. A weighed soil sample is dried and then put through a nest of sieves where each succeeding sieve has smaller holes than the one preceding it. Then each fraction is weighed, and the percentage of each particle size range can be computed.

Scanning electron microscopy (SEM) has become a useful tool for many crime laboratories and has been applied to the analysis of soils. Using an electron microscope, an analyst can visualize very tiny particles that cannot be seen with a light microscope. In addition, the elemental composition of an individual particle can be determined at the same time using **energy dispersive X-ray analysis (EDX)**.

CHEMICAL PROPERTIES

Chemical analysis of soil is performed less frequently than physical analysis. One method receiving increased attention is high performance liquid chromatography. A soil sample is extracted with an organic solvent and then subjected to HPLC. The resulting chromatogram gives a profile of many of the organic substances found in the soil. This

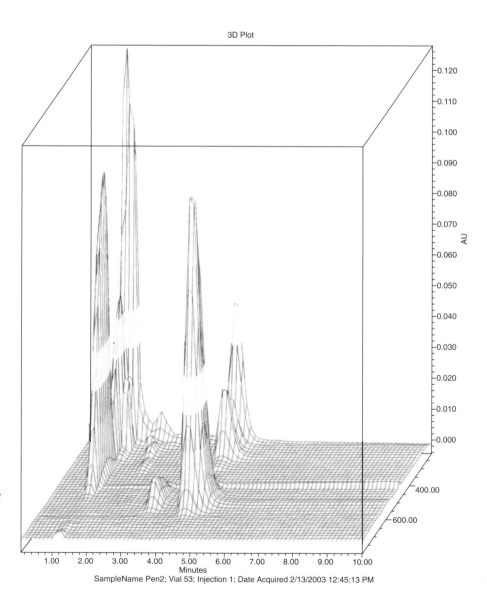

Figure 17.3. HPLC *chromatogram of a soil sample. The peaks are due chiefly to the organic components of soil, including the so-called "humic acids."*

provides excellent comparative data for known and unknown samples. Infrared spectrophotometry of soils is also a useful technique. It is possible, for example, to obtain the infrared spectrum of a bulk soil sample, then extract the organic fraction and obtain its infrared spectrum, and then by spectral subtraction, obtain the IR of the rest of the mostly inorganic fraction. A liquid chromatogram of a soil sample is shown in Figure 17.3. The pseudo three-dimensional appearance is due to simultaneous measurements of time, absorbance wavelengths, and intensity of absorption.

Other tests are sometimes done on organic fractions of soil. These tests include oxygen bioavailability and DNA analysis on bacteria or other microbes in the soil. The latter is a relatively new technique and is quite complex and expensive.

BACK TO THE CASE: SOILS

Remember that soil is not individual evidence. It cannot be associated with a particular location. It must be acknowledged that a soil sample could have come from a number of locations. If a forensic scientist has a single type of unknown soil that has similar characteristics to a soil sample of known origin, this is of limited probative value. If, however, the forensic scientist has the situation presented in the "Trouble with Harry" type cases, in which several different types of soil are found as evidence of a crime and all of them can be associated with soils from several independent locations, then the value of the evidence is magnified greatly. The more locations there are, the more impact the evidence has.

In the case presented at the beginning of the chapter, the biggest challenge is to isolate the various types of soil on the body so they can be analyzed independently. This can sometimes be done by using color or particle size distribution characteristics to separate the soils. If all of the soils are mixed together and cannot be physically distinguished, then it may be necessary to mix together all of the known samples and do a single bulk comparison. This has less value than separate analyses but may be the only alternative.

In a case like this, the laboratory would perform a particle size distribution analysis by sieving and then perhaps liquid chromatography. If the forensic scientist has some skills in mineral identification, then a microscopic mineral assay may be performed.

THE CASE: GLASS

A man is walking across the street with a load of packages at dusk. Suddenly, from his left, comes a truck with a drunk driver. He hits the man and runs him over and keeps on driving. The point of impact at the truck is the right front headlight, which is shattered. It was not turned on even though visibility is poor at that time. Pieces of the headlight are strewn about the scene, and some tiny shards of glass are lodged in the man's coat. Some pieces are left in the headlight frame. A witness gives the police a description of the truck, and a few minutes later, the truck is stopped by a patrol car. After noting what appears to be fresh damage to the front of the truck and traces of what appear to be blood in that area, the police officer arrests the driver on suspicion of vehicular homicide. The truck is impounded, and a search reveals the damaged front headlight. The headlight is removed and sent to the lab, as are the victim's clothes and sweepings of glass from the scene of the incident. The forensic scientist is asked to compare the glass fragments and render an opinion.

WHAT IS GLASS?

Glass is sometimes defined as an **amorphous solid**. This is a hard, brittle material that is usually transparent, but which lacks the ordered arrangement of atoms (**a crystal lattice**) that is found in most solids. Instead, common glass is made up largely of oxides of silicon that have been doped with other materials to give it its familiar properties. The silicon oxides come chiefly from sand. The silicon dioxide matrix and the methods used to produce glass impart properties of both solids and liquids. Such substances are sometimes called **fluids**. If you have ever bumped into a plate glass window or been hit with a glass bottle, you know that glass is hard like a solid. The amorphous structure of glass is shown in Figure 17.4.

GLASS MANUFACTURE

Glass is manufactured by melting sand and the other desired ingredients and then allowing it to cool without crystallizing. The cooling process may take place in a mold, or the glass may be injected into a particular shape or may be cooled in such a way that it is perfectly flat. For example, molten glass can be cooled on top of a bath of molten tin. This allows the glass to be very flat so it can be used in windows and other similar applications. This type of glass is called **float glass**.

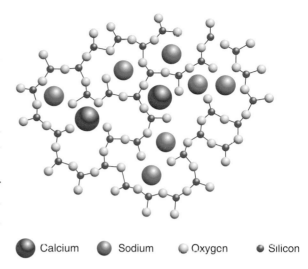

Calcium   Sodium   Oxygen   Silicon

Figure 17.4. The *chemical structure of basic glass—silicon dioxide. Source:* **http://www. pilkington.com /pilkington/ corporate/ english/ education/ chemistry/ default.htm**

Basic silicon glass is seldom used in a pure form. Instead, while it is in the melted state, it is "doped" with measured amounts of various impurities that alter its properties in a predictable way. For example, when soda (sodium carbonate, Na_2CO_3) is added, the glass will melt at a lower temperature and flow more easily, making it easier to work with. Other materials such as lime (calcium oxide, CaO) can be added to stabilize the glass and make it less soluble in solvents. When both calcium oxide and sodium carbonate are added, the product is called **soda-lime glass**. If boron oxide (B_2O_3) is added to glass, it becomes more heat resistant. It is then called **borosilicate glass** (Pyrex©). This type of glass is used in cookware, thermometers, and automobile headlights.

Glass may be strengthened by **tempering**. This is a process whereby the glass is heated and cooled rapidly, producing deliberate stress in the surface. When this glass breaks, it shatters into small, solid pieces rather than sharp shards. This type of glass is used in car windows. In the United States tempered glass is not used in front windshields. Instead, a laminate consisting of two layers of glass with a layer of plastic between is used. If the windshield is broken, the plastic sheet helps keep shards of glass from flying around the car and injuring the passengers.

Three major types of glass are encountered as evidence in forensic cases: sheet or flat glass, container glass, and glass fibers. Flat glass is used to make windows and windshields and can also be shaped to form light bulbs, headlights, and other materials. Container glass is used to make bottles and drinking glasses. Glass fibers are used to make fiberglass and fiber optic cables as well as glass-plastic composite materials. Less frequently, optical glass is used to make eyeglass lenses, and similar materials may be encountered in forensic cases.

FORENSIC EXAMINATION OF GLASS

More than 700 types of glass are in use today in the United States. Obviously, some are more common than others. One would expect to find much more bottle and container glass as well as window glass in a given environment than optical or specialty glass. Because many types of glass are mass-produced, individual glass objects ordinarily do not possess any characteristics that are so unique that a piece of glass from this object could be individualized to it. As a result, small pieces of glass are considered to be class evidence.

THE MECHANICAL FIT (FRACTURE MATCH)

In one set of circumstances glass can be individualized to a particular object. This is called the **mechanical fit** or **fracture match**. This occurs where a piece of glass breaks into relatively large pieces that have at least one good, intact edge that can be fitted to the edge of another piece from the same source. Since glass is hard and brittle, it doesn't deform when broken. Since it is amorphous, there are no lattice points along which the glass would fracture when subjected to force, so fractures would be random events, and no two pieces of similar glass would be expected to break in the same manner. Thus, if there is a good mechanical fit to two pieces of broken glass, it can be concluded with certainty that they were once joined. This fit is often aided by the presence of stress marks along the broken edge. These marks are also randomly generated and are caused by the application of force at the breaking point. They can be seen only with the aid of a microscope. Even if a broken edge is relatively featureless, the stress marks will align where the break occurs. In Figure 17.5A, a glass fracture match can be seen. Figure 17.5B shows stress marks that appear in glass where it is broken.

EXAMINATION OF SMALL GLASS PARTICLES

The vast majority of cases involving glass consist of particles that are too small to be physically matched by a mechanical fit. Such particles

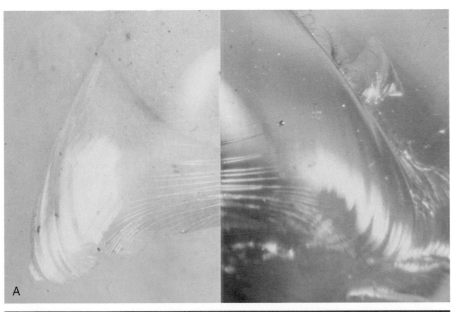

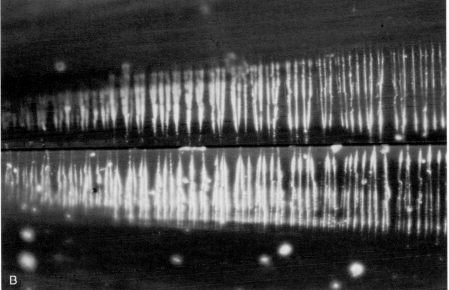

Figure 17.5. Λ) *A glass fracture match. B) Stress marks produced when glass is broken.*

generally do not contain any unique characteristics that would permit them to be individualized to a particular source. Thus, they are class evidence. A number of tests can be performed on glass particles to characterize and compare them. The most important of these tests are identification, **density**, and **refractive index**. Other tests include color, thickness, flatness, surface features, and fluorescence. Some of these tests are discussed in the following sections.

Identification

Before testing of glass particles is undertaken, it must be shown that the particles actually are glass and not some other material. This can be done by measuring its hardness, structure, and behavior when exposed to polarized light. If glass is examined under crossed polars, it will disappear because it is isotropic: It behaves the same at any orientation of a polarizer. Glass can be differentiated from plastic by pressing it with a needle point. Most plastics will show an indentation from the needle, whereas glass will not. Substances like table salt can be differentiated by their shape; they have a regular shape owing to an ordered chemical structure, whereas glass is amorphous. Table salt, for example, exhibits cubic crystals. Some minerals can be differentiated from glass using a polarizing microscope. Many minerals will show different colors and brightness than glass when exposed to polarized light.

Preliminary Tests

Prior to a comparison of glass fragments, it is advisable to do some preliminary testing to show that all of the pieces of glass in one exhibit could come from one single object. These preliminary tests include color, surface characteristics, flatness, thickness, and fluorescence. These tests are also valuable in comparing known and unknown samples.

Density

The density of an object is its mass divided by its volume (see equation 17.1).

$$\text{Equation 17.1} \qquad D = M/V$$

For example, if a cubic block of wood weighs 250 grams and is 10 cm on each side, its volume would be 1,000 cubic centimeters or cc (10 cm × 10 cm × 10 cm), and its density would be 250 g/1,000 cc, or 0.25 g/cc.

Most objects can be easily weighed on a common balance. If an object has a regular shape such as a cube, whose volume can be easily measured, then its density is easily determined. If the object has an irregular shape but is fairly large, its volume can be measured by dis-

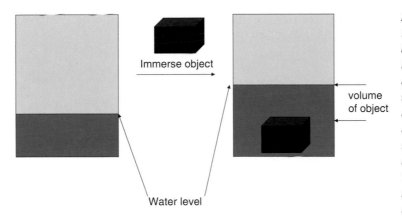

Immerse object

Water level

volume
of object

placement of a liquid. A measured amount of liquid, such as water, is put into a container and the object is added. The volume of the liquid is then measured, and then the volume of the liquid alone is subtracted, leaving the volume of the object. Recall the old tale about the thirsty bird that could not reach a small amount of water in a deep hole. He dropped pebbles into the hole, thus raising the level of the water until he could reach it.

If an object is irregular in shape and too small to measure its volume by displacement, then indirect methods must be used. This is shown graphically in Figure 17.6.

The easiest way to compare the densities of two small pieces of glass is by the sink/float method. In this method, two miscible liquids (liquids that can mix with each other) are chosen such that one is less dense than glass and the other is denser. Bromoform and bromobenzene are often used, although they are both toxic, so caution must be observed. A tube is filled with a mixture of the two liquids, and pieces of the glass are inserted. In the glass case that is described earlier in the chapter, pieces of glass from the broken headlight, fragments from the crime scene, and fragments taken from the man's clothes could be compared this way. The particles are allowed to settle in the liquid mixture; then a determination is made to see if they all settle at the same level. Then the liquids are heated slightly using a hair dryer. Heat decreases the density of liquids, so the glass particles should then move lower. If they all move to the same spot, then they have the same density. It should be noted that a glass object, such as a headlight, will exhibit slight variations in density within the object. The sink/float method of density determination is very sensitive, and two pieces of glass sometimes may show slight differences in density but may actually arise from the same object. This problem may be minimized by using several particles from each source.

Refractive Index

In Chapter 5, "Spectroscopic Techniques," the property of **refraction** was discussed. Recall that refraction occurs when light passes through a transparent medium or object. The light bends away from its path and slows down. Refraction is much more pronounced in solids than liquids or gases. Glass is an excellent medium for exhibiting refraction. The amount of refraction caused by glass is an important physical property for the comparison of known and unknown exhibits.

Refractive index (**n**) is the ratio of the velocity of light in a vacuum (or air) to the velocity as it passes through the medium (see equation 17.2).

$$\text{Equation 17.2} \qquad n = \frac{\text{Velocity of light in vacuum}}{\text{Velocity of light in medium}}$$

Refractive index is always greater than one because light travels fastest in a vacuum. For glass, the range of refractive indices is usually between 1.4 and 1.7. Different types of glass have different refractive indices, so this property can be valuable in determining what type of glass is present and for comparing glass fragments.

There are two other important properties of refractive index. One is its variation with temperature. Refractive index is inversely proportional to temperature (see equation 17.3).

$$\text{Equation 17.3} \qquad N \propto 1/T$$

This means that, as the temperature is raised, the refractive index decreases. Over a range of perhaps 30° this effect is quite pronounced with liquids, but solids such as glass exhibit almost no variation. The other important property of refractive index is its variation with wavelength of light. It is also inversely proportional to wavelength (see equation 17.4).

$$\text{Equation 17.4} \qquad N \propto 1/\lambda$$

How is refractive index measured in a forensic science laboratory? Clearly, it is not possible to measure the refractive index directly because it would be impractical to try to measure the speed of light as it passes through a piece of glass. Instead, an indirect method must be used. A number of indirect methods are used in forensic science labs. The most popular are **immersion methods**, and the most commonly used of these is the **Becke line method**.

The Becke Line Immersion Method

The human eye can detect transparent objects in air, such as glass fragments, because they refract light. If two objects have the same refractive index, then our eyes would not see any difference in the light that passed through them. If a piece of glass is immersed in a liquid whose refractive index matches that of the glass, then the glass should be invisible because light that passes through the glass would have the same refractive index as the liquid, and our eyes would not be able to detect any difference. In reality, if a piece of glass is immersed in a liquid with the same refractive index, the glass usually does not completely disappear. The reason is that there are other effects caused by light passing through glass. These effects are more pronounced when the glass is thick. The result is that we can still see faint borders of the glass even if they have the same refractive index.

When a piece of glass is immersed in a liquid of different refractive index and observed under a microscope with transmitted light, a bright halo in the shape of the glass will appear to surround the glass. This halo, called the Becke line, is caused by the difference in refraction by the glass and the liquid. If the glass and liquid have the same refractive index, the Becke line will disappear even if the glass does not. The Becke line can be clearly seen in Figure 17.7.

In practice, the Becke line method takes advantage of several properties of refractive index. Again, referring to the glass case discussed earlier in the chapter, suppose the glass examiner wishes to determine the refractive indices of the glass from the headlight, the crime scene, and the victim's clothes. The glass chemist would need to have a set of liquids that are made to have an accurately determined refractive index. A liquid is chosen that has a slightly higher refractive index than the glass being examined. A glass particle is immersed in the liquid on a microscope slide and observed under a microscope outfitted with a **hot stage**. This device allows the slide to be heated under controlled conditions. The slide is slowly heated while the Becke line is observed. Recall that, as a liquid is heated, its refractive index decreases, whereas temperature has very little effect on solids such as glass. At some increased temperature, the refractive index of the liquid will decrease until it is the same as that of the glass, and the Becke line will disap-

Figure 17.7. *A piece of glass immersed in a liquid of different refractive index. The Becke line can be seen as a bright halo around the glass.*

pear. Commercial hot stages contain a **thermocouple** (temperature-measuring device). The hot stage is also connected to a computer that keeps track of the change in temperature. The maker of the refractive index liquids determines how much the refractive index decreases with each degree rise in temperature. This data permits the exact calculation of the refractive index of the liquid at the elevated temperature and thus the glass. An example of a refractive index calculation follows: A microscope outfitted with a hot stage is shown in Figure 17.8.

A piece of glass is immersed in a liquid whose refractive index at 25°C is 1.520. The Beckc line is plainly visible. This particular type of glass is known to have a refractive index of less than 1.520. The temperature is raised to 45° at which time the Becke line disappears, indicating that the glass and the liquid are now at the same refractive index. The bottle label indicates that the refractive index of the liquid drops 0.0003 units for every degree increase in temperature. Over this limited temperature range, we may consider the refractive index of the glass to remain constant. When the Becke line disappeared, the temperature had risen 20° (25–45). This corresponds to a decrease of .006 refractive index units (.0003 × 20). This means that at 45° the refractive index of the liquid and glass are 1.514 (1.520–.006).

The Becke line method is very accurate and precise, but it is always advisable to get as much data as possible before rendering an opinion

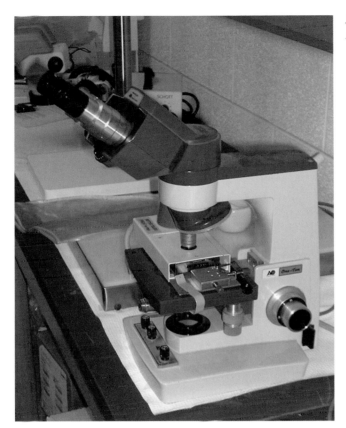

Figure 17.8. A hot stage and microscope.

about the association of evidence. The Becke line experiment is usually performed using light at 589 nm (the sodium D line). If a different wavelength of light is used, the refractive index of the liquid and the glass will be different so the experiment can be repeated at several different wavelengths. Then a plot of wavelength versus refractive index can be constructed for each glass particle. This is called a **dispersion plot** (see Figure 17.9). If two pieces of glass have the same dispersion curve, this is good evidence that they could have come from a common source.

Elemental Analysis of Glass

Glass may contain trace amounts of elements that get there either from contamination during the manufacturing process or on purpose when it is desirable to impart certain properties to the glass. Measurement of the types and quantities of these trace elements can help in determining whether two pieces of glass were ever part of the same object.

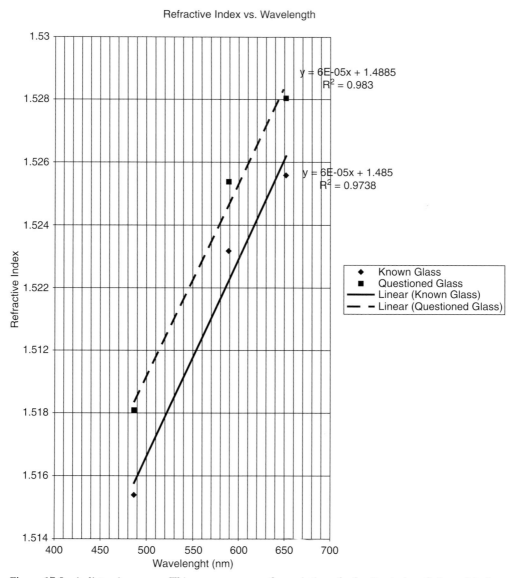

Figure 17.9. *A dispersion curve. This curve measures the variation of refractive index of glass with the wavelength of light used to measure it.*

The problem here is to be able to digest the glass so as to free the trace elements. This is not easy to do. Inductively coupled mass spectrometry (ICP/MS) identifies the trace elements in the glass. The glass can be dissolved by laser ablation or by using hydrogen fluoride. Electron microscopy can also be used to determine the presence and amounts of trace elements.

THE EFFECTS OF PROJECTILES ON GLASS

When a piece of glass such as a window is struck by a projectile, a number of things happen that can leave evidence of the direction and angle of impact. What actually transpires depends on the thickness of the glass and the velocity and size of the projectile. The observations that can be made include where the projectile ends up, the formation of a crater in the glass, the formation of a cone of glass, and the types and positions of cracks that appear in the glass.

When a high-speed projectile, such as a bullet, passes through a piece of glass, a crater will form in the glass that is larger on the exit side of the glass. This, by itself, may reveal the direction of impact of the projectile. The crater may show some asymmetry that can yield information about the angle of impact. In addition, **radial cracks** will form on the side of the glass opposite the side of the impact, as shown in Figure 17.10. These cracks radiate out in all directions from the point of impact. In many cases, there will also be **concentric cracks** on the side of the direction of impact. In addition, the orientation of stress marks in the glass at the point of the break or penetration can help determine the direction of impact. Some of the marks will form a right angle at the point of impact. The direction of the angle will indicate the direction from which the projectile came.

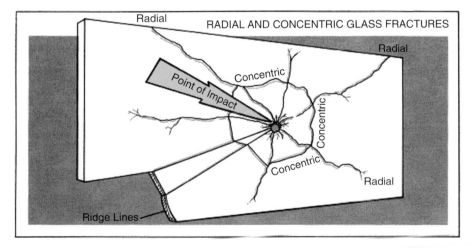

Figure 17.10. Radial and concentric cracks in glass and multiple penetrations in a piece of glass. Note that the radial cracks made by the second penetration stop at a crack made by the first penetration. This shows how it can be determined which penetration was made first.

Source: **http://www.adtdl.army.mil/atdls.html**

Figure 17.11. A
headlight filament that
was on when the lamp
was broken. Courtesy of
Christopher Bommarito,
Michigan State Police
Forensic Science
Division.

LAMP ANALYSIS

Once again, consider the hit-and-run case described earlier in the chapter. Suppose the facts of the case are altered so that the incident took place at dusk and a question arose as to whether the headlights were on at the time. If a headlight breaks during the impact, it is usually possible to determine its status at the time of impact. As shown in Figure 17.11, if a light filament is hot and then the vacuum in the light is broken, then oxygen will be available. This will cause a reaction between the tungsten in the filament and oxygen in the air, resulting in the formation of oxides on the surface of the filament. If the filament was not on, then no oxides will be seen.

BACK TO THE CASE: GLASS

An all too typical hit-and-run case was described earlier in this chapter. Glass can play a pivotal role in the investigation of this incident. It will be necessary to collect all of the glass around the crime scene as well as the broken headlight and the victim's clothes. All of this will be taken to the lab. At the lab, the scientist will carefully brush the clothes to recover tiny glass particles. The glass from the scene will have to be extensively examined to determine if any of it could have come from the headlight. There may be glass from many sources on the street, including headlights from other cars if other accidents occurred there. Most likely, these extraneous pieces of glass would differ in one or more ways from the glass fragments that were thrown off from the headlight of the truck involved in this hit-and-run.

One of the most important examinations would be to determine if there exists a mechanical fit between pieces of glass from the victim's clothes and/or the scene with the glass remaining in the headlight. This would be possible only if the fragments were large enough to match.

Fragments of glass from all of the sources will be compared for shape, thickness, surface characteristics, color, and other characteristics. Assuming that they are all in order, then the analyst would most likely perform a refractive index test such as described earlier and may also do a density test. If the results of these tests were similar for glass from the victim and the headlight, and perhaps from the scene, then the scientist would conclude that the glass from the victim's clothes could have come from the headlight. Owing to the mass production of headlights, it would not be possible to conclude that the glass from the victim's clothes came from that particular headlight, nor would it be possible to state the probability that it did or did not originate from that source. If the results of the testing were very different for the known and unknown glass fragments, then the analyst would be justified in eliminating the headlight on the suspect car from being the source of the glass taken from the victim. This is in keeping with the characteristics of class evidence: It is possible to absolutely eliminate a particular source but not to associate it as the only possible source.

CHAPTER SUMMARY

Glass and soil can be very important trace evidence in criminal cases.

Soil is virtually always class evidence. It is easily transported from place to place, so soil samples can differ significantly when they are only a few meters apart vertically or horizontally.

Soil consists of organic and inorganic fractions. The organic material consists chiefly of decaying and decayed animal and vegetable material. The inorganic part is mostly crushed minerals. Soil can range from nearly 100% organic to 100% inorganic. Various tests can be performed on the organic and/or inorganic fractions. Mineral microscopy (petrography) is frequently carried out on known and unknown soils for comparison. Liquid chromatography can be used to separate organic fractions. Density and size gradients can also be determined on the soil sample as a whole.

There have been many cases in which the whereabouts of an automobile can be traced by the layers of soil that collect with time in protected areas such as the wheel wells.

There are many types of glass present in our environment and they are often found at crime scenes, especially automobile accidents and home and business invasions. Glass is a material somewhat like a solid and a liquid. When glass fragments are large enough to be pieced together like pieces of a jigsaw puzzle, the evidence can be individualized to a particular source. In most cases, the glass fragments are too small to be fracture matched, so class characteristics are emphasized. Most commonly, this involves measuring the density and refractive index of the glass fragments. Electron microscopy or ICP-MS can also be used in the analysis of elemental composition of glass.

Test Your Knowledge

1. What is soil? What are its major components?

2. What is forensic geology? How does it differ from conventional geology?

3. Where does humus come from?

4. In the case that was presented at the beginning of the chapter, where was the soil collected that was used for comparison with various locations around the country?

5. How is glass defined?

6. What properties does glass have that are like those of a solid? A liquid?

7. What is density? What are its units?

8. What is refractive index? What are its units?

9. Glass is sometimes called an amorphous solid. What does this mean?

10. What is float glass? How is it made?

11. What is tempered glass?

12. What are radial cracks? How do they help determine the direction of impact of an object on glass?

13. What are concentric cracks? What is their role in determining the direction of impact of an object on glass?

14. What is dispersion? How are dispersion curves constructed?

15. What is a Becke line and how does it form?

16. If a piece of glass is immersed in a liquid and a Becke line is seen under a microscope, how can you tell which has the higher refractive index?

17. What is the relationship between refractive index and temperature?

18. What is the relationship between refractive index and incident light?

19. Under what conditions can a piece of glass be individualized to a particular source?

20. How can you tell if a headlight were on or off at the time it is broken?

Consider This . . .

1. Soil cannot generally be individualized to a particular source. Explain why this is so. (Keep in mind the variations of soil horizontally and vertically within one location, the transportability of soil, and the forensic taxonomy of soil, if any.)

2. Suppose that, in the glass case given earlier in this chapter, there are only tiny fragments of glass found at the hit and run scene, instead of large pieces. Since one could accomplish a fracture match, what tests would you do on the glass fragments? What known samples would you obtain? Assuming that the results for the tests are the same for both knowns and unknowns, what would your conclusion be about the source of the tiny glass fragments?

3. When multiple tests are performed on scientific evidence, they have the most value when they are entirely independent of one another; that is, they are based on entirely different principles. If density and refractive index comparisons are done on known and unknown evidence, does this principle hold? Are density and refractive index completely independent of one another? (Hint: What happens when you try to walk under water compared to walking through air on land?)

BIBLIOGRAPHY

Caddy, B. (Ed.) (2001). *Forensic examination of glass and paint: Analysis and interpretation.* London: Taylor and Francis.

Houck, M.M. (Ed.) (2004). Trace evidence analysis: More cases in mute witnesses. Amerstam: Elsevier.

Miller, E.T. (1982). *Forensic glass comparisons.* In R. Saferstein (Ed.) *Forensic science handbook, vol. 1.* Englewood Cliffs, NJ: Prentice Hall.

Fires and Explosions

CONTENTS

KEY TERMS

Accelerant

Accidental

Activation energy

Active adsorption

Adsorption-elution

Arson

Blast pressure

Bomb seat

Charring

Combustion

Deliberate

Detonation velocity

Electron capture detector

Endothermic

Exothermic

Explosion

Explosive train

Fire tetrahedron

Fire trail

Flame point

Flash point

Flashback

Fragmentation

Headspace

High explosive

High order explosion

Ignition temperature

Incendiary

Incomplete combustion

Initiating (primary) high explosive

Instantaneous combustion (detonation)

Low burning

Low explosive

Low order explosion

Material distortion

Natural

Negative controls

Negative pressure phase

Noninitiating (secondary) high Explosive

Oxidation

Passive adsorption

Point of origin

Selective ion monitoring (SIM)

Solid phase micro-extraction

Soot and smoke staining

Spalling

Spontaneous combustion

Target compound

Thermal (heat) effects

Vapor pressure

Vapor trace analyzer

V-pattern

Weathering

THE CASE: ARSON

A physician has a small, solo medical practice in an area of downtown that is losing population; thus, his practice is starting to suffer. Clearly, he is facing a dire financial predicament. The office consists of a waiting room and a long hall with examining rooms at intervals on each side of the hall. Late one night, a witness sees someone enter the office through the front door with no apparent force being used to gain entry. (S)he was carrying what looked to be a large metal can. A few minutes later the office becomes engulfed in flames. The witness calls the fire department but never sees the person leave the office by the front door. The fire extensively damages the office. A fire trail is seen that travels from the front waiting room, down the hall, and into each of the examining rooms, all of which show extensive damage. The most intense burning appears to have taken place in the front waiting room. The fire scene investigator is suspicious about the cause of the fire, and the insurance company calls for a thorough investigation before paying the doctor's claim on the fire insurance. Ultimately, the physician is arrested and charged with arson. A warrant is issued to search his home and office. A can containing gasoline is recovered from his home and sent to the forensic science lab. Debris from the fire is sent to the forensic science laboratory for analysis for the presence of an accelerant. This debris includes samples from the front office, the hall, and all of the examining rooms. Unburned samples of carpeting are also sent for comparison.

INTRODUCTION

Virtually all fires and explosions are the result of a chemical reaction known as combustion. Combustion is simply the reaction of a fuel with oxygen. The products of a complete combustion are carbon dioxide, water, and energy (see equation 18.1). An example of combustion is the reaction of natural gas (methane) with oxygen. The energy produced by this reaction is used every day by millions of homes and businesses for heat and other purposes.

Equation 18.1
$$CH_4 + 2O_2 \rightarrow CO_2 + 2H_2O + energy$$
Methane   Oxygen   carbon dioxide   water Δ

A chemical reaction that releases energy as one of its products is described as **exothermic**. All combustion reactions are exothermic. A reaction that requires the input of energy in order for the reaction to take place is called **endothermic**. In reality, all chemical reactions require an input of energy to get them started. This is the **activation energy** (see equation 18.2). So an exothermic reaction is one in which energy is produced in excess of the energy put in as activation energy.

Equation 18.2

activation energy $+ CH_4 + 2O_2 \rightarrow CO_2 + 2H_2O +$ excess energy

The most familiar combustion reactions are fires or explosions. Whether a fuel burns or explodes has to do with the nature of the fuel and how close the oxygen and the fuel are to each other during the reaction.

Everyone is familiar with fire, the various types of fires, and the many ways that fires can start. From the forensic science standpoint, it is necessary to know the cause of a fire because, in many cases, fires are deliberately set with criminal intent. These fires are classified as **arson**. Deliberately setting fires constitutes a criminal act and the perpetrators must be punished.

A chemical explosion results from the same type of chemical reaction as a fire. Fires take place by a slow or ordinary combustion wherein the fuel and the oxygen are physically and chemically separated; the oxygen is obtained from the air that surrounds the fuel. The oxygen in air is supplied in a molecular form, O_2 (see equation 18.3).

Equation 18.3   part of activation energy $+ O_2 \rightarrow O + O$

In order for the fire to take place, the oxygen molecules must be broken up into atoms, and they must get close to the fuel molecules. This takes energy and time, so this type of combustion is relatively slow.

If a fuel such as gasoline is confined to a closed space, such as a cylinder in an automobile engine, the fuel and oxygen are compressed, raising their temperature. The spark from the spark plug causes a very rapid combustion to take place. This is called an **explosion**. The difference, then, between an explosion and a fire in this case is the speed of the reaction. Explosions can also be made to occur without confining the fuel. For example, if the fuel and the oxygen are physically mixed and the oxygen is combined with another element instead of itself, it can be made to explode. An example of this is ANFO: ammonium nitrate (NH_4NO_3) and fuel oil. Pellets of ammonium nitrate are coated with fuel oil. The ammonium nitrate supplies the oxygen, which is easily released and very close to the fuel. The resultant combustion is very rapid and is an explosion. ANFO is classified as a **low explosive** because the velocity of the explosion is not as powerful as in the case of more energetic explosives.

It is also possible to combine the oxygen and fuel into a single molecule. In this case, the oxygen and fuel are chemically combined. This is the most advantageous situation for combustion to take place. Such materials undergo **instantaneous combustion** or **detonation**. Of course, the combustion is not really instantaneous; there is always a time lag, but it is even more rapid than in an explosion. The speed of a detonation is referred to as the **detonation velocity**, and it can range from 9000 fps to over 25,000 fps, which works out to be almost 5 miles in one second. These are **high explosives**. Examples include trinitrotoluene (TNT) and nitroglycerine (NG).

FIRES

The analysis of fire scene evidence requires that the investigator be knowledgeable about all aspects of fires. This action provides answer to five fundamental questions about fire. They are listed below.

- What is a fire and what is necessary to start one and to keep it going?
- What are the types of fires?
- How are fire scenes investigated?
- What are fire residues and how do they arise?
- What is the role of the forensic science laboratory in fire scene investigation?

CONDITIONS FOR A FIRE

A simple way of looking at the conditions necessary to have a fire is the **fire tetrahedron**, as shown in Figure 18.1.

The fire tetrahedron depicts the elements that must be present in order to have a fire: a source of heat or energy, a fuel, a source of oxygen, and a chain reaction between the fuel and oxygen. The source of energy is necessary to elevate the fuel and oxygen molecules into an excited state so that they can undergo chemical reactions. This is the aforementioned activation energy. The temperature necessary to do this varies with the fuel and is called the **ignition temperature**. Once this temperature is reached, a fire can continue on a self-sustaining basis. If any of these three elements are eliminated, the fire will not continue. This is the basis of fire extinguishers. Some merely use water to lower the temperature of the fire below the ignition temperature. Others smother the fire, depriving it of oxygen. Still others coat or disperse the fuel.

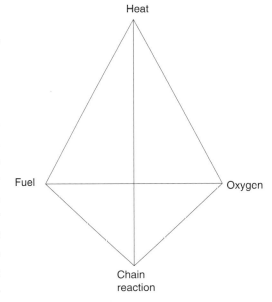

Figure 18.1. *The fire tetrahedron.*

In most cases, the fuel must be a vapor. Often it is necessary to supply energy to change a liquid or solid fuel to a vapor. This would be true, for example, in the case of gasoline. There must be sufficient energy present to convert enough of the liquid gasoline to a vapor to support combustion. This temperature varies with the nature of the fuel and is called the **flash point** of the fuel. Thus, the flash point is the lowest temperature that will allow a liquid to produce a flammable vapor. Even if a fuel is heated to its flash point, it must still be ignited, so a source of ignition is still needed. In addition, the flash point of a fuel will allow combustion but will not sustain it. This requires a higher temperature known as the **flame point**.

Flash Points of Some Common Liquids

Flammable and combustible liquids are divided into the following classes, based on flash points and boiling points. Flammable liquids are defined as those with flash points below 100°F and combustible liquids have flash points at or above 100°F. Flammable and combustible liquids are further subdivided into the following classes:

- **Class IA:** Flash point below 73°F (22.8°C) and boiling point below 100°F (37.8°C). Examples include acetaldehyde, diethyl ether, pentane, ethyl chloride, ethyl mercaptan, hydrocyanic acid, and gasoline.

- **Class IB:** Flash point below 73°F (22.8°C) and boiling point at or above 100°F (37.8°C). Examples include acetone, benzene, carbon disulfide, cyclohexane, ethyl alcohol, heptane, hexane, isopropyl alcohol, methyl alcohol, methyl ethyl ketone, toluene, petroleum ether, acetonitrile, and tetrahydrofuran.
- **Class IC:** Flash point at or above 73°F (22.8°C) and below 100°F (37.8°C). Examples include glacial acetic acid, acetic anhydride, cyclohexanone, and dichloroethylether.
- **Class II:** Flash point at or above 100°F (37°C) and below 140°F (60°C). Examples include kerosene, diesel fuel, hydrazine, and cyclohexanone.
- **Class IIIA:** Flash point at or above 140°F (60°C) and below 200°F (93.4°C). Examples include aniline, cyclohexanol, phenol, o-cresol, naphthalene, nitrobenzene, and p-dichlorobenzene.
- **Class IIIB:** Flash point at or above 200°F (93.4°C). Examples include diethyl sulfate, diethylene glycol, and p-cresol.

Besides flash point and flame point, there are other energy considerations if the fuel is a liquid or solid. Energy must be supplied to convert a solid to a vapor and a liquid to a vapor. All of this heat must be present before the fire will start and continue.

There is a well-known expression: "Where there is smoke, there is fire." Observation of a fire also leads to the opposite observation: "Where there is fire, there is smoke." What is smoke? Smoke occurs when there is **incomplete combustion** in a fire. Equation 18.1 shows the products of the reaction of methane with oxygen when the conditions are right for complete combustion. The limiting factor here is usually the presence of oxygen. If there is not enough oxygen to completely combust the molecules of fuel, then some of the combustion will be incomplete. This will yield products such as carbon particles (soot), unburnt and partially burnt gases. Together these comprise smoke. Sometimes a fire will occur in a building where the oxygen supply is limited, and as the oxygen is used up, more smoke is formed. If this fire is then suddenly ventilated, the increased oxygen will cause an explosive fire. This phenomenon, familiar to all firefighters, is called **flashback**.

Accelerants

Because the fuel must normally be in the vapor phase in order to sustain a fire, many times it is difficult to start a fire or to make one burn quickly. If someone wanted to start a fire in a room where there is only wooden furnishings, it would be very difficult to do with just a match or lighter because there would not be enough heat available to provide the activation energy needed to start and sustain the fire. In such cases, arsonists often turn to the use of **accelerants**, fuels that are easily vaporized and support combustion and are highly exothermic. These liquids are poured around the area that is to be burned and then ignited. They burn easily, and the heat they give off vaporizes the solid materials in the room, causing them to be involved in the fire. From the arsonist's standpoint, the problems with accelerants are that they often leave a residue behind even if burned, and procuring the accelerant may bring unwanted attention from citizens or law enforcement agents. This is probably why gasoline is by far the most common accelerant in the United States. Other common ones are kerosene, charcoal lighters, and some paint thinners. These are common consumer products that many people keep in their homes.

The Bureau of Alcohol, Tobacco and Firearms has developed a classification scheme for common ignitable liquids. All of the common accelerants used in fires today are found in this scheme. Table 18.1 contains an abbreviated list of the common accelerants in this scheme.

TYPES OF FIRES

There are a number of ways of classifying fires depending on the use of the classification system. From the forensic standpoint, there are just

Table 18.1. Ignitable Liquid Classification System.

CLASS #	CLASS NAME	PEAK SPREAD	EXAMPLES
1	Light Petroleum Distillates	C_4–C_{11}	Lighter fluids Camping fuels
2	Gasoline	C_4–C_{12}	Gasoline
3	Medium Petroleum Distillates	C_8–C_{12}	Some charcoal lighters and paint thinners, mineral spirits
4	Kerosene	C_8–C_{17}	#1 Fuel oil, Jet-A Some charcoal lighters and paint thinners
5	Heavy Petroleum Distillates	C_{10}–C_{23}	Diesel fuel #2 Heating oil

three types of fires: **natural**, **accidental**, and **deliberate**. Of course, fire scene investigators are most interested in the last type—deliberate. A fire that is deliberately set is also called an **incendiary** fire. If an incendiary fire is set with the intent to illegally destroy a structure or evidence that may lie within it, or to cover up another crime, then the fire is classified as arson, as was stated earlier in the chapter. Not all incendiary fires are arson. It is possible for someone to set a fire on purpose but not have the intent to destroy something illegally. Brush fires may be set deliberately in a forest to minimize the danger of an uncontrolled accidental fire, but there would be no intent to illegally burn down the forest. Such a fire would be incendiary but not arson.

Natural Fires

Although there are others, the chief natural cause for a fire is lightning. In many places in the United States, there are more lightning fires than all other types combined. Lightning may also be the proximal cause of a fire even if it doesn't actually start the fire itself. For example, lightning can strike the electrical lines leading into a building, causing an overload that in turn may cause an electrical fire. See Figure 18.2 for damage from a lightning fire.

Figure 18.2. *The gaping hole in this house was caused by lightning. When it struck the house, it caused a major fire and extensive damage. Such direct hits on buildings, especially those with lightning rods, are relatively rare. Courtesy of John Lentini.*

Accidental Fires

Accidental fires may arise from any of a number of different sources and may sometimes be difficult to distinguish from deliberate fires. For example, a simple malfunction of a furnace may cause a fire. This would ultimately be ruled an accidental fire. However, if someone deliberately tampered with the furnace that led to a malfunction that in turn caused the fire, this would be arson. If the furnace is badly damaged in the fire, it may not be possible to tell if the malfunction was accidental or deliberate. In many of these cases, the fire scene investigator would call in experts such as electricians, plumbers, or heating contractors to help determine if the appliance were tampered with. Even so, the investigator may not be able to determine if the fire was deliberate.

Deliberate Fires

Determination that a fire has been deliberately started with malicious intent involves a number of steps. First and most important is that all possible natural and accidental causes of the fire must be eliminated. If this is done, then the only other possibility is that the fire was deliberately set.

Once this determination has been made, then the investigator will seek to determine if the fire is arson. Remember that it is possible to have a fire that is started deliberately but without malicious intent to destroy or damage. Determination of arson can be aided by finding residues from an accelerant or other evidence such as multiple points of origin, fire trails, etc. This will be discussed further in the section on fire scene investigation.

Other Ways of Classifying Fires

From the fire scene investigator's standpoint, it is important to not only be able to determine the type of fire as explained in the preceding sections, but to determine the source of ignition or cause of the fire. Each of the more common causes of fire will be discussed in the following sections.

SELF-IGNITION

Self-ignition is sometimes called **spontaneous combustion**. This is a misnomer because this phenomenon is not at all spontaneous. A familiar example of this effect occurs in compost heaps or haystacks. Bacterial action can cause an increase in temperature within the biomass.

Although the maximum temperature achievable from bacterial action is not high enough to cause combustion, internal heating can promote other oxidation processes that can raise the temperature high enough to cause self-ignition. This will occur only if the rate of heat generation exceeds that of heat loss via convection, conduction, or radiation. Certain naturally occurring drying oils such as linseed oil can also undergo self-ignition through oxidation and polymerization, both of which are exothermic. This is the cause of the self-ignition of oily rags containing these materials. Finally, finely divided coal dust can self-ignite owing to its huge surface area where oxidation reactions can take place.

DIRECT IGNITION

Direct ignition is the most basic of all of the causes of fire. It involves direct application of a spark or flame source to the fuel. It can be as simple as using a match or flint or lighter or as complicated as using a time-delay mechanism. In the latter case, a mechanical or chemical or physical process is used to delay the application of the flame or spark so the fire setter is not around when the fire starts. Everything from candles to clocks to mousetraps to exothermic chemical reactions have been used in time-delay mechanisms. This is one of the few types of fire for which determining intent is easy. A time-delay mechanism is almost never used unless the fire is arson.

ELECTRICAL FIRES

Electric appliances, wires, components, and connections are all capable of giving off sparks or overheating given the proper set of conditions. Sometimes this may be due to malfunction, whereas other times it may be a natural part of the functioning of the object. If the proper mixture of fuel and oxygen is present under the right set of conditions, then a spark that contains enough energy to raise the mixture beyond its flash point can cause a fire. An appliance may be rigged to fail and thus overheat or spark so that it can cause a fire. Other times, it may wear out or malfunction because of mistreatment or age. It may be difficult for the fire scene examiner to determine which happened in this case, thus making it difficult to reach a conclusion about the cause of the fire. An example of how electrical overloads may be deliberately created can be seen in the movie *She-Devil*, in which Roseanne Barr, having been jilted by her husband, exacts her revenge by overloading electrical circuits, plugging in many high-current appliances into the same socket. This causes fires that eventually burn down the house.

WEATHER-RELATED FIRES

As mentioned previously, lightning strikes are surprisingly common. They are responsible for hundreds of forest fires annually. Lightning striking a dead tree in a forest can provide more than enough energy to vaporize some of the wood or resin and raise it above its flash point, thus starting a fire. There are also many incidents of lightning striking homes and other buildings. Most tall buildings are protected by lightning rods, but others are not and a lightning strike can cause both structural damage and fire. Finally, it is possible for the sun to cause a fire. Many children have used a magnifying glass to focus the sun's rays on a leaf or grass, causing it to smolder and even burn. Any object, such as a glass or vase, could conceivably act as such a lens and cause a fire, although actual incidents of this are very rare.

MECHANICAL FIRES

Many types of machines can overheat either through misuse or incorrect placement. For example, a shaft or wheel that relies on bearings to reduce friction can overheat if the bearings become damaged or worn. The localized heating can cause a lubricant to ignite and a fire to spread. This may be especially dangerous in cars, where many parts operate at high temperatures and gasoline is present. Overheating of an engine or catalytic converter or muffler can cause a fire if fuel leaks in the wrong place. Even appliances that give off heat and are not given proper ventilation could overheat and cause a fire.

RECOGNITION AND COLLECTION OF FIRE SCENE EVIDENCE

INVESTIGATION OF FIRE SCENES

Fire scenes are among the most difficult places to investigate. In many cases, a building may be completely destroyed by fire, and material from upper floors may have collapsed onto the lower parts of the building. The fire department may have attended the fire and attempted to suppress it. This usually involves using many thousands of gallons of water to cool and douse the fire. The scene is usually dark because the electricity has been cut off. Fire scenes can be exceedingly dangerous.

There may be smoldering embers buried in the debris. The structure of the building may be weakened, so walking in the building can be hazardous. There is also the danger of parts of the building collapsing on the investigators. At the same time it is necessary to avoid disturbing the scene as much as possible so as to not dislodge or contaminate potential evidence.

Fire scenes can be properly investigated only by highly trained fire scene investigators. They must be aware of the causes of fires, burning patterns, how different materials react to fire, the characteristics of points of fire origin, how fires normally proceed through a structure, unusual fire characteristics, and the effects of fire suppression on the scene. In addition, many experienced investigators are knowledgeable about appliances, especially furnaces and hot water heaters, and how they fail or can be tampered with.

The main duty of the fire scene investigator is to determine the cause of the fire. The most important piece of evidence in this determination is the point of origin of the fire. Finding the point of origin is crucial to determination of the cause because it will permit the investigator to determine if a fire was accidental or incendiary.

Proper investigation of a fire scene involves many important processes. The investigator must proceed in an orderly, methodical way and must make accurate, thorough records of the investigation through still or video photography and good note taking. Like many criminal examinations, the investigation of a fire scene starts with a general examination of the scene and gradually focuses on the room or origin and then the point of origin. If the fire scene is a building, then the investigation normally starts with the exterior of the building and works toward the point of origin inside the building. If the fire is outside, such as a forest, then the investigation proceeds from outside the damaged area in toward the area of origin. In some cases fire investigator will make use of specially trained dogs. This is explained in "In More Depth: Arson Dogs."

Point of Entry/Point of Exit

Conventional wisdom is that the most important piece of evidence about a fire is the point of origin or ignition of the fire. As will be shown later, this is crucial evidence, but in the case of a deliberate fire, the perpetrator may have started the fire from within the building, rather than remotely. In this case, there would be a point where this person entered the building—the point of entry. In order to leave the

IN MORE DEPTH:
Arson dogs

One of the most effective tools for searching fire scenes for accelerants is the arson dog. These specially trained dogs can sniff out trace evidence of hydrocarbon accelerants. Dogs have an extremely refined and sensitive sense of smell that can be exploited at a fire scene, where they can smell hydrocarbons even where there has been extensive burning. Arson dogs are used to locate possible sources of accelerants so that they can be collected by the fire scene investigator for laboratory analysis. They enable the search of a fire scene to proceed much faster and more efficiently.

Research has shown that there is a higher level of positive findings for an accelerant by lab scientists in cases in which arson dogs have located the accelerant first. The major disadvantage to the use of arson dogs is that they do not know what they have located. They cannot discriminate between a real accelerant and a hydrocarbon that is part of some object and was released by the fire. If the laboratory cannot confirm the presence of the accelerant, then the dog's reaction cannot be used as evidence that an accelerant was present.

arsonist a route of escape, there may also be a remote point of exit. Of course, the points of entry and exit may be the same. These locations may be very important in determining who started the fire because they will likely be the source of trace evidence left by the perpetrator upon entering or exiting the fire scene. Even if the fire is severe, these locations are often remote from the point of origin and the most severe damage from the fire. The reason is that the perpetrator would not want to be detected when starting the fire, so he or she would start it in an area away from the point of origin. Likewise, the perpetrator would want to provide a safe mode of exit after the fire starts, so the point of entry would be remote from the point of exit. In such cases, the trace evidence may be well preserved.

Physical evidence that is found at points of origin and exit of fire scenes includes fingerprints, shoe prints, hairs and fibers, soil, and even blood. These items can provide circumstantial or direct evidence of the identity of the arsonist and must not be overlooked.

Point of Origin

Certainly, the points of entry and exit can provide important clues about **who** may have started an incendiary fire. The most important evidence about **how** a fire was started is the **point of origin** or ignition. This is the location where the initial ignition took place. If accelerants were used in a fire, they are most likely to be here. As a general rule, the most intense burning and damage are found in the area around the point of origin. Deviations from this can occur for a number of reasons including wind direction, efforts at fire suppression, locations of fuels and/or accelerants, drafts, etc.

Locating the Point of Origin

Generally, a number of characteristics are present at the point of origin of a fire. They include the following:

- **Low burning:** Fires generally start in a low area of a building. Arson fires are seldom started at a high place because the perpetrator may not have a safe point of exit, and the damage will generally not be as great since fires burn in an upward direction.

- **V-patterns:** If the point of origin is near a wall or corner of a room, smoke damage on the wall(s) usually occurs in a "V" shape. This is not universally true, and there may be other areas in the building where V patterns occur.

- **Charring of wood:** The depth of charring of wood depends on the intensity of the heat near the wood and the time of exposure. Often, wood near the point of origin of the fire will have charring to a greater depth than elsewhere in the building, although this is not always true because there may be other points where the fire burns hotter or longer than at the point of origin.

- **Spalling of plaster or concrete:** Spalling is the destruction of a surface due to heat or other factors. In the case of concrete, the spalling may be explosive owing to trapped moisture and expansion of the concrete. Spalling usually occurs most where the heat is most intense.

- **Material distortion:** Metal and glass may melt or distort owing to high heat. Since melting points of many of these materials are well known, such destruction may indicate the approximate temperature of the fire at that point. If the fuels that are supporting the fire are not capable of reaching that temperature during burning, an accelerant may be suspected.

- **Soot and smoke staining:** The amount of soot present in a fire may indicate the point of origin and the direction of travel of the fire. If there are indications that soot was first deposited on a surface and then burned further, this may be good evidence of the point of origin.

The point of origin of a fire can be seen in Figure 18.3.

Figure 18.3. Point of origin of a fire. The point of origin is on the right side of the couch as you look at the picture. Courtesy of James Novak, Novak Investigations.

Indications of an Arson Fire

In order to definitely know the cause of a fire, the investigator must find the fuel that was first ignited to start the fire, the source of the heat that got the fire started, and how the the two came together. Finding the point of origin of a fire is usually a necessary condition of determining if a fire was deliberately set for malicious purposes, but it may not be sufficient. A number of factors **may** be present that would indicate that a fire was arson:

- **The presence of an accelerant:** If an accelerant is present at or near the point of origin of the fire, it usually, but not always, means that the fire is arson. There may be some cases in which a can of gasoline is stored in a building and the fire starts by some other means near the can and it subsequently becomes involved in the fire. The investigator would find the accelerant, but in this case, a finding of arson on that basis alone would not be correct.

- **Elimination of natural or accidental causes of a fire:** This is a necessary condition of determination that a fire is arson. If the point of origin is found and there is no evidence that the fire was started by natural or accidental causes, then the fire must have been incendiary. This may be difficult to determine at times. For example, a furnace may malfunction, leading to a fire. The furnace, being at the point of origin, will be damaged or destroyed. It must be determined if the furnace malfunctioned accidentally or if it was due to deliberate tampering. This determination can be very difficult if there is extensive damage to the furnace.

- **Fire trails:** In order to enable a fire to travel rapidly in particular directions, a **fire trail** may be employed. This can be accomplished by pouring an accelerant along a floor in the desired direction. The result will be an uneven, intense burn along the fire trail. This would not be seen in an accidental or natural fire; thus, fire trails of this nature are always arson. The investigator must be careful not to mistake a fire trail pattern caused by a burnt, plastic carpet runner or wear in a carpet or floor. Usually, such "fire trails" are much more regular in shape than those caused by pouring an accelerant. Figure 18.4 shows a fire trail.

- **Multiple points of origin:** Investigation of a fire such as the case described at the beginning of this chapter would show a fire trail and what would appear to be several points of origin. Pouring some accelerant in each room and then starting multiple fires, one in each room, could also accomplish this. In either case, the investigation would show multiple points of origin, so it would be easy to classify this fire as incendiary. This is usually, but not always, the case with fires that have more than one apparent point of origin. There are exceptions. For example, suppose the electrical wiring inside a building was faulty and perhaps an overload occurred at an outlet. This might lead to overheating of the wiring and insulation at several points inside the walls of the house. Fires might start at some or all of these points of overheating. The investigation of the fire would show multiple points of origin, but the fire would not be arson (unless the original overload was made to happen deliberately). A good fire scene investigator can usually tell if electrical wiring burned because it overheated or because it came in contact with fire that started elsewhere. Aside from such exceptions, multiple points of origin are generally indicative of incendiary fires.

PRESERVATION OF FIRE SCENE EVIDENCE

Trace Evidence

Trace evidence such as hairs and fibers, fingerprints and shoe prints, soils, blood, documents, etc., can be found anywhere at a fire scene.

Figure 18.4. A fire trail. Courtesy of James Novak, Novak Investigations.

Finding the evidence can be more difficult at a fire scene than other types of scenes because of fire suppression activities and the condition of the scene (e.g., dark, smelly, dangerous footing). As a result, the fire scene investigator will usually concentrate his or her efforts to find trace evidence in the areas where a perpetrator would most likely have been at the start of or during the fire. These would be the points of entry, exit, and origin. As explained previously, points of entry and exit tend to be remote from the point of origin of the fire and thus more likely to have trace evidence that has been relatively shielded from the fire. The same precautions need to be obeyed when collecting trace evidence from fire scenes as with any other scene. The additional problem is that contamination with combustion products, fuels, and water makes it more likely that evidence will be adulterated or destroyed at fire scenes.

Accelerant Evidence

The presence of an accelerant such as gasoline, especially around the point of origin of a fire, is generally indicative of the fire being incendiary. This must be put in context, however. A fire scene investigator will make a determination about the type of fire based on observation of all of the factors that have been listed previously, including the presence of an accelerant. Merely finding gasoline at a fire does not mean that the fire was arson or even incendiary. The gasoline could have become involved in the fire incidentally rather than purposefully. On the other hand, the absence of an accelerant where the conditions of the fire scene would seem to indicate its presence does not rule out arson as a possible cause. Depending on the duration and intensity of the fire, suppression efforts, and the nature of the accelerant employed, there may not be enough accelerant residue available to detect. Also, owing to extensive damage from a fire and collapse of debris from higher to lower areas of the building, the fire scene investigator may not be able to collect the residues that may be present. Thus, a fire scene investigator may have ample evidence that a fire was arson, yet there may not be a finding of an accelerant by the laboratory.

If an accelerant was present, it will most likely be found at the point(s) of origin or along fire trails. A number of methods are used to detect possible accelerants. The most sensitive method is the use of hydrocarbon-sniffing dogs. These dogs are specially trained to sniff out the smallest traces of hydrocarbons (the main constituents of fuels and accelerants) at fire scenes even after extensive burning and fire suppression. They are especially useful in finding traces of accelerant in large volumes of fire scene residue. There are also analytical devices that can detect small amounts of hydrocarbons in a large amount of debris. These are essentially stripped-down gas chromatographs with a gas sensor. They are also quite sensitive.

In jurisdictions that do not have dogs or hydrocarbon sniffers, the investigator relies on experience and observation to determine which fire residues are most likely to have trapped accelerant residues. For example, cloth materials such as bedclothes, clothing, carpeting, and upholstery are usually good sources of evidence because accelerants will soak into these materials and, in some cases, may be recovered intact even though there has been considerable fire damage. Hard, non-porous items such as flooring or wall boards are generally poor sources of evidence because accelerants will not soak into them and are thus easily burned or evaporated away. On the other hand, if there are seams in wooden flooring, for example, some of the accelerant may seep inside and be protected from combustion.

When a determination is made of what samples to collect, it is very important that sufficient sample quantity be taken and that **negative controls** are also collected. In general, the investigator cannot collect too much sample, but there are, of course, practical considerations, so the rule is to collect as much sample as is practical and likely to be fruitful. An example of the importance of using negative controls is shown in the case in "In More Depth: The Importance of Controls."

A negative control in this context is essentially the matrix where the accelerant residue is being collected. For example, if a partially burned carpet is suspected to contain accelerant residues, then some of the unburned carpet should be collected far from the burned area to make sure that there is no unburned accelerant. If possible, some of the burned carpet that is known not to contain any accelerant should also be collected. Collecting such evidence is especially impor-

IN MORE DEPTH:
The importance of controls

Many years ago, the author was involved in an arson case where a negative control was critical to figuring out what happened. This case involved a fire at a barracks at a military base in Virginia. Some of the bedding and furniture was piled up in a corner of the barracks and set on fire. There was extensive damage to the barracks. One of the soldiers went AWOL shortly after the fire started, so he was immediately suspected. Evidence brought to the forensic science laboratory included charred remains of the fire at the probable point of origin. Chemical analysis by gas chromatography determined that there was a hydrocarbon-based material present that resembled kerosene in the pattern of peaks, but the positions (retention times) of the peaks were all displaced from where they should be if kerosene was present. The peaks were very strong, so it was clear that there was something present. The criminal investigators were asked to go back to the barracks and retrieve some flooring and wall material that had not been involved in the fire (a negative control). Upon examination of the flooring, the cause of the peaks in the GC trace became clear. There was a large buildup of floor wax on the floor, and this produced a similar peak pattern, albeit displaced to later times as that produced for kerosene. If there was kerosene or any other common accelerant present, it was swamped by the huge amounts of floor wax in the debris.

tant when synthetic textiles are encountered because some of these textiles may interfere with the chemical analysis of the accelerant residues. The analyst must have a control sample of this material to aid in the analysis of the evidence.

Packaging of Evidence

If there is one rule about packaging of fire scene evidence that is to be tested for accelerant residues, it is that **airtight containers must be used**. Accelerants are generally volatile substances; that is, they evaporate easily. If airtight packaging is not used, then some or all of the accelerant may evaporate before the analysis is completed. Some crime laboratories will store all of their fire scene evidence in one cabinet or locker, usually an explosion-proof one. If some of the packages are leaking, then much of the evidence may be contaminated with these vapors. If this takes place over a long enough period of time, false positive conclusions may be the result. In addition, the leaking evidence container may lose so much accelerant to evaporation that, when it is finally examined, little or no accelerant may be found.

Over the years a number of containers have been used for storage of fire debris suspected to contain flammable residues. The most popular have been metal containers. Many fire scene investigators employ unused paint cans. These make excellent containers for fire residues because they are made to be airtight when sealed properly. They can be heated without danger of breaking and are generally rugged and easy to transport. Holes are easily punched in the top for access to the evidence without removing the top. They usually have a protective coating on the inside to retard rust. If this gets scratched, then the inside may rust quickly, especially when wet materials are stored there.

Glass jars have also been used as containers for fire residues. Generally, these are smaller than paint cans, which limit them to small samples. Either new Mason jars or used peanut butter or vegetable jars work. They are fragile and must be heated carefully to avoid breaking. The metal tops are suitable for punching holes in to get access to the evidence.

Bags have also been used for packaging fire residue evidence. Paper bags are totally unsuitable because they can disintegrate if exposed to water and, more importantly, they "breathe"; that is, they will allow hydrocarbon or other flammable vapors to escape. Some plastic bags have also been used, albeit with mixed success. Polyethylene bags are very common, but not suitable. They can be sealed easily but may be reactive to some hydrocarbon vapors and some solvents. They are not impervious to hydrocarbons, and thus, sample will be lost. Contamination of the fire residues inside the bag may occur if the outside of the bag is exposed to hydrocarbons. On the other hand, polyvinylidine bags are quite suitable for storage, being impervious to solvents and flammable materials such as those found in fire residues. They come in various sizes and can accommodate many different types of evidence; however, they do not lend themselves well to hole punching for access to the evidence.

Of course, as with all evidence, once the fire residues have been packaged, they must be sealed and properly labeled. They should be transported to the laboratory as soon as possible so that they can be analyzed with a minimum of sample loss due to evaporation of volatile materials.

ANALYSIS OF FIRE SCENE RESIDUE EVIDENCE

The analysis of fire scene residues includes two major steps. The first is to isolate the accclerant, usually an ignitable liquid or residue from the matrix of charred or unburned material. This usually involves some sort of extraction but can, under certain conditions, be a direct capture. The second step is to determine the nature of the accelerant residue. This most often involves determining the type of ignitable liquid present, such as gasoline, a kerosene-based material, etc. This typically involves gas chromatography. An increasing number of laboratories are using mass spectrometry as a detector for the GC. FTIR can also be used for characterizing these residues.

ISOLATION OF ACCELERANT RESIDUES

A number of methods are commonly used for the isolation of accelerant residues in the analysis of fire scene residues. The one(s) chosen will depend on personal preference, available equipment, and the nature of the exhibits being processed. Following is a list of the more commonly encountered types of exhibits that are encountered:

- **Neat ignitable liquid:** Occasionally, residues will contain some intact, unburned accelerant mixed with debris from the fire. Even if the exhibit is wet, hydrocarbons are not miscible with water, so they would form a separate layer. It may be possible to pour off the liquid from the residue, filter it to remove solid particulates, separate the hydrocarbon from the water, and make a direct injection into the gas chromatograph.

- **Partially burned accelerants:** A much more common occurrence in fire scene residues is accelerants that have been partially burned. The major change that these substances undergo is evaporation of the most volatile components, leaving the higher-boiling components behind. Usually, such exhibits must be extracted from the matrix in order to be concentrated.

- **Nearly completely burned accelerants:** If an accelerant has been subjected to extreme heat for a significant period, nearly all of the substances present will evaporate or burn. The best that can be hoped for in such cases is that there will be some non-volatile residue left that can be extracted. Identification of these residues can be difficult owing to a lack of characteristic chromatographic information.

Four methods and some variants are used for isolation of accelerant residues. They are described in the following sections.

Headspace Methods

Consider fire residue containing some small amount of liquid accelerant mixed in. This residue is put in a sealed metal container. Some of the accelerant will vaporize, while the rest will remain a liquid. The amount of the liquid that becomes a vapor depends on the **vapor pressure** of the substance and the temperature. The higher the temperature, the greater percentage of vapor there will be. Eventually, equilibrium will be established. Once the equilibrium is established, then some of the vapor above the matrix, the **headspace**, can be sampled with a gas-tight syringe and injected into a gas chromatograph. The amount of heating that the container can be subjected to is limited. If too much heat is applied, then the ensuing increase in pressure in the container can cause the top to come off a can or the glass to break in a jar. Typically, a container will be heated to no more than about 60°C. Figure 18.5 is a diagram of a can showing the headspace.

Adsorption Methods

Although headspace methods are quite useful when volatile substances are present in the accelerant residue, the equilibrium condition and

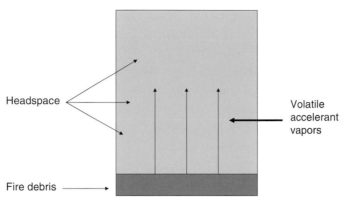

Headspace

Volatile accelerant vapors

Fire debris

Figure 18.5. *The headspace in a can. The fire debris is at the bottom. The headspace is the air layer on top of this that contains accelerant vapors.*

the limitations on heating can be drawbacks to the method. Another approach to isolating and concentrating accelerant residues is using adsorption methods. This group of methods utilizes the ability of charcoal (finely divided pure carbon) or synthetic materials such as Tenax to adsorb large quantities of hydrocarbons onto its surface. There are two variants of the adsorption method. The first is **passive adsorption**. In this variation, a small container of charcoal or Tenax or a plastic strip coated with one of them is placed or suspended inside the container. It can either be left overnight at room temperature or for a shorter time while being heated. Sometimes the fire scene investigator will put a charcoal strip into the container with the evidence at the crime scene. This means that the container won't have to be opened to insert the strip at the lab, thus minimizing the loss of accelerant vapors.

In **active adsorption**, sometimes called **adsorption-elution**, two tubes containing charcoal or Tenax are inserted partway into the container through holes in the top. Then air is pumped through one of the tubes into the container. This causes air to flow from inside the tube out through the other tube. As the container is heated, more accelerant evaporates into the headspace. It is swept through the outlet tube along with the air and is trapped or adsorbed onto the charcoal or Tenax. This upsets the vapor-liquid equilibrium in the container, and the consequence is for more of the liquid accelerant residue to evaporate. This continues until there is no more accelerant in the container; it has all been trapped in the outlet tube. In some laboratories a vacuum is applied to one of the tubes, drawing in air from the outside through the other one. The accelerant vapors are trapped in the tube

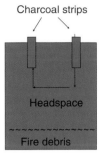

Charcoal strips

Headspace

Fire debris

Figure 18.6.
Absorption-elution.
The paint can has
two tubes that can trap
accelerants. The
vacuum pulls air
through one tube
into the can. Any
contaminants in the air
that might interfere with
the analysis are trapped
in the first tube. The
headspace vapors are
pulled out through the
other tube and trapped
in the charcoal in the
tube.

where the vacuum is applied. The result is the same. See Figure 18.6 for a diagram showing absorption-elution.

Once the accelerant has been adsorbed onto the charcoal or Tenax, then it is eluted off using a suitable solvent. Carbon disulfide (CS_2) has been used for many years for this purpose, but it is toxic and highly flammable. Other solvents have been tried, including butane and pentane, but they are less satisfactory because they are also constituents of many accelerants.

Another variant of the adsorption methods is **solid phase microextraction**. In this method, a fiber made from fused silica is coated with an adsorbent such as charcoal or Tenax. This is inserted into the heated fire residue container. After adsorption is complete, the fiber can be inserted directly into the inlet of a gas chromatograph, where the high heat of the injector zone rapidly elutes the accelerant into the mobile phase stream for analysis. The advantages of this technique are extreme sensitivity and removing the necessity of a separate elution step.

Solvent Extraction

Solvent extraction is a very simple and sensitive technique, usable with a wide range of accelerants. The evidence container is opened, and a small quantity (depending on the amount of debris in the container) of a suitable solvent is added. Carbon disulfide is the most popular solvent for this process. The solvent is then poured off and filtered and then evaporated to a small volume, leaving behind the accelerant residue. This solution can then be introduced into a gas chromatograph.

Disadvantages of solvent extraction are, first, that the solvent will also dissolve unwanted pyrolysis products, matrix materials and other substances, some of which may interfere with the subsequent analysis; and second, that the evaporation of the solvent may also cause evaporation of some of the volatile components of the accelerant residues.

Steam Distillation

Steam distillation is the oldest technique for isolation of accelerant residues. Some of the accelerant residue is put in a distillation apparatus with some water, which is then boiled and distilled. The steam will heat and carry over accelerant residues. Those that are immiscible with water will form a layer on top of the distilled water. If water-soluble residues are suspected to be present, then the first aliquot of water must be collected and analyzed. Steam distillation is not very sensitive, and relatively large quantities of matrix are need. This technique is not as subject to contamination interferences as is solvent extraction, but it does favor high boiling fractions. It is also the most complicated to run.

ANALYSIS OF FIRE SCENE ACCELERANT RESIDUES BY GAS CHROMATOGRAPHY

GC is almost universally employed in crime labs for the analysis of fire scene accelerant residues. Today, most laboratories use capillary GC columns for increased sensitivity and efficiency. Increasingly, mass spectrometry has been mated with the GC to identify certain components of the residues.

All of the hydrocarbon accelerants including gasoline, kerosene-based materials, fuel oils, and other consumer products have many components. Gasoline, for example, is made up of more than 300 substances. In basic GC analysis of accelerant residues, the result will be a pattern of peaks that is characteristic of the accelerant type. Thus, an analyst can identify an accelerant as gasoline or as a kerosene-based product, for example, but it is generally not possible to identify a specific product or manufacturer by this method.

The key to effective analysis of accelerants by GC is to have a comprehensive library of chromatograms that are obtained preferably on the same instrument as the analysis of unknowns, or at least taken under the same conditions. This library would include not only the various products, brands, and types of accelerants, but also their various forms. For instance, an accelerant may appear in its neat, unburned form or partially burned or almost totally consumed. The chromatograms of these materials will be quite different, and it would be difficult to tell what is present unless there are good standards for comparison. Figures 18.7A and 7B show chromatograms of pure gasoline and kerosene. Figure 18.8 shows the chromatogram of gasoline headspace.

Mass spectrometry has added flexibility and refinement to GC analysis of fire scene evidence. Individual components of residues can be unequivocally identified. In addition, mass spectrometry provides some features that make it especially valuable in identifying ignitable liquids in the presence of contaminants or in mixtures containing multiple ignitable liquids. The first of these is called **selective ion monitoring** (SIM), whereby the mass spectrometer looks for particular ions that are characteristic of particular types of flammables. The other enhancement is called **target compound** analysis. In this technique, a

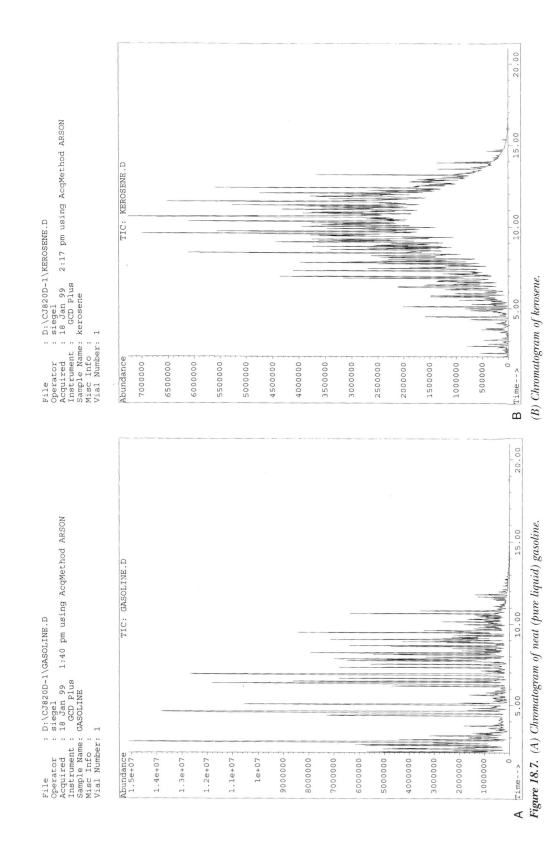

Figure 18.7. (A) *Chromatogram of neat (pure liquid) gasoline.* (B) *Chromatogram of kerosene.*

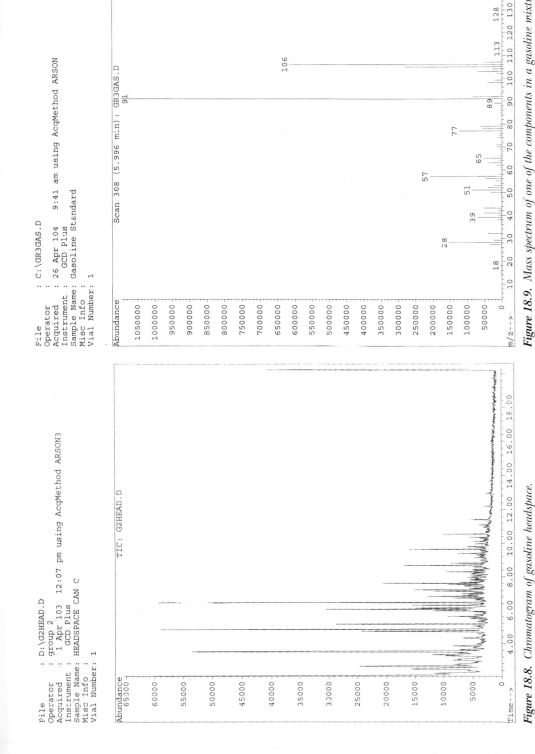

Figure 18.8. Chromatogram of gasoline headspace.

Figure 18.9. Mass spectrum of one of the components in a gasoline mixture.

profile of compounds that are present in each type of accelerant, such as gasoline, is monitored by the mass spectrometer. These compounds can be easily identified even in complex mixtures. The mass spectra of individual components of a material such as gasoline can be easily displayed. The mass spectrum of one of the compounds found in gasoline is shown in Figure 18.9.

INTERPRETATION AND ASSOCIATION

In a fire scene investigation there are two major goals: determining the type of fire (e.g., accidental, natural, incendiary) and, if the fire was deliberately started, who did it. In most cases it is not particularly difficult to determine the type of fire. It is much more difficult to determine who committed the crime. Of course, the presence of evidence such as fingerprints, DNA, or trace evidence can be quite useful in determining the identity of the perpetrator. In the absence of such evidence, there is not much else that can provide identification.

Consider the case at the beginning of this chapter. There is certainly circumstantial evidence of the doctor's involvement and no other readily identifiable suspects. If there is no direct evidence of identification, what else is left? The facts indicate that the perpetrator brought a can of accelerant into the office and spread its contents around. Suppose the can had a few drops of the ignitable liquid in it that was used to start the fire? Would it be possible to first identify the accelerant at the fire and then associate it with the remains in the can? This would provide a link between the fire, the can, and perhaps the perpetrator. Unfortunately, it is not possible to individualize such fire residues to a particular source.

When the forensic scientist extracts some of the accelerant residue from the matrix using the methods presented in the preceding sections, the resulting chromatogram seldom looks exactly like the standards made from pure samples of various ignitable liquids. Remember that the accelerant has been in a fire, and heat may have evaporated some of the more volatile substances away. There may be contamination by the matrix or other materials present. If the scientist is to report that a class or type of accelerant is present, then there must be a clear majority of characteristic peaks present with few, if any, unexplainable peaks. For example, examine the chromatogram of gasoline liquid in Figure 18.8. If a real fire residue sample had some of the early peaks missing, that would be explainable by evaporation caused by heat. But if all of the peaks were present at the beginning and end of the chromatogram but some were missing in the middle, then that would be a circumstance that might necessitate a finding of inconclusive or even negative in the report. A negative finding may also be reached even if there are some indications that an accelerant was used. This can occur when there is not enough material to give a good quality chromatogram or when there is too much contamination to identify the necessary peaks. The term **weathering** is used to describe the degradation of an accelerant due to heat or other environmental factors.

It is also important to note that a negative finding in a report does not preclude the presence of an accelerant at a fire. As was pointed out previously, there are several reasons why an accelerant might not be found even if it was actually used. It could have been burned to such an extent that there is not enough left to detect, or perhaps the evidence that contained the accelerant residues may have been overlooked. Fire scene investigators always try to take these factors into account when reaching conclusions about fire causation.

BACK TO THE CASE: ARSON

The case described at the beginning of this chapter is fairly typical of arson fires. There is ample evidence that the fire was deliberately started. The unusual fire trail patterns into each room are indisputable. The difficulty in investigating a fire such as this is identifying the perpetrator. There is circumstantial evidence that the doctor was involved. His dire financial condition and the opportunity to collect insurance money provide evidence of motive. There is also the issue of who else would benefit from targeting this particular office to destroy. The finding of a container of gasoline in the doctor's home is also of some evidentiary value, especially if the contents of the can are similar to the accelerant residues obtained from the fire scene. It is also true, however, that many people keep gasoline and other ignitable liquids in their homes. A thorough search of the fire scene might pick up some evidence that the doctor was there, but that must also be tempered by the fact that he worked in that office and his fingerprints and trace evidence of his presence would be expected to be present everywhere in the office.

THE CASE: EXPLOSIVES

In the Oklahoma City bombing of the Edward Murrah building, a truck containing hundreds of pounds of ANFO, ammonium nitrate and fuel oil, was parked on the street in front of the building. These explosives were detonated remotely by a device that created a spark or electrical arc that caused detonation of the explosive. Since the explosives were confined to the inside of the truck, tremendous pressure was built up from the superheated gaseous products of the explosion. These gases finally burst through the walls of the truck and were hurled at thousands of feet per second into the building. The resulting shock and heat did extensive damage to the building and killed many people.

EXPLOSIVES AND EXPLOSIONS

In the following sections, the nature of the chemical reactions that result in explosions, the different types of explosives, and how they are detected and identified will be explored.

EFFECTS OF EXPLOSIONS

The effects of an explosion can all be explained by understanding what happens when an explosion or detonation takes place. Solid and/or liquid fuels combine with oxygen to form gaseous products such as carbon dioxide and other products that are converted to gases from the heat of the combustion. These very hot gases expand rapidly away from the origin of the explosion (the **bomb seat**). These rapidly moving gases create three primary effects: **blast pressure**, **fragmentation**, and **thermal** or **heat effects**.

Blast Pressure

Escaping gases can travel as much as 8,000 miles per hour and exert hundreds of tons per square inch of pressure. This compresses the gases and the surrounding air. The wave that is created by this blast will shatter anything that gets in its way. The damage decreases with distance as the wave loses power. As the blast wave travels away from the bomb seat, it creates a partial vacuum because the air itself has been displaced. When the blast wave dissipates, the vacuum must be filled. The compressed air and gases now rush back toward the bomb seat. This causes another blast effect, the **negative pressure phase**. This is not as powerful as the positive pressure blast phase, but it is capable of doing additional serious damage to objects that have already been damaged by the initial blast. The two phases of a blast can be seen in Figure 18.10.

Fragmentation Effects

Fragmentation damage from a bomb can occur in several ways. First, the bomb casing itself can shatter, and the pieces can be propelled away from the bomb seat with great force. Second, the bomber may wrap nails or other pieces of metal around the bomb to create shrapnel that will cause fragmentation damage. Finally, the blast may break up objects in its way that may also fragment and be propelled.

Thermal Effects

Thermal or temperature effects are generally the least damaging of the effects of an explosion. At the instant of detonation a large ball of fire or flash is produced at the bomb seat. This will be very hot and very brief if a high explosive is used and will be longer in duration but not as hot in the case of low explosives. The flash usually dies very quickly, and no further effects will be seen unless there is combustible material nearby the blast, in which case secondary fires may be ignited.

TYPES OF EXPLOSIVES

Explosives are commonly categorized by the velocity of the explosion or detonation. This gives rise to two types of explosives: **low explosives** and **high explosives**.

LOW EXPLOSIVES

By definition, low explosives have detonation velocities below 3,280 fps. In low explosives the oxygen is physically mixed with the fuel, and the explosion takes place at a slower rate than would be the case if the fuel and oxygen were chemically combined. As a result, the main effect of low explosives is to push rather than to shatter. Such explosives are

***Figure 18.10.** Positive and negative phases of blast pressure during an explosion.*

Positive Pressure Phase of an Explosion

Negative Pressure Phase of an Explosion

often used as propellants in guns that fire bullets or shot pellets. Thus smokeless powder is used in weapons to propel a bullet away from the cartridge and out of the barrel of the gun without causing the weapon to blow apart. Low explosives are also often used in blasting operations when it is desired to push earth or other material out of the way.

Low explosives such as black powder are also used in pipe bombs. Here, the explosive is confined inside a metal or plastic pipe. As detonation takes place, pressure builds up inside the bomb until it shatters. Low explosives can be easily detonated using a flame, a spark, or chemicals such as acids.

Some examples of low explosives are black powder, smokeless powder and, as previously mentioned, ANFO. Black powder is a mixture of potassium nitrate, charcoal (carbon), and sulfur. Smokeless powder can contain nitrocellulose and/or nitroglycerine. It occurs in pellets of varying shapes and sizes. Some so-called low explosives such as urea nitrate can still cause considerable damage. An example is the World Trade Center bombing discussed in "In More Depth: The World Trade Center Bombing."

IN MORE DEPTH
The World Trade Center bombing

On February 26, 1993, just after noon a homemade explosive device was detonated on the second level of the parking garage in the World Trade Center in New York. The bomb was made from a commercial fertilizer known as urea nitrate. It weighed about 1,400 lbs. The initiator for the explosive was lead azide. This was ignited using a burning fuse with a 20-minute delay. Also incorporated into the bomb were three pressurized tanks of hydrogen gas, which is extremely flammable. This bomb had been placed in the cargo area of a Ford van, which was then left under Building One of the Trade Center.

The explosion produced a huge crater that extended five floors deep in the garage and was approximately 150 feet in diameter. The garage structure was mainly steel-reinforced concrete and was at least one foot thick in most places.

Six people were killed by the blast and more than a thousand were injured. The entire World Trade Center complex was evacuated, more than 50,000 people.

The crime scene was one of complete devastation. In addition to the thousands of tons of rubble, there was no lighting to see by and several water and sewer lines were ruptured, pouring millions of gallons of water and sewage into the crater. It took more than a week for 300 law enforcement agents from around the country to sift through the rubble and piece together the cause of the explosion.

The damage done by the explosive was characteristic of a heaving or pushing rather than shattering. It was surmised that the escaping gases had a velocity of around 15,000 fps. There are a number of explosives that have these characteristics, and some fertilizer-based devices are among them.

The FBI and ATF sent chemists to the scene, and a makeshift lab was set up. Fragments were found that led to the identification of the van that carried the explosive and the explosive itself. Ultimately, four men were convicted of this terrorist bombing.

HIGH EXPLOSIVES

High explosives have detonation rates above 3,280 fps. Some dynamites, for example, have rates as low as 6,000 fps, whereas some military explosives approach 28,000 fps. These explosives are designed to shatter objects and destroy them. The mechanism by which high explosives detonate is quite different than for low explosives. The latter are generally granular, and at ignition, the burning travels from one particle to the next. Most high explosives, on the other hand, require a severe shock to get them to detonate. This can be accomplished using a blasting cap or a primary or **initiating** explosive. When these explode, they create a strong shock wave that shatters the chemical bonds that hold molecules of fuel and oxygen together. Detonation takes place and travels from molecule to molecule, picking up velocity along the way so that the result is practically instantaneous detonation.

There are two types of high explosives: **initiating (primary)** and **non-initiating (secondary).**

Initiating High Explosives

Initiating high explosives are usually very powerful and very sensitive. Even the slightest shock or spark can be enough to cause detonation. For this reason, they are used only in very small quantities, usually to detonate less sensitive explosives in **explosive trains**. Examples are PETN, RDX, and mercury fulminate. Figure 18.11 shows chemical structures of some common initiating high explosives.

Non-initiating High Explosives

Non-initiating high explosives are not sensitive, and it usually takes a good deal of effort to cause detonation. These explosives can be easily transported and used without fear of accidental detonation. They are generally so insensitive that it takes a major shock, such as that supplied by a nearby initiating explosive in an explosive train, to get them

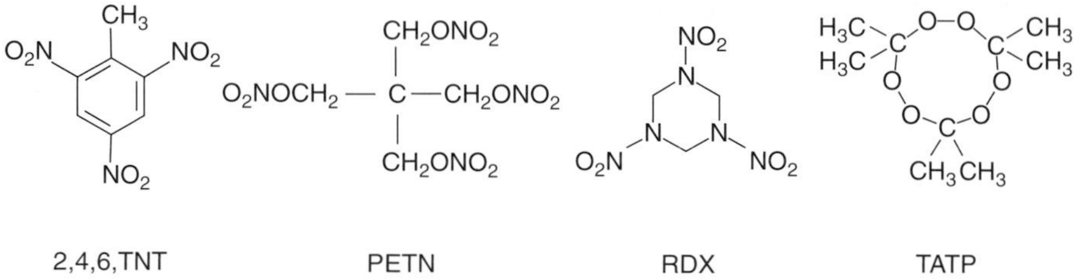

Figure 18.11.
Structures of TNT, PETN, RDX, and TATP.

2,4,6,TNT   PETN   RDX   TATP

to detonate. Examples include many dynamites and the military explosive C4.

HIGH AND LOW ORDER EXPLOSIONS

Two other terms that describe explosions sometimes cause confusion. **Low** and **high order explosions** have nothing to do with a type of explosive, but instead describe the efficiency of a particular explosion. A high order explosion is one that occurs at or near its maximum theoretical detonation velocity. It is the explosion that you get if everything works out right!

A low order explosion, on the other hand, is one that takes place at less than optimal efficiency. This can be due to any of a number of factors, which include

- Old, out-of-date explosive
- Explosive that is subject to excessive moisture or humidity
- Improperly constructed explosive device
- Improper placement of the device

EXPLOSIVE TRAINS

Sometimes it is necessary to use one explosion to set off another. Other times it may be desirable to have a series of explosions take place in a particular order. These situations may require the use of an **explosive train**. Explosive trains may contain as few as two steps or up to four or more. They are classified as low or high, depending on whether the final explosive in the train is a high or low explosive. Examples of low and high explosive trains will be considered in the following sections.

LOW EXPLOSIVE TRAINS

Low explosive trains are usually two-step trains. One example would be a pipe bomb wherein a fuse made of black powder is used to detonate smokeless powder inside a pipe bomb. This would be classified as a low explosive train because the final explosive, the smokeless powder, is a low explosive.

*Figure 18.12. An
explosive train.*

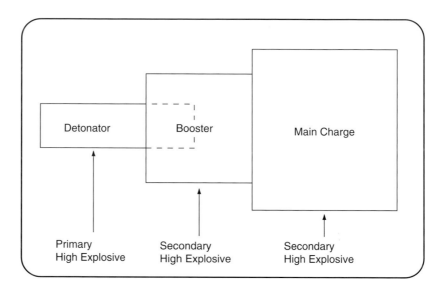

HIGH EXPLOSIVE TRAINS

In high explosive trains, the final explosive is usually a secondary explosive. The detonator may be a blasting cap or other suitable primary explosive. In between there may be other secondary high explosives that act as boosters. See Figure 18.12 for an example of an explosive train.

ANALYSIS OF EXPLOSIVES

Two major types of explosive residues are encountered in bombs: undetonated explosive and exploded residues that are products or side products of the explosion. In addition, the bomb seat area may yield parts of the device that was used to hold the explosive and set it off. Those cases in which undetonated explosive is found are the easiest to analyze because the exact explosive can be identified. In those cases in which no exploded residue can be located, the analytical situation is more complex because it may be difficult to distinguish explosive residue from materials that are present in the normal environment. For example, in the case of black powder, various nitrate-containing compounds would be the products of the explosion. Many of these are found naturally in soil, and it may be difficult to determine their origin.

This requires care in interpretation of a finding of explosive residues where no intact explosive is found.

Finding bomb device parts can be a huge advantage in reconstructing the explosion. From surprisingly few pieces of device, an explosives examiner may be able to tell the type of device, how it was detonated, and perhaps what part of the world it came from and who made it. This capability has obvious advantages in terrorism situations.

THE VAPOR TRACE ANALYZER

To aid sifting through evidence to find explosive residues from the scene, a **vapor trace analyzer** or other, similar detector is often employed. The VTA is a specialized gas chromatograph that is optimized for explosives. The VTA utilizes a type of head space analysis wherein vapors from the explosive residue are introduced into a collector, and the polar explosive residues are isolated and chromatographed. A special detector, called an **electron capture detector**, is used to detect the presence of explosive residues.

VISUAL EXAMINATION

In order to analyze explosive residues, they must be isolated from the matrix in which they are found at the crime scene. The best way to do this is to manually remove them under a low-power microscope. This often tedious and time-consuming activity can pay dividends later. Isolated residues can be more easily characterized without the analyst's having to be concerned about the presence of impurities. More important, the forensic scientist can testify that actual particles of explosive were recovered from the debris and were chemically identified.

If no large particles of undetonated explosive can be isolated, then it may be necessary to dissolve microscopic particles with a suitable solvent and remove them from the debris. This is a lot faster than manual sifting but suffers from several disadvantages. First, other substances may also dissolve, so the explosive residue is not really being purified. Second, many explosives have components that are ionic, such as potassium nitrate (KNO_3). When this material is dissolved, it dissociates into potassium and nitrate ions. The resulting analysis indicates that the ions were present but not the actual compound. It could be argued that these ions could have arisen from any of a number of sources, and this doesn't prove that an explosive is present.

INSTRUMENTAL AND OTHER METHODS OF ANALYSIS

Once an explosive residue has been isolated from bomb debris, then it should be further characterized and, ultimately, identified. A variety of techniques can be used depending on the amount and type of explosive available. Thin layer chromatography and infrared spectrophotometry are widely used for this purpose.

Thin Layer Chromatography

Many explosives can be conveniently separated by thin layer chromatography. They include many low and high explosives including smokeless powder, dynamites, TNT, etc. Known explosives are normally spotted along with the unknowns so that the identifications can be more certain. A number of specific stationary and mobile phase combinations have been used for various classes of explosives. Several visualizing reagents have also been employed. For example, **Greiss** reagent has been employed extensively because it will color most nitrate-containing species bright red. Most explosives have nitro groups, making Greiss reagent very versatile.

High Performance Liquid Chromatography

HPLC is also useful for the analysis of explosive residues. It has advantages over TLC in that it can be used for quantitative analysis, should this be called for, and it doesn't require the use of visualizing reagents. It is preferred over GC because high concentrations of explosives can cause problems with the high temperatures used in GC.

Capillary Electrophoresis

CE is similar in some ways to HPLC but operates on somewhat different principles. It is a relatively new technique that is now being exploited to a greater extent in forensic science. It has some advantages over HPLC, the major one being the amount of sample required. CE is very sensitive, and only tiny amounts of material are needed for analysis. This can be a real advantage when only trace amounts of explosive residue are present.

Infrared Spectrophotometry

Because chromatographic methods provide only a tentative identification, a confirmatory test, IR, is often used. A variety of sample types can be used for IR including solids, liquids, mulls, and solutions.

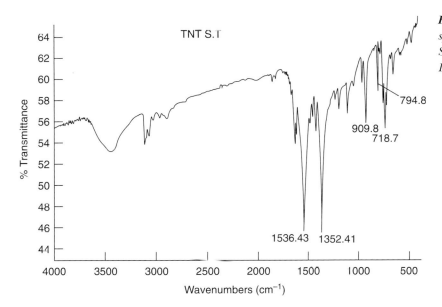

Figure 18.13. *Infrared spectrum of TNT. Source: ATF National Laboratory.*

Similar compounds such as TNT and DNT (dintrotoluene) can be differentiated by IR. See Figure 18.13 for the infrared spectrum of TNT.

BACK TO THE CASE: EXPLOSIVES

The Murrah building explosion, discussed earlier in this chapter, and the World Trade Center bombing, also discussed, are examples of terrorist bombings using home-made devices. There are several similarities as well as differences in the two crimes. Both involved large amounts of explosive made of materials that are readily available to a determined terrorist. Both involved fertilizer and a fuel. In the Oklahoma City bombing, it was ammonium nitrate fertilizer pellets coated with fuel oil. The Murrah building was much smaller than the WTC and didn't have the extensive underground parking levels, so planting the explosives underground in a truck may not have been a viable option. The evident plan of the terrorists who detonated the bomb in the WTC was to weaken the foundation of the building in the hope that it would collapse. In Oklahoma City, the truck was parked in front of the building, and the blast caused extensive damage above ground.

In both cases, the investigators had to plow and sift through thousands of tons of rubble in order to find evidence of the crime. This job was somewhat easier in the case of the Murrah building because the vehicle containing the explosive was parked away from the building, and the investigators did not have to uncover it from the rubble. In both cases, no direct evidence was available to identify the perpetrators. This is very common in major explosions. The perpetrators are identified by collateral evidence such as tracing the rental vehicle back to the renters.

Finding evidence of unexploded residue is also very difficult to do in major cases such as this. Painstaking sifting through tons of rubble is often necessary. This can be a tedious and dangerous job and must be done while being aware of the possible presence of bodies or parts of bodies that must also be preserved.

CHAPTER SUMMARY

Fires and explosions are both the result of combustion reactions, where a fuel and oxygen react, sometimes violently and instantaneously to give off large amounts of energy. In fires, the combustion takes place relatively slowly because the fuel and the oxygen are separated, the oxygen being supplied by the air surrounding the fuel. The oxygen is in molecular form, and these bonds must be broken before the oxygen can be used. Also, the fuel must be a vapor before it will burn. This also takes time and energy. A fire can be made to simulate an explosion by confining it to a closed space, as is the case with the combustion of gasoline in the cylinders of an internal combustion engine.

In explosions, the fuel and oxygen are more intimately mixed. In low explosions, the oxygen is present in molecules such as potassium nitrate. This oxygen is more readily available than in molecular oxygen form. In high explosions, the oxygen is actually chemically incorporated into the fuel, and combustion is practically instantaneous.

In order to determine if a fire is arson (deliberately set), all accidental and natural causes of the fire must be eliminated. This is done by finding the point of origin of the fire and looking for particular characteristics. Many arson fires are set using an accelerant, and one hopes, some of these residues will also be recovered. The nature of the accelerant is determined by gas chromatography after the accelerant is isolated by extraction.

In the investigation of an explosion, the point of origin, or bomb seat, must be located. This is where residues of the explosive are most likely to be found as well as parts of the device used to set off the bomb.

Test Your Knowledge

1. What is the fire tetrahedron? How is it important in explaining the elements necessary to have a fire?

2. What is an oxidation reaction? Give an example.

3. What is a combustion reaction? Give an example.

4. How do fire extinguishers work in general?

5. What does "exothermic" mean? Give an example of an exothermic reaction.

6. What is a flashpoint?

7. What is an accelerant? Give an example.

8. One of the possible types of fire is accidental. What are the others?

9. In order to determine that a fire is arson, what must a fire scene investigator be able to do?

10. Why is finding the point of origin of a fire so important in determining the cause of the fire?

11. What is the crucial difference between a fire and an explosion?

12. How is a detonation defined? Why do some explosives detonate, whereas others do not?

13. What is an initiating explosive? A non-initiating explosive?

14. What is an explosive train? When is it used?

15. What is the difference between a high explosive and a high order explosion?

16. What characteristics of fuel and oxygen give rise to a high explosive?

17. What is smokeless powder? What type of explosive is this and where is it used?

18. How are explosive residues collected? Where is the most important place to look?

19. Why is it so important to be able to recover intact residues of unexploded material?

20. What is a Greiss reagent? What types of explosives is it used to help analyze?

Consider This . . .

1. Both fires and explosions involve the same types of chemical reactions. They can be differentiated by the velocity of the gases that escape from the point of origin. Explain how the arrangements of the various reactants help dictate the power of the reactions and thus the velocity of escaping gases.

2. In the explosion that took place at the Murrah building, it was determined that ANFO was the explosive used. Why would someone use this as an explosive? Why wouldn't the perpetrators have used dynamite or nitroglycerine? What could have been done to the truck containing the explosive to maximize the damage to the building?

3. The most popular methods of concentrating accelerant residues from a fire involve either active or passive adsorption onto charcoal. Explain why these methods are preferred over headspace, distillation, or solvent extraction methods.

BIBLIOGRAPHY

DeHaan, J. (1991). *Kirk's fire investigation (3rd ed.).* Englewood Cliffs, NJ: Prentice-Hall.

Redsiker, D., & O'Connor, J. (1997). *Practical fire and arson investigation (2nd ed.).* New York: CRC Press.

Siegel, J. (Ed.) (2000). Fire investigation. In *Encyclopedia of Forensic Science.* London: Academic Press.

Urbanski, T. (1964). *Chemistry and technology of explosives, Vols 1–3.* Oxford: Pergamon Press.

Yinon, J., & Zitrin, S. (1991). *The analysis of explosives.* Oxford: Pergamon Press.

Yinon, J., & Zitrin, S. (1993). *Modern methods and applications in analysis of explosives.* Chichester: John Wiley.

PART FIVE

Physical Sciences

Friction Ridge Examination

KEY TERMS

Accidental

Amido Black

Aqueous Amido Black

Aqueous Leucocrystal Violet

Arches

Automated fingerprint identification system (AFIS)

Bertillionage

Biometrics

Central pocket loop

Core

Delta

Dermal layer

DFO

Double loop

Fingerprint powder

Gentian Crystal Violet

Glue fuming

Individualization

Iodine

Latent prints

Level 1 detail

Level 2 detail

Level 3 detail

Loop

Minutiae

Ninhydrin

Partial prints

Patent prints

Physical developer

Plain arch

Plain whorl

Point counting standard

Primary classification

Primary friction ridges

Radial loop

Secondary friction ridges

Small particle reagent

Sudan Black

Tented arch

Type lines

Ulnar loop

Vacuum metal deposition

Whorls

INTRODUCTION

From the early days of complicated body measurements to today's sophisticated biometric devices, the identification of individuals by their bodies has been a mainstay of government and law enforcement. Fingerprints are the current leader in identification markers, especially in forensic science. Recent court challenges, however, have brought fingerprinting into the spotlight again and may force some changes if it is considered to be a science.

THE NATURAL BORN CRIMINAL

Cesare Lombroso's theory of *l'umo delinquente*—the criminal man—influenced the entire history of criminal identification and criminology. Lombroso, an Italian physician in the late 1800s, espoused the idea that criminals "are evolutionary throwbacks in our midst . . . [and] these people are innately driven to act as a normal ape or savage would, but such behavior is considered criminal in our civilized society" (Gould, 1996; p. 153). He maintained that criminals could be identified because of the unattractive characteristics they had, their external features reflecting their internal aberrations. While normal "civilized" people may occasionally commit crimes, the natural-born criminal could not escape his or her mark.

Lombroso's comparison of criminals to apes made those of the lower classes and "foreigners" most similar to criminals: The "nature" of criminals was reflected in the structure of Lombroso's society. His list of criminal "traits" sounds laughable to us today: Criminals were said to have large jaws, large faces, long arms, low and narrow foreheads, large ears, excess hair, darker skin, insensitivity to pain, and an inability to blush! It's easy to see the racial stereotypes of Lombroso's description, how society's "others" were automatically identified as criminal.

The idea of identifying "natural born killers" caught the attention of many anthropologists and law enforcement officials in the late 1800s, and even though Lombroso's work was later repudiated (many of his assertions were not supported by objective data), it spawned a great deal of activity in the search for real, measurable traits that would assist the police in identifying criminals. One of them, a French police clerk named Alphonse Bertillion (Ber-TEE-yin), devised a complex system of anthropometric measurements, photographs, and a detailed description (what he called a *portrait parlê*) in 1883; it was later to be called **Bertillionage**, after its inventor. At that time, the body was considered to be constant and, as Lombroso's work then maintained, reflective of one's inner nature. Bertillion's system was devised to quantify the body; by his method, Bertillion hoped to identify criminals as they were arrested and booked for their transgressions. Repeat offenders, those whom we would today call career criminals or recidivists, were at that time considered a particular problem to European police agencies. The growing capitals and cities of Europe allowed for a certain anonymity, and criminals were free to travel from city to city, country to country, changing their names along the way as they plied their illegal trades. Bertillion hoped that his new system would allow the identification of criminals no matter where they appeared and, thus, help authorities keep track of undesirables (Cole, 2001).

Bertillionage was considered the premier method of identification for at least two decades—despite its limitations. The entire Bertillionage of a person was a complicated and involved process requiring an almost obsessive attention to detail. This made it difficult to standardize and, therefore, replicate accurately. Bertillion often lamented the lack of skill he saw in operators he himself had not trained. If the way the measurements were taken varied, then the same person might not be identified as such by two different operators. The portrait parlé added distinctive descriptors to aid the identification process, but here, again, the adjectives lacked precise objective definitions. "Lips might be 'pouting,' 'thick,' or 'thin,' 'upper' or 'lower prominent,' with 'naso-labial height great' or 'little' with or without a 'border,'" writes Simon Cole, quoting from Bertillion's own instruction manual (2001; p. 39). What was meant by "pouting", "prominent", or "little" was better defined in Bertillion's mind than in the manual.

Bertillionage was used across Britain and in its colonies, especially India. The officials in the Bengal office were concerned with its utility, however. They wondered if Bertillionage could distinguish individuals within the Indian population. Another concern the Bengali officials had with Bertillionage was the inconsistency between operators. There were variations in the way operators took the measurements: Some rounded the results up and some rounded them down, and some operators even decided which measurements were to be taken and which ones could be ignored. Staff in the Bengal office even attempted to solve the variance problem by mechanizing the system! All of these variances made searches tedious, difficult, and ultimately prone to error, defeating the point of using the method. The problem became so extreme that the Bengal office dropped Bertillionage entirely except for one small component of the system: Fingerprints.

Maintaining this part of the system begged for a way to classify fingerprints systematically, and this was the limiting factor in the adoption of any identification system. Bertillionage was too cumbersome and finicky to systematize for quick sorting, as were photographs. Additionally, with the growing number of individuals who were being logged into police records, any system of identification had to be able to handle hundreds, thousands, and eventually thousands of thousands of records quickly, correctly, and remotely.

FINGERPRINTING IN THE UNITED STATES

The first known systematic use of fingerprint identification in the United States occurred in 1902 in New York City. The New York Civil Service Commission faced a scandal in 1900 when several job applicants were discovered to have hired better educated persons to take their civil service exams for them. The New York Civil Service Commission therefore began fingerprinting applicants to verify their identity for entrance exams and to prevent better qualified persons taking tests for unscrupulous applicants. The first set of fingerprints was taken on December 19, 1902, and was the first use of fingerprints by a government agency in the United States (Cole, 2001).

Also in 1902, officials from the New York State Prison Department and the New York State Hospital traveled to England to study that country's fingerprint system. The following year, the New York state prison system began to use fingerprints for the identification of criminals; the use of fingerprinting increased even more when the United States Penitentiary in Leavenworth, Kansas, established a fingerprint bureau. This was the first use of fingerprints for criminal identification in the United States. During the 1904 World's Fair in St. Louis, John K. Ferrier of Scotland Yard taught the techniques and methods of fingerprinting to the public and law enforcement. Because of the notoriety of the fair and novelty of fingerprints as a "modern" method, the public and professional awareness of fingerprinting was greatly enhanced in the United States (Wilson and Wilson, 2003).

The first U.S. criminal conviction using fingerprint evidence occurred in Chicago, in the case of Thomas Jennings. Charles Hiller had been murdered during a burglary, and Jennings was charged and tried for the crime and ultimately convicted in 1911. The International Association for Identification (IAI) was formed in 1915, initially as a professional association for "Bertillon clerks," but as fingerprinting grew and eventually replaced Bertillionage, the focus of the IAI also changed. *The Finger Print Instructor* by Frederick Kuhne was published in 1916 and is considered the first authoritative textbook on fingerprinting in the United States.

The growing need for a national repository and clearinghouse for fingerprint records led to an Act of Congress on July 1, 1921, that established the Identification Division of the FBI in Washington, D.C., in 1924. A boost to the non-criminal use of fingerprinting came in 1933 when the United States Civil Service Commission (now the Office of Personnel Management) submitted over 140,000 government employee and applicant fingerprints to the FBI's Identification Division; this prompted the FBI to establish a Civil Identification Section, whose fingerprint files would eventually expand well beyond the Criminal Files. In 1992, the Identification Division was renamed the Criminal Justice Information Services Division (CJIS) and is now housed in Clarksburg, West Virginia. The increasing use of biological identification methods, or biometrics, continues to expand the use of fingerprinting. See "Issues in: Biometrics and Forensic Science" for more information.

ISSUES IN . . .
Biometrics and forensic science

The terms **"biometrics"** and "biometry" have been used since early in the twentieth century to refer to the field of development of statistical and mathematical methods applicable to data analysis problems in the biological sciences, such as the analysis of data from the yields of different varieties of wheat or data from human clinical trials evaluating the relative effectiveness of competing therapies for a disease. Recently, the term "biometrics" has also been used to refer to the emerging field of technology devoted to automated methods for authentication of individuals using physiological and behavioral traits, such as retinal or iris scans, fingerprints, hand geometry, face recognition, handwriting, and gait.

Forensic science and biometrics both apply various identification sciences, some the same and some unique to the particular discipline, although they do so for very different reasons (see Figure 19.1). Biometrics uniformly applies to a *pre-event situation*, such as gaining access, surveillance, or verification. In this way, biometrics chooses which mode of identification will be used. Forensic science, however, applies to *post-event situations*; as a historical science, forensic science reconstructs past criminal events to assist adjudication. Some of the identification sciences may be used by both forensic science and biometrics (blue), some may be used by only one or the other (yellow and red), or await application (purple). Because forensic scientists never know which mode of identification will be used ahead of time (DNA, fingerprints, dentition, etc.), they must sort through all of the information to discern significant clues. This highlights another important difference between the two disciplines: The results of a forensic science report may ultimately end up in court, whereas those of a biometric analysis rarely do.

For more information about biometrics, see Woodward, Orlans, and Higgins (2003).

Figure 19.1. *Biometrics versus forensic science.*

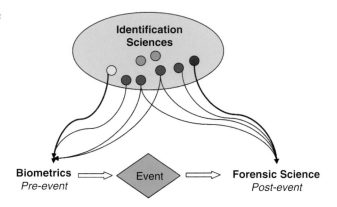

WHAT ARE FRICTION RIDGES?

Friction ridges appear on the palms, soles, and the ends of the fingers and toes (see Figure 19.2). These ridges are found on the palms and soles of all primates (humans, apes, monkeys, and prosimians); in primates with prehensile tails ("finger-like" tails, such as spider monkeys), friction ridges also appear on the gripping surface of the tails. All primates have an arboreal evolutionary heritage: Trees have been and continue to be the primary habitat for most apes and monkeys, and humans share this arboreal heritage. Primates' hands and feet show adaptations for locomotion and maneuvering in the branches of trees. The opposable thumb provides a flexible and sturdy means of grasping branches or the food that hangs from them. Primates, unlike other mammals such as squirrels or cats, have nails instead of claws at the distal end of their phalanges. Claws would get in the way of grasping a branch (imagine making a fist with 2″ nails) and would provide insufficient structure to hold an animal with a high body weight (a 1 lb. squirrel is highly maneuverable in a tree, but a 150 lb. jaguar is not). The ridges on the palms and soles provide friction between the grasping mechanism and whatever it grasps. Without them, it would be nearly impossible to handle objects in our environment.

Friction ridges begin forming when in the 9th or 10th week of fetal development. These **primary friction ridges** develop deep in the **dermal layer** of the skin, as shown in Figure 19.3. At about 14 weeks of gestation, sweat glands and sweat ducts begin to form, proliferating from the primary friction ridges. They infiltrate into the dermis and

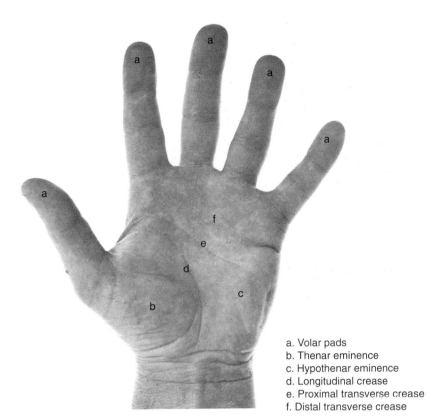

Figure 19.2. *The terminology for the palms is important for the proper identification and description of friction ridges.*

a. Volar pads
b. Thenar eminence
c. Hypothenar eminence
d. Longitudinal crease
e. Proximal transverse crease
f. Distal transverse crease

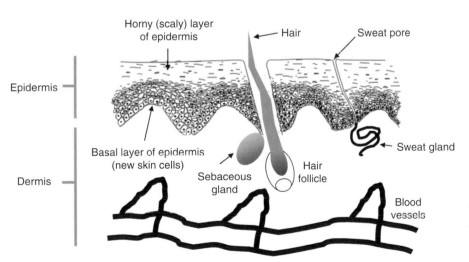

Figure 19.3. *The primary friction ridges develop deep in the dermal layer of the skin.*

develop into mature ducts and glands. The primary friction ridges pro-liferate until about the 15th or 17th weeks of gestation; at this point, the primary friction ridges stop proliferating and **secondary friction ridges** appear.

Secondary friction ridges develop from week 17 and mature by week 24. The interface or margin between the epidermis and the dermis provides a template of the configuration of the friction ridges on the surface. Numerous factors may affect the patterning and arrangement of friction ridges, including the fetus' genetics, environ-mental factors, drugs, disease, and perhaps even the shape of the volar pad itself.

Friction ridges develop *in utero* and remain the same throughout life, barring some sort of scarring or trauma to the epidermal-dermal margin of a friction ridge area. This interface between the epidermis and the dermis acts as a template for the configuration of the friction ridges seen on the surface of the skin. Although we grow and increase in size, the friction ridges on our bodies, which became permanent and fixed in their patterns from about 17 weeks of embryonic development, do not change like other parts of our bodies (Carlson, 2003).

WHAT'S A FRICTION RIDGE PRINT MADE OF?

A friction ridge print is a representation of a friction ridge pattern in some medium. Friction ridge prints can be classified as either **patent**, if they are visible with the unaided eye, or **latent**, if they require some sort of assistance to make them visible. Patent prints can appear because of some transferable material on the ridge pattern, such as liquid blood, liquid paint, or dust, or because the ridge pattern was transferred to a soft substrate that had "memory" and retained the impression, like clay, fresh paint, or putty.

Latent prints are composed of the sweat and oils of the body that are transferred from the ridge pattern to some substrate where they persist for some time until found by one of numerous visualizing tech-niques. The most familiar visualizing technique is the use of **fingerprint powder**—colored, fluorescent, or magnetic materials that are very finely ground—which is brushed lightly over a suspected print to produce contrast between the background and the now-visible print. More latent print visualization techniques are described later in this chapter.

COLLECTING PRINTS AT A CRIME SCENE

Friction ridge prints, especially fingerprints, can be left on a wide variety of surfaces and may persist for quite some time. Our homes, cars, and offices can be littered with friction ridge prints, but only a very few directly relate to a specific incident. Friction ridge patterns suitable for comparison can be obtained from a variety of surfaces, including glass, painted surfaces, plastics, ceramics, paper, and books. Just because the culprit may have worn gloves doesn't mean that no prints were left. The glove could have slipped, allowing a partial palm print, or the criminal may have taken the gloves off for some reason and touched something. Even leather gloves can leave prints of the cowhide patterns—don't assume and forget to collect potentially important evidence!

FRICTION RIDGE PATTERN VISUALIZATION TECHNIQUES

Not all friction ridge patterns are patently obvious, and some require physical, chemical, or optical enhancements to make them visible. The oldest and still most common method is to use one or more fingerprint powders to create contrast between the ridge pattern and the background. These powders typically are available in black, white, and other colors, including metallic. Black is the most popular color because it creates the most contrast on a white card, commonly used for filing and recording friction ridge prints. This card provides a uniform medium for the comparison of black ridges of the questioned print to the black inked ridges of the known print.

For photography, however, other colors are of considerable use, especially against the variety of colored backgrounds found in homes, offices, and cars. Once photographed, the prints can be lifted and placed on a black background card for further contrast. Some companies produce powders with two contrasting colors in them, like black and silver, to provide contrast regardless of the background. Magnetic powders, finely ground magnetic metals, work best on coated or shiny surfaces. Additionally, many fluorescent powders have been developed that fluoresce at specific wavelengths for easy visualization with tunable light sources and special film.

Figure 19.4.
Fingerprint powders are
applied with brushes
that have very fine,
long bristles.

The powders are applied with a soft fiberglass brush that has long, very fine bristles, as shown in Figure 19.4; some examiners use brushes with natural bristles (usually squirrel or camel hair), but they are rare today. The brush is dipped into the powder and gently applied to the latent print with a light touch; it is very easy to stroke too hard and remove latent print evidence. Magnetic powders require a special magnetic applicator that picks up a "fuzzy ball" of magnetic powder, which is then "brushed" over the print. The print is then photographed and/or lifted with frosted or clear tape for mounting on a contrasting background card. Information about the print, where, when, and how it was lifted and by whom is also recorded on the card, along with an identifier for chain of custody and reference purposes.

Many other chemicals and processes have been developed for the visualization of latent prints, some of which are listed in Table 19.1.

Visual detection methods have the advantage of being non-destructive and, therefore, these techniques do not prevent applying conventional fingerprint development methods afterward. Three main types of lasers have been used to detect fingerprints: the argon, the copper vapor, and the neodymium-yttrium/arsenide/gallium (Nd:YAG,

Table 19.1. *Various methods of visualizing latent prints and precautions.*

VISUALIZATION METHOD	USE	LIMITATIONS
Amido Black	Protein dye sensitive to blood, turning a blue-black color in its presence. Treatment with physical developer may be done after Amido Black to improve the developed print.	It will not stain the normal constituents in a latent print. Amido Black should not be used as a presumptive test for blood because it reacts to more than just blood.
Aqueous Amido Black	Protein dye solution sensitive to blood, turning a blue-black color in its presence. Treatment with physical developer may be done after Amido Black to improve the developed print. Can be washed over any non-porous surface; the item may also be submerged in the solution.	It will not stain the normal constituents in a latent print. Amido Black should not be used as a presumptive test for blood because it reacts to more than just blood. May permanently stain some surfaces. Presumptive tests for blood should be done **before** using Aqueous Amido Black. It is corrosive and will damage metal surfaces if not washed off quickly.
Aqueous Leucocrystal Violet	Enhances and develops latent prints stained with blood on porous or non-porous surfaces. Best applied by washing or submersion.	It will not stain the normal constituents in a latent print. May permanently stain some surfaces.
Gentian Crystal Violet	A protein dye that stains the fatty portions of sebaceous sweat a deep purple color; it also works on bloody prints. GCV will visualize latent prints on the adhesive side of all tapes. Fluoresces at 525, 530, and 570 nm (use red goggles); also at 485 and 450 nm (use orange goggles).	May permanently stain some surfaces.

Table 19.1. *Continued*

VISUALIZATION METHOD	USE	LIMITATIONS
DFO (1,8-Diazafluoren-9-one)	A Ninhydrin analogue that reacts to the amino acids present in body proteins; especially good for paper evidence. Once DFO is applied, the evidence should be heated for 10 minutes at 100°C (212°F). Using orange goggles, best fluorescence is seen at 450, 485, 525, and 530 nm for most papers. For brown and yellow papers, DFO fluorescence occurs at 570 to 590 nm. When DFO, Ninhydrin, and physical developer are each going to be used in the processing of a specimen, DFO must be used as the first process if there is to be any fluorescence.	**Not** recommended for spraying. Special conditions apply for photography.
Glue Fuming	Fumes from cyanoacrylate ester adhesives (Super Glue® and similar products) will develop latent prints by binding the proteins in the prints. The cyanoacrylate ester adhesive is heated in the presence of water to create the fumes. The developed prints may then be dusted to enhance their details; fluorescent materials may be incorporated into this process.	The fumes from cyanoacrylate ester adhesives are irritating but nontoxic.
Iodine	Fumes from iodine crystals develop latent prints on surfaces that are impractical for traditional dusting or have residue such as grease. The FBI has developed a method for spraying iodine solutions on large surfaces, such as walls.	Latent prints developed with iodine are visible for only a few hours.

Table 19.1. *Continued*

VISUALIZATION METHOD	USE	LIMITATIONS
Ninhydrin	Develops latent prints on porous surfaces like paper by reacting with amino acids in latent print residue. In a fume hood, the specimens are submerged in the Ninhydrin solution and then air dried. Ninhydrin may be applied after DFO and before physical developer.	Avoid contact with the powder and solution form of Ninhydrin. Any source of heat or spark should be avoided.
Physical Developer	A silver-based liquid reagent that reacts to lipids, fats, oils, and waxes present in the print residue. It is good for porous objects but should be the last process in the chemical sequence.	Numerous safety precautions are required for physical developer. Paper with a pH above 7 (like thermal fax paper) is not suitable for processing with physical developer.
Small Particle Reagent Molybdenum Disulfide, MoS_2)	A physical development technique in which small black particles adhere to the fatty substances left in print residue and is useful on many different surfaces. Well known for its ability to develop prints on wet surfaces and even under water.	Numerous safety precautions are required for Small Particle Reagent. Developed prints should be photographed before lifting is attempted.
Sudan Black	Working best on glass, metal, or plastic materials that are greasy or sticky, Sudan Black is a dye that stains the fatty components of sebaceous secretions. Sudan Black also works well on the inside of latex gloves. Specimens must be glue fumed prior to applying Sudan Black.	Stains many surfaces. Should not be used on porous or absorbent surfaces.
Vacuum Metal Deposition	This is reported to be the most effective technique for most smooth, non-porous surfaces. The process evaporates gold or zinc in a vacuum chamber, which coats the specimen surface with a microscopic layer of metal.	The equipment is expensive.

for short) lasers. These lasers have been shown to work well on metal surfaces, skin, and some plastics. The item of evidence should be illuminated with different wavelengths of light while the object is observed through filtered goggles designed for the wavelengths used.

PRESERVING PRINTS FOR ANALYSIS

Friction ridge prints should be photographed as soon as they are found at the crime scene or in the laboratory. This emphasis on preserving latent print evidence has numerous advantages, including showing the object where the print was found and leaving the print untouched for further examination. Photographing friction ridge prints is not as easy as photographing other types of evidence at a crime scene. The friction ridge photographer must be skilled in various methods of lighting, exposure, filters, and latent print enhancement. The final image of the print should be 1:1—the real size of the print—to facilitate the eventual comparison.

If the item has a plastic print or is a difficult surface to process where it's found, like a knob or switches, it should be removed, packaged properly, and submitted to the laboratory.

PRINCIPLES OF FRICTION RIDGE ANALYSIS

Although Francis Galton was not the first person to propose the use of fingerprints for identification, he was the first to study them scientifically thereby laying the foundation for their use in criminal cases, biometrics, and anthropology. Galton, a dilettante who studied a wide variety of disciplines including anthropology, genetics, geology, and statistics, was influenced by his cousin Charles Darwin and collaborated with Karl Pearson. *Fingerprints*, the first scientific text on the subject, was published in 1892 by Macmillan; Galton went on to publish additional works on fingerprints in 1893 (*Decipherment of Blurred Finger Prints*) and *Fingerprint Directories* (1895). In his 1888 paper for the Royal Institution, Galton estimated the probability of two persons having the same fingerprint and studied the heritability and racial differences in fingerprints. Galton's work on fingerprints summarized common pattern in fingerprints and devised a classification system that is still used to this day. The method of identifying criminals by their finger-prints had been introduced in the 1860s by William Herschel in India, but their potential use in forensic work was first proposed by Dr. Henry Faulds in 1880 (Wikipedia, **www.wikipedia.com**).

Since Galton's time, friction ridges have been considered unique; that is, no individual's friction ridges are identical to anyone else's. The concept of uniqueness is typically associated with the philosopher Gottfried Wihelm Leibniz, who stated, "For there are never in nature two beings that are perfectly alike and in which it is not possible to find a difference that is internal or is founded on an intrinsic denomenation" (Rescher, 2001; 64). While it is one thing to understand all people and things are separate in space and time, it is quite another to prove it.

Galton was the first to attempt to calculate the likelihood of finding two friction ridge patterns that are the same. Numerous researchers have recalculated this probability over the years by various calculations based on differing assumptions (see Table 19.2). But they all indicate that the probability of any one particular fingerprint is somewhere between 0.000000954 and 1.2×10^{-80} (0.0 with 78 zeros and 12) all very small numbers indeed. Technically, even infinitesimal probabilities such as these are still *probabilities* and do not represent true uniqueness (which would be a probability of 1 in ∞), but the values are such that latent fingerprints, with sufficient minutiae, have been considered unique by the vast majority of forensic scientists and the courts. The values in Table 19.2 also demonstrate the value of finding as many points of comparison as possible; more similarities—with no significant differences—leads to a lesser probability of a coincidental match (false positive).

Under low-power magnification, friction ridge patterns are studied for the kind, number, and location of various ridge characteristics or **minutiae**. As with many other types of forensic evidence, it is not merely the presence or absence of minutiae that makes a print unique: It is the *presence, kind, number*, and, especially, *arrangement* of those characteristics that create a one-of-a-kind pattern. When two or more prints are compared, it is a careful point-by-point study to determine if enough of the significant minutiae in the known print are present in the questioned print, with no relevant differences. This comparison process is demonstrated graphically in Figure 19.5.

Figure 19.5 shows the comparison of two fairly complete prints; in reality, the majority of prints that are identified, resolved, and compared are **partial prints**, representing only a portion of the complete

Table 19.2. *Comparison of probability of a particular fingerprint configuration using different published models for 36 minutiae and 12 minutiae (matches involve full, not partial, matches). Data from Table 8.2 in Maltoni, Maio, Jain, and Prabhakar (2003, p. 267).*

AUTHOR	PROBABILITY VALUE FOR A LATENT PRINT WITH *36* MINUTIAE	PROBABILITY VALUE FOR A LATENT PRINT WITH *12* MINUTIAE
Galton (1892)	1.45×10^{-11}	9.54×10^{-7}
Henry (1900)	1.32×10^{-23}	3.72×10^{-9}
Balthazard (1911)	2.12×10^{-22}	5.96×10^{-8}
Boze (1917)	2.12×10^{-22}	5.96×10^{-8}
Wentworth and Wilder (1918)	6.87×10^{-62}	4.10×10^{-22}
Pearson (1930, 1933)	1.09×10^{-41}	8.65×10^{-17}
Roxburgh (1933)	3.75×10^{-47}	3.35×10^{-18}
Cummins and Midlo (1943)	2.22×10^{-63}	1.32×10^{-22}
Trauring (1963)	2.47×10^{-26}	2.91×10^{-9}
Gupta (1968)	1.00×10^{-38}	1.00×10^{-14}
Osterburg, et al. (1977)	1.33×10^{-27}	1.10×10^{-9}
Stoney (1985)	1.20×10^{-80}	3.5×10^{-26}

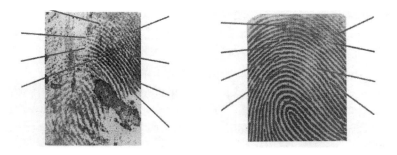

Figure 19.5. *The presence, kind, number, and arrangement of minutiae create the pattern used in a fingerprint comparison. The points are studied side by side with a magnifying lens.*

print pattern. A friction ridge print scientist must then determine if a partial print is suitable for comparison, that is, if the print has the necessary and sufficient information to allow a proper comparison. A partial print, or even a complete print for that matter, may be identifiable as such but be smudged, too grainy, or too small for the scientist to make an accurate and unbiased comparison. Often this is the crucial step in a friction ridge print examination that is dependent on the scientist's experience, visual acuity, and judgment.

One of the ongoing debates among forensic scientists is how many points of comparison are necessary and sufficient to reach a conclusion of identification. For years, many agencies had a **point counting standard**, which dictated how many points of comparison were required before a positive conclusion could be reached. The number of points varied from 8 to 16 to even 20 in some agencies. A concern with point counting, however, is that no scientific or statistical basis has been established that would indicate that 8 is not enough but 10 might be, or 16, or more. The frequency of individual kinds of minutiae (deltas, bifurcations, crossings, etc.) in any population is not known and so begs the question of a numerical standard's significance. Another concern is the question of the threshold limit: If your agency requires 10 points for an identification, what if you find 9 very clear points? Could you possibly make out a 10th or must you just claim an exclusion? What if it's a small partial print and 9 very clear points are all you can find? At an agency with a threshold of 8, this would be a match, and the problem is obvious.

Many, if not most, agencies have now adopted a "no-point" standard, summarized in 1973 by the International Association for Identification, a professional association for forensic scientists involved in identification techniques, as "no valid basis exists for requiring a predetermined minimum number of friction ridge characteristics which must be present in two impressions in order to establish positive identification" (1973; p. 1). The threshold then becomes one of a sufficient number of characteristics necessary to make a conclusion of identification, however many that might be. A scientist's experience and judgment become central to the process of a quality examination; this, ultimately, is derived from proper and comprehensive training coupled with a mentoring process of practical experience. This does not absolve scientists of the obligation to be able to articulate the points of comparison, their significance, and why they lead to a conclusion of identification. Two or more experts may disagree, but they need to be able to offer cogent arguments as to why and how they reached their different conclusions.

CLASSIFYING FINGERPRINTS

The patterning and permanency of friction ridges allows for their classification. As discussed earlier, the fact that fingerprints could be systematically sorted and cataloged was a main reason for their widespread adoption among government agencies. The history of the classification of fingerprints is central to understanding the widespread adoption of fingerprinting; see "History: Classifying Fingerprints" for more information. It is important to keep in mind that it is the general patterns, and not the individualizing elements, that allow for this organization.

Today, all fingerprints are divided into three classes: loops, arches, and whorls. **Loops** have one or more ridges entering from one side of the print, curving back on themselves, and exiting the fingertip on the same side, as shown in Figure 19.6. If the loop enters and exits on the

HISTORY:
Classifying fingerprints

The first person to describe a taxonomy of fingerprints was Dr. Jan Purkyně, a Czech physician and one of the historical giants in the field of physiology. In 1823, Dr. Purkyně lectured on friction ridges in humans and primates and described a system of nine different basic ridge patterns. In 1880, Dr. Henry Faulds, a Scot who worked in a Tokyo hospital, researched fingerprints after noticing some on ancient pottery; Faulds even used "greasy finger-marks" to solve the theft of a bottle of liquor. He published his research on the use and classification of fingerprints in a letter to the scientific journal *Nature*. The publication of Fauld's letter drew a quick response from William Herschel, a chief administrator from the Bengali British government office in India, who claimed that he, Herschel, and not Faulds had prior claim to the technique of fingerprints.

Herschel had been using finger and palm prints to identify contractors in Bengal since The Indian Mutiny of 1857, employing a simplistic version of the system that was eventually instituted some 40 years later. In fact, it may not have been Herschel's own idea to use prints for identification: The Chinese and Assyrians used prints as "signatures" at least since 9,000 years before the present, and the Indians had probably borrowed this behavior. Sir Edward Henry had tried to institute fingerprinting as the primary means of identification across all of India; his supervisor thought otherwise and Herschel's work languished until Fauld's letter was published. The argument about who was first—Fauld or Herschel—would continue into the 1950s (Thorwald, 1964; Cole, 2001).

side of the finger toward the little finger, it is called an **ulnar loop**; if the loops enters and exits on the side toward the thumb, it is termed a **radial loop**. All loops are surrounded by two diverging ridges called **type lines**; the point of divergence is called a **delta** because of its resemblance to a river delta and the Greek letter Δ (*delta*). The central portion of the loop is called the **core**.

LOOP

Figure 19.6. Loops have one or more ridges entering and exiting from the same side of the print, looping back on itself in the middle. Ulnar loops exit and enter the side of the finger toward the little finger, whereas radial loops are on the side toward the thumb.

Arches are the rarest of the three main classes of patterns. Arches are either **plain** (see Figure 19.7), with ridges entering one side of the finger, gradually rising to a rounded peak, and exiting the other side; or **tented**, which are arches with a pronounced, sharp peak. A pattern that resembles a loop but lacks one of the required traits to be classified as a loop can also be designated as a tented arch. Arches do not have type lines, cores, or deltas.

Whorls are subdivided into **plain whorl** (see Figure 19.8), **central pocket loop, double loop**, and **accidental**. All whorls have type lines and at least two deltas. Central pocket loops and plain whorls have a minimum of one ridge that is continuous around the pattern, but it does not necessarily have to be in the shape of a circle; this ridge can be an oval, ellipse, or even a spiral. Plain whorls are located between the two deltas of the whorl pattern, whereas central pocket loops are not. This difference can be easily determined by drawing a line equidistant between the two deltas: If the line touches the circular core, then the whorl is a plain whorl; if not, it is a central pocket loop (see Figure 19.9).

A **double loop** is made up of two loops that swirl around each other (see Figure 19.10). Finally, an **accidental** is a pattern that combines two or more patterns (excluding the plain arch) and/or does not clearly meet the criteria for any of the other patterns.

The relative appearance of loops overall is 60–65%; whorls, 30–35%; and arches, 5%.

Figure 19.7. Arches enter one side of the fingertip, peak, and then exit the opposite side. Arches are either plain or tented.

CLASSIFICATION

The modern system of fingerprint classification is based on Henry's original design, which could process a maximum of 100,000 sets of prints, with modifications by the FBI to allow for the huge number of entries that have accumulated over the years. The FBI Criminal Justice Information Section (CJIS) currently has over 80 million fingerprints stored in its files.

PLAIN
ARCH

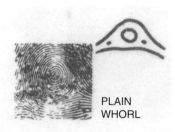

PLAIN
WHORL

Figure 19.8. *All whorls have type lines and at least two deltas. Whorls are classified as plain, central pocket loops, double loops, and accidentals.*

The modern fingerprint classification consists of a **primary classification** that encodes fingerprint pattern information into two numbers derived as follows: All arches and loops are considered "non-numerical" patterns and are given a value of zero. Whorls are given the values depending on which finger they appear, as shown in Table 19.3.

The values are summed, with one added to both groups, and the resulting primary classification is displayed like a fraction, as shown in equation 19.1:

$$\text{Equation 19.1} \quad \frac{\text{R index} + \text{R ring} + \text{L thumb} + \text{L middle} + \text{L little} + 1}{\text{R thumb} + \text{R middle} + \text{R little} + \text{L index} + \text{L ring} + 1}$$

If, for example, all of an individual's fingers had whorls, the formula would look like $16 + 8 + 4 + 2 + 1 + 1/16 + 8 + 4 + 2 + 1 + 1 = 32/32$; if all of this person's fingers had arches or loops instead, the formula would be $0 + 0 + 0 + 0 + 0 + 1/0 + 0 + 0 + 0 + 0 + 1 = 1/1$.

In and of itself, a primary classification is just that: class evidence. The primary classification was originally devised to sort individuals into smaller, more easily searched, categories; this, of course, was when fin-

Figure 19.9. *To determine whether a print is a plain whorl or a central pocket loop, draw a line between the two deltas. If the line touches the core, it's a plain whorl; if not, then the print is a central pocket loop.*

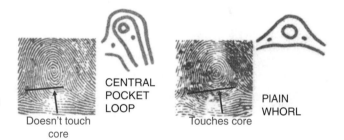

CENTRAL
POCKET
LOOP

Doesn't touch
core

PLAIN
WHORL

Touches core

Table 19.3. *Values for fingers in the Henry classification system.*

Right thumb, right index	16
Right middle, right ring	8
Right little, left thumb	4
Left index, left middle	2
Left ring, left little	1

gerprints were searched by hand. Additional subdivisions of the classification scheme may be used, but they still serve only as a sieve through which to organize and efficiently search through filed prints. Comparison of minutiae and higher level details is the only method for fingerprint identification.

The problem with storing and sorting fingerprints using only the Henry-FBI classification system is that, while the system stores all 10 prints as a set, rarely are full sets of fingerprints found at a crime scene. To search through even a moderately sized database of 10 print sets for an individual print would take too long and be too prone to error. Many agencies used to keep single-print files, which contained the separate fingerprints of only the most frequent locally repeating criminals.

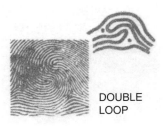

DOUBLE LOOP

Figure 19.10. *Double loops have two loops that swirl around each other; accidentals combine two or more*

AUTOMATED FINGERPRINT IDENTIFICATION SYSTEMS (AFIS)

The advent of computers heralded a new age for many forensic sciences, and among the first to utilize the technology was the science of fingerprints. The process of capturing, storing, searching, and retrieving fingerprints via computer is now a standard occurrence among police agencies and forensic science laboratories. Automated fingerprint identification systems, or AFIS (pronounced "AY-fis"), are computerized databases of digitized fingerprints that are searchable through software—essentially, a computer and a scanner hooked to a network-type server computer. An AFIS can store millions of print images that can be searched in a matter of minutes by a single operator. The core of this electronic system is a standard format developed by the FBI and the National Institute of Standards and Technology (NIST), with the advice of the National Crime Information Center (NCIC), which provides for the conversion of fingerprints into electronic data and their subsequent exchange via telecommunications and computers. Although the data format was a standard, the software and computers that operate AFIS are not, and several vendors offer products to law enforcement and forensic science agencies. The drawback was that these products were not compatible with each other, precluding the easy exchange of information between systems.

This situation began to change in 1999 when the FBI developed and implemented a new automated fingerprint system known as the

Integrated Automated Fingerprint Identification System, or IAFIS (pronounced "EYE-ay-fis"). Although IAFIS is primarily a 10-print system for searching an individual's fingerprints like a standard AFIS, it can also digitally capture latent print and 10-print images and then do the following:

- Enhance an image to improve its quality.
- Compare crime scene fingerprints against known 10-print records retrieved from the database.
- Search crime scene fingerprints against known fingerprints when no suspects have been developed.
- Automatically search the prints of an arrestee against a database of unsolved cases.

Other advances are being made to solve the problem of non-compatible AFIS computers. The Universal Latent Workstation is the first in a new generation of interoperable fingerprint workstations. Several state and local agencies, the FBI, NIST, and AFIS technology manufacturers are developing standards to provide for the interoperability and sharing of fingerprint identification services. The Workstation, which is part of that program, assists agencies and manufacturers understand and develop the concept of "encode once and search anywhere." The Workstation allows agencies to enter data into the format of the system they purchased and use but also to share that data with other, previously incompatible systems. Agencies will eventually be able to use this type of Workstation to search local, state, and neighboring systems and the FBI IAFIS system, all with a single entry.

IDENTIFICATION

The final identification decision in a fingerprint comparison is reached when sufficient quality and quantity of corresponding Level 1, 2, and 3 friction ridge details are present. **Level 1 detail** includes the general ridge flow and pattern configuration. It is not sufficient for individualization but can exclude an individual. Level 1 detail may include information enabling orientation, core and delta location, and distinction of finger versus palm.

Level 2 detail includes formations, defined as a ridge endings, bifurcations, dots, or combinations of these features; Level 2 detail enables individualization. It is not the presence or absence of these features that can individualize a print, although they play a major role: It is the *relationship* of these features across the print that defines the uniqueness of the print. This is analogous to each of us having a nose, two eyes, a mouth, and ears but our face being much more than a laundry list of those parts.

Finally, **Level 3 detail** includes all attributes of a *ridge*, such as ridge path deviation, width, shape, pores, edge contour, incipient ridges, breaks, creases, scars, and other permanent minutiae. Level 3 detail obviously can lead to individualization as well and, it has been argued, when a fingerprint examiner looks at a print, he or she automatically takes Level 3 detail into account.

Some jurisdictions, both in the United States and abroad, require a number of points of comparison before an identification can be made and confirmed. Fingerprint examiners who argue against this approach maintain that friction ridge examination is much more holistic than just "counting points," and the entire print is considered and examined. Those who support point counting standards claim that they *do* take the whole print into account, but some *objective* standard must be met for the good of the profession. The recent Llera-Plaza case is a good example of the challenges that fingerprint examiners, and possibly other forensic sciences, face in the future. See "In More Detail: Is Forensic Fingerprint Examination Science?"

HOW LONG DO FRICTION RIDGE PRINTS LAST?

> **ON THE WEB:**
> www.usao-edpa.com/invest.htm
> www.onin.com

Plastic prints will last as long as the impressed material remains structurally intact. A print left in some medium, such as blood or dust, is quite fragile and does not last very long. Latent prints, however, if in the proper environments, can last for years. Currently, the age of a set of fingerprints is almost impossible to determine.

ELIMINATION PRINTS

As with any other type of evidence, obtaining known samples for elimination purposes can be of great assistance to the forensic scientist.

IN MORE DETAIL:

Is forensic fingerprint examination a science?

In a federal drug and murder prosecution case, *U.S. v. Llera Plaza* (2002), the defendants moved the court to suppress the offered fingerprint evidence because, in their opinion, fingerprint evidence is not a science. The United States countered that the fingerprint evidence should be admitted at trial. The prosecutors also requested the judge to take official notice that "fingerprints offer an infallible means of personal identification" (www.fbi.gov).

Judge Pollak agreed with the defendants and ruled on January 7, 2002, that the FBI's experts would not testify at trial that the evidentiary fingerprints "matched" those of the defendants. After additional testimony, however, Judge Pollak changed his ruling. On March 13, 2002, he reversed his decision and ruled that the FBI's experts could testify to the fingerprint evidence.

Judge Pollak's initial ruling has been interpreted as "trashing" fingerprint evidence. This overstates his ruling: Judge Pollak ruled that FBI fingerprint examiners could testify to the processing of the latent fingerprints, give demonstrative evidence to the jury in the form of enlarged comparison prints, and indicate the points of comparison among them. What the judge disallowed was the experts' opinion that the fingerprints were the same. In Judge Pollack's view, that was the sole right of the jury.

In coming to this decision, Judge Pollak relied upon two United States Supreme Court opinions: *Daubert v. Merrill Dow Pharmaceuticals, Inc* (113 S. Ct. 2786 [1993]) and *Kumho Tire Co. Ltd. V. Carmichael* (119 S. Ct. 1167 [1999]). The Supreme Court held that scientific evidence must meet four factors to be admitted under the *Daubert/Kumho* guidelines:

- The scientific technique must be testable.
- The technique must have been subjected to peer review and publication.
- The technique's known or potential error rate must be known, and standards for using the technique must exist and be followed.
- The technique must be generally accepted.

Judge Pollak determined that fingerprint comparison, if it was a science, failed the first three points—thus, the surprise when Judge Pollak reversed his decision. Part of the explanation for this was that the attorneys and the judge restricted their evidence in the first hearing to a two-year-old transcript of another case. Extensive evidence about the scientific reliability of fingerprints had been introduced by both the defendants and the prosecutors. Judge Pollak requested to hear additional evidence based on the prosecution's motion to reconsider his first ruling. Part of the evidence presented this second time was live testimony from one witness who had testified in the previous case—his "old" testimony had simply been read by Judge Pollak. The court also heard additional testimony from additional experts that ultimately swayed his decision.

Does this mean fingerprints are now "off the hook"? Probably not, by anyone's guessing. Forensic scientists must be scientists in the laboratory *and* in the courtroom despite the change in basic rules between those venues. Any forensic scientist who steps into a courtroom unprepared to provide supporting research, standards, and protocols for his or her methods is asking for trouble. More research is needed in all aspects of forensic science—no one study will solve a legal problem and make it "go away." Science is a search for understanding and knowledge, and forensic scientists are as bound by this search as any other scientific discipline. In their recent book, Champod, Lennard, Margot, and Stoiloric (2004) note that

> When it comes to identification issues, it is clear that the criminal justice system is approaching fingerprint evidence with a much more critical eye than in the past. Certainly, the highly debated introduction of DNA evidence and its systematic comparison with fingerprint evidence has promoted such renewed critical interest. We welcome this regain of interest, as it will force the profession to analyze its foundations critically. (p. 204)

These samples may not only eliminate individuals from an investigation's focus, but they can also demonstrate a proper scientific mindset through a comprehensive series of comparisons. If these elimination knowns are incorporated into a trial presentation, they can create confidence in the mind of the trier of fact that, not only do the defendant's known prints match, but the other potential subjects' prints do *not* match. Displaying what is and is not a match can clarify the forensic scientist's process of identification and comparison to the layperson.

CHAPTER SUMMARY

Friction ridge examination is one of the bedrock disciplines in forensic science and has been employed for over 100 years. It is considered the preeminent method of individualization in forensic science. That status, however, has recently been upset by court challenges and high-profile misidentifications. Friction ridge examination will no doubt continue as a mainstay of forensic science, but changes in methodology and interpretation may be on the horizon. How the discipline weathers these changes depends on the willingness of the profession to critically examine its foundations, procedures, and methods.

Test Your Knowledge

1. Who was Cesare Lombroso?
2. What was a *portrait parlé*?
3. What is Bertillionage?
4. When was the first systematic use of fingerprints in the United States?
5. When was the first U.S. conviction using fingerprints?
6. What is CJIS?
7. Why are friction ridges important?
8. When do friction ridges begin to develop? When are they complete?
9. What is fingerprint powder?
10. List four methods of visualizing latent prints, how they are used, and their limitations.
11. How are lasers used with latent prints?
12. What are minutiae?

13. What is a point counting standard?

14. Name the types of fingerprint patterns and describe them.

15. What is AFIS? What's the difference between AFIS and IAFIS?

16. What are Level 1, Level 2, and Level 3 detail?

17. What are elimination prints?

Consider This . . .

1. Is the discipline of forensic friction ridge analysis a science? What are the hallmarks of a science? Do they apply to fingerprints?

2. How would you present evidence in an admissibility hearing under the *Daubert* guidelines to support forensic friction ridge analysis as meeting those criteria? How would you counter those arguments if you were an opposing expert?

3. Why would you not want to use a point counting standard in fingerprints? What's the positive aspect and what are the negatives?

BIBLIOGRAPHY

Ashbaugh, D. (1999). *Quantitative-qualitative friction ridge analysis: An introduction to basic and advanced ridgeology.* Boca Raton, FL: CRC Press.

Carlson, B. (2003). Human Embryology and Developmental Biology Updated Edition. Amsterdam, Elsevier.

Champod, C., Lennard, C., Margot, P., & Stoilovic, M. (2004). *Fingerprints and other ridge skin impressions.* Boca Raton, FL: CRC Press.

Cole, S. (2001). *Suspect identities.* Cambridge, MA: Harvard University Press.

Gould, S. (1981). *The mismeasurement of man.* New York: W.W. Norton.

Federal Bureau of Investigation. (1984). *The science of fingerprints.* Washington, D.C.: U.S. Government Printing Office.

International Association for Identification Standardization Committee (1973). *Report of the standardization committee,* Mendota Heights, MN, International Association for Identification.

Kuhn, F. (1916). *The finger print instructor.* New York: Munn & Company.

Maltoni, D., Maio, D., Jain, A., & Prabhakar, S. (2003). *Handbook of fingerprint recognition.* New York: Springer-Verlag.

Pankanti, S., Prabhakar, S., & Jain, A. (2001, December 11–13). On the individuality of fingerprint. Proceedings of Computer Vision and Pattern Recognition (CVPR), Hawaii.

Rescher, N. (2001). *Philosophical reasoning.* Malden, MA: Blackwell Publishing.

Thomas, R. (1999). *Detective fiction and the rise of forensic science*. Cambridge, UK: Cambridge University Press.

United States v. Llera Plaza, 181 F. Supp. 2d 414 ED Pa. (2002).

Wilson, C. and Wilson, D. (2003). *Written in blood: A history of forensic detection*. New York City, Carroll and Graf Publishers.

Woodward, J.D., Orlans, N.M., & Higgins, P.T. (2003). *Biometrics*. New York: McGraw-Hill/Osborne.

Questioned Documents

KEY TERMS

Abbreviated signature

Abrasive erasure

Charred document

Chemical erasure

Document alteration

Document dating

Electrostatic Detection Apparatus (ESDA)

Exemplar

Forged signature

Formal signature

Indented writing

Informal signature

Laser desorption

Non-requested writing

Obliteration

Questioned document

Requested writing

Scribbled signature

Stylistic signature

INTRODUCTION

When you hear the term "questioned document," what do you think of? Most people think of a piece of paper that has been handwritten and that contains some important writing. People believe that the forensic questioned document examiner must compare this document with a sample of someone's handwriting to determine if that person was the author. In fact, questioned document examiners do a good bit of handwriting comparison. But there is a lot more to questioned documents than simply handwriting on paper. Sometimes the document is a piece of paper, but questioned documents have been written on mirrors, walls, large placards, and other objects. Document examiners are often called upon to compare typewritten or computer printer-generated documents, analyze inks and papers, determine the age of a document, uncover credit card forger-

THE CASE

A very wealthy, elderly woman died in 1999, leaving her vast fortune to the "Rocking Pebbles," a somewhat aged rock music group that was very popular in the 1960s. The will that directed the distribution of the estate was dated March 19, 1970, and was purportedly handwritten by the deceased. The alleged will was not witnessed or notarized. A few days after the woman's death, her surviving male child, aged 50 and long estranged from his mother, came forward with another will, dated August 21, 1972, also apparently handwritten and signed by the deceased, but also not witnessed nor notarized. In this will, all of the money in the estate was to go to the son. The executor of the original will sued the son, he counter-sued, and the case ended up in court. In cases such as this that involve disputed wills, two major issues arise: Which wills, if any, were written by the woman, and were they written at the time that they were dated?

Both of these issues will ultimately be decided with the help of questioned document examinations. In order to explain and answer the issues, a number of materials must be obtained. They include writings that can be definitely ascribed to the deceased (authenticated), and that were written around the time of each of the wills. It would also be helpful to obtain the writing instruments used to write each of the wills, but this is often not possible.

ies and currency fraud, and reconstruct charred or obliterated writing. In this chapter you will learn how questioned document examiners are trained; what their duties are; and how questioned documents are collected, preserved, and analyzed.

WHAT IS A QUESTIONED DOCUMENT?

A questioned document need not be a piece of paper; it can be any object. To be a document, it must contain linguistic or numerical markings that are put there by handwriting, typewriting, copying, computer printing, or by other means. If there is doubt about whether the document is authentic (e.g., a real draft card or passport) or who the author is (as in the wills in the case at the beginning of this chapter), then it is a questioned document. For example, in a case in Virginia that occurred about 25 years ago, a man wrote a threatening note in large letters using spray paint on the outside wall of a woman's house. By the broad definition given above, the wall is a questioned document. Usually, however, questioned documents are bills, wills, letters, checks, contracts, passports, lottery tickets, driver's licenses, etc.

THE QUESTIONED DOCUMENT EXAMINER

The first questioned document examiners were photographers. Interestingly enough, photography is an area that questioned document examiners must be very familiar with, but being an expert photographer is by no means sufficient to qualify someone as a questioned document examiner. One of the most notorious cases in which a photographer acted as an expert in questioned documents was the Dreyfus treason case. Alphonse Bertillion, a photographer and developer of the science of anthropometry, testified in court that Alfred Dreyfus authored the pivotal document in the case. Later, it was determined that Dreyfus was innocent, thus discrediting both Bertillion and the budding science of questioned document examination.

Until the twentieth century, questioned document analysis and testimony were not readily accepted in court. Under English Common law, handwriting standards that were not already in evidence were not admissible. This prevented the comparison of questioned documents with known samples since they could not be introduced into court. Since much of early U.S. law was based on the Common law, this rule had the effect of delaying the use of forensic document examination in the United States until 1913, when through the efforts of the famous document examiner Albert Osborne, Congress passed a rule that allowed document standards into evidence in court.

TRAINING AND EDUCATION OF QUESTIONED DOCUMENT EXAMINERS

There are no formal college-based education programs in questioned document analysis in the United States, and it is unlikely that there will be in the future. Some college courses in questioned document analysis can provide a theoretical foundation for the field, but questioned document analysis, like firearms, tool marks, and fingerprints, is essentially a classical apprenticeship field. The trainee may spend several years studying with an accomplished professional in the field. After passing a number of tests and exercises, the trainee may then become a journeyman questioned document examiner. A certification program in the United States for questioned document examiners is administered by the American Society of Questioned Document Examiners (ASQDE, http://www.asqde.org/).

A typical training program for a questioned document examiner trainee would consist of formal coursework, if any is available; reading and studying the relevant basic and advanced books and journals that are concerned with all aspects of document examination; and study and examination of actual questioned documents under the supervision of the trainer. There are also quality assurance and control materials including blind tests and then mock trial exercises that must be successfully completed. The trainee would also be expected to learn the relevant statutes and legal considerations that govern the examination, reporting, and expert testimony of questioned documents.

HANDWRITING COMPARISONS

When children first learn to print or write, they are given exemplary samples of each letter and are told to copy them as neatly as possible. Then they are taught to put the letters together and make words. During this period, the handwriting of most children would be very similar. After learning how to form words, the children then put these skills to use in writing, and teachers begin to focus on content. This means that the writers begin to focus more on what they are writing and less on how the letters are formed. Writing habits become internalized, and the act of writing becomes subconscious; thus, individual characteristics develop, many of which the writer is unaware of. With time, handwriting becomes a sort of unconscious habit; however, this habit is not a static one. Handwriting changes with time. Over a period of years, a person's handwriting undergoes gradual changes that result in different characteristics. This can have a profound influence on the collection of known handwriting samples. Handwriting can also change as a result of changes in physical and emotional condition. Depression, drug abuse, and physical illness can all have effects on handwriting.

The ability of a questioned document examiner to identify the author of a handwritten document by comparison of unknowns with

exemplars (known, authentic writing samples) depends on two factors: First, there must be sufficient individual characteristics present in the unknown sample, and second, there must be sufficient samples of the purported writer's authentic handwriting. If both of these conditions are met, then it is possible to individualize an unknown handwriting exhibit to a particular person. If either of the above conditions is not met, then an equivocal conclusion may be necessary—that the suspect's writing could not be eliminated as a source for the unknown. Of course, it is also possible to use class or individual characteristics to eliminate a suspect as being the source of questioned handwriting.

As with other types of comparative evidence such as shoe prints, bullets, fingerprints, etc., there are no standards as to the minimum number of characteristics of handwriting that must be present in known and unknown handwriting samples in order to reach a conclusion of individuality. The minimum number depends on the size of the questioned and exemplar samples, the nature of the characteristics, and the experience of the examiner. If a handwriting sample contains many unusual characteristics, then a questioned document examiner may need to find fewer of them than would be the case with more "normal" handwriting. In any case, the individual examiner must find enough points on the questioned document and exemplar to be convinced to a degree of reasonable, scientific certainty that the questioned document was written by the same hand as the exemplar. Under no circumstances is it permissible to state a percent probability that the questioned document was written by the suspect. That is, it would not be permissible to state "I am 95% sure that the suspect wrote the will." This is an important concept that also applies to other types of evidence such as fingerprints and toolmarks. In order for an examiner to be able to cite numerical probabilities about the likelihood that known and unknown evidence have a common source, there must be studies that yield information about how common a particular set of characteristics is in a given population. Although this works well for DNA, where there is a finite set of types of DNA, it is not possible to do this for handwriting, where characteristics do not fall into discrete groupings.

HANDWRITING COMPARISON CHARACTERISTICS

A large number of handwriting characteristics, both class and individual, can be used for comparisons of questioned and exemplar specimens. A few of the more common ones are as follows:

- Spacing between letters
- Spacing between words
- Relative proportions between letters and within letters
- Individual letter formations
- Formations of letter combinations
- The overall slant of the writing
- Connecting strokes
- Pen lifts
- Beginning and ending strokes
- Unusual flourishes
- Pen pressure

Unlike other pattern evidence such as fingerprints, handwriting changes throughout life. There may also be changes in a person's handwriting due to such factors as stress, physical and emotional condition, illness, influence of drugs or alcohol, and the conditions under which the writing is made, including type of writing instrument and paper, level of comfort, and lighting. Figure 20.1 shows how handwriting characteristics can be used to compare known and unknown samples of handwriting. A description of the document comparisons in the famous "Clifford Irving" case is given in "History: The Clifford Irving Case."

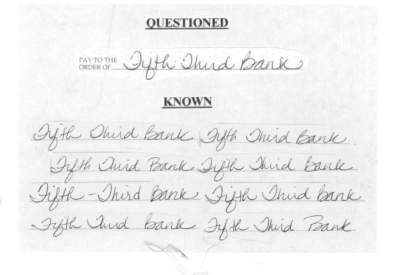

Figure 20.1. *A handwriting comparison. The questioned document is a bank check. Courtesy of Robert Kullman, Speckin Forensic labs.*

COLLECTION OF HANDWRITING EXEMPLARS

Although the questioned document examiner or the investigator has little or no control over the quality and quantity of the questioned writing sample, he or she does have or should have control over the exemplars. Known writing samples are critical to a successful handwriting comparison. The quantity of exemplar material is important so that the full range of variation in the subject's writing can be obtained. Just as important, however, is the quality and type of exemplars. There must be some similarities between the known and unknown writings in order to make meaningful comparisons. This will be explained later in this section.

There are two types of exemplars: requested and non-requested writings. **Requested writings** are writing samples taken from someone for the purpose of comparison with a questioned document. These writings are generally taken by multiple dictations of a passage that has words and phrases similar to the questioned document. This type of data collection minimizes the chance that the subject is deliberately trying to alter her handwriting style. The writing often consists of a few generic sentences or passages followed by all or part of the exact questioned document. Requested writings are often preferred because the questioned document examiner has a good deal of control about the content, paper, writing implement, etc., used in the writing and can request enough writing sample so as to minimize the chances that the writer will be able to disguise her writing. The major disadvantage of requested writings is temporal: If a great deal of time has elapsed between the time the questioned document was written and the collection of the requested writing, there is a chance that the questioned document examiner may not be able to make a valid comparison because the handwriting could have changed dramatically during the period between the two samples.

Proper Collection of Requested Writings

Successful comparison of known and unknown handwritings depends in part on the quality and quantity of the exemplars. Following are some of the more important considerations that must be kept in mind when collecting these known samples:

- The most important consideration is that there must be a sufficient amount of writing to ensure that the normal variations that are present in everyone's handwriting are represented. There is no standard amount of writing that would suffice, but some examiners believe that 10–20 samples of the comparable writing should be sufficient.

- Each writing sample should be on a separate sheet of paper and should be removed from the sight of the writer before collecting the next sample.

- Requested writings should be collected by dictation, not copying. Dictation minimizes the chance of deliberate alteration of handwriting because the subject must concentrate on listening to the dictation.

- Dictated passages should be long. This will also help uncover attempts to deliberately alter handwriting because the longer someone is forced to write, the harder it is to make deliberate alterations to what is essentially a subconscious process.

- The requested writing should contain some words and phrases that are present in the questioned document. If there are misspelled words in the questioned document, they should also be given in the exemplar. The same holds true with mistakes in punctuation.

- To the extent possible, the subject should be supplied with the same type of writing instrument and paper used in the questioned document. The subject should also be made as comfortable as is practical, and there should be adequate lighting.

The temporal limitation of requested writings can sometimes be overcome by using **non-requested writing** exemplars. These are examples of the subject's writing that are taken in the normal course of business or personal transactions. They might include checks, bills, a diary, deeds, etc. Every effort should be made to obtain non-requested writings that were written around the same time as the questioned document. The major problem with non-requested writings is establishing their authenticity. In order for such writings to be admissible in court, there must be proof that the subject wrote them. This may be accomplished in a number of ways including having a witness who saw the subject write the exemplar, having the subject available to testify that the subject wrote the exemplar, or getting an exception to the hearsay rule under the business records exception. There have also been situations in which a questioned document examiner has compared non-requested writings with admitted writings and shown them to be written by the same hand. In most questioned document examinations, it is preferable to have both request and non-request exemplars.

HISTORY
The Clifford Irving Case

Howard Hughes was an eccentric, reclusive billionaire. Many people were fascinated with his life, but few people knew much about him. In 1971 the publisher McGraw-Hill and *Life* Magazine announced that they were going to publish an authorized autobiography of Hughes. This caused a sensation in the media, and it was predicted that the book would be an instant best-seller. The autobiography was said to have been written with the help of a freelance writer, Clifford Irving. Irving claimed that he had recorded Hughes' own words about his life during more than 100 meetings.

After the announcement, Howard Hughes came out of seclusion to lambaste Irving as a fraud and the supposed autobiography as fiction, worthy of a Hollywood script. Irving appeared on television claiming that the book was genuine and that Howard Hughes was just being his unpredictable, irascible self.

Several handwriting experts were retained to examine hundreds of pages of handwriting that Irving claimed were written by Hughes to Irving. These pages set out various aspects of Hughes' life and provided the basis for much of the autobiography. The consensus among responsible questioned document examiners who looked at the evidence was that the Irving papers were very good forgeries. Irving was convicted of fraud and sentenced to 2 1/2 years in prison. Since his release, he has written a number of fiction and nonfiction books, including a presumably authorized biography of the famous art forger, Elmyr de Hory.

Source: www.crimelibrary.com/gangsters_outlaws/cops_others/clifford_irving/9.html

SIGNATURES

Signatures can present special problems for questioned document examiners. In many cases, a person's signature does not represent typical handwriting, nor does it always contain the same individual characteristics as normal handwriting. A single signature may be the sole handwriting on the entire questioned document, giving rise to problems that the sample may be insufficient for a definite conclusion. Finally, signatures tend to be very sensitive to context. Consider the many situations in which a person is called upon to furnish a signature. Most people would sign an official document such as a will or property deed carefully, so there would be no doubt about the name. This would be a **formal signature**. An **informal signature** would be used in routine correspondence such as personal letters and other documents where the writer wants the reader to recognize the signature. Finally, there is the **abbreviated** or **stylistic signature** that would be used in signing checks, credit card receipts, etc. This can also be used to describe the physician's illegible signature on a prescription. It is often highly stylistic.

Because of the sensitivity to context of signatures, there are rules for the collection of exemplars of signatures. The circumstances under which the exemplar is collected should be as similar to the way that the questioned document was written as possible. Therefore, signatures are always collected in context. For example, if the questioned document is a forged check signature, the subject might be asked to fill out 10–20 blank checks for varying amounts and sign them.

Forged Signatures

Signatures are very often the subjects of forgery attempts. Depending on the circumstances, the forgery may be accomplished in a number of ways. If the signature is to be furnished on the spot as in a stolen check, the forger would attempt the forgery in one of two ways. Either she would practice it from a copy of the authentic signature beforehand and then try to duplicate the check-owner's signature on the check, or she would just write the check freehand, with no attempt to forge the signature. In this case the forger hopes the merchant won't pay much attention to the signature. Figure 20.2 shows how signatures are compared.

The forgery of a document that does not have to be signed in front of someone else may be done in different ways. It may be traced, for

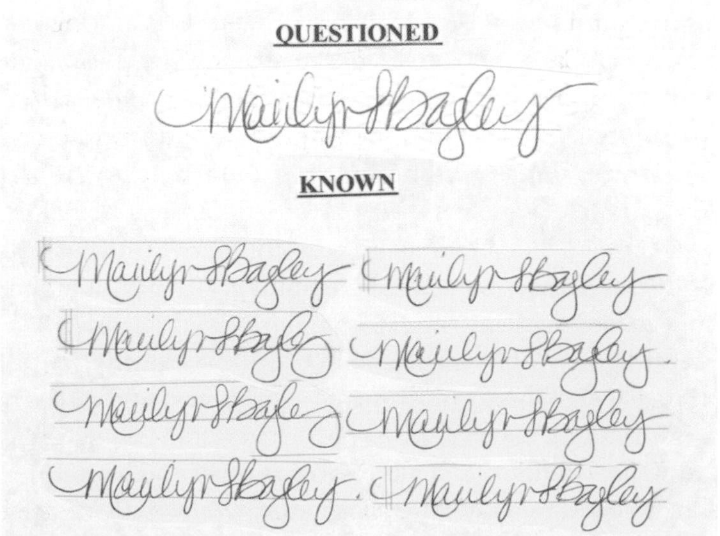

Figure 20.2. A *signature comparison. Courtesy of Robert Kullman, Speckin Forensic labs.*

example. This could be done using transmitted light or tracing by making an indented writing of the signature in a piece of paper and then retracing it. Most tracings of either type have characteristics that make them look artificial, and the tracing may be apparent to the trained questioned document examiner. As with all questioned documents, the key to discovering a **forged signature** is collecting sufficient numbers of exemplars.

PRINTED DOCUMENTS: TYPEWRITERS, COMPUTER PRINTERS, ELECTROSTATIC COPIERS

Printed documents are subject to different considerations than handwritten ones. Except under unusual circumstances, mass production of machines such as typewriters, printers, and copiers prevents individualization of a document to a particular machine. The only exception to this rule occurs when there is a defect in the printing or copying mechanism that results in the repeated appearance of an unusual or unique characteristic, or preferably several such characteristics. Typewriters are more likely to show these defects than other types of printing instruments.

TYPEWRITERS

Forensic document examiners are normally asked two questions about typewritten documents: Can the document be traced to a particular machine, and can the make and model of the typewriter be determined? The answer to the first question requires some knowledge about how typewriters operate and how unique characteristics are likely to arise. Older style typewriters generally work using a standard keyboard. When a key is struck, the corresponding raised, metal character mounted on a long stem strikes a ribbon that then leaves an inked imprint of the letter on the paper. As time passes, the metal character may become worn, chipped, bent, or misaligned. This gives rise to reproducible defects that can serve to identify the particular typewriter. Newer model typewriters, exemplified by the IBM Selectric models, employ a metal sphere with all of the characters raised on its surface. When a character is struck on the keyboard, the sphere is aligned so that the portion containing that character is lined up with the ribbon.

The sphere strikes the ribbon, leaving an inked impression on the paper. Because the characters are all on one sphere and are more rigidly held than if they are on individual stems, these Selectric typewriters are less prone to developing wear or misalignments in the typefaces, thus making individualization more difficult. This problem is compounded by the fact that the spheres are easily replaced, so a comparison of a typewritten document with a particular machine requires the original sphere that typed the document.

Determining the make and model of the typewriter that made a document requires that the examiner have a complete library of sample writings of every make and model of typewriter that is available. The font type and size are generally characteristic of a particular manufacturer's models of typewriters, so a questioned, typewritten document can be compared with library entries to determine the make and model used. A typewriting comparison is shown in Figure 20.3.

Typewriter Exemplars

When a questioned document is written by a typewriter, the best policy is to get the actual, suspect machine into the hands of the questioned document examiner. A number of variables with typewriters can affect the appearance of the type on paper, and they need to be controlled by the document examiner. For example, many typewriters permit the typist to adjust the pressure of the keystrokes to make the typed image darker or lighter. In addition, having the typewriter gives the examiner the opportunity to examine all of the typefaces as well as the ribbon. In the case of Selectric models, the typeface ball that was alleged to have made the questioned document must also be submitted.

If the typewriter cannot be submitted, then it will be necessary to get a complete set of all characters at all settings of pressure. It is also advisable to type the exact questioned document on the typewriter so the examiner can make direct comparisons. If possible, the ribbon should also be submitted for examination.

LASER PRINTERS AND COPIERS

Laser printers and copy machines that use Xerography© do not use ink to make characters on the paper. Instead, the characters are made using a toner made from finely divided carbon powder and binders

Figure 20.3. *A
typewriter comparison.
The arrows point to
unusual features of
certain letters made
by the typewriter in
question. The features
show up in both the
known and questioned
typewritings. Courtesy
of Robert Kullman,
Speckin Forensic labs.*

QUESTIONED

```
Do not call help
There is bomb in bank set
to go off at 5:45.
If we do not see or hear
police.  Will call location of bomb
at 5:40.
Put all 100's, 50's, 20's, 10's & 5's
in contain & send out.
My partner is right behind me.
Act natural.

P.S. I now know your name.
```

KNOWN

```
Do not call help
There is bomb in bank set
to go off at 5:45.
If we do not see or hear
police.  Will call location of bomb
at 5:40.
Put all 100's, 50's, 20's, 10's & 5's
in contain & send out.
My partner is right behind me.
Act natural.

P.S. I now know your name.
```

and a laser. There is virtually no chance that characters will become deformed over time. If the cartridge that contains the toner starts to run out, then the print quality may become irregular, but this will change with time and may not be of much help. In some situations, however, extraneous marks or blotches of toner may show up repeatedly in the same location until the machine is cleaned or repaired. If the questioned document and exemplars both show these same markings in the same location on the copy, this may provide individual information about the source of the questioned document. It would not be necessary for the examiner to have the printer or copier. Instead, whoever investigates the incident should make sure that an adequate number, at least a dozen, of exemplars are taken to show the degree of consistency of the extraneous markings.

If there are no reproducible, extraneous markings on the paper, and the source of the copy is not known, it may still be possible to determine the make of the copier or printer by chemical analysis and comparison of the toners with known samples. This evidence is not individual, however; it is normally not possible to determine with certainty that a particular machine was the source of a questioned document.

INK-JET PRINTERS

Ink-jet printers generally use water-based ink that is sprayed onto the paper for character and picture formation. As with laser printers and copiers, there is no issue of a user-defined pressure or darkness of the characters, and misalignment of particular characters is not normally an issue. In addition, as with laser printers or copiers, there is the chance that extraneous, reproducible markings may appear on the paper that may help in pinpointing the source of the copy. As the cartridge runs out of ink, the typescript may become uneven and fade. This is especially pronounced with color cartridges wherein the loss of one color ink will distort the colors that appear on the page. Again, however, this problem continues with time until there is no more ink or until the user changes cartridges.

If there are no reproducible imperfections from the printer, it may still be possible to pinpoint the printer manufacturer from chemical analysis of the ink. Of course, other class characteristics such as font type can be used to help associate a document with a computer and printer.

FAX MACHINES

Fax machines are similar to the machines that were discussed in the preceding sections with respect to ink or toner and the possible presence of extraneous, individual markings on the paper. In addition, however, facsimiles also possess a special header that describes some of the characteristics of the fax, such as originating and recipient fax numbers, date and time, etc. This header is called a Transmitting Terminal Identifier or TTI. The TTI can be a very important characteristic in the comparison of known and unknown facsimiles. It will usually be in a special font that is different from the text font, and attempts to forge TTIs and place them on a document are usually detectable by document examiners.

OTHER EXAMINATIONS PERFORMED BY DOCUMENT EXAMINERS

In addition to the comparison of handwritings and printed writings, questioned document examiners are sometimes called upon to perform a variety of other related examinations. One broad category is termed **document alterations**. They include obliterations, erasures, additional markings, indented writings, and charred documents. Another major category of analysis involves tools such as inks, papers, and pencil leads. Finally, there is the emerging issue of the age of documents, especially those that are handwritten using ink. This technology involves considerable skill and knowledge of chemistry and is presently being performed by only a handful of document examiners worldwide, and research into new methods of ink dating is presently being carried out by a number of investigators.

DOCUMENT ALTERATIONS

Obliterations

Obliteration is the overwriting of a sample of writing or printing with another writing instrument. It may be accidental or deliberate. The document examiner may be called upon to discover what is contained in the writing beneath the obliteration. Obliterations can be made with pens or pencils or other types of writing instruments.

The method used to discover what is written under an obliteration depends on the original writing and the means used. It may be possible to dissolve all or part of the obliteration or to use a light source that will be transmitted through the obliteration but not the original writing. This will permit the examiner to read the original writing underneath the obliteration. In some cases, it is possible to treat the paper with an oil that changes its refractive index so that the original writing can be viewed from the back of the paper. In some cases, a strong back-lit light source may penetrate the obliteration enough to read the underlying writing. A sample of obliterated handwriting is shown in Figures 20.4A and B.

Recently, several cases were encountered in which sections of printed documents were obliterated using a black marker. In one case,

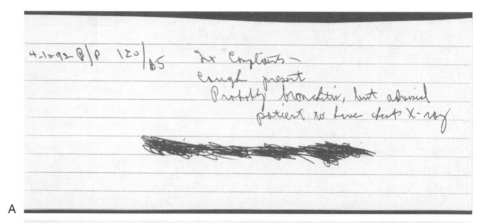

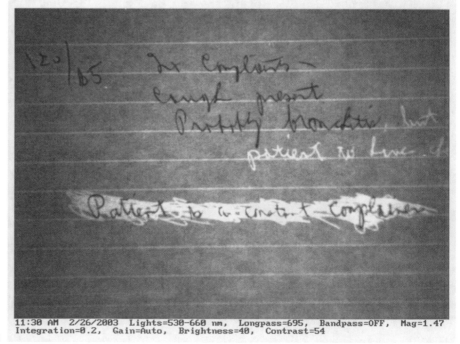

Figure 20.4.
Obliteration. (A) A portion of a patient's medical record with an entry obliterated. (B) The same record photographed with light of wavelengths 530–660 nm. Courtesy of Robert Kullman, Speckin Forensic labs.

soaking the document in methyl alcohol removed enough of the marker to see through it without damaging the printing below. In another case, ethyl alcohol did the same thing. Such soaking techniques may not be successful if the original writing is made with ink as opposed to toner because the solvent used to dissolve the obliterating agent may also dissolve the ink.

The use of infrared or ultraviolet light may enable the examiner to read underlying handwriting that has been obliterated. The wavelength of light is chosen so that it is transmitted through the obliteration to the ink below. If the ink then absorbs the light, it will appear dark. Sometimes the light will cause the ink to fluoresce in the visible or IR region, enabling it to be seen with the naked eye or a camera and special film or filters.

Erasures

Erasures can occur in a number of ways. The familiar **abrasive erasure** involves removing writing (usually that made with a pencil) with an abrasive eraser material. **Chemical erasures** involve dissolving or bleaching ink so that it is no longer visible. Typewriter erasures involve the use of a ribbon that lifts typewritten images from the paper. All of these erasure types can, in principle, be detected.

Abrasive erasures are the easiest to detect. They virtually always involve disturbance of paper coatings and fibers at the point of the erasure that can be seen with a low-power microscope. Chemical erasures may be detected by differences in shades of color in the paper from bleaching effects or by behavior of the chemicals in UV or IR light. Typewriter erasures can be detected by observing the indentations made by the typewriter in the paper. This can be done using oblique-angle photography and a low-power microscope.

Even though it may be relatively easy to detect that an erasure has taken place, it is usually more difficult to determine what was erased. An efficient erasure may be impossible to overcome, and the writing may never be reconstructed. Partial erasures may enable the examiner to read some or all of what was erased, as shown in Figures 20.5A, B, and C.

Indented Writings

In many situations a document is written on the top sheet of a pad of paper using a writing instrument, such as a ballpoint pen, that exerts pressure on the paper. This may result in an image of the writing being formed on one or more pages below. There are a number of ways of viewing this **indented writing**. The simplest and most popular method of viewing indented writings is with oblique lighting. If a light is directed across the surface of the page at an angle, the indentations may cast enough shadow on the paper to reveal the contents of the writing. This technique may work for several pages below the top sheet, especially if a high-resolution digital camera is used to capture the image. Sometimes TV or movies will show indented writing being visualized by rubbing the indented area with the side of a pencil. This will not only fail to visualize the indented part of the writing, but will destroy the evidence so that workable methods cannot be used.

There is also an instrumental method for capturing indented writings. It is called the **Electrostatic Detection Apparatus** (ESDA). This apparatus works a bit like an electrostatic copy machine, in that it uses a toner which collects within the indented writings so that they can be visualized. Figure 20.6 illustrates how ESDA is used.

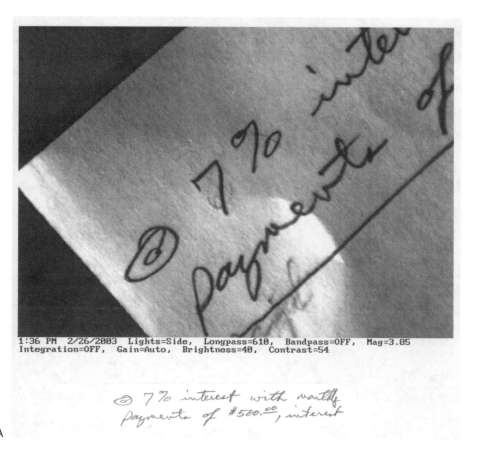

Figure 20.5. *Erasures.*
Original documents
read "6% interest . . . "
(A) A pencil erasure.
(B) A typewriter erasure.
(C) A chemical erasure.
Courtesy of Robert
Kullman, Speckin
Forensic labs.

A

B

2:52 PM 2/26/2003 Lights=Side, Longpass=OFF, Bandpass=OFF, Mag=8.18
Integration=OFF, Gain=Auto, Brightness=40, Contrast=54

@ 7% interest with monthly
payments of $500.00, interest

C

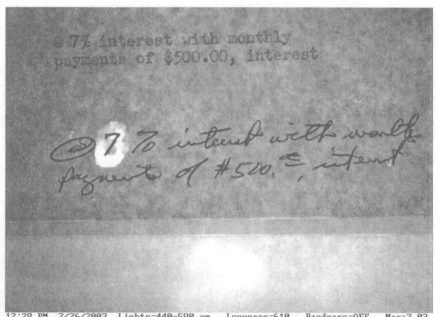

12:28 PM 2/26/2003 Lights=440-580 nm, Longpass=610, Bandpass=OFF, Mag=2.03
Integration=0.2, Gain=Auto, Brightness=40, Contrast=54

@ 7 % interest with monthly
payment of #500.⁼ interest

Figure 20.5.
Continued

Figure 20.6. *The use of an Electrostatic Detection Apparatus (ESDA). This is a note given to a bank teller during a robbery. ESDA reveals what was written on the piece of paper above the note in the pad of paper on which the note was written. Courtesy of Robert Kullman, Speckin Forensic labs.*

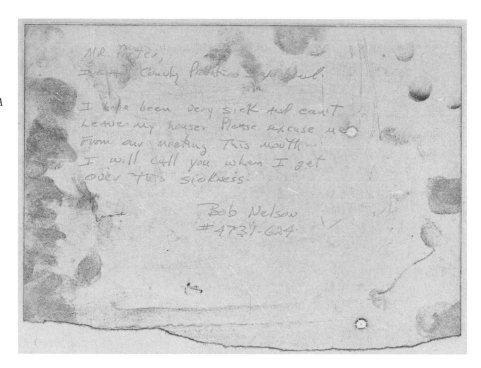

Charred Documents

In many civil infractions or crimes, the perpetrator tries to destroy documentary evidence that would otherwise be incriminating. This can be conveniently accomplished by burning the documents. If the documents are recovered before being completely destroyed, there is a chance that some of the writing can be identified. The reason is that, although the paper may become charred, the ink or pencil may not char as easily.

One of the problems with **charred documents** is that they are very fragile. Sometimes they can be strengthened by misting them with a lacquer or alcohol or water. This may provide enough strengthening to enable the examiner to preserve the document between sheets of glass or plastic.

Analysis of charred documents is carried out somewhat like that for obliterations. The key is that the ink or toner is preserved, so the goal is to provide as much contrast as possible between the darkened, charred paper and the ink. If the ink fluoresces with UV or IR light, then photography using specialized films may be used along with the proper lighting. Even intense incandescent light sources have some-

Figure 20.7. *Charred writing. Through the use of infrared light, the figures made with the ink fluoresce against the charred background.*

times provided the needed contrast to visualize the writing. See Figure 20.7 for an example of the analysis of charred writing.

INK ANALYSIS

In recent years forensic document examiners have become more comfortable performing chemical analyses as part of their routine work. This has been aided by research into the composition of inks that has led to the development of easier methods of characterizing and comparing inks. Understanding the composition of an ink sample and the chemical changes that it undergoes as it dries can be very important in several types of document cases. For example, in a number of questioned document cases an examiner may be called upon to determine if two documents could have been written using the same writing instrument, such as a ballpoint pen. In other cases a question arises as to whether a document could have been altered by means of adding writings after the original document was written, for example, a check that was altered by adding zeros to the amount of the check. Other cases involve a question of whether a document could have been altered by adding further writings and then backdating them to make it seem as if they were put in at an earlier date.

The first two of these questions involve the chemical analysis of an ink sample and comparison of two or more writings. Although it is not possible to individualize an ink sample to a particular writing instrument, it is possible to show that a suspect pen, for example, could be included in the population of pens that could have written a document, or that it could not have possibly written a document. The last question involves determining the age of the ink sample either relative to other writings on the document or by estimating the actual age of the ink. This involves knowledge of what happens to the ink as it dries.

Analyzing and Comparing Ink Samples

There are many different types of inks as evidenced by the myriad types of pens available today. They range from the familiar ballpoint pens, roller ball pens, fiber or porous tip pens, gel pens, and the venerable India ink pens and fountain pens. There are also ink-jet inks used in some computer printer cartridges. The inks for each of these pen and printer types are formulated especially for the ink delivery system.

Inks can be quite complex materials. For example, modern ballpoint pen inks contain a solvent such as ethylene glycol and dyes. There may also be drying agents and other additives in an ink formulation. When the composition of an ink sample is determined, it can help determine what type of writing instrument it came from.

SAMPLING

Sampling ink on a questioned document presents unique problems because the legibility of the handwriting or printing must be maintained so that the handwriting characteristics are not destroyed. This means that the samples of ink taken for analysis must be very small. Sampling is typically accomplished by using a blunt-point syringe needle that takes tiny plugs from individual letters in the writing. Many plugs may be taken from a writing sample as long as there are sufficient writings available to maintain legibility. The number and size of the plugs will, of necessity, dictate the amount and type(s) of testing that can be done.

THIN LAYER CHROMATOGRAPHY

The most popular and one of the easiest methods of ink analysis and comparison is TLC. Normally, about 10 plugs of ink are dissolved in a minimum amount of a solvent such as methanol and then spotted at the bottom of the plate. When the plate is developed, there will usually

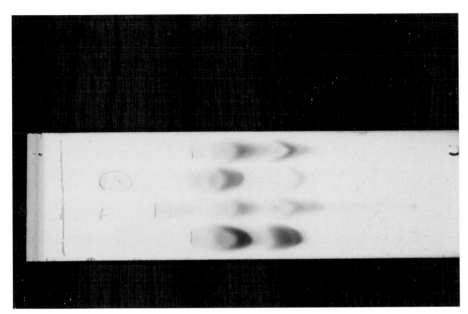

Figure 20.8. *Thin layer chromatography of inks. Lanes one and three are standard black ballpoint pen inks. Lanes two and four are different black inks. Courtesy of Jamie Dunn, Michigan State University, Department of Chemistry.*

be several spots from the dyes in the ink, since there may be several dyes used to formulate that particular color of that ink. No visualization of the plate is necessary since the dyes are already colored. In addition to the dye peaks, there may be other peaks from some of the other materials present in the ink. These would not be colored, so some sort of visualization would be necessary. Figure 20.8 shows a TLC of some ink samples.

OTHER TYPES OF CHROMATOGRAPHY

Both gas chromatography and liquid chromatography have been successfully employed in the analysis and comparison of inks. Their advantage over TLC is that they yield quantitative as well as qualitative data, and they are both generally more sensitive than TLC, which means that they require fewer plugs of ink. In addition, GC-MS is often used, allowing identification of each of the components in the ink.

A new technique in chromatography, capillary electrophoresis, has also been applied to ink analysis, although the research has been very limited thus far. Capillary electrophoresis is similar to HPLC but uses tiny columns for separation so as little as one or two plugs may be used for analysis. This has great advantages in cases in which the sample is very limited.

INFRARED SPECTROPHOTOMETRY

Infrared spectrophotometry has also been used for characterizing inks. This technique shows absorption peaks for all of the components of the ink at one time, including the solvent, dyes, and additives. Because of this, IR can be very helpful in comparing two ink samples to see if they could have originated from the same source. One of the disadvantages of IR is that it requires more sample than do chromatographic methods in general. It is difficult to use microplugs for sampling in IR. Various sampling methods have been tried, including using a microscope to focus on microplugs, with mixed success.

MASS SPECTROMETRY

Mass spectrometry, with and without gas chromatography, has been used for the analysis of inks. One new type of MS, called **laser desorption**, uses a laser to remove ink from the surface of paper and then analyze it. An advantage of this type of mass spectrometry is that the ink does not have to be removed from the paper first. A piece of paper with ink writing on it can be directly introduced into the instrument. It is also essentially non-destructive. Laser desorption can be used to track an ink dye as it ages. When dyes age, they undergo chemical degradation. One popular ballpoint pen ink dye, Methyl violet, degrades by losing CH_3 groups, replacing them with hydrogen atoms. This results in the loss of 14 mass units from the molecule. Figures 20.9A and B show how Methyl violet degrades and an aging study on a typical ballpoint pen.

MICROSPECTROPHOTOMETRY

One of the more difficult problems in ink analysis is to determine if a specimen of writing is the same color as the ink from a suspect pen. The human eye is an excellent discriminator of color but obviously cannot give objective data about the color of an ink. The exact color of an ink (or any other colored material) is defined by the wavelengths of visible light that are either absorbed or transmitted by the dyes in the ink. If an examiner is working with very small samples of ink, then a visible microspectrophotometer is the ideal instrument. It determines the absorption or transmission spectrum of the dyes. If two inks are the same color, their visible spectra will be the same.

Document Dating

In addition to analyzing and comparing inks, questioned document examiners are sometimes called upon to determine if the age of a doc-

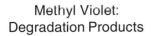

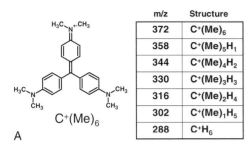

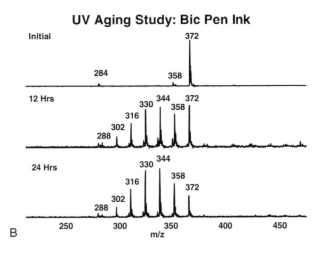

A

B

Figure 20.9. (A) The degradation of Methyl violet. Each degradation causes the replacement of a Methyl group with a hydrogen atom, resulting in the loss of 14 mass units. (B) An artificial aging study of Methyl violet dye in a blue ballpoint pen, by tracking with laser desorption mass spectrometry.

ument is consistent with what it is purported to be. For example, in the case cited at the beginning of the chapter, one of the questions that needs to be addressed is whether the will was written at a time consistent with its circumstances. Another example involves a situation in which some writing on a document is not written at the time it is dated. These cases involve determining the age of the document. The most common way of doing this is by determining the age of the ink on the document. For example, a few years ago there was a case in Michigan in which a physician was accused of entering notes about a particular medical procedure in a patient's file after the patient had died. A questioned document examiner was called upon to determine if the notes were written at the time that the date on them indicated.

There are a number of ways of estimating the age of an ink sample. The most common method is to examine its behavior as it dries. When ink dries, a number of chemical processes are going on at the same time; for example, the solvent in the ink is constantly evaporating as the ink dries. As this happens, some of the components of the ink may form polymers (e.g., resins). In addition, the dyes in the ink will change in chemical composition over time. One way of tracking the drying of ink with time is to determine how easily it will redissolve in a suitable solvent. The drier the ink, the harder it will be to get it to redissolve. This process can be tracked by thin layer chromatography. Infrared spectrophotometry can also track this drying process. As the solvent evaporates, some IR peaks will decrease. As polymers form in the ink, some new peaks will appear.

As the dyes in inks age, their chemical compositions change. This process can be tracked by mass spectrometry or some types of chromatography. For example, as the composition of a dye changes, its molecular weight will change, and this will be reflected in its mass spectrum.

These techniques can be used to determine if one set of writings on a document is appreciably older or younger than others or if a document as a whole is as old as it is purported to be. It should be kept in mind that these processes that age documents take place over a long period of time, and measuring these changes may not be entirely accurate. It may take several weeks or months before a change in the ink is reliably detectable.

ARTIFICIAL AGING OF INKS

Since inks age slowly over time it is necessary to have a reliable method of speeding up the aging process so that drying processes may be conveniently studied. The only method that has been studied to any great extent is to heat the document in an oven. It has generally been shown that heating at 100°C for a few minutes or hours can age a document by several months.

Since heating may cause unwanted interactions between ink and paper, other methods of artificial aging are also being studied. They include using UV light and using an oxygen-rich atmosphere. Whatever process is used, it must age the ink in the same way that natural aging takes place; that is, the same chemical processes must take place.

BACK TO THE CASE

In the case cited at the beginning of the chapter, there were two wills. One of the issues is which, if any, were written by the diseased. In order to settle this issue, the questioned document examiner would have to have exemplars (known writing samples) that were written at around the same time that each will was alleged to have been written. Since the wills were supposed to have been written just two years apart, one set of known writings would most likely suffice here. It should be kept in mind, however, that a person's handwriting will be affected by the general health of that person. If the woman's health had deteriorated badly between the times that the two wills were purported to have been written, her handwriting may have changed significantly. It must also be noted that the exemplars will be **non-requested**; that is, they will be writings that she made during her normal course of life, and the conditions under which they are obtained cannot be controlled. For this reason, it is necessary to collect as much exemplar writing as possible.

The other issue is whether the wills, especially the second, were written back in 1972, or were written after the woman's death in order to collect her money. Ink dating techniques can be useful here in determining if the ink is as old as it would have to be if the document were written back in 1972.

CHAPTER SUMMARY

A questioned document is any piece of writing or printing whose source or authenticity is in doubt. Questioned document examiners undergo years of apprenticeship training before being certified to take their own cases. Questioned document examination involves many activities, including handwriting comparisons to determine the source of handwriting, typewriting or printing comparison, analysis of paper and inks, and restoration of altered documents. Examiners also determine if a document has been forged or fraudulently produced.

Handwriting comparisons are the most familiar of the tasks of the document examiner. Many characteristics of handwriting are characterized, and it is necessary to have sufficient, timely, representative known samples of the subject's handwriting for comparison. There is no standard number of characteristics that must be found in the known and unknown samples, in order for the examiner to determine that a particular person wrote a document. There is a similar need for reliable known samples of typewriting, computer printing, or electrostatic copiers.

Inks and papers are being given more attention in recent years. This especially applies to inks, where much research is being carried out to identify dyes in the inks. This analysis can also be used to determine the age of a writing sample by tracking the chemical aging of the dyes.

Document examiners also examine charred and obliterated documents and documents in which one line is written over the top of another one.

Test Your Knowledge

1. Why is it more difficult to distinguish handwriting among a group of third graders than among adults?

2. Explain the precautions that should be taken when obtaining requested handwriting samples to check against a questioned document.

3. Why would it be easier over time to individualize a sample of typewriting to a particular typewriter than it would be to do so with a computer printer page to a printer?

4. What are some of the major characteristics of forged handwriting?

5. Give an example of a handwritten questioned document that is not written on paper (or similar materials). What special procedures might have to be used in such a case?

6. What characteristics of ink are most useful for comparing known samples with unknowns?

7. How can chemical erasures be detected on a document?

8. How can mechanical erasures be detected on a document?

9. Suppose someone tries to alter a check written for $100 by adding a zero so that it reads $1000. What are some of the ways this can be detected?

10. How should requested writings be taken if the entire questioned document is a check with an allegedly forged signature?

11. What is the definition of a questioned document in the broadest sense?

12. What is "ink dating"? What is its purpose?

13. How many points of identification are necessary for a questioned document examiner to be able to declare that a questioned document was written by a particular person?

14. What is the difference between requested writings and non-requested writings?

15. What is ESDA? What is it used for?

16. What is the best way of deciphering indented writings? Why isn't rubbing the indented writing with the side of a pencil a good idea?

17. What are some ways that an examiner could determine some pen handwriting that has been obliterated with another pen?

18. What is a charred document? How can the writing on a charred document be deciphered?

19. How does a person become a questioned document examiner? Why is this considered to be an apprenticeship field?

20. What type of certification is there for questioned document examiners?

Consider This . . .

1. How does handwriting vary over time? At what time of life is the change the greatest? How is this taken into account when comparing questioned documents to exemplars?

2. What is the basis for the conclusion that a handwritten questioned document can be matched to a particular writer? In what way is the status of handwriting comparison changing forensically?

3. What is "graphology" (or graphoanalysis)? What relationship does it bear to questioned document analysis? Many questioned document examiners do not trust or agree with the principles of graphoanalysis. What problems has this caused?

BIBLIOGRAPHY

Ellen, D. (1997, August). *The scientific examination of documents: Methods and techniques* (2nd ed.) New York: Taylor & Francis.

Hilton, O. (1992, June). *Scientific examination of questioned documents.* Boca Raton, FL: CRC Press.

Hilton, O. (1997, March). *Detecting and deciphering erased pencil writing.* Springfield, IL: Charles C Thomas Pub Ltd.

Robertson, E. W. (1991, July). *Fundamentals of document examination.* Chicago: Nelson-Hall Co.

www.crimelibrary.com/gangsters_outlaws/cops_others /clifford_irving/9.html

Firearms and Tool Marks

KEY TERMS

Bore

Breech block

Breech face

Breech marks

Broach

Bullet wipe

Caliber

Cartridge

Choke

Ejection marks

Extraction marks

Firing pin impression

Gauge

Grains

Grooves

GSR particles

Gunshot residues

Lands

Modified griess test

Muzzle-to-target distance

Primer

Rifling

Rifling button

Smokeless powder

Sodium rhodizinate

Striations/striae

Trigger pull

Twist

INTRODUCTION

Firearms examination is one of the key services a forensic science laboratory provides; even smaller laboratories with only a few employees will probably have a firearms examiner. Many crimes are committed with a firearm, to coerce cooperation or directly harm, and society has judged this implied or actual violence to be a severe crime. Firearms examination is complex, technical, detailed—and experiencing a renaissance with the development and growth of automated database searches. This computerization promises to revolutionize the nature of firearms examination and, perhaps, forensic science.

In 1863, Confederate General Stonewall Jackson was fatally wounded on the battlefield during the U.S. Civil War. The deadly projectile was excised from his body and, through examination of its size and shape, determined to be .67 caliber ball ammunition. This was not the .58 caliber minie ball used by the Union army but ammunition typical of the Confederate forces—Jackson had been shot by one of his own soldiers. In 1876, a Georgia state court allowed the testimony of an expert witness on the topic of firearms analysis. These are the first examples of firearms analysis and testimony in the United States. (Thorwald, 1964; Wilson and Wilson, 2003).

In a report issued by the Bureau of Alcohol, Tobacco, and Firearms nearly 125 years after Jackson's death (ATF, 2000), over 84,000 guns were trafficked illegally in the United States, and over 1,700 defendants were charged with illegally trafficking guns. The National Crime Victimization Survey (2002) found that in that same year, 533,470 victims of serious violent crimes faced an offender who had a firearm. While the number of crimes committed with firearms has stabilized in the past few years, it is still over 350,000 (see Figure 21.1). In 2000, about 66% of murders, 41% of all robberies, and 18% of all aggravated assaults that were reported to the police were committed with a firearm. Based on data from the National Center for Health Statistics, in 1999, of the deaths that resulted from firearms injuries, 38% were homicides, 57% were suicides, 3% were unintentional, and 1% were of undetermined intent (see Table 21.1). From these and other data, it is obvious that the analysis of firearms plays a central role in a forensic science investigation.

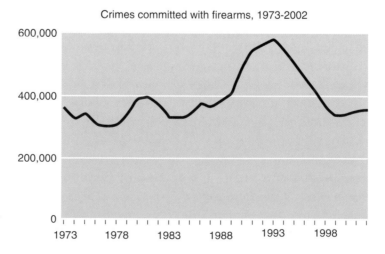

Figure 21.1. Over 350,000 crimes are committed with handguns each year in the United States. Crimes include the Uniform Crime Report index offenses of murder, robbery, and aggravated assault. Source: FBI, The Uniform Crime Reports (UCR).

Table 21.1. *Of all the deaths from firearms, about 96% are either homicides or suicides; over a third of all deaths by firearms are homicides.*

		NUMBER OF FIREARM DEATHS				PERCENT OF ALL FIREARMS DEATHS			
	ALL	UNINTENTIONAL	SUICIDE	HOMICIDE	UNDETERMINED	UNINTENTIONAL	SUICIDE	HOMICIDES	OTHER
1991	38,317	1,441	18,526	17,986	364	4	48	47	1
1992	37,776	1,409	18,169	17,790	408	4	48	47	1
1993	39,595	1,521	18,940	18,571	563	4	48	47	1
1994	38,505	1,356	18,765	17,866	518	4	49	46	1
1995	35,957	1,225	18,503	15,835	394	3	51	44	1
1996	34,040	1,134	18,166	14,327	413	3	53	42	1
1997	32,436	981	17,566	13,522	367	3	54	42	1
1998	30,708	866	17,424	12,102	316	3	57	39	1
1999	28,874	824	16,599	10,828	324	3	57	38	1

Source: National Center for Health Statistics, retrieved from www.cdc.gov/nchs.

FIREARMS

The field of forensic firearms examination is sometimes referred to as "ballistics" or "forensic ballistics." This terminology is not wholly accurate: Ballistics is the study of an object in flight and is under the domain of physics. "Forensic ballistics" may be somewhat more accurate but does not capture what forensic firearms examiners do in their job. They certainly are not analyzing the trajectories of bullets *while* they are in flight! Many of the principles, equations, and methodologies of ballistics are used, for example, to reconstruct a shooting incident. But the discipline of forensic firearms science is more than that and encompasses the study of firearms; their manufacture, operation, and performance; the analysis of ammunition and its by-products (such as muzzle-to-target distance and gunshot residue); and the individualizing characteristics that are transferred from firearms to bullets and cartridge cases.

The examination of tool marks is related to firearms in many ways but is also very different in others. It requires an understanding of the way in which tools are made and used; additionally, it includes the restoration of serial numbers that a criminal has attempted to obliterate.

TYPES OF FIREARMS

Very generally, firearms can be divided into two types: handguns and shoulder firearms. Handguns include revolvers and pistols, whereas shoulder firearms are more diverse, encompassing rifles, shotguns, machine guns, and submachine guns. Broad knowledge and familiarity with the various types, makes, models, and styles of firearms are crucial to being a successful forensic firearms scientist. This knowledge and familiarity should cover not only new products as they emerge on the market but also older models and the history of manufacturers and their products. Each year, at the Sporting, Hunting, and Outdoor Trade Show and Conference (SHOT Show), hundreds of exhibitors, many of them firearms related, display their new products and give out product information; in 2002, over 11,000 people attended the show (see Figure 21.2). The SHOT Show is an excellent source of information about firearms and related products.

Figure 21.2. The largest sporting goods show of its kind, the Sporting, Hunting, and Outdoors Tradeshow (SHOT Show), is a good source of information about firearms and their manufacturers. From **www.shotshow. org**, *with permission by the National Shooting Sports Foundation.*

Handguns are firearms designed to be fired with one hand. These firearms appear in two major types: revolvers and (semi)automatic pistols (see Figures 21.3A and B). A revolver is a handgun that feeds ammunition into the firing chamber by means of a revolving cylinder. The cylinder can swing out to the side or be hinged to the frame and released by a latch or a pin for loading and unloading. A single-action revolver requires that the hammer be cocked each time it is fired; a double-action revolver can be cocked by hand or by the pulling of the trigger, which also rotates the cylinder.

A (semi)automatic pistol, on the other hand, feeds ammunition by means of a spring-loaded vertical magazine. Although the term "automatic" is often applied to pistols fed by magazines, they are not truly automatic in their firing. An automatic firearm is one that continues to fire ammunition while the trigger is depressed; a semiauto-

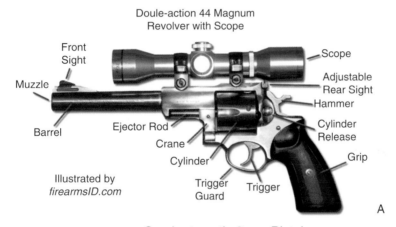

Figure 21.3. (A) A revolver is a handgun that feeds ammunition via a revolver cylinder (hence the name), whereas (B) a pistol feeds ammunition through a spring-loaded magazine. Courtesy of www.FirearmsID.com, artwork by Erik Dahlberg.

matic firearm fires one bullet for each pull of the trigger. When fired, semiautomatic pistols use the energy of the recoil and the sliding of the **breech block** (slide) or the recoil of the cartridge to expel the empty cartridge from the firearm and load a live round into the firing chamber. Springs are used to store the energy and expend it.

Shoulder arms consist of rifles, automatic rifles, machine guns, and shotguns. Rifles are designed to be fired from the shoulder with two hands (see Figures 21.4A and B). Rifles may be single-shot, repeating, semi-automatic, or automatic. A single-shot rifle must be loaded, fired, the cartridge extracted, and then reloaded; these rifles were common as a young boy's first firearm just after the turn of the century but are almost non-existent now. Repeating rifles fire one bullet with each pull of the trigger, but the expended cartridge must be expelled, cocked, and reloaded from a magazine manually. Repeating rifles may be bolt-action (like the M1 from war movies or many hunting rifles) or lever-action (made popular by cowboy movies). Semi-automatic rifles use the energy of the fired ammunition to expel the empty cartridge, cock the firing mechanism, and reload a live round; thus, one pull of the trigger fires one round, and this may be done sequentially until the magazine

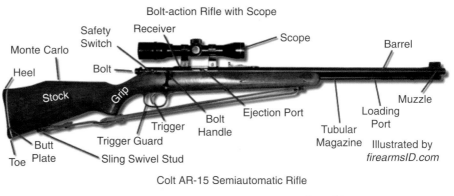

Figure 21.4. Rifles are firearms designed to be fired with two hands— one to pull the trigger, the other to stabilize the barrel for aiming. Rifles can be single-shot (A), repeating, or assault rifles (B). Courtesy of www.FirearmsID. com, artwork by Erik Dahlberg.

is empty. Assault rifles, like the AK-47 or M-16, can be fired either like semi-automatic rifles or in automatic mode: Pull the trigger, and the firearm will fire ammunition continuously until all the ammunition is gone. A machine gun is a fully automatic firearm and therefore is fed ammunition from a high-capacity belt or box. Because of their size and the strength of the recoil, machine guns are meant to be fired from a tripod or other mounted/fixed position. A submachine gun is a machine gun meant to be fired while held in the hands.

FIREARM BARRELS

The interior surface of the barrels of the firearms discussed so far (handguns and rifles, but not shotguns) are **rifled** with a series of ridges and valleys, called **lands** and **grooves**, respectively, that spiral the length of the barrel (see Figure 21.5). The lands dig into the bullet surface as it travels down the barrel, imparting spin to stabilize the bullet's flight once it leaves the barrel. This creates land and groove impressions on the bullet surface as well as impressions of the microscopic imperfections of the interior barrel surface called **striations** or **striae** (see Figures 21.6A and B).

Figure 21.5. Spiral grooves are cut into the inner surface of a firearm barrel to impart spin to the bullet as it leaves the barrel, stabilizing its flight. The raised portions between the grooves are called lands. Courtesy of Richard Ernest, Alliance Forensics, Inc.

Figure 21.6.
Imperfections in the surface of the tool (A) that cuts the rifling grooves are transferred to the barrel's inner surface (B). These striations, or striae, are then transferred to a bullet's outer surface when it is fired. Striations are considered to be unique to a particular barrel, through manufacturing and use. These are on lead bullets fired from a revolver. Courtesy of www.FirearmsID. com, *artwork by Erik Dahlberg.*

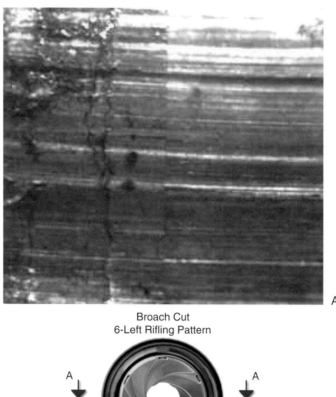

A

Broach Cut
6-Left Rifling Pattern

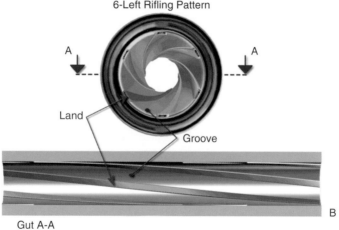

Land

Groove

Gut A-A

B

During manufacture of a barrel, a hole is drilled down the length of a steel bar of the proper size for the intended firearm. The grooves are cut into the barrel by either a large segmented tool, called a **broach**, or a **rifling button**, a stiff metal rod with a flanged tip, which is run down the length of the hole. When the grooves are cut, they are cut in a spiral of a certain direction or **twist** (right-handed/clockwise or left-handed/counterclockwise); this is what spins the bullet and creates a stable flight path. Some manufacturers produce barrels with four

Pump-Action Shotgun

Heel · Comb · Stock · Safety Switch · Receiver · Bolt · Double Action Bars · Barrel · Front Sight · Muzzle · Grip · Trigger · Trigger Guard · Butt Plate · Sling Swivel Stud · Toe · Release for Action Bar Lock · Magazine · Magazine Loading Port · Fore-End · Magazine Cap · Illustrated by firearmsID.com

Figure 21.7. *A shotgun is essentially a rifle that doesn't shoot bullets. Instead, it fires many small, round pellets or a single large slug. Therefore, shotgun barrels do not have rifling and have a limited effective firing range. Courtesy of* www.FirearmsID. com, *artwork by Erik Dahlberg.*

grooves, some with five or six, depending on the design and desired performance of the firearm.

The interior or **bore** diameter of a rifled barrel is the diameter of a circle that touches the tops of the lands. The **caliber** of a firearm used to mean the same thing as bore diameter but now refers mostly to the size of a particular ammunition cartridge; firearms are still referred to in their nominal caliber, however. A barrel's internal diameter is an exact measurement, whereas caliber is an approximation; the barrel of a .38 firearm may actually measure between 0.345 and 0.365 inches (also note that calibers do not use the zero before the decimal). The caliber of American and British ammunition is typically measured in inches, and all others are measured in millimeters (a Smith and Wesson .32 versus a Beretta 9 mm).

Shotguns can fire numerous projectiles, called pellets or "shot," of varying sizes (see Figure 21.7, and Table 21.2); they may also fire single projectiles called "slugs." A single-barrel shotgun can be either single-shot (manually loaded) or repeating-shot in design (with a spring-loaded auto-feeder or manual pump feeder with a reservoir of three to five shells). The interior of a shotgun barrel is smooth so that nothing deflects or slows down the pellets as they traverse its length. The muzzle of a shotgun barrel may be constricted by the manufacturer to produce a **choke**, which helps to keep the pellets grouped longer once they leave the barrel. The influence of choke on the shot pattern increases with the distance the pellets travel; the range of a shotgun is, compared to rifles, short, but the choke can improve the chance of hitting targets at near-to-mid ranges (see Table 21.3). The choke may also be modified by barrel inserts.

The diameter of the shotgun barrel is called **gauge** and is the number of lead balls with the same diameter as the barrel that would

Table 21.2. *The size of pellets is organized numerically, except for the two largest, "BB" and "00" (or "double ought") buckshot.*

PELLET SIZE	DIAMETER (INCHES)
9	0.08″
8	0.09″
7	0.10″
6	0.11″
5	0.12″
4	0.13″
2	0.15″
1	0.16″
BB	0.18″
00 Buck	0.33″

Table 21.3. *Choke is the measure of constriction of a shotgun barrel, intended to group the pellets and produce a tighter pattern at impact. Some shotgun barrels may have their choke modified by a removable insert.*

CHOKE	PELLETS THAT FALL WITHIN A 30-INCH CIRCLE AT 40 YARDS
Full-choke	65–75%
Modified choke	45–65%
Improved cylinder	35–45%
Cylinder bore	25–35%

Table 21.4. *The size of a shotgun barrel is measured in gauge, except for the smallest, which is labeled a "410" ("four-ten") because the barrel's internal diameter is 0.410 inches wide.*

GAUGE	INCHES	MILLIMETERS
10	0.775″	19.68 mm
12	0.729″	18.52 mm
16	0.662″	16.82 mm
20	0.615″	15.62 mm
410	0.410″	10.41 mm

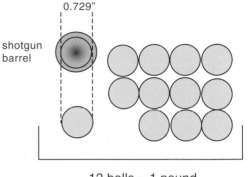

12 balls = 1 pound
12 gauge shotgun

weigh one pound. For example, 12 lead balls, which together weigh one pound, have the same diameter as the interior of a 12-gauge shotgun (about 0.729″). The exception to this system is the so-called 410-gauge shotgun, which has its bore diameter measured in inches (0.410″) (see Table 21.4).

ANATOMY OF AMMUNITION

Ammunition is what a firearm fires; it is typically a self-contained **cartridge** that is composed of one or more projectiles, propellant (to act as fuel), and a primer (to ignite the propellant). As with firearms, ammunition comes in two major types: cartridges, for handguns and rifles; and shells, for shotguns (see Figures 21.8A–C).

Bullets, the first type of projectile, can be classified as lead (or lead alloy), fully jacketed, and semi-jacketed. Lead (alloy) bullets are lead hardened with minute amounts of other metals (such as antimony) and formed into the desired shape. Although hardened, they are too soft to use in most modern firearms other than .22 rifles or pistols. A fully jacketed cartridge has a lead core, which is encased in a harder material, usually copper-nickel alloys or steel. A semi-jacketed cartridge has a metal jacket that covers only a portion of the bullet with the nose often exposed. Because the nose of the bullet is softer than the surrounding jacket, the tip expands or "mushrooms" on impact, transferring its energy to the target. A hollow-point cartridge is a

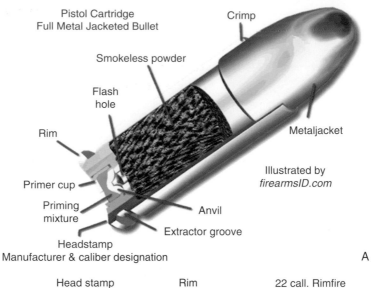

Pistol Cartridge
Full Metal Jacketed Bullet

Crimp

Smokeless powder

Flash hole

Rim

Metaljacket

Illustrated by
firearmsID.com

Primer cup

Priming mixture

Anvil

Extractor groove

Headstamp
Manufacturer & caliber designation

A

Head stamp
Manufacturer & caliber designation

Rim

22 call. Rimfire

Brass case

Lead bullet

Primer
Mixture

Smokeless powder

Illustrated by
firearmsID.com

B

Figure 21.8.

Ammunition comes in two basic types: bullets and shells. (A&B) Bullets have a single solid projectile, a propellant, a primer to ignite the propellant, and a casing to hold the components together until fired. (C) Shells are similar to bullets, except that the pellets or shot must also be held together and wadding is inserted between them and the propellant. The wadding maintains an even pressure on the pellets, pushing them all out of the barrel at about the same time. Courtesy of www.FirearmsID. com, *artwork by Erik Dahlberg.*

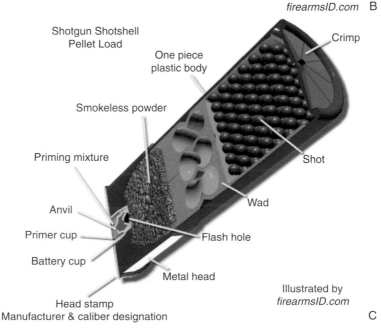

Shotgun Shotshell
Pellet Load

One piece
plastic body

Crimp

Smokeless powder

Shot

Priming mixture

Anvil

Wad

Primer cup

Flash hole

Battery cup

Metal head

Head stamp
Manufacturer & caliber designation

Illustrated by
firearmsID.com

C

semi-jacketed bullet that has a hollowed-out tip to increase this effect. Some semi-jacketed cartridges leave the base exposed but cover the tip; they have a greater penetrating power due to the hardness of the tip material and tend to pass through the target.

Shotguns, as noted previously, can fire pellets or slugs. Dozens of varieties of projectiles, from explosive bullets to "safety" ammunition consisting of pellets in a small sack to disable airline hijackers, are currently available and may be encountered in casework.

The propellant is the fuel that propels the projectile down and out of the firearm's barrel. Black powder, the first propellant to be used in firearms, was invented across numerous cultures at various times. The most common formulation for black powder is 15 parts charcoal, 75 parts potassium nitrate, and 10 parts sulfur, but there are others. Black powder is now pressed into cakes and broken up as needed (this prevents the separation of the mixed components); the size of the pieces, or **grains**, controls the rate of burning, with the smallest burning fastest. The grains are sorted by size and small grains are used for handguns, medium grains for shotguns and small rifles, and large grains for larger rifles. Because it is still in use today, mostly with black powder hunting enthusiasts and Civil War re-enactors, forensic firearms scientists must be familiar with this oldest of propellants.

The more common by far propellant is **smokeless powder**, which was developed in response to the huge plumes of smoke that black powder produces upon ignition. Smokeless powder is composed of nitrocellulose combined with various chemicals to stabilize the mix and modify it for safe manufacture and transport.

The **primer** is what ignites the propellant. It consists of a small metal cup containing a percussion-sensitive material (it explodes on impact) that, when struck, creates enough heat to ignite the propellant. The small cup is set in place at the rear of the cartridge, where it is struck by the firing pin. Modern primer materials consist of lead styphnate, antimony sulfide, barium nitrate, and tetracene. Because of the concerns of toxicity over long-term exposure to law enforcement officers, many primers are now made from organic primers that are lead-free.

WHAT HAPPENS WHEN AMMUNITION IS DISCHARGED?

When the hammer strikes the primer cap on a live round chambered in a firearm, the primer explodes and ignites the propellant. The burning of the propellant generates hot gases that expand and push the bullet from its cartridge case and down the barrel. The propellant is designed and the ammunition constructed so as to continue to burn; if the propellant stopped burning, the friction between the bullet and the rifling of the barrel would cause the bullet to stop. The friction between the bullet and the rifling also transfers the pattern of lands and grooves to the bullet's exterior. More importantly, it also transfers the microscopic striations—themselves transferred to the barrel's inner surface from the tool used to cut the lands and grooves—and these are used by the forensic firearms scientist in the microscopical comparison of known and questioned bullets.

If the firearm retains the spent cartridge, a revolver, for example, then the only marks to be found on the cartridge that could be used for comparison would be the **firing pin impression**, the mark made by the firing pin as it strikes the primer cap. Firearms that expel the spent cartridge, however, may produce a variety of marks indicative of the method of cartridge extraction (**extraction marks**) and ejection (**ejection marks**) from the chamber. Other common marks left on a cartridge case during discharge are **breech marks**. The discharge of a firearm creates recoil, forcing the cartridge case backward into the **breech face** of the firearm; the breech face holds the base of the cartridge case in the chamber. Recoil causes the cartridge base to smack against the breech face and receive an impression of any imperfections in the breech face (see Figures 21.9A and B).

As the bullet leaves the muzzle of the barrel, it is followed by a plume of the hot gases that forced it down the barrel. This plume contains a variety of materials, such as partially burned gunpowder flakes, microscopic molten blobs of the primer chemicals, the bullet, and the cartridge. As these materials strike, or come to rest on, a surface, they transfer potential evidence of that surface's distance from the firearm's muzzle and other materials that may indicate that surface's association with the firing of a firearm or one that has been fired.

COLLECTION OF FIREARMS EVIDENCE

You're watching TV, the detective finds a handgun at the crime scene, picks it up by sticking a pencil down the barrel (or with bare hands, or with gloved hands by the grip, or . . .) and says to her partner, "Hey, Charlie, I think this is what we're looking for. . . ." There is hardly a

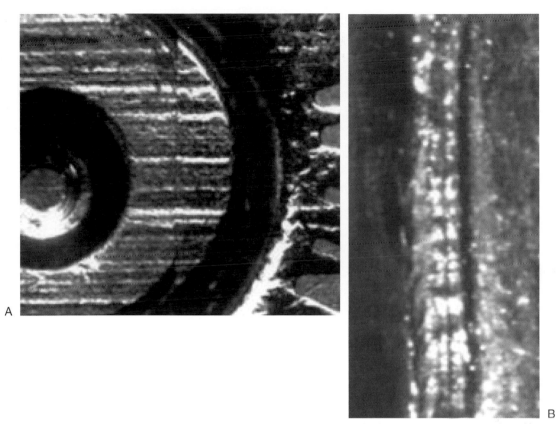

Figure 21.9. Various impressions can be left on the cartridge casing by the discharge of ammunition, such as (A) firing pin impressions and breech marks, (B) extraction marks, and ejection marks. Courtesy of www.FirearmsID.com.

more enduring, or inaccurate, image in the visual lexicon of police dramas. Although TV and movie dramas are hardly the place to learn the proper methods of evidence collection, watching them can be a good way to learn what *not* to do.

A firearm is a durable piece of evidence, subject to analyses beyond the standard forensic firearms examinations, such as latent prints, fibers, and hairs. Additionally, safety is a primary concern when collecting firearm evidence because any firearm could be loaded. After the firearms examiner photographs and documents the location of all firearms, they should be secured in packaging that prevents shifting during transit and that locks the trigger into place (see Figure 21.10).

In shooting reconstructions, it is vital to locate, photograph, and measure the location of all bullets, bullet holes, and spent cartridges.

Figure 21.10. *It is important to properly package firearms when submitting them to the laboratory. Companies that sell crime scene materials usually offer special packaging for firearms that prevent them from accidentally discharging during shipment or transport. This type of packaging is critical to preserve evidence and keep forensic professionals safe. Only unloaded firearms should be shipped. Courtesy of TriTek, Inc.*

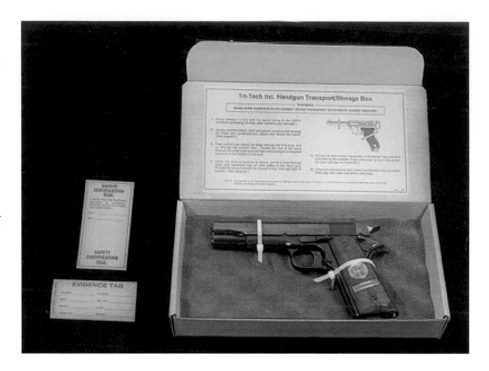

This data will be used later to generate three-dimensional data about the shooting, possibly for court demonstrations (see Figures 21.11A–C).

FIREARMS ANALYSIS

SAFETY AND OPERATIONS TESTING

Firearms, especially those collected as evidence, are inherently dangerous. It is of paramount importance that a firearm be checked prior to any testing or examination: A firearm with a live round in the chamber should not be transferred or stored as evidence unless there is an important reason to do so. A person trained in the safe handling of firearms should check any weapon to see if it's loaded and, if it is, the chamber should be cleared. Proper precautions should be taken to ensure the integrity of any evidence on, in, or removed from a firearm.

Information that is important to note in a preliminary firearms examination is the manufacturer, the caliber, the type of firearm, the model, the ammunition capacity, the barrel length, and the serial number. Criminals will sometimes attempt to obliterate the serial numbers to avoid

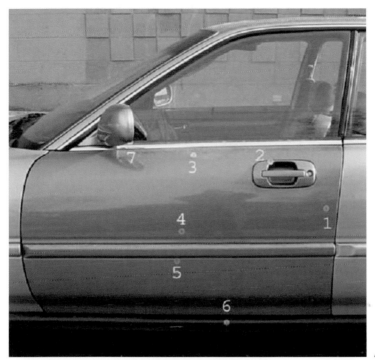

A

Figure 21.11. In shooting reconstructions, it is vital to locate, photograph, and measure the location of all bullets, bullet holes (A), and spent cartridges. These provide the data for reconstructing the bullet pathways (B and C). Courtesy of Alliance Forensic Science Consultants.

B

Figure 21.11.
Continued

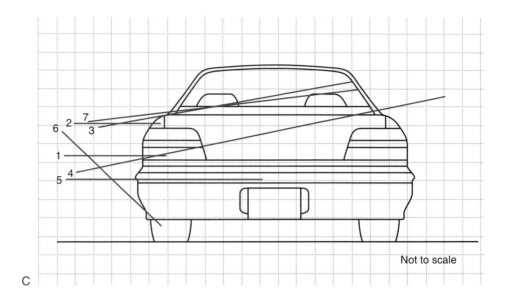

C

Not to scale

being traced through serial number registration; all firearms' serials numbers and their purchasers must be recorded by retailers.

Often the first step in a forensic firearms examination is to determine whether the firearm in question operates properly. The action, the safety, the **trigger pull**—the force required to pull the trigger to the firing position—and other typical functions of the firearm should be tested and recorded. The ability of a weapon to fire may be important in the investigation. These questions may sound mundane, but their answers could be the difference between an accidental death and a homicide.

BULLET COMPARISONS

Many published studies have demonstrated that no two firearms produce the same unique marks on fired bullets and cartridge cases; this is even true with firearms of the same make and model. The machining of the manufacturing process, combined with the use of the firearm, leaves surface marks on the metal parts of the firearm that are not reproducible in other firearms. These marks are transferred to the bullets and casings when discharged from the firearm.

Because there is no practical method of comparing the striations on the inner surface of a rifled weapon with the striations on a fired bullet, reference bullets of the same make, style, and caliber must be created by firing them from the questioned firearm. Not only would cutting the barrel open be impractical, but the comparison would then be between positive (the barrel) and negative (the questioned bullet) impressions. The known fired bullets must be captured and preserved, however, so that they are as "pristine" as possible and not deformed or damaged. Firearms are typically discharged into a water tank where the water slows and eventually stops the bullet without altering its striations; other bullet recovery systems are used, from the simple (a bucket filled with rubber shavings) to the high tech (sandwiched layers of specialized materials) (see Figures 21.12A and B). The known bullet is then recovered, labeled, and used as a reference in the comparison; multiple known bullets may be created, if necessary.

The questioned and known bullets are first examined with the naked eye and slight magnification. The number of lands, grooves, their twist, and the bullets' weights are recorded. Because these are higher order class characteristics, any deviations from the known bullet indicate that the two bullets were fired from different barrels. If the lands, grooves, and direction of twist all concur, then the next step is microscopical comparison of the striations on the bullets.

The comparison is performed on a comparison stereomicroscope (see Chapter 4, "Microscopy," for more information) with special stages that facilitate positioning the bullets in the focal plane and allow for rotation of the bullets on their long axis (see Figure 21.13). The bullets are positioned on the stages, one on each, both pointing in the same direction, and held in position with clay or putty; this allows for easy repositioning, and the soft material will not mark the bullets' surfaces. The known bullet is then positioned to visualize a land or groove with distinctive striations. The questioned bullet is then rotated until a land or groove, respectively, comes into view with the same striation markings (see Figure 21.14). The lands and grooves of the two bullets must have the same widths. More importantly, the two bullets must not be merely similar but must have the same striation patterns with no significant differences. This last point is critical: Not only must the forensic firearms scientist see the positive correlation between the significant information on the bullets' surfaces, he or she must also not see any unexplained differences. Each rifled barrel is unique: No two of them will have identical striation patterns. This is true even of barrels that have been rifled in succession, one after the other. It takes education, training, and mentoring to train a person's eye and judgment on the subtleties of bullet striation patterns.

Figure 21.12. *Test firing bullets into a water tank preserves the striae on the bullets but also slows the bullets down so they can be safely discharged and retrieved. Water tanks can be difficult to maintain, however, and other methods have been devised. Some are simple, like a 5-gallon bucket filled with rubber shavings (A), and some are technologically complex, like the bullet recovery system made by Ballistics Research, Inc., of Rome, Georgia (B). The system uses two specialized types of material sandwiched in a series of alternating layers inside a caster-mounted metal box. Projectiles come to rest within the series of layers, where they are easily recovered by hand. (A) Courtesy of Richard Ernest, Alliance Forensics, Inc. (B) Courtesy of Ballistics Research, Inc.*

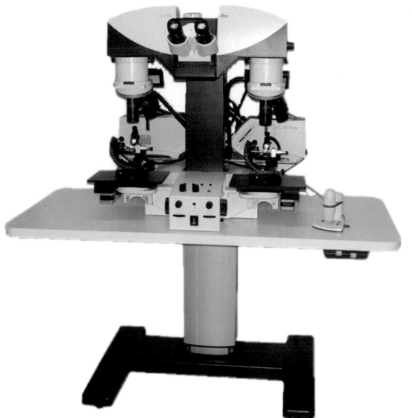

Figure 21.13. *Much like the comparison microscope used for hairs and fibers work, the firearms comparison microscope optically joins two stereomicroscopes. This allows the forensic scientist to view two objects simultaneously side by side. Courtesy of* www.FirearmsID. com.

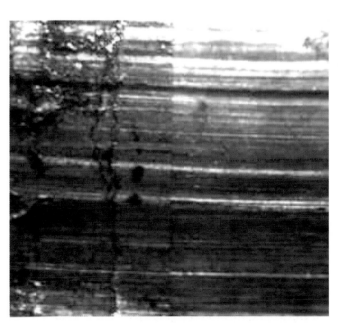

Figure 21.14. *A positive association between bullets must have the same land and groove patterns, widths, and striation patterning with no significant negative correlations. A negative association between bullets would demonstrate numerous significant misalignments of striations with no significant correlation of striae. Courtesy of* www.FirearmsID. com.

Figure 21.15. The lead bullet has deformed and separated from the copper jacket. It would be difficult, if not impossible, to find useful suitable striations on a corresponding known bullet. Courtesy of www.FirearmsID.com.

Bullet striation comparisons are difficult enough with intact, clean bullets; in reality, most bullets recovered from crime scenes are mangled, deformed, and dirty. Often only a small portion of a bullet may have useful striations for comparison (see Figure 21.15). The barrel may not have been cleaned recently, and rust, grit, and built-up residues may have been transferred to the bullet when it was fired.

FIREARM DATABASES AND AUTOMATED SEARCH SYSTEMS

Whether a firearm is used by the same criminal or shared between members of a criminal enterprise, firearm evidence can link a person or persons to multiple crimes. The problem in doing so is the difficulty of searching and comparing numerous bullets or casings. If the crimes were committed across multiple jurisdictions, then the task becomes even more involved.

Two automated search systems were developed in the 1990s, one by the Federal Bureau of Investigation, called DRUGFIRE, which analyzed cartridge casing markings; and one by the Bureau of Alcohol, Tobacco, and Firearms, called the Integrated Ballistic Identification System (IBIS), which primarily analyzed bullet striations but could also work with cartridge casings. The systems integrated digital imaging, novel data collection, computerized databases of images, and communications technology. Unfortunately, the systems were not compatible with each other, and specialized hardware and software were needed for each one.

IN MORE DETAIL:
Fifteen leads for the Boston police department

The Boston, Massachusetts, Police Department is aggressive in the use of advanced technology to combat illegal firearms and firearms violence. Departmental regulations require that all recovered evidence relating to firearms be submitted to the laboratory for entry into its NIBIN unit. The power of ballistic imaging technology, and Boston's thorough approach to its deployment, has enabled the department to find links undetectable by other means. On September 9, 2000, in Boston, several subjects were apprehended and found to be in the possession of three handguns. (The public possession of firearms is in itself a criminal offense in Boston.) The subjects were arrested and charged with the possession offense; the three handguns—a .25 caliber, a .40 caliber, and a 9 mm—were all seized as evidence, test fired, and entered into IBIS.

Correlation of the test firings returned several promising similarities. Examiners from the department's firearms laboratory viewed the correlation results and then examined the recalled evidence. The following criminal offenses were positively connected to the test fired weapons:

- On June 2, 1999, in Boston, shots were fired, but no victim was identified. Several 9 mm cartridge casings were recovered at the scene.
- On October 28, 1999, also in Boston, shots were fired, but no victim was identified. More 9 mm cartridge casings were recovered.
- On April 3, 2000, in Boston, one victim was wounded by gunfire; 9 mm cartridge casings were collected at the scene.
- Also on April 3, 2000, in Boston, shots were fired, but no victim was identified; 9 mm cartridge casings were recovered at the scene.
- On April 19, 2000, in Boston, one victim was wounded by gunfire. In the area, 9 mm cartridge casings were recovered.
- On April 23, 2000, in Boston, shots were fired, but no victim was identified; 9 mm cartridge casings were collected at the scene.

- On May 9, 2000, in Boston, shots were fired, but no victim was identified; 9 mm cartridge casings were recovered at the scene.
- On June 8, 2000, in Boston, four victims were shot; 9 mm cartridge casings were collected at the scene of this violent crime.
- On June 15, 2000, in Boston, a victim was assaulted with a firearm; 9 mm cartridge casings were recovered.
- Providence, Rhode Island, is located about one hour south of Boston by car. On June 19, 2000, Providence police responded to the scene of a shooting and found there a large amount of blood and several 9 mm cartridge casings, but no victim. The firearms evidence recovered was entered into Providence's RBI unit, which communicates with the Boston DAS.
- On June 25, 2000, in the city of Brockton, about 25 minutes south of Boston, an assault with a firearm took place. The 9 mm casings recovered at this scene were transported to the Boston PD for entry into Boston's IBIS unit.
- On July 6, 2000, in the city of Randolph, just south of Boston, three victims were wounded in a shooting; 9 mm cartridge casings were recovered.
- On July 7, 2000, in Boston, shots were fired, but no victim was located; 9 mm cartridge evidence was collected at the scene.
- On July 20, 2000, in Boston, shots were fired, but no victim was located. Cartridge casings from a .40 caliber firearm were recovered.
- A routine arrest for firearms possession charges resulted in the discovery of links among these shooting incidents spread over several police jurisdictions in two states. As a result, each agency involved now has a wealth of information to use in its investigation, including the identities of the possessors of the guns. (In all of the cases, investigation is ongoing.) Without Boston's participation in NIBIN, these crimes would likely not have been linked.

On the Web: www.nibin.gov

In January of 1996 the ATF and the FBI acknowledged the need for IBIS and DRUGFIRE to be compatible. This meant the systems had to capture an image according to a standard protocol and with a minimum quality standard and exchange these images electronically so that an image captured on one system could be analyzed on the other. In June 1996, the National Institute of Standards and Technology (NIST) issued the minimal specifications for this data exchange. In May of 1997, the National Integrated Ballistics Information Network (NIBIN) was born.

In 2002, the NIBIN program had expanded to 222 sites. When complete in all 16 multistate regions, NIBIN will be available at approximately 235 sites, covering every state and major population centers. Since the inception of this technology, over 5,300 "hits" have been logged, providing investigative leads in many cases where none would otherwise exist. For an example of how this technology links cases, see "In More Detail: Fifteen Leads for the Boston Police Department."

DISTANCE OF FIRING DETERMINATION

GUNPOWDER RESIDUES

When a firearm is discharged, the bullet is not the only object expelled from the weapon (see Figure 21.16). The violent chemical reaction of the primer and accelerant result in a cloud of molten metals, partially burned gun powder flakes, smoke, and other microscopic debris. This residue may be found on the person who discharged the firearm, on an entrance wound of a victim, or on other surfaces. The discharge of a firearm, particularly a revolver, can deposit residues 3′ or more from the hand of the shooter, so interpretations about who fired the gun can be problematic. Some of these **gunshot residues** may be used to make determinations about the location of the discharged firearm in relation to its surroundings and its target.

The patterning of gunshot residues on a target is indicative of the distance from the muzzle to the target (see Figure 21.17). The patterning and density of the gunshot residues will vary with the firearm and ammunition used. Therefore, the patterns must be empirically generated by discharging a questioned firearm in order to make a comparison with a questioned gunshot residue pattern. The distances tested are typically contact (where the muzzle is against the target), 6,

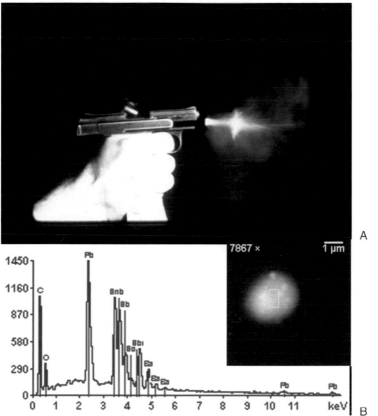

Figure 21.16. When a firearm is discharged, the projectile is not the only thing that leaves the barrel. The residues of completely or partially burned propellant, volatilized metals from the projectile and the casing (if any), and wadding (in the case of shotguns) are also ejected.

12, 18, 24, 30, and 36 inches; other distances may be tested based on the case circumstances.

Contact or near-contact bullet entrance holes demonstrate severe damage to a textile or garment. Bullets that strike an object before hitting their final target tend to have uneven edges. Typically, the greater the damage to the textile in a contact gunshot, the higher the velocity of the ammunition.

Firearms discharged more than about 3′ from the target will not impart any residues other than a **bullet wipe**. A bullet wipe is a residue of lead, primer materials, carbon, and other materials from the barrel that are transferred ("wiped") onto the outermost surface of the target by the bullet as it passes through. For the sake of clarity and standardization, the questioned weapon is discharged onto a 1′ × 1′ piece of white cloth. If the firearm is not recovered as evidence, then the range of distance estimates will be greater (1′–3′, for example, instead

Figure 21.17. GSR materials expand over distance and can leave a transfer pattern on any objects between the barrel and the target. When a firearm with similar ammunition is test fired, a range of patterns can be established and compared with the crime scene pattern. This leads to an estimate of the muzzle-to-target distance. This pattern was made by a 2″ muzzle to target distance Federal ammunition .357 magnum from a Smith & Wesson Magnum Revolver with a 6″ barrel. Courtesy of Alliance Forensic Science Consultants.

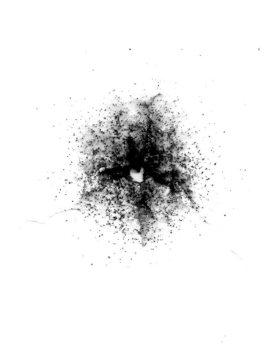

of 1.5′–2.5′) to reflect the uncertainty. Additionally, many circumstances may affect the distance estimate, including position of the shooter and target, weather, and intervening items.

To make the comparisons, often the gunshot residue pattern on the target must be visualized by some means. Infrared photography may reveal the residue pattern when the clothing is dark or heavily patterned (see Figures 21.18A and B). The first test that should be used to reveal residues is the **Modified Griess Test**. In the Modified Griess Test, a piece of desensitized photographic paper is treated with a mixture of sulfanilic acid in distilled water and alpha-naphthol in methanol. Photographic paper is desensitized by exposing the paper to a hypo solution, desensitizing it to light but making it reactive to nitrite residues. The evidence is placed target-side down on the photographic paper and pressed with a steam iron filled with a dilute acetic acid solution. The resulting residues appear as orange dots on the photographic paper. This variant of the Greiss Test was developed and published by Scott Doyle of the Kentucky State Police Crime Laboratory (www.firearmsid.com), who has enjoyed great success with it.

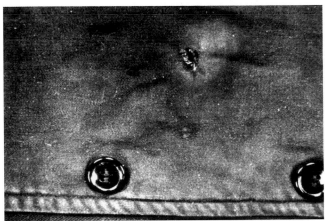

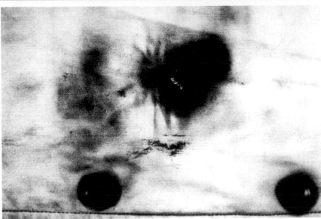

Figure 21.18. *(A) When a shooting victim's clothing is dark or heavily patterned, it may be difficult to see the pattern of gunshot residues. (B) Infrared photography may reveal the gunshot residue pattern. Courtesy of Wisconsin State Police.*

The last chemical treatment used to visualize gunshot residue is to spray **sodium rhodizonate** on the surface and then treat that area with a series of acid sprays. The residues turn pink and then purplish-blue and are easily seen (see Figure 21.19).

When a scene or body is examined for gunshot residue, it is necessary to remember that lead residues may look like GSR. Lead residues may be found up to 30′ from the muzzle and are present on the opposite side of a penetrated target.

SHOTGUN DISTANCE DETERMINATION (SHOT PATTERNS)

The determination of **muzzle-to-target distance** of a shotgun is similar to the method used for other firearms except it is the pattern of pellets

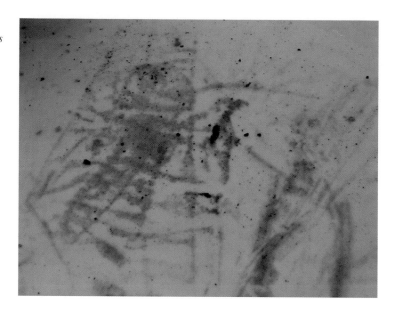

Figure 21.19. A color-test that visualizes the pattern of gunshot residues. *Courtesy of Alliance Forensic Science Consultants.*

that is measured. As the pellets leave the shotgun barrel, they begin to spread based on the distance they travel and the choke of the barrel. Their pattern of spreading is indicative of the distance between the end of the barrel and the target. Because of the variations in choke, gauge, and ammunition materials, it is important, just as it is with other firearms, to use the same weapon and ammunition as suspected in the crime.

PRIMER RESIDUES

Primer residues may also land on the hands of a shooter. The residue is mostly microscopic blobs of the molten metals from the primer cap, the primer compound, the casing and other metallic components (containing copper [Cu], zinc [Zn], nickel, aluminum [Al], among others). The major primer elements are lead (Pb), barium (Ba), and antimony (Sb); typically, all three are found in GSR. The minor elements include aluminum (Al), calcium (Ca), chlorine (Cl), potassium (K), silicon (Si), sulfur (S), and tin (Sn). As the blobs fly through the air, they condense into heterogeneous spheres of various sizes (sub-micron up to 50 or more microns, with most in the 2–10 micron range).

Several tests have been developed over the years, including the dermal nitrate test (which also tested positive for fertilizers) and atomic absorption (AA). Still used in some laboratories, AA suffers from a number of limitations, including collection problems (swabbing the suspect's hands with a mild acid solution), a lengthy analysis time and, most importantly, lack of specificity. The result of an AA analysis yields quantities of the elements tested for but not their distribution. Because GSR particles are aggregates of compounds, no information about the form of the sample is known with AA. Fireworks, matches, and other common objects could yield a positive result with AA.

A method was published in 1976 (Nesbitt, Wessell, and Jones) for detecting **GSR particles** using a scanning electron microscope (SEM) outfitted with an energy-dispersive spectrometer and imaging system (EDS). The particles were collected from the hands of a suspected shooter, placed on a carbon-coated aluminum mount, and viewed in the SEM with a detection mode that relates brightness to atomic number. The bright particles on a dark background of carbon were those that were most likely GSR. The EDS detector would then analyze the individual spheres to detect the presence of antimony, barium, and lead—the three main components found to be in GSR but not other high-atomic number particulate matter. The shape and elemental content of the particles defined them as GSR. It is important to note that the primers in .22 ammunition differs from other primers and may not be detected by a system screening for PbSbBa particles.

This method gained greater acceptance when the technology allowed for the detection and analysis of the method to be automated for unattended operation. Multiple samples could be loaded into the SEM, the software calibrated, and then left on its own to run samples overnight. The human operator still needed to verify any positive "hits" because other materials could produce elemental signatures that were overlapping with or that confounded the signatures of true GSR particles, such as automotive brake pads.

The only conclusion that can be drawn from a GSR test is that the subject discharged a firearm, was near a firearm when it discharged, or handled a recently discharged firearm. These factors, plus the potential for contamination during arrest from GSR-rich environments, like police officers' hands, patrol car seats, and handcuffs, make for a limited application of GSR analysis. Many laboratories continue to offer GSR analyses, and some perform a high volume of casework (>500 cases) annually.

Figure 21.20. *The machining of metals leaves microscopic striations that can be transferred to a softer surface, such as a screwdriver prying open a wooden window casing. The class characteristics can also be useful in identifying (but not individualizing) the tool used. Courtesy of Wisconsin State Police.*

TOOL MARK COMPARISONS

The potential sources for tool marks are many, and some are surprising. Metal tools are made by a variety of methods, but most are finished in a way that leaves microscopic striations on their working surfaces. A tool may leave class characteristics that may help to identify what kind of item it is: An ice pick will leave different markings than would a flat-head screwdriver, for example (see Figure 21.20). The types of markings left will not only depend on the type of tool but also how it was used, the angle of contact, the force of contact, what was contacted, among other factors. It is important for the tool mark examiner to have a foundational knowledge about how various tools are made and machined.

Bullets travel down the barrel of a firearm in nearly the same way each time; the same is not true of a tool mark. The forensic tool mark examiner must pay close attention to the potential orientations of the questioned tool; otherwise, the test marks may not be comparable with the questioned marks (see Figures 21.21A and B). Like bullets, however, the questioned tool cannot be compared directly to the mark it may have left. Crime scene personnel or the tool mark examiner may

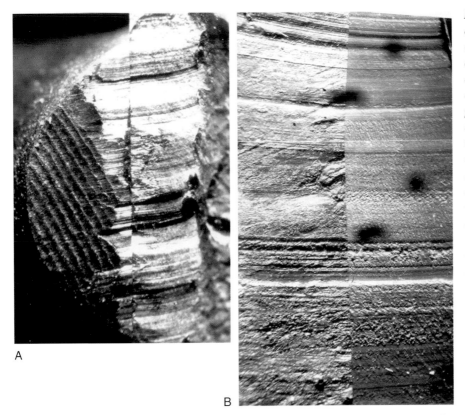

A

B

Figure 21.21. Examples of tool mark striations. (A) A pair of bolt cutters was used to cut a lock shackle. The comparison micro-photograph shows the evidence lock shackle on the left side and the test cut produced by the bolt cutters in a piece of lead on the right. (B) An unusual example of a tool mark comparison is shown on the left, where a knife edge from a suspect's knife was compared and matched to a piece of rib cartilage from a victim. On the left is a silicone rubber cast of the cut in the victim's rib cartilage compared with a cast of a test cut produced with the suspect's knife using a plastic material (Dip-Pak). Courtesy of Alliance Forensic Science Consultants.

take casts of the tool mark(s) for comparison purposes. These casts are lightweight, easy to handle, and easy to store; fine-grained polymer materials are sold by most forensic science supply companies. Dental stone is also a favored medium for making tool mark casts.

SERIAL CHARACTER RESTORATION

Most mass-produced items have some sort of identifier on them, such as a part or serial number. Criminals may hope to evade capture by attempting to obliterate the serial numbers from metal parts from automobiles, firearms, motorcycles, and other products by grinding, filing, or scratching.

Serial numbers (which may include letters) are transferred to the metal part by a stamping die made of a material harder than the sub-

Figure 21.22. Serial number restoration is part of the repertoire of the firearms examiner because many criminals try to obscure the identifying information on stolen guns. Other metallic parts with serial numbers, like automotive parts, may also be subject to this test.

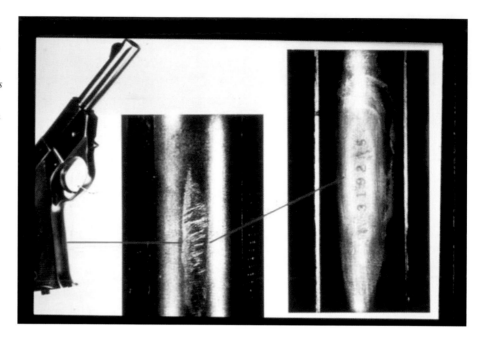

stance to be impressed. This die is stamped onto the metal part with sufficient force to impress the numbers into the metal part. The crystallography of the metal in the area of the characters is permanently changed, and this alteration allows the forensic scientist to raise obliterated characters (see Figure 21.22). The characters are restored by applying an etching chemical on the obliterated area. The stamped, weakened area dissolves faster than the surrounding unstamped area, and this reveals the pattern of the characters. If the stamped area has been ground below this zone, however, the characters may be lost permanently. It is also possible to apply too much etching agent and dissolve the stamped zone, so caution and patience are recommended.

This method works best on a clean surface, free of dirt and oil. The etching chemical is applied with a cotton swab while the affected area is observed; different "recipes" are used for different materials. A stereomicroscope may also be helpful with this technique. Different chemicals are required for different metals, such as steel, copper, brass, etc. This process should always be carried out with a camera at the ready because in many cases, a number will appear from treatment with the etching reagent and then disappear forever.

CHAPTER SUMMARY

Because of the ubiquity of firearms in the United States and their primary violent purpose, firearms examination is a central function of most forensic science laboratories. The exams performed, however, are more than just comparing bullets and cartridges. A wide range of knowledge is required to be a qualified firearms examiner, from chemistry to physics to computers and digital imaging. The field has advanced considerably since its battlefield beginnings and will continue to do so for some time to come.

Test Your Knowledge

1. What is the difference between a revolver and a pistol?
2. List four differences between rifles and shotguns.
3. Why is a pistol not a rifle?
4. What is wadding?
5. What is gauge?
6. How is caliber determined?
7. Name the parts of a cartridge.
8. What does a primer do?
9. What are lands and grooves?
10. Why don't shotguns have lands and grooves?
11. What materials are used to estimate muzzle-to-target distance?
12. What is in gunshot residue?
13. Do striations appear only on bullets? Why or why not?
14. Are more firearms involved in homicides or suicides?
15. What determination can be made from a GSR analysis?
16. How are bullets compared?
17. How are serial numbers restored?
18. What is sodium rhodizonate used for?
19. What is "bullet wipe"?
20. How many firearms-related crimes occur in the United States each year?

Consider This . . .

1. Firearms, like any other mechanism, are made up of parts; one of these parts is the barrel. The striations in each barrel are unique because of

manufacturing and use. How could this information affect the comparison of questioned bullets with test-fired bullets from a suspected gun? Knowing this, what would you do differently, if anything, in your analysis?

2. One bullet is fired from a firearm in much the same way as the next bullet; this makes generating comparison (known) bullets easy. Why is this not the case for tool marks? What implications does this have for a tool mark comparison?

BIBLIOGRAPHY

Biasotti, A. (1959). A statistical study of the individual characteristics of fired bullets. *Journal of Forensic Sciences, 4,* 35–50.

DiMaio, V. (1985). *Gunshot wounds.* New York: Elsevier.

Nesbitt, R.S., Wessel, J.E., & Jones, P.F. (1976). Detection of gunshot residue by use of the scanning electron microscope. *Journal of Forensic Sciences,* 21, 595–610.

National Center for Health Statistics, retrieved from **www.cdc.gov/nchs**.

National Crime Victimization Survey 2002 (2002). Washington, D.C.: United States Department of Justice, Office of Justice Programs, Bureau of Justice Statistics.

Rowe, W. (1988). Firearms identification. In R. Saferstein (Ed.), *Forensic science handbook, volume II,* pp. 393–449. Englewood Cliffs, NJ: Prentice-Hall.

Thorwald, J. (1964). *The century of the detective.* New York City: Harcourt, Brace & World.

U.S. Bureau of Alcohol, Tobacco, Firearms, and Explosives (2000). *Following the Gun: Enforcing Federal Laws Against Firearms Traffickers.* Washington, D.C.: U.S. Bureau of Alcohol, Tobacco, Firearms, and Explosives.

Wilson, C., & Wilson, D. (2003). *Written in blood: a history of forensic detection.* New York City: Carroll and Graf Publishers.

www.firearmsid.com, maintained by Scott Doyle of the Kentucky State Police.

Impression Evidence

KEY TERMS

Cast	Footwear impression	Shoe print
Dental stone	Lift	Tire tread stance measurement
Electrostatic lifting device	Serial numbers restoration	Wheelbase measurements

INTRODUCTION

When a donor object or a material is pressed against a recipient object or material and some force is applied, the donor may leave an impression in the recipient. If you have read the chapter on fingerprints (Chapter 19, "Friction Ridge Examination"), you may recall that a fingerprint may be left in putty. This is an impression. Likewise, in Chapter 21, "Firearms and Tool Marks," you learned that a wire cutter may leave an impression of its cutting edge in a piece of wire. Many other types of impressions occur in criminal and civil cases. Some are fairly common, such as footwear and tire treads, whereas others are quite rare. An example of a rarely found impression is fabric impressions on paint.

Impression evidence can, under certain conditions, be quite powerful in its ability to associate the donor with the recipient. Consider a brand new, left foot, size 10, men's tennis shoe of a particular brand and model. Suppose you went to a shoe store and obtained two of these shoes right out of the box. Careful examination of the soles of these

THE CASE

At approximately 2 a.m. on December 3, a break-in occurred at a suburban home. The burglar jimmied a locked window outside the living room on the first floor of the home. In doing so, he left a thumbprint impression in the window putty on the window, which had been replaced the day before after being broken. To gain entrance through this window, the burglar stepped in a flower garden in front of the window, leaving shoe prints from his sneakers. He broke in and took whatever he could find of value.

When the burglary was discovered the next morning, the police were called. They noted pry marks in paint from a screwdriver used to pry open the window. They also noted the fingerprint impression in the putty and the shoe prints in the soil. Soil was also found on the living room floor.

The suspect was apprehended when he tried to "fence" the stolen merchandise. One of the articles in the haul was a rifle. The burglar had filed off the serial number to make tracing the weapon more difficult. He intended to keep the rifle, and it was seized when he was apprehended.

The rifle was sent to the lab. The shoe print was photographed, and a **dental stone cast** was made. This was sent to the lab along with a pair of sneakers seized from the suspect's apartment pursuant to a search warrant. Likewise, the window frame with the pry marks and putty containing the fingerprint and a screwdriver found in the suspect's car were also sent to the lab.

shoes would indicate very little difference between them. If either one left an impression at a crime scene, it would not normally be possible to determine which one it was. Now, suppose that you bought two pairs (left and right feet) of these tennis shoes, and they were worn by two different people for a period of months. If the left soles were examined again, an examiner would most likely see significant differences between them. Parts of the soles would become worn, pitted, cracked, or broken, and these things would happen in a random way because each person would travel in shoes differently on different types of surfaces for different lengths of time. These random imperfections would soon accumulate to the point where each sole would be measurably different than any other sole. The evidence would become individualized, so it would be possible to show that an impression could be traced back to one and only one shoe. Several types of impressions lend themselves to this type of analysis. You have already learned about fingerprint and tool mark impressions. In this chapter we will discuss footwear, tire treads, and serial numbers, as well as some less common types of impression evidence. There is an important distinction between fingerprint comparsions and **shoe print**, **tire tread**, or other impressions. Fingerprints remain the same throughout life, whereas most other impression evidence changes with time. Over several months new impressions may appear in shoe prints or tire treads that are not on the exemplar, and comparisons of these impressions must take that into account. A single, major disparity between a known and unknown fingerprint may eliminate the known, whereas that would not necessarily be the case with other impressions.

TYPES OF IMPRESSION EVIDENCE

Remember that the definition of impression evidence involves a donor and a recipient. The donor contains some three-dimensional markings, and the recipient is made of a material that can form and hold a negative image of the donor markings. Common donors that occur as evidence in crimes include shoe soles and heels, tire treads, fingerprints and other friction ridges such as footprints and lip prints (discussed in Chapter 19), tools that leave markings on the objects on which they are used (discussed in Chapter 21), metal dies that are used to make serial numbers, ribbing and texture in fibers, etc. Common recipients include soft plastics, soil, putty, paint, dust, metals, plastics, wood, and other, similar materials. With all of these possibilities, there is a variety of impression evidence available. Consider the case presented at the beginning of the chapter. How many different types of impression evidence are created by the burglary?

SIGNIFICANCE OF IMPRESSION EVIDENCE

As you are now aware, much of the time spent by forensic scientists in the lab is to try to associate evidence with its source. Sometimes this can be done with a great deal of certainty, whereas in other cases, the association is more equivocal. As you have also learned, the factor that determines if one object can be associated with another is the presence of unique characteristics in the donor. This means that the three-dimensional pattern or markings must contain some characteristics that are unique to that object. For example, a shoe print must contain some wearing characteristics such as pits or cracks that would make it unique. The same would hold true with a tire tread. These kinds of unique characteristics come about as the object is used. When a shoe or tire is brand new, it does not contain unique characteristics; one tire that comes off the assembly line looks pretty much like all of the others that come off the same line. It is only as the tire is used on a car that it picks up unique characteristics. The reason is that the processes that give rise to these characteristics are random in nature. No two tires or shoes will wear the same way, so each will change with time and become unique. When there are enough of these unique features present, then this evidence can be individualized to a particular tire or shoe. Thus, these types of evidence can be very valuable in associating a particular shoe or vehicle with a crime scene. In the case discussed at the beginning of the chapter, it might be possible to associate the footprint left in the soil of the garden outside the window to the sneaker worn by the suspect.

Serial numbers restoration presents a different picture. Here, the goal is to restore an obliterated serial number on an object so the number can be used to identify that particular object. The serial number itself is usually unique and makes the object unique by virtue of being the only object with that particular serial number.

Some other types of impressions such as the one made when a car strikes a pedestrian so hard that an imprint of the fabric pattern in the victim's clothes is impressed in the car's paint do not lend themselves to individualization. It would not be possible to individualize an impression to a particular garment, but it wouldn't normally be necessary. There is almost always other evidence that can link a motor vehicle to a hit and run. The fabric pattern is class evidence that adds to the total picture.

FOOTWEAR IMPRESSIONS

GENERAL CHARACTERISTICS

When you think about it, most people wear footwear (in the broadest sense, this includes shoes, boots, and sandals) when they are outside of their home except perhaps at the beach or in certain Oriental restaurants. Certainly, most people wear shoes when they commit a crime. In the hypothetical burglary discussed at the beginning of the chapter, the burglar would have left footwear prints coming into the crime scene (the house), while at the crime scene, and in leaving the scene. In many cases, these impressions are difficult to locate, especially if they are latent or invisible. In addition, a suspect's footprints may be mixed with those of other people, including police investigators, paramedics, and crime scene technicians. Many crime scene technicians are not familiar with the best methods for visualizing and preserving footwear impressions, so they tend to overlook them or not bother to search for them.

Information That Can Be Derived from Footwear Impressions

Footwear impressions can indicate the type, manufacturer, model and, often, the exact size of the footwear. If enough unique characteristics are present, the impression can be matched to the actual footwear that made the impression. These impressions can indicate the route(s) taken into and away from the crime scene. They can also indicate some of the activities that took place during the crime. The number of people, and perhaps suspects, that were at the scene may be determined. Even if the footwear cannot be identified, characteristics of the walking (or running) gait of the wearer may be uncovered. Although this would have little value in identifying the person, gait has been used successfully in tracking criminals, illegal aliens, missing persons, kidnap victims, and others.

When Footwear Touches the Ground . . .

A number of things can happen when footwear touches the ground. First, a static electricity charge can be applied to the impression if the shoe is clean and dry. This charge dissipates after a short time but is useful because it helps in the transfer of trace residues and dust to the impression.

If the surface is soft or pliable, pressure exerted by the foot will cause the surface to deform and take on the contours of the surface of the footwear. This may be permanent as would be the case of dirt and snow, or it may be temporary as with grass or carpet. Even in cases in which the impression is temporary, trace residues may be transferred to it. As mentioned previously, if the footwear surface contains unique characteristics, they may be transferred to a pliable surface if it has sufficient resolution to capture small features such as would occur on shoe surfaces as they wear. Figure 22.1 shows a shoeprint that was made in soil.

An imprint is made when there is enough residue on the footwear to leave an impression on the recipient surface. This would be a positive impression because the residue is on the surface of the footwear that touches the surface. Positive impressions are the most common type of footwear imprints. If the shoe sole is clean and the recipient

Figure 22.1. *A shoe print that was left in soil at a crime scene. Courtesy of Cheryl Lozen, Michigan State Police, Forensic Science Division.*

surface contains a lot of dust or residue, then a negative impression can form. Here, the parts of the shoe that touch the floor remove the residue, leaving behind a negative impression. Imprints also result when someone tracks through blood, wet paint, or grease. Depending on how much of this material is present, a negative impression may be left in the liquid, and a positive impression may be deposited further away on a clean surface from the residue picked up by the shoe.

FOOTWEAR IMPRESSIONS AT THE CRIME SCENE

Detection

The first problem with footwear evidence is finding it. This may involve a systematic search that should include route of entry and exit, as well as the scene itself. It must be remembered that impressions may be latent or invisible, and the scene investigators must develop strategies based on likely locations for impressions. Oblique lighting and physical methods of development, similar to those used for fingerprint residues, may be useful for discovering hard-to-see images. The search must encompass both two- and three-dimensional impressions. At this point, the purpose is just to locate impressions. Preserving them will come later.

General Treatment of Footwear Impressions at the Scene

Once footwear impressions have been detected at a crime scene, there is a set of relatively universal procedures for processing them. The most important consideration is to avoid altering an impression until examination-quality photographs have been taken.

As with any crime scene, this first investigative activity is a complete visual record. Increasingly, digital still and video photography are replacing classical film and tape methods. No matter what method is used, photography will provide a permanent record of the position of all footwear impressions and their general conditions. As with all crime scenes, those containing impressions must be immortalized with careful, complete notes and sketches that further document exact locations and circumstances. These will also help associate photos, casts, and sketches with each other.

For impression evidence of any type, it is important to take photographs that can be used for examination of the individualizing characteristics. They must be close-up photos that have sufficient resolution and lighting to be used on their own for comparison, even if casts will also be made.

The next step is to make a decision about how to best preserve and/or enhance the impression. This will depend on where the impression is, how easy it is to remove the medium and the impression from the scene, and whether the impression is two or three dimensional. If at all possible, the impression and its substrate should be physically removed and transported to the laboratory where there are usually better facilities for additional photographic or other treatments. Even if carpeting or flooring has to be cut, it should be removed. If removal is not practical or possible, then a cast should be made if the impression is three dimensional or it should be lifted if two dimensional. If lifting isn't possible, then the impression should be enhanced to the maximum degree possible and more examination photographs taken.

CASTING THREE-DIMENSIONAL FOOTWEAR IMPRESSIONS

The popularity of casting footwear impressions has varied greatly over the years and has depended on the quality of photography at a given time. Early on, photography was rather crude with uneven lighting and low-resolution film. This resulted in photographs that often did not show sufficient detail for comparison purposes. Thus, casting was heavily used. At this time the major method for casting was to use plaster of Paris. This is a dense material, and sometimes several pounds were required for each cast. In addition, plaster of Paris is relatively slow in drying and, when dry, is not very hard. Thus, when impressions were taken in dirt, the cast was often damaged during the cleaning process to remove the dirt. Figure 22.2 shows a plaster cast of a shoe print.

In recent years, photography has improved greatly. New types of lights and lighting techniques have been developed. Higher resolution films have been introduced that show more impression detail. This has had the effect of decreasing the use of casting of impressions.

At present, even with the development of digital photography and video, casting has made somewhat of a comeback. This is due to improved casting materials such as **dental stone**, which is less dense, dries faster, and shows more detail than does plaster of Paris. Experts recognize several advantages of modern casts over photography. For example, even the best photography requires a level, two-dimensional

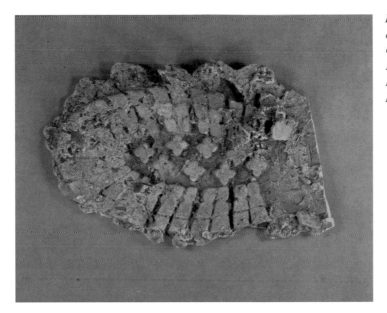

Figure 22.2. *A plaster cast of a shoe print. Courtesy of Cheryl Lozen, Michigan State Police, Forensic Science Division.*

subject to be most effective. If an impression is imbedded deeply into a substrate such as sand, it will be difficult to gain a proper perspective photographically. It will also be difficult to properly locate a measuring device and to even get a proper measurement of the size of the shoe that made the impression. In addition, if oblique lighting is used to photograph the impression, some of the most important details may be partially obscured in shadow. None of these are problems with a properly made cast.

Casting Materials

Plaster of Paris has been the most popular casting material for all types of impressions. Its major drawback is its softness even after drying. It also is made up of relatively large particles that may cause the loss of crucial detail. Today dental plasters and stones are more often used. They are more dense and have more uniform, smaller particle size than plaster of Paris. They dry quickly and show more detail.

The softness of plaster of Paris necessitated that reinforcing materials such as pieces of wood be used within the cast. This required that casts be at least two inches thick. In order to make a cast that thick, a form had to be used. With modern dental stones and plasters, the increased density and hardness of the dried casting material means that the cast doesn't have to be as thick and, therefore, often doesn't require a form. Forms are still sometimes used if the impression is on a steep slope or is of very different depths throughout.

FOOTWEAR IMPRESSIONS IN SNOW

Many people are surprised that it is possible to make a cast of a **footwear impression** in snow. They visualize pouring plaster of Paris onto a snow print, having the print melt or collapse under the weight of the plaster—and indeed it would! Today the most popular and successful method for making casts of snow print impressions is by the use of snow print wax. This comes in a spray can and is sold in bright red and brown. It is sprayed on the snow print and dries in a few minutes. This thin cast shows excellent detail but is fragile. After the wax cast dries, it is filled with cold dental casting mixture. This adds strength and bulk to the stone. When snow print wax is used, precautions must be taken to keep direct sunlight away from the cast because the dark colors of the wax absorb light and might cause the print to melt. Once the wax and stone cast is made, it should be covered with a box or other container to hasten drying. Other materials such as paint thinner, spray paints, paraffin, and sulfur can be used to make snow print castings. Figure 22.3 shows the cast of a shoe print that was made in snow.

LIFTING IMPRINTS

An imprint in a material like dust or one that has been visualized using a powder technique analogous to fingerprint techniques can be lifted from a surface in a number of ways. An examiner can use large pieces of tape just as would be done on a smaller scale with fingerprints or palm prints. There are also gelatin materials that are made to be used for lifting prints. The most popular method used today, however, involves one of a number of **electrostatic lifting devices**. The principle behind these techniques is that a large, static electricity charge will strongly attract dust and other fine powders. A low-current, high-voltage charge is put across a film that attracts the particles from the impression, thus effecting a transfer. A contrasting color film can be used as the transfer surface. The transferred image can then be easily photographed. Figure 22.4 shows an electrostatic **lift** of a shoe print that was made in dust.

Some imprints are difficult or impossible to lift. They include those in grease or oil or in blood. Figure 22.5 shows a shoe print that was

Figure 22.3. *A cast of a shoe print that had been left in snow. Courtesy of Cheryl Lozen, Michigan State Police, Forensic Science Division.*

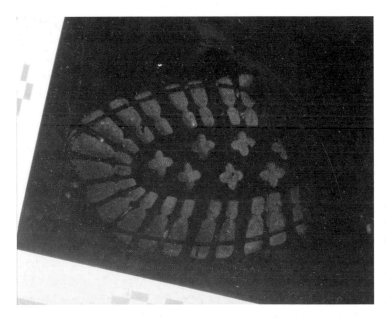

Figure 22.4. *An electrostatic lift of a shoe print left in dust. Courtesy of Cheryl Lozen, Michigan State Police, Forensic Science Division.*

made in blood at a crime scene. There are also impressions made in materials that deform when impressed but then bounce back and lose the impression, such as cushions or carpeting materials. The three-dimensional image would be lost when the substrate regains its shape, but residues imbedded in the impression that forms an imprint may, in some cases, be lifted.

Figure 22.5. *A shoe print left on a floor after the wearer stepped in blood at a crime scene. Courtesy of Cheryl Lozen, Michigan State Police, Forensic Science Division.*

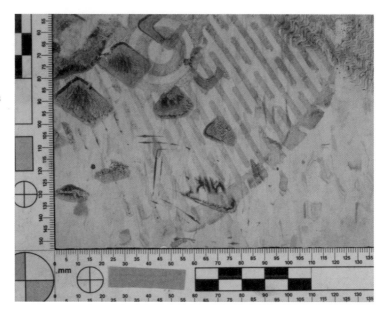

COMPARISON OF FOOTWEAR IMPRESSIONS

Footwear impressions can be individual evidence. There must be sufficient, unique characteristics present in the impression. The impression must be of good enough quality to have these characteristics, and the lifting or casting technique must be able to faithfully record them for comparison with the actual footwear. It is important to note that the impression will be a negative of the footwear; that is, raised areas on the footwear will be depressions in the impression. Figure 22.6 shows a comparison of a shoe print and a cast made of the shoe print at the crime scene.

There is no set number of unique characteristics that must be present on an impression in order for a forensic scientist to reach the conclusion that the impression was made by a particular shoe. Each case must be considered on its own merit and will have its own facts and circumstances. The forensic scientist must never be put into the position of making a conclusion that it always takes a certain minimum number of characteristics to declare a match because there is no database that would support one, single number. Generally, once a shoe has been worn to the point where it starts gaining unique characteristics, there will be plenty to choose from. Footwear comparisons are illustrated by the OJ Simpson case described in "Issues in Footwear Analysis."

Figure 22.6.
Comparison of a shoe print with a plaster cast of the shoe. Courtesy of Cheryl Lozen, Michigan State Police, Forensic Science Division.

ISSUES IN FOOTWEAR ANALYSIS:
Shoe print investigation of the Simpson–Goldman murders

Following the murders of Nicole Brown Simpson and Ronald Goldman in the summer of 1994, photographs depicting shoeprint impressions in blood from the Brentwood, California, crime scene were delivered to the FBI Laboratory. The Los Angeles Police Department requested the determination of the brand and size of footwear that made the impressions. The impressions submitted were mainly from a path adjacent to the home of one of the victims, Nicole Brown Simpson. However, other partial impressions were on the victims' clothing.

By examining these impressions and researching the FBI reference and standards files, an FBI examiner was able to positively link some of the crime scene impressions to size 12 Bruno Magli™ Lorenzo shoes. The examiner issued a report directly to the judge in the case and was subsequently called to testify. Although the shoe prints from the crime scene could be positively linked to a particular brand and size of shoe, at the time of the criminal trial no evidence was available that defendant O. J. Simpson owned such shoes. In the interval between the criminal and civil trials, pictures depicting Mr. Simpson in such shoes were discovered. The shoes became evidence in the civil trial, when the examiner restated his testimony.

Source: **http://www.fbi.gov/hq/lab/fsc/backissu/april2001/held.htm**

TIRE IMPRESSION EVIDENCE

DEFINITION

Tire impressions are similar in some ways to footwear impressions. They both have the same purpose: to increase friction and minimize slippage. This can be more important in tires than in shoes because tires travel at much higher speeds, under all sorts of weather conditions, and must be able to start and stop more rapidly than shoes. The part of the modern tire that is in contact with the road is called the **tread**. Today treads have many intricate designs that serve several functions. Like footwear and some other types of impressions, treads wear with time and pick up unique characteristics because of the random nature of the wearing. The two- or three-dimensional impressions that treads can transfer to a medium can be individualized in the same ways that footwear impressions can.

TIRE TREADS

The first recorded patent for something approaching the modern vehicle tire was granted in England in 1846 to Robert Thompson. His "aerial wheel" went unused and unappreciated until 1888, when the pneumatic (air-filled) tire was reinvented by John Dunlop (of Dunlop tires). There was no tread on these tires; they were bald. This caused problems that became evident when the first roads were built around the beginning of the twentieth century. It soon became clear that some sort of traction mechanism for tires would have to be developed because the roads were in such poor condition (McDonald, 1993).

In 1907, Harvey Firestone (of Firestone Tires) developed a traction design in the tread for the first time. He took the words "Firestone" and "Non-skid" and carved them into the tire so they were alternately raised and lowered in the tread. An impression of this tire revealed these words. This was actually a crude form of advertising. Today, tread design has become a science unto itself. Treads are designed not only for traction, but for channeling away water to prevent hydroplaning, for noise reduction, and for comfortable driving. Many tread designs are quite intricate in order to be able to accomplish the goals of the tire. Sidewalls of tires contain a number of markings that describe the characteristics of the tire. These are explained in "In More Detail."

IN MORE DETAIL:
What do all those numbers mean on the sidewall of a tire?

Take a look at the sidewall of a modern tire. There are lots of letters and numbers stamped or embossed on the tire. Some describe the company and model of the tire. These are pretty easy to figure out. But what about the mysterious combination of letters and numbers such as

P235/75 R 15

The "P" means that the tire is built for a passenger car. If the vehicle were a pickup truck, the tire would be designated "LT." The "235" is the cross-section width or diameter of the tire in millimeters measured from sidewall to sidewall. Since tires can be mounted on different size rims and this would affect the diameter, the designated diameter is that when the tire is mounted on the rim that it was built for. The "75" is called the **aspect ratio**. This number is derived from the height of the tire, measured from the bead (where the tire seals to the rim) to the top of the tread. The actual number is the percentage of the tire width, so the 75 means that the height of the tire is 75% of its width. In our example, the height would be 176 mm, which is 75% of 235 mm.

The "R" designates how the tire is manufactured. The most common method is radial. Other tires can be designated "D" for diagonal bias or "B" for bias belted. Finally, the "15" is the diameter in inches of the rim that the tire was designed for.

Source: McDonald, P. (1993). *Tire imprint evidence*. Boca Raton, FL: CRC Press.

TIRE IMPRESSIONS AS EVIDENCE

It is surprising to learn that more than two-thirds of major crimes in the United States involve an automobile, if only as the "getaway car." It is also true that a tire impression is the most effective way of positively identifying a motor vehicle that has been at a crime scene. Many crime scene investigators, however, do not look for or record tire impressions nearly as often as should be the case with such potentially important evidence.

As with footwear impressions discussed previously, there are three methods for recording tire impressions at a crime scene. Tire impressions may be three dimensional or two dimensional and may be negative or positive, depending on how they are produced. As with any other crime scene, photographs and drawings are the best methods of faithfully recording the overall scene, and this should be done before examination-quality castings are made of tire impressions. As with footwear impressions, both recording the impression photographically and casting for three-dimensional impressions should be done. The main advantage of a cast is that all three dimensions can be easily seen. With tire impressions in particular, there is often a need to make a

Figure 22.7. A plaster cast of a portion of a tire tread. Courtesy of Cheryl Lozen, Michigan State Police, Forensic Science Division.

three-dimensional cast at the scene because, unlike footwear impressions, tire impressions often cannot be taken up and moved to the laboratory for further analysis. Figure 22.7 shows a portion of a tire tread cast in plaster.

There are disadvantages of casting tire impressions relative to photography. Some of these are not found in footwear impressions. First, it can be difficult to make a cast on a steep incline because the casting material may tend to flow downhill and the top part of the cast may be too thin and fall apart. Second, unlike footwear impressions, which are usually about 6–12 inches long, tire impressions may be many feet long; this requires very large casts, which can be bulky and unwieldy. It would be much easier to take a series of photographs of a long impression. Finally, there is the problem with three-dimensional impressions being negative; the raised areas of the tread become depressions in the cast. Negative impressions should never be compared to a positive image. To correct this problem, a photograph negative would have to be taken of the tire tread, which adds time and expense to the project.

Not surprisingly, tire impressions are made from the same materials and are done in the same way as footwear impressions. Dental stone is the preferred medium for casts in soil, etc., and snow casting wax is

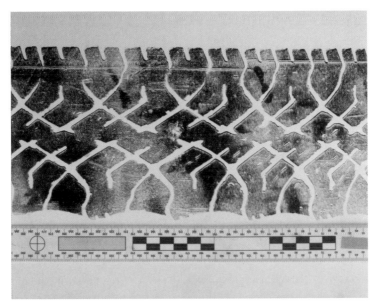

Figure 22.8. *A portion of the tread of a known tire that has been inked and rolled out on paper to provide points of comparison with an unknown tire tread. Courtesy of Cheryl Lozen, Michigan State Police, Forensic Science Division.*

best for impressions in snow. A suitable measuring instrument should be placed in all photographs.

Lifts of two-dimensional tire impressions are made in the same way as with footwear impressions. Because of the great length of many tire impressions, several lifts of the same impression may have to be made in order to get the whole impression although, in some cases, a roll of Mylar film the full length of the impression can be used to lift it all at once. In many cases, this situation means that photography would be preferred to the impression being made. Figure 22.8 shows a portion of a tire tread that was made in ink for comparison with an unknown tread.

In addition to the tread patterns, other data can be derived from tire impressions. For example, the Michigan State Police Forensic Science Laboratory collects data on **wheelbase** and **tire tread stance measurements**. The wheelbase of a motor vehicle is the distance from the center of the front hub to the center of the rear hub. The stance is the distance from the centerline of the right tire to the centerline of the left tire. The Michigan State Police Laboratory keeps a database of automobiles organized by these three measurements, and from this data and the tire tread design, an examiner may be able to determine the make and model of a car or truck.

SERIAL NUMBERS RESTORATION

Many consumer products and commercial items are identified by a unique serial number. In some cases this is required by law; firearms and certain auto parts are examples. If the object is metal, the serial number is often stamped into the surface of the metal using a set of alphanumeric dies. Like footwear and tire treads, serial numbers of this type are considered to be impressions. With serial numbers, however, the impression is not the evidence. If an object is stolen, the perpetrator may attempt to obliterate the serial number in order to foil attempts to trace or identify the object. In the burglary case described at the beginning of the chapter, the burglar attempted to obliterate the serial number on the rifle he took from the residence. Most people, when faced with the task of wiping out a stamped serial number on metal, will use a file or grinder. They will consider the job to be a success if they can no longer read any of the numbers or letters. If so, they may be making a grave mistake! It is possible for a forensic chemist to restore an obliterated serial number impression. In order to understand how this is done, it is necessary to learn a bit about how the chemistry and physics of metals change when the metal is subjected to a stress.

Metals, like most solids, have a definite crystal structure and therefore an ordered arrangement of chemical bonds between atoms (called "metal-metal bonds"). When a serial number is stamped into a metal, two things happen to the metal under the number. First, it is compressed, making it more dense than the surrounding metal. Second, the metal-metal bonds in that area are disrupted and the metal structure becomes weakened. When someone tries to remove a serial number impression by abrasion with a file or grinder, the metal surrounding the stamped number is removed. Once the metal surface surrounding the numbers becomes level with the stamped numbers, then they cannot be seen anymore. However, the compressed, deformed metal UNDER the numbers is still there (unless the perpetrator continues the grinding process beyond where the numbers disappear).

To restore the serial number, the metal surface that has been abraded is polished with a fine abrasive and then slowly treated with a corrosive acid. The acid slowly dissolves the metal. However, the metal that is under the serial numbers behaves differently toward the acid than the surrounding metal, which had not been disturbed by the stamping process. There are two possible ways that the stamped metal can behave differently. First, remember that it has become more dense when compressed. Thus, this metal would be expected to dissolve more slowly than the less dense metal surrounding it. As the metal dissolves, the serial number would be seen to be raised above the faster dissolving, surrounding metal surface.

Second, remember that the metal-metal bonds of the stamped metal have been disrupted by the stamping process, and thus weakened. This would be expected to cause the stamped metal to dissolve more quickly than the surrounding metal. The serial number would thus appear to be pressed into the metal once again.

How do we know which mechanism is more important in the restoration of serial numbers? Well, the best way is to observe the process using a low-power stereomicroscope. You will be able to easily see that the serial numbers are lower than the surrounding metal surface as the numbers are restored. This means that the weakened bond theory must be most responsible for the dissolving process of the acid.

The actual process of dissolving the metal to restore an obliterated serial number must be done carefully. Once a serial number is restored, it will eventually disappear and then will be gone forever. It is good practice to have a camera ready to take pictures of each number as it is restored so there will be a permanent, visual record of the restoration. The acid is generally applied with a cotton swab. When a number appears, the acid is washed off quickly to minimize further dissolving while the operator views the restored number and photographs it.

Various metals can have serial numbers stamped into them. Each type of metal requires different acids and conditions. Acidified, aqueous copper chloride solutions are used to restore serial numbers in iron and steel. The copper chloride acts to oxidize the iron so it will dissolve. The copper ions are reduced to metallic copper, which will deposit on the metal surface. If the metal surface is stainless steel, a more powerful acidic solution is needed. Acids are too strong for aluminum surfaces, which would dissolve almost immediately. In such cases, mild alkaline solutions are used. Figures 22.9 A and B show a serial number that was stamped into the door of a farm tractor. The number had been ground off and then restored in the lab.

Today, serial numbers are stamped into many objects other than metals. Is it possible to restore obliterated numbers in other surfaces?

Figure 22.9. A restored serial number on the inside of the metal door of a tractor. Figure 22.9A shows where the serial number has been obliterated by grinding. Figure 22.9B shows how the number has been restored.

A

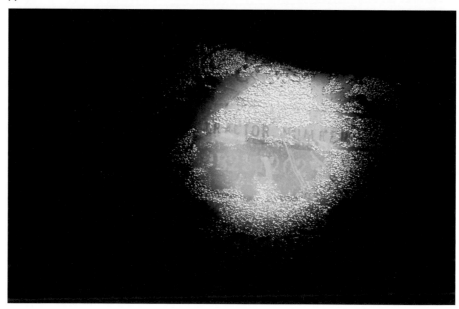

B

Plastics present difficult problems. They are generally polymers. These are very stable substances that are generally insoluble in most solvents, so it would be extremely difficult to restore serial numbers in most plastics. An unusual numbers restoration case involving old wood is described in "In the Lab."

IN THE LAB:
An unusual case of numbers restoration

Can worn serial numbers be restored from wood? This actually occurred in a case. An old wooden ladder was used to gain entry into a second-floor business that was then burgled. The prosecutor wanted to try to trace the origin of the ladder, which had been brought to and then left at the crime scene. There were some remnants of a serial number that had been stamped into the ladder when it was new, but then had worn away with time. The ladder was sent to the author's lab, where a number of restorative methods were attempted, all ending in failure. Finally, a deceptively simple way was discovered that actually worked. Can you guess what the method is?

The area of the wood containing the serial number was wetted with water to increase its contrast with the surrounding wood. Then a strong light source was aimed at the serial number at a steep angle so that the light was almost parallel to the ladder surface. The indented serial numbers created faint shadows in the wood and were then photographed and read.

BACK TO THE CASE

Did you spot how many impressions were generated by the burglary described at the beginning of the chapter? There was the thumbprint in putty, the screwdriver tool mark left in the window frame, the footwear impression in the soil outside the window, and the serial number that had been filed off the stolen rifle. The putty, the window frame, and the rifle can be easily transferred to the laboratory for analysis. The shoe print in the soil would have to be photographed and a dental stone cast made. Of course, it would not be advisable to try to dig up enough soil containing the shoe print and transfer that to the lab. The shoe print, fingerprint, and tool mark could all be individualized to the particular object or person from which they arose.

CHAPTER SUMMARY

Impressions can be two or three dimensional. The most common impression evidence types are footwear, tire treads, and serial numbers. Other types of evidence may also be in the form of impressions. An example of this would be a fingerprint left in putty. Impression evidence can be individualized to one particular object if there are sufficient unique characteristics present. These characteristics arise from the random wearing of the shoe wear or tire tread.

Preserving impressions is very important because they often cannot be transported to the forensic science lab intact. Proper, high-resolution photography is commonly done, with digital photography becoming more popular. A suitable measuring instrument must be in the picture to facilitate scale determination. The measuring instrument must be a ruler or other device that actually measures distance. Ordinary objects such as coins or a cigarette pack that could provide perspective but not measurement should not be used. Dental stone has become the casting material of choice in many impressions because of its ease of use and high definition.

Serial numbers that have been stamped into metal can be restored after scraping off if the scraping hasn't gone too deep into the metal. The fatiguing of the metal under the stamped numbers will dissolve faster than surrounding metal when subjected to oxidizing acids.

Test Your Knowledge

1. What is "impression evidence"? What types of evidence are included?
2. What are the differences between two-dimensional and three-dimensional impressions?
3. Under what conditions can impression evidence be individualized?
4. How are imprints "lifted"?
5. How are three-dimensional impressions "lifted"?
6. How are casts made of impressions? What are the best casting materials?
7. How many characteristics are necessary in order to individualize a footwear or tire tread impression?
8. What are the major class characteristics present in footwear impressions?
9. What are the major individual characteristics in footwear impressions?
10. What are the major differences between footwear and tire tread impressions?

11. What is tire tread stance?

12. What is the wheelbase on a car? How is it measured?

13. What are the major reasons tires have treads?

14. Write an equation that shows how copper chloride reacts with iron.

15. What is metal-metal bonding? What is its importance in serial number restoration?

16. Some serial numbers are "embossed" on a surface. Could an obliterated, embossed serial number be restored in the same way that stamped ones are?

17. Why can stamped serial numbers normally not be restored in plastic?

18. What would happen if you used acid to restore a serial number set and just left the acid on the object and went away for a while?

19. Under what conditions would fingerprint evidence be considered to be impression evidence?

20. Why are firearms and tool marks considered to be impression evidence?

Consider This . . .

1. Impression evidence can be either class or individual. Under what conditions does it change from class to individual? How do probabilities enter into the conclusion that such evidence is class or individual?

2. Describe how you would preserve and photograph a tire tread impression that was left in grease on a garage floor. How would you handle a tire tread impression left in dirt?

3. Besides the types of evidence described or mentioned in this chapter, are there other types of impression evidence? What are they?

BIBLIOGRAPHY

Bodziak, W.J. (1993). *Footwear impression evidence.* Amsterdam: Elsevier.

McDonald, P. (1993). *Tire imprint evidence.* Boca Raton, FL: CRC.

PART SIX

Law and Forensic Science

Legal Aspects of Forensic Science

KEY TERMS

Admissibility of evidence

Attorney-client privilege

Chain of custody

Cleric-penitent privilege

Criminal investigation

Daubert v. Merrill-Dow

Deposition

Demonstrative evidence

Discovery

Doctor-patient privilege

Due Process Clause

Expert witness

Falsifiability

Frye v. United States

Hearsay

Marital privilege

Materiality

Peer review

Probativeness

Real evidence

Reciprocal discovery

Relevant

Rules of evidence

Novel scientific evidence

Self-incrimination

Subpoena

Subpoena *Duces Tecum*

Technical evidence

Tort

Trier of fact

Unreasonable search and seizure

Voir dire

INTRODUCTION

One of the major differences between practicing science and practicing forensic science is that forensic science is done within a legal context. The results of scientific examination of evidence in the forensic science arena often end up in a courtroom. Forensic scientists must be familiar not only with the scientific principles that guide their analyses, but also the rules and regulations that govern their conduct in a civil or criminal court.

In the U.S. legal system the admissibility of all evidence is guided by **rules of evidence**. These rules determine what evidence may be admitted, for what uses, and under what conditions. This is true of scientific evidence as well. In addition, as we shall see, scientific evidence also has an additional layer of rules that must be obeyed in order for it to be admitted.

The culmination of a forensic scientist's efforts is to write a report on the examinations performed and then, if needed, testify to those results in a court of law. The word "forensic" means "applied to public or legal concerns" and is as important as the word "science." If someone were the greatest scientist on earth but could not effectively testify to what he or she did, that person would be a failure as a forensic scientist. Professional forensic scientists must be able to explain the theories, methods, procedures, analyses, results, and interpretations of the scientific examinations he or she performed. And the scientist must do this without being an advocate for either side in a case: Impartiality is the hallmark of science, and this is especially true of forensic science.

FORENSIC SCIENCE IN THE CRIMINAL JUSTICE SYSTEM

As our world has become more complex and increasingly driven by science and technology, the nature of crime and the evidence that helps convict the guilty and exonerate the innocent have also become more technical. This, in turn, has increased the responsibility and importance of forensic science and the scientists who practice it. Forensic science is widely considered to be part of the investigation process. From the moment that a crime or civil infraction is discovered, the investigation process begins. This process includes such important processes as developing and interviewing witnesses interrogating suspects, and collecting and processing evidence. This chapter covers the role that forensic scientists play in the investigation process with emphasis on the criminal investigation process. For the most part, the same considerations hold if there is a civil infraction.

THE CRIMINAL INVESTIGATION PROCESS

When a criminal action begins as the result of activity by the police, prosecutor, or grand jury, the **criminal investigation** process also begins. This involves discovering who committed a crime or, if someone is arrested for the crime, searching for evidence that helps convict or exonerate that suspect. Criminal investigation is a continuous process. It begins when the crime is first discovered and may continue up to the time of a preliminary hearing, a grand jury hearing, or a trial and beyond. Forensic scientists become involved in this process in a number of ways:

- **Response to the crime scene**: It is the policy of many law enforcement agencies to have a forensic scientist, or perhaps several, attend crime scenes whenever there is a homicide suspected or in cases in which there may be a high-profile or otherwise notorious crime committed. No one knows better than a forensic scientist how to recognize, collect, and preserve evidence within his or her area of responsibility. Because of time constraints and the need to preserve human resources, forensic scientists are usually sent only to the most important or high-profile crime scenes.

- **Analysis of evidence**: This is, of course, the primary task that forensic scientists perform. This is an ongoing job during a criminal investigation process. Investigators may bring in evidence to the crime lab at many points during the investigation, and the results of the analyses can have profound effects on the course of the investigation and the outcome of the case.

- **Depositions**: At one time, **depositions** were used almost exclusively in the civil justice system, but recently more states have begun to use them in criminal cases. A deposition is a proceeding in which testimony is given under oath but not in court or before a judge. A reporter or recorder authorized to administer oaths is present to take down the testimony. The main purpose of the deposition is to preserve testimony for trial, but it may also be used as a discovery tool to gain information about what the opposition knows. A witness giving a deposition is usually questioned in the same manner as at a trial; the proponent questions first, followed by cross-examination by the opponent counsel. One major difference is that there is no one present to rule on objections to questions. In such cases, the objection is lodged and the witness will usually answer the question, but the challenged testimony may not be used at trial until the court has ruled on the objection.

- **Court testimony**: Along with the analysis of evidence, court testimony is the heart of the job of a forensic scientist. It is the most effective way to impart highly technical and complicated facts and opinions to the judge or jury. Forensic scientists may testify at preliminary hearings and trials. They do not ordinarily offer testimony in a grand jury proceeding.

Legal Constraints on the Criminal Investigation Process

The U.S. Constitution, the U.S. Congress, and state legislatures have created rules that govern the criminal justice system, including the criminal investigation process. Some of these rules apply to all of the states, whereas some represent law that is valid only in the state where it was legislated. A few of the more important rules that sometimes have bearing on forensic scientists will be discussed in the following sections.

DISCOVERY

Discovery is a process whereby one side in litigation seeks to find out something known by the other side. Historically, the rule in the United States has been that the defendant in a criminal case is not entitled to inspection or disclosure of evidence that is in possession of the prosecution. The modern trend is to broaden discovery in criminal cases, both in favor of the defendant and the prosecution. This is known as **reciprocal discovery**.

In the Federal Code Title 28, Federal Rules of Civil Procedure, Rule 16 governs the discovery pretrial in the federal courts, and many states have similar rules. Rule 16(F) of the Federal Rules of Criminal Procedure states that: "[t]he government must permit a defendant to inspect and to copy or photograph the results or reports of any physical or mental examination and of any scientific test or experiment" subject to certain conditions. Rule 16(G) states that the government is obliged to "give to the defendant a written summary of any testimony that the government intends to use" that could constitute expert testimony as defined by the Federal Rules of Evidence.

The Jencks Act permits the defense in a criminal case to obtain any written and oral statements made by a prosecution witness, but only after the witness has testified at trial. A reasonable recess is allowed for inspection of the written statement. It may then be used by the defense for purposes such as impeachment.

Finally, the Supreme Court ruled, in *Brady v. Maryland*, that the government has a continuing burden in a criminal case to turn over to the defense any evidence that can reasonably be construed as favorable to the defense. The evidence must be disclosed at a time when it would be valuable to the defense. Exculpatory lab reports fall within the purview of the Brady doctrine. Failure to comply with this doctrine can result in a mistrial or even dismissal of the charges against the defendant.

SEARCH AND SEIZURE

The Fourth Amendment of the U.S. Constitution states:

> *The right of the people to be secure in their persons, houses, papers, and effects against unreasonable searches and seizures shall not be violated and no warrants shall issue, but upon probable cause, supported by oath or affirmation, and particularly describing the place to be searched and the person or things to be seized.*

The familiar Fourth Amendment to the U.S. Constitution prohibits **unreasonable searches and seizures**. What actually constitutes an unreasonable search or seizure has been the subject of numerous court decisions during the past 200 years. The Supreme Court has held that the Constitution expresses a clear preference for searches conducted only after judicial authorization, that is, after a search warrant has been obtained. The authority to search for and seize also may be limited by other laws enacted by Congress and state legislatures.

There may be times when forensic scientists are called upon to attend crime scenes and help in the collection of physical evidence. They should understand the legal limitations on this collection process so as to avoid improper seizure of important evidence. There are numerous instances in which evidence has been collected and analyzed and shown to be highly important in the prosecution of the defendant, only to be stricken because the evidence collector ran afoul of the prohibition against unreasonable searches. If there is any question about the reasonableness of a search or seizure, a search warrant should be obtained. This is the responsibility of the criminal investigators and the prosecutor.

SELF-INCRIMINATION

A portion of the Fifth Amendment of the U.S. Constitution reads in part:

> *Nor shall any person be compelled in any criminal case to be a witness against himself.*

Statements made by a suspect confessing to all or part of a crime, and later used against that suspect to prove guilt, constitute self-incrimination. **Self-incrimination** is not prohibited. Only *compelled* self-incrimination falls within the scope of the Fifth Amendment. Self-incrimination is allowed subject to certain limitations. It applies to both the verbal and written statements of a suspect. The protection of objects, property, personal papers and effects, and so forth, is the subject of the Fourth Amendment.

The Fifth Amendment also contains the so-called **Due Process Clause**. This clause protects people against outrageous government behavior, such as the case in which a person's stomach was pumped to get to illicit drugs. Obtaining exemplars of hair, blood, etc., can constitute a seizure requiring a warrant under the Fourth Amendment; however, evidence of this nature does not constitute "incrimination" that falls within the scope of the Fifth Amendment. Other identifying data such as fingerprint specimens and photographs generally do not require a warrant.

PRODUCTION OF EVIDENCE

Subpoenas

The U.S. Constitution guarantees the accused in a criminal case the right to confront all witnesses against him. A forensic scientist who examines physical evidence and reaches conclusions that are inculpatory is such a witness. There exists a mechanism by which a witness can be compelled to offer testimony in a criminal or civil case. It is known as the **subpoena**. Subpoena comes from the Latin and means "under penalty." A subpoena is an order issued under the direction of a court commandng the presence of a witness at a specific time and place to give testimony or other evidence. If a witness ignores a subpoena, he or she may be charged with being in contempt of court.

In most cases a subpoena must be delivered to the subject in person, although sometimes a subpoena for a forensic scientist is mailed or delivered to the laboratory where the scientist retrieves it. Usually, a forensic scientist will receive a certain type of subpoena called a **Subpoena *Duces Tecum***. This type of subpoena commands not only the presence of the witness in court but also any documentation or evidence in his or her possession that is material to the case. Thus, the forensic scientist must bring in all lab reports as well as other related charts and graphs. This type of command carries with it the same potential penalties for disobeying as do other types of subpoenas.

THE RULES OF EVIDENCE

Evidence in a criminal case can be defined as anything that will help prove or disprove a material fact. It helps the judge or jury reach conclusions about the guilt or innocence of the defendant in a case. There are two major types of evidence: real and testimonial. Real evidence is physical; it consists of things that help link a suspect to a crime or explain the circumstances of the incident. It may be **real evidence** such as fingerprints, fibers, blood, weapons, or **demonstrative evidence**, which consists of supporting materials such as crime scene photos or sketches. Demonstrative evidence does not arise directly from the commission of the crime but is generated by observation and documentation of the crime scene.

Testimonial evidence is the oral recitation of facts and sometimes impressions or opinions by a witness under oath. Lay witnesses' testimony is usually limited to first-hand observations and impressions. Lay witnesses usually are not permitted to offer opinions; they may only state facts. **Expert witnesses**, on the other hand, are often called upon to state conclusions and opinions based on their examination of evidence or observation of the crime scene or parties to the crime. These conclusions and opinions must be within the purview of their expertise. Forensic scientists and other experts are required to state an opinion in some cases when they do not wish to. It is part of their responsibility as scientists.

AUTHENTICATION OF EVIDENCE: THE CHAIN OF CUSTODY

Virtually all real evidence is subject to authentication. There must be a showing that the evidence is the same and in the same condition from the moment it was seized at the crime scene until it is used in court. It must have "sponsors" who can identify it and follow its trail. The only exception to rigorous authentication of real evidence occurs when it has some unique characteristic that makes it differentiable from all other objects. This might include a weapon with a unique serial number that has been noted by a police officer when the weapon was first seized.

The most commonly accepted method for authentication of evidence is the **chain of custody**. The chain of custody is both a process and a document that memorializes the transfer of evidence from the custody of one person to another. The process of authentication starts at the crime scene or anywhere evidence is seized. Each item of evidence that doesn't have an obvious unique label such as a visible serial number is given a unique identifier. Each piece of evidence is packaged separately in a tamper-evident container and sealed. The official who packages the evidence affixes his or her signature or initials and the date to the evidence container. Every time someone, such as a forensic scientist, opens the container to examine the evidence, this person must do so, in a manner that doesn't disturb the already affixed seals. He or she must also reseal and label the evidence with unique identifiers so it can be easily seen who opened it.

The evidence is also accompanied by a chain of custody form, as shown in Figure 23.1. This document contains a description of the evidence and a place for the signatures of everyone who handles the evidence. That person signs for the evidence when it is received and then signs it over to the next person. Each signature is accompaniedby the date and time. This way, it is easy to tell who had custody of the evidence at any time and that person can be called to testify what condition the evidence was in, what was done to it, and how it was stored.

A substantial break in the chain of custody, either the process or the documentation, can result in the evidence being excluded from admission to court. In such cases, the opponent must show that the break in the chain could have reasonably resulted in the evidence being adulterated or otherwise tampered with.

THE ADMISSIBILITY OF EVIDENCE

The major questions that must be answered about evidence in a courtroom setting are: What types of evidence may properly be brought to the attention of the trier of fact (the judge or the jury)? And what uses may the trier of fact make of the evidence that it is permitted to consider?

This is the essence of the **admissibility of evidence**. A number of rules determine what evidence may be admitted and under what conditions. Most of the rules of evidence govern admissibility. There are fewer rules that govern how a judge or jury may use evidence that is admitted properly. The consideration of what uses a jury can make of admitted evidence is beyond the scope of this book. We will consider only the question of admissibility here. The rules of admissibility of evidence provide a framework to help preserve the integrity of the evidence.

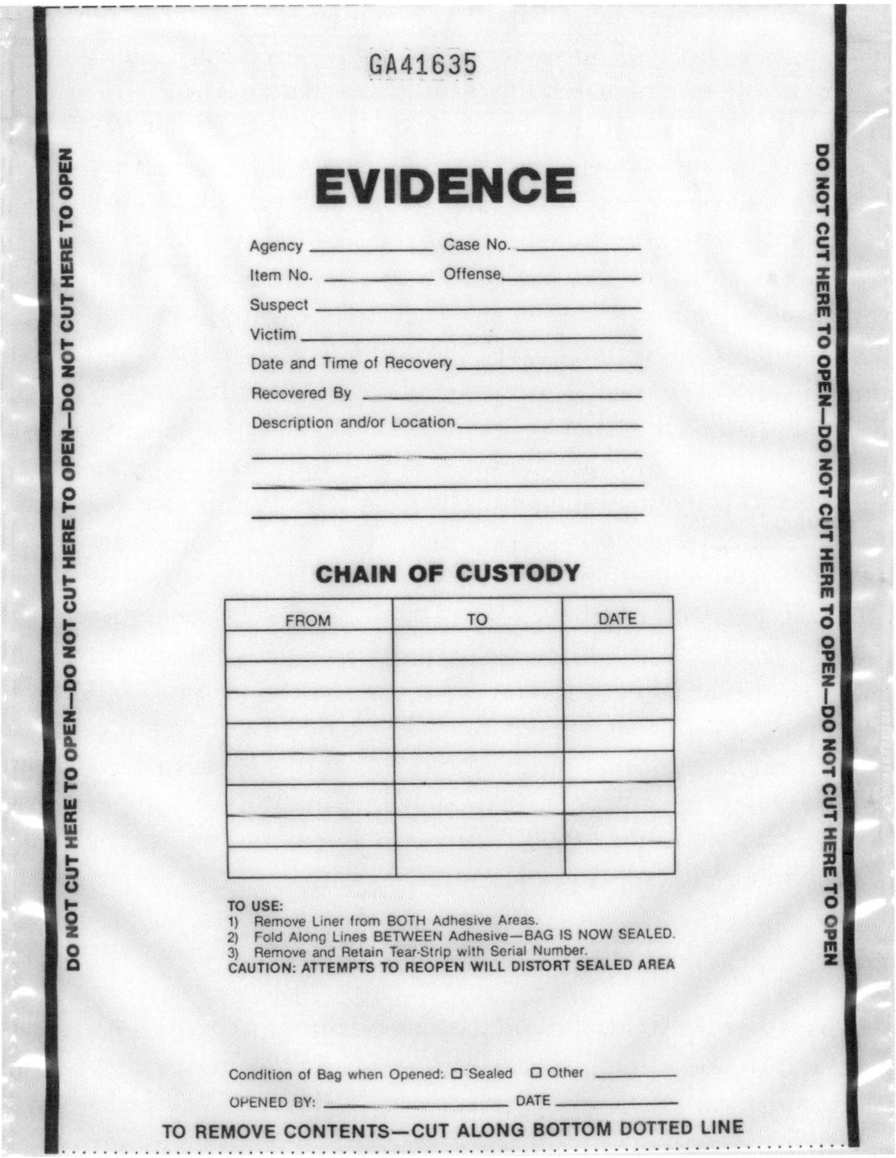

GA41635

DO NOT CUT HERE TO OPEN—DO NOT CUT HERE TO OPEN—DO NOT CUT HERE TO OPEN

EVIDENCE

Agency ______ Case No. ______
Item No. ______ Offense ______
Suspect ______
Victim ______
Date and Time of Recovery ______
Recovered By ______
Description and/or Location ______

CHAIN OF CUSTODY

FROM	TO	DATE

TO USE:
1) Remove Liner from BOTH Adhesive Areas.
2) Fold Along Lines BETWEEN Adhesive—BAG IS NOW SEALED.
3) Remove and Retain Tear-Strip with Serial Number.
CAUTION: ATTEMPTS TO REOPEN WILL DISTORT SEALED AREA

Condition of Bag when Opened: ☐ Sealed ☐ Other ______
OPENED BY: ______ DATE ______
TO REMOVE CONTENTS—CUT ALONG BOTTOM DOTTED LINE

DO NOT CUT HERE TO OPEN—DO NOT CUT HERE TO OPEN—DO NOT CUT HERE TO OPEN

Figure 23.1. *Chain of custody form. The form is on the front of a secure evidence package.*

The Rules of Admissibility

In general, any evidence is admissible if it is **relevant**.

RELEVANCE

Relevance has two components: **materiality** and **probativeness**.

RELEVANCE = MATERIALITY + PROBATIVENESS

If evidence is material, it means that it applies to a matter dealing with the case at hand, and not some other case.

As an example of materiality, assume that a man is on trial for injuring another person with a knife during a street robbery. Evidence that the accused had a stash of stolen guns at his home at the time might indicate that he was violent or that he liked to sell guns, but it would not be material to the robbery charge.

The other component of relevance is probativeness. Evidence is probative if it tends to make a fact or issue more or less probable than if the evidence were not present. Another way of expressing this is: Does the evidence tend to prove something about a fact at issue in this case? Let's take the example of a man who is arrested on suspicion of distribution of methamphetamine. A search warrant is issued to search his home. No methamphetamine is found, but the police uncover a large number of plastic bags, a postal scale, a large bag of a material that is commonly used as a diluent for drugs, and a book that describes in detail how methamphetamine is made. Is this evidence probative? Most certainly it is. Although it doesn't prove by itself that the accused is a drug dealer, it tends to make this assertion more probable.

So we can see that the test for relevance is two-pronged. The evidence at issue must be both material and probative in order for it to be considered relevant. But this is only half the story. Even if evidence is highly relevant, it may still be inadmissible because of other issues. These are described below.

OTHER CONSIDERATIONS THAT AFFECT THE ADMISSIBILITY OF EVIDENCE

- **Prejudice**: Prejudicial evidence may be highly relevant but also has the effect of putting the accused in a bad light. It can impugn the accused's reputation or turn the jury against him or her. Consider some examples. In many jurisdictions, color photographs of an autopsy of the victim of a violent homicide may not be admissible because their graphic and sickening nature may cause the jury to focus on the gore and blood they depict, at the expense of their value as evidence. The jurors may view the defendant as being guilty because of the photographs of the victim. Unless the sponsoring pathologist can make a case that these color photographs are necessary to illustrate certain probative facts about the killing, these photos may not be admissible.

Prejudice is also one of the major reasons why a defendant's prior criminal record is not admissible evidence of guilt in a new charge because the jurors may feel that because this person committed similar crimes before, he or she did it again, and they will assume that because this person may have committed this type of crime in the past that he or she has done it again. Thus, they may not properly consider the facts of the present case.

- **Time wasting**: Although attorneys are generally given wide latitude to prosecute and defend in criminal cases, judges will generally not tolerate evidence that wastes time because it is unnecessarily cumulative or repetitive or lacks relevance. This has the effect of distracting and perhaps even confusing the jury.

- **Unreliable**: This is one of the criteria that governs the admissibility of certain scientific (and perhaps pseudoscientific) evidence. Evidence must be reliable if the jury is to be able to weigh it properly in reaching a decision about guilt or innocence. This is also the reason that much eyewitness testimony is discredited in court. Research has shown that, when witnesses experience a startling or surprise event, their recollections can differ widely from one person to the next. Each person reacts to such events differently, and their reactions will bias their perceptions about what they experienced.

Another form of unreliable evidence is **hearsay**. Hearsay is defined as a statement made by someone outside of a courtroom and not under oath and that is now being used inside court to prove what it asserts. An example of hearsay would be if person A witnessed a crime committed by person C and tells person B what he saw, and then person B offers to testify in court what person A told him about the crime. If the testimony by person B is being used to prove that C committed the crime, this would be hearsay. Some types of hearsay may be considered unreliable because it is difficult or impossible to effectively cross-examine someone who did not witness something but is only repeating something that he or she heard.

There are numerous exceptions to the hearsay rule. For example, suppose that a forensic scientist analyzed some plant material and concluded that it was marijuana. She then wrote a report that communicated these findings. At trial, the prosecutor wishes to introduce the report as evidence to prove that the plant material was marijuana. The report may be admissible under certain circumstances even if the scientist is not present, even though it would be technically considered to be hearsay. The difficulty of cross-examining a lab report can be appreciated, but it may be admissible if properly attested and agreed to by both sides. The lab report may also be important when a long period of time has elapsed between the time the scientist analyzed the evidence and wrote the report, and the time of the trial. The scientist may not remember doing the analysis in this specific case, so the lab report then stands as the most reliable evidence of what was done to the evidence.

- **Improper procedures**: Courts generally do not allow surprise witnesses to testify without the attorneys giving notice to the other side. Evidence is also generally inadmissible if it is offered out of turn or after the proponent has already rested his or her case. Attorneys are not allowed to present testimony during an opening or closing argument. These prohibitions against improper evidence are all in place to protect the jury as well as the rights of the accused.

- **Existence of privileges**: In the legal context, a privilege is a protection given someone to protect that person from having to offer testimony against another person. Certain privileges that exist in many locales have been created by legislation. They include the **attorney-client**, **doctor-patient**, **cleric-penitent**, and **marital** privileges. All of these privileges in some way are designed to protect sensitive or intimate or otherwise special communications. The attorney-client privilege, for example, protects communications between an attorney and a client by allowing the attorney to refuse to testify to such communications. Likewise, a priest is protected from testifying about the contents of confessions made by a member of the congregation. There are two marital privileges: one protects one partner in a marriage from testifying against the other, and the other permits one partner to silence the other partner who wishes to testify. There are, of course, many exceptions to the privileges mentioned here.

- **Constitutional constraints**: Certain provisions of the Constitution, discussed earlier, provide for the exclusion of evidence and/or testimony where violations have occurred. They include evidence seized in violation of the fourth Amendment and testimony of self-incrimination.

In summary, evidence will be admissible in court only if it is both relevant and doesn't violate the other considerations described above. This applies to all evidence in a criminal case. Next, we will turn to the question specifically of the admissibility of scientific or expert evidence. We will see that because of its technical nature, it is treated somewhat differently than non-scientific evidence.

Admissibility of Novel Scientific and Technical Evidence

Scientific and **technical evidence** differ from ordinary evidence in a number of significant ways. First, such evidence is usually given by people who have been qualified as experts by the court. This permits them to offer opinions and conclusions about such evidence that would be beyond the knowledge of the average person. Experts are, in effect, interpreters of technical evidence for the jury. Second, because technical evidence is given by scientists and experts, it sometimes has an aura of truth and infallibility about it, which necessitates ensuring that the evidence is valid and reliable when it is presented to the jury. Most jurors and judges do not have the knowledge to determine if the testimony about technical evidence they are receiving is correct or reliable or valid, so there must be legal protections in place to ensure or at least promote reliability. These considerations are most important for evidence involving new or novel scientific techniques or instruments. There must be some assurance that the new technique has been sufficiently tested and validated so that the jury or judge may rely on its conclusions.

The history of the admissibility of scientific testimony involves two very instructive and interesting cases that illustrate the difficulties that courts have had in determining the standards for admissibility of **novel scientific evidence**.

THE FRYE CASE

Prior to 1923 in the United States, most courts treated scientific evidence the same as any other type. The rules governing the admissibility of evidence were derived from the Common Law. There was no codification of specific rules. In 1923, the landscape changed for novel scientific evidence, owing to a murder case in Washington, D.C. James Frye was on trial for murder. As part of his case, he sought to have introduced the results of a test that utilized a machine that could be considered the forerunner of today's polygraph. He claimed that the results of the test helped to prove that he was innocent. The prosecution objected to the admission of this novel evidence, and the judge agreed. On appeal, the court upheld the trial judge's decision. In effect the appeals court stated that, with respect to novel scientific evidence, not only must it meet the relevancy standard, but an additional hurdle must be overcome as revealed in its ruling:

> Just when a scientific principle or discovery crosses the line between the experimental and demonstrable stages is difficult to define. Somewhere in this twilight zone the evidence force of the principle must be recognized, and, while courts will go a long way in admitting expert testimony deduced from a well-recognized scientific principle or discovery, *the thing from which the deduction is made must be sufficiently established to have gained general acceptance in the particular field to which it belongs.* (emphasis added). (**Frye v. United States**, 293F. 1013, 1014 [D.C. Cir. 1923])

Thus, the standard for novel scientific or technical evidence that came out of this decision was that before a *new scientific technique* could be introduced in court, the underlying principle that governed it must have achieved general acceptance within the particular scientific field to which it belongs. One important issue was not decided by the court: what constitutes general acceptance. In fact, this issue has never been clearly decided. It has come to mean, more or less by default, that the technique and principles have been published in a peer-reviewed journal or other equivalent exposure to the field. This implies that **peer review** for a journal and publication means that a technique will be generally accepted. There are numerous examples in all scientific endeavors where this has not been borne out. Many valid and reliable scientific principles have never been published, and there are numerous examples of techniques that have been published and were later shown to be unreliable.

Over the next 70 years, the federal courts and about half of the states used the *Frye* case as the yardstick to evaluate the admissibility of new scientific techniques. During that time a number of novel scientific techniques were subject to "Frye challenges" in various courts. They included voiceprint spectrography, blood spatter pattern analysis, polygraph analysis, and even DNA typing techniques.

On January 2, 1975, the Congress, for the first time, approved an evidence code. This had been proposed by the U.S. Supreme Court in a preliminary draft form in 1969. Its effective date was July 1, 1975. The initial set of rules of evidence contained a specific article dealing with expert and opinion testimony (Article VII) that contained individual rules, which have since been amended. Under those rules, specifically Rule 702, the proponent of expert testimony had the burden of demonstrating that the expert was qualified and that the opinion evidence would have been helpful to the fact finder (the judge or jury).

After the new evidence code was adopted by Congress, federal and many state courts became divided as to whether *Frye* or the new Federal Rules should be used to determine the admissibility of novel scientific evidence. The question was addressed and settled by the U.S. Supreme Court in ***Daubert v. Merrill-Dow***.

DAUBERT V. MERRILL-DOW

The plaintiff in *Daubert v. Merrill-Dow*, heard in Federal District Court, was a pregnant woman who took Bendectin, a Merrill-Dow product that had been prescribed for many years to relieve nausea that occurred during pregnancy. After her baby was born with birth defects, she sued Merrill-Dow, claiming that Bendectin caused the birth defects. This type of civil harm is called a **tort**. Since the biochemical causes of birth defects are not fully understood, there was no direct way for Daubert to establish directly that Bendectin was the cause of the defects. Instead, the plaintiff had to resort to epidemiology, the study of the cause and effects of disease on large populations. The plaintiff and defendant both retained statisticians to determine whether the instances of birth defects among women who took Bendectin during pregnancy were statistically greater than birth defects in the general population. The plaintiff's experts concluded that there was a significant increase in birth defects among Bendectin users' babies, whereas the defendant's experts concluded that any increase was not statistically significant. The defendant also argued that the plaintiff's experts did not use methods that were *generally accepted by the scientific community* in reaching their conclusions; that is, they argued that the plaintiff had failed to meet the *Frye* standard for technical evidence. The court agreed and, upon a motion for summary judgment, found for the plaintiff, Merrill-Dow. Daubert appealed and eventually the case reached the U.S. Supreme Court. The Court ruled that the trial judge had used the wrong standard of admissibility in reaching his ruling. The Supreme Court concluded that the Federal Courts could not use the *Frye* rule any more in deciding questions about the admissibility of scientific or technical evidence, and that the doctrine of general acceptance was not the proper yardstick. Instead, the courts must use the Federal Rules of Evidence relevance standard when evaluating the admissibility of novel scientific or technical evidence. The Court drew particular attention to FRE 702, reproduced here:

> *If scientific, technical, or other specialized knowledge will assist the trier-of-fact to understand the evidence or to determine a fact in issue, a witness qualified as an expert by knowledge, skill, experience, training or education, may testify thereto in the form of an opinion or otherwise.*

In interpreting the Federal Rules including 401, 402, 403, and 702, the Court indicated that the judge must be the "gatekeeper" who decides when novel scientific evidence is admissible. In doing so, the court must use rational criteria for determining whether evidence is reliable and valid. The Court went so far as to suggest several criteria that a judge could use in the gatekeeper role. These criteria were not meant to be exhaustive, but only suggestions:

- **Falsifiability**: If the underlying theory or principle behind a novel technique has been repeatedly tested to see if it is false and in all cases the theory is verified, this can be a good measure of validity. This is not amenable to all principles, and proper research designs must be implemented for this to be a valid criterion.

- **Knowledge of error rates**: If the error rates of the results of a technique are known or can be estimated, then a judge could presumably make some determination as to the reliability and validity of a technique. For some techniques however, there is little or no quantifiable data available to determine an error rate.

- **Peer review**: Certainly, a technique or method or principle that has survived the peer review process and has been found worthy of publication has demonstrated some level of scientific validity. This is tempered, however, by the issue of the quality and scholarliness of the journals in the applicable field.

- **General acceptance**: The Supreme Court never meant to discard general acceptance as an acceptable criterion for determining scientific validity. The Court concluded that this should not be the only criterion and that there exist other, perhaps better ones. The Court did not, however, seek to define what it meant by general acceptance.

In addition, the Court's decision mandated that novel scientific techniques must be based on scientific principles, not speculation, and that the scientific basis for the principles had to be demonstrated.

Admissibility of Scientific and Technical Evidence Today: Fallout from Daubert

Since the *Daubert* decision in 1993, most states in the United States have adopted the principles of the decision in whole or in part. The rest still use the *Frye* rule.

There has also been interesting fallout from the *Daubert* decision. The mandate for having a demonstrable scientific basis for introducing novel scientific or technical techniques in court has caused the legal and forensic scientific communities to take a fresh look at forensic scientific disciplines that were heretofore assumed to be proper and correct from a scientific basis. For example, there have been some recent challenges to the admissibility of testimony by expert questioned document examiners where there is a definite conclusion of authorship of a handwritten document. The basis for the challenges is that there has been little or no demonstrated scientific research that proves that handwriting comparisons are valid techniques for establishing definite authorship.

At least one case has had a challenge to morphological human hair comparisons. In this case, the forensic scientist was going to testify that the defendant was one of an indeterminate population of people that could have been the source of the hairs found at the crime (rape) scene. The judge excluded the evidence as being too speculative and not scientific enough for the jury to consider.

There have also been some recent challenges to the validity of "matching" partial latent fingerprints with known prints of a suspect. As in the cases with handwriting comparisons, the issue is the scientific basis (or lack of it) for concluding that fingerprints are unique and can be matched reliably. There is little doubt that the courts will be dealing with additional issues of the sufficiency of the science that underlies scientific and technical methods, processes, and techniques.

Recently, as a result of the findings in *Daubert*, the Congress changed some of the rules regarding the admissibility of novel scientific evidence. For example, language was added to FRE 702 at the end:

> *if (1) the testimony is based upon sufficient facts or data, (2) the testimony is the product of reliable principles and methods, and (3) the witness has applied the principles and methods reliably to the facts of the case.*

WRITING REPORTS

Forensic science laboratory reports vary widely in their particular formats but all should contain the following information:

- The name of the examiner who conducted the tests
- The agency where the examiner works
- The date the report was issued
- The case identification information (laboratory number, case number, etc.)
- Contact information for the examiner
- The items examined
- The methods and instrumentation used to examine and analyze the submitted items
- The results of the examinations and/or analyses
- Any interpretations or statistics that are relevant to the results
- A statement of the disposition of the evidence
- The signatures of the examiner and any reviewers of the report

The format of the report should roughly follow that of a standard scientific paper, that is, Introduction, Materials and Methods, Results, Conclusions, and Discussion. It is important to remember, however, that, unlike a scientific paper in a peer-reviewed journal, a forensic science laboratory report is *not* intended for other scientists. Most of the readers of a forensic science laboratory report are law enforcement officers, attorneys, and judges, all of whom may have little to no training in science. This requires a special effort to make the reports readable, intelligible, and concise, while retaining the necessary information to maintain its scientific rigor. To this end, forensic science laboratory reports should be *summations* of analyses, not complete and definitive examples of scientific research results. Figure 23.2 shows an example of a forensic science report.

EXAMPLES OF ANALYSIS AND REPORTS

Several examples of analyses and reports follow, starting with a relatively simple drug analysis, moving through increasingly complex interpretations (arson, DNA), and ending with a difficult-to-interpret trace evidence case. These examples will show how the structure, wording, and content of a report must reflect the nature of the analyses it sum-

CERTIFICATE OF ANALYSIS
INDIANA STATE POLICE

Figure 23.2. An example of a forensic science report.

State Police Laboratory
8500 East 21st Street
Indianapolis, IN 46219

Telephone: (317) 899-8521
Fax: (317) 899-8298

May 30, 2005

TO: INVESTIGATING OFFICER
POLICE AGENCY
123 NORTH STREET
ANYTOWN, IN 46219

Lab File # 05Q1234

Agency Case #: 123-4567

Evidence Submitted by: Officer John Smith

Received by Laboratory: 05/10/05 at 9:00

Item 1 Sealed plastic bag containing a ziploc plastic bag containing a white powder.

Results:

Item 1 was found to contain Cocaine, a controlled substance.
The net weight was 2.54 grams.

Donna K. Roskowski
Forensic Scientist 1

DKR

Lab File #05Q1234 Page 1 of 1 Reviewed by 5656

marizes. In many of his lectures and presentations to scientists and non-scientists, Jose Almirall (personal communication) has used this approach to great effect and it is used here with his permission.

Drug Analysis

A packet of white powder is seized from a suspect and submitted to the laboratory. The powder (Item Q1) is analyzed by GC-MS and is identified as cocaine (see Figure 23.3).

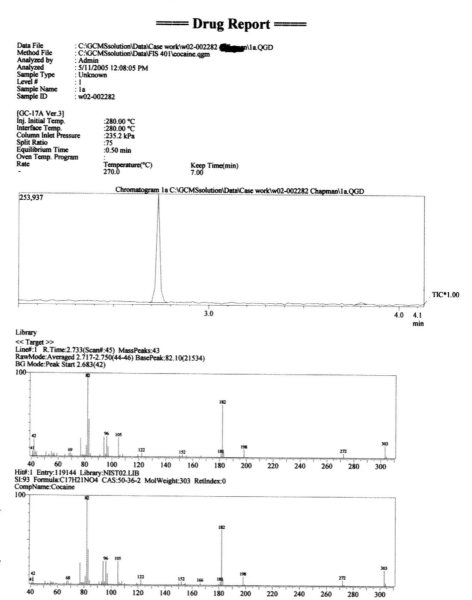

Figure 23.3. A gas-chromatography-mass spectrometry spectrum of cocaine. By itself, this is a qualitative analysis because the weight of the cocaine has not been determined.

As it stands, this is a qualitative analysis because the amount (weight) of the cocaine has not been determined. Federal law requires a quantification of the substance to determine its purity. Total weight of the drug is reported (15 g) along with a statement of the uncertainty of the measurements (± 0.1 g). The case file includes all of the methodology, data, and quality control information as required by the laboratory's protocols and in accordance with the relevant laws.

Figure 23.4 shows typical wording for this kind of analysis.

Arson Analysis

A sample of nylon carpeting from a burned house is submitted to determine if accelerants were used to start the fire. The carpeting is subjected to an extraction process to remove any accelerants from the carpet and its backing. The instrumental analysis provides a qualitative identification of the compounds (see Figure 23.5), these compounds are classified according to the ASTM standards (see Figure 23.6).

The analyst must determine which compounds are present and if any of them indicate the presence of an accelerant. Volatiles from the carpet, its backing, and any other substances that may have been present must be identified and cataloged. The interpretation of the data indicates that gasoline is present, along with many other compounds. Can the analyst reasonably explain the presence of the other compounds? How can the analyst be sure that the other compounds aren't being confused for gasoline? Are the compounds masking another or a different accelerant? These questions may come up in court, and the forensic scientist must be prepared and able to answer them. The case file must contain all of the methodology, data, and quality control information as required by the laboratory's protocols.

DNA Analysis

In a suspected homicide case, the victim was found in his apartment, apparently strangled. There were signs of a violent struggle, and some broken glass was found. The victim was wearing a shirt that had some spots of blood on it. The suspect had been to the hospital emergency room the night of the incident to get a gash in his arm stitched up.

The analyst will test the blood from the shirt as well as DNA samples from the victim and the suspect. Currently, forensic science labs use a panel of 13 loci to characterize DNA. The known DNA samples and the unknown blood stains would all be tested. The result DNA types

Figure 23.4. An example of a drug report. Courtesy of Donna Rostowski, Indiana State Police.

CERTIFICATE OF ANALYSIS
INDIANA STATE POLICE

State Police Laboratory
8500 East 21st Street
Indianapolis, IN 46219

Telephone: (317) 899-8521
Fax: (317) 899-8298

May 30, 2005

TO: INVESTIGATING OFFICER
POLICE AGENCY
123 NORTH STREET
ANYTOWN, IN 46219

Lab File # 05Q1235

Agency Case #: 123-4568

Evidence Submitted by: Officer John Smith

Received by Laboratory: 05/10/05 at 9:00

Item 1	Sealed brown paper bag containing a pair of tennis shoes from John Doe.
Item 2	Sealed brown paper bag containing a red shirt from John Doe.
Item 3	Sealed brown paper bag containing a pair of blue jeans from John Doe.
Item 4	Sealed brown paper bag containing a red carpet standard from 123 First St.
Item 5	Sealed white pill box containing a glass standard from 123 First St.

Received by Laboratory: 05/11/05 at 15:30

Item 6	Sealed brown paper bag containing a 15" crow bar recovered from 321 Main St.

Results:

Examination of the tennis shoes "from John Doe" (item 1) and the blue jeans "from John Doe" (item 3) both revealed the presence of red nylon carpet fibers with similar color, chemistry and microscopic characteristics in comparison to the red carpet standard "from 123 First St." (item 4).

Examination of the tennis shoes "from John Doe" (item 1) and the 15" crow bar (item 6) both revealed the presence of glass fragments with similar physical and optical characteristics in comparison to the glass standard "from 123 First St." (item 5).

Examination of the red shirt "from John Doe" (item 2) did not reveal the presence of red carpet fibers or glass fragments for comparison purposes.

Damon W. Lettich
Forensic Scientist 1

DWL

Lab File #05Q1235 Page 1 of 1 Reviewed by 5656

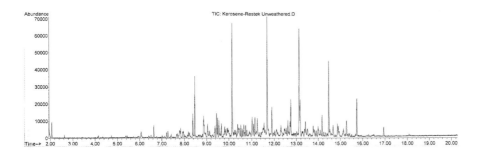

Figure 23.5. A mass chromatogram of material suspected of being an accelerant used in an arson case.

E 1618

TABLE 1 Ignitable Liquid Classification Scheme[A]

Class	Light (C$_{4-9}$)	Medium (C$_8$-C$_{13}$)	Heavy (C$_8$–C$_{20+}$)
Gasoline-all brands, including gasohol	Fresh gasoline is typically in the range C$_4$-C$_{12}$		
Petroleum Distillates	Petroleum Ether Some Cigarette Lighter Fluids Some Camping Fuels	Some Charcoal Starters[B] Some Paint Thinners Some Dry Cleaning Solvents	Kerosene Diesel Fuel Some Jet Fuels Some Charcoal Starters
Isoparaffinic Products	Aviation Gas Specialty Solvents	Some Charcoal Starters Some Paint Thinners Some Copier Toners	Some Commercial Specialty Solvents
Aromatic Products	Some Paint and Varnish Removers Some Automotive Parts Cleaners Xylenes, Toluene-based products.	Some Automotive Parts Cleaners Specialty Cleaning Solvents Some Insecticide Vehicles Fuel Additives	Some Insecticide Vehicles Industrial Cleaning Solvents
Naphthenic Paraffinic Products	Cyclohexane based solvents/products	Some Charcoal Starters Some Insecticide Vehicles Lamp Oils	Some Insecticide Vehicles Lamp Oils Industrial Solvents
N-Alkanes Products	Solvents Pentane Hexane Heptane	Some Candle Oils Copier Toners	Some Candle Oils Carbonless Forms Copier Toners
De-Aromatized Distillates	Soem Camping Fuels	Some Charcoal Starters Some Paint Thinners	Some Charcoal Starters Odorless Kerosenes
Oxygenated Solvents	Alcohols Ketones Some Lacquer Thinners Fuel Additives Surface Preparation Solvents	Some Lacquer Thinners Some Industrial Solvents Metal Cleaners/Gloss Removers	
Others-Miscellaneous	Single Component Products Soem Blended Products Some Enamel Reducers	Turpentine Products Some Blended Products Various Specialty Products	Some Blended Products Various Specialty Products

[A]The products listed in Table 1, in the various classes are illustrations of known commercial uses these ignitable liquids have. These examples are not intended to be all-inclusive. Reference literature materials may be used to provide more specific examples of each classification.

[B]As can be noted there are products found in multiple classifications such as "charcoal starters". Therefore, many of the examples can be preface by the word "some", as in "some charcoal starters."

Figure 23.6. Standards for identification of accelerant compounds from ASTM, International, standard E-1618, Table 1, "Ignitable Liquid Classification Scheme."

at the 13 loci would be reported. If the DNA from the blood matched the DNA from the suspect, then the odds of a chance match would be computed and reported.

Trace Evidence Analysis

A witness sees a man break a window of a dry cleaner and steal the cash register. A suspect is arrested 30 minutes later, and his clothing is submitted to the laboratory, along with known samples of the window glass. The forensic scientist is asked to examine the suspect's clothing and, if glass is found, to compare the glass with the known glass from the window. Eight fragments of glass were recovered from the suspect's clothing, and all eight were analytically indistinguishable from the known window samples. Based on the experience, training, and education of the analyst, the conclusion is that the glass very likely originated from the window.

A lot of information is bundled into this last statement! The analyst knows from the relevant literature, casework experience, and perhaps research he or she has performed, that he or she would not expect to find glass on the clothing of an individual selected at random who was not involved in this incident. Further, it would be highly unusual to find eight loose fragments of glass on clothing that are analytically indistinguishable from the window glass if the suspect were not associated to the incident (see, for example, Coulson, Buckleton, Gummer, & Triggs, 2001). Therefore, the results provide strong evidence to support the interpretation that the suspect was involved with the glass breaking.

Can the analyst testify that the suspect *did* break the glass? Of course not. Can the analyst testify with a percentage of certainty reflecting his or her confidence (say, 80%) that the suspect broke the glass? No. If asked, "Is it likely that the suspect did **not** break the glass?" however, the analyst would have to answer "no" to that question also: Valid evidence exists that associates the suspect to the breaking. Most of the time, trace evidence interpretation lies in this middle ground between exclusion and definite identity. Attorneys, juries, and judges all expect numbers (statistics or probabilities) to be part of the answer and, when they aren't, these people may lose confidence. Although trace evidence can be exceedingly strong evidence, it is rarely quantifiable in the way DNA is.

TESTIMONY

Getting into Court

Not every case will go to trial. In fact, opportunities for testifying may be infrequent and irregular. A summons to testify may not be issued for a number of reasons, such as the defendant may plead guilty, a plea arrangement may be made for a lesser sentence, the attorney may decide that evidence isn't needed for trial, or the charges may be dropped. Only one thing is definite about testifying: Sometime during their careers all forensic scientists will have to go to court to testify to their results.

IN MORE DETAIL:

What's the difference between civil and criminal cases?

Civil cases usually involve private disputes between persons or organizations. Criminal cases involve an action that is considered to be harmful to society as a whole.

Civil Case

A civil case begins when a person or entity (such as a corporation or the government), called the plaintiff, claims that another person or entity, called the defendant, has failed to carry out a legal duty owed to the plaintiff. Both the plaintiff and the defendant are also referred to as parties or litigants. The plaintiff may ask the court to tell the defendant to fulfill the duty, or make compensation for the harm done, or both. Legal duties include respecting rights established under the Constitution, under federal or state law, or by prior agreement of the parties.

Civil suits are brought in both state and federal courts. An example of a civil case in a state court would be if a citizen (including a corporation) sued another citizen for not living up to a contract.

For example, if a lumberyard enters a contract to sell a specific amount of wood to a carpenter for an agreed-upon price and then fails to deliver the wood, forcing the carpenter to buy it elsewhere at a higher price, the carpenter might sue the lumberyard to pay the extra costs incurred because of the lumberyard's failure to deliver; these costs are called damages. If these parties were from different states, however, then that suit could be brought in federal court under diversity jurisdiction if the amount in question exceeded the minimum required by statute ($75,000).

Individuals, corporations, and the federal government can also bring civil suits in federal court claiming violations of federal statutes or constitutional rights. For example, the federal government can sue a hospital for overbilling Medicare and Medicaid, a violation of a federal statute. An individual could sue a local police department for violation of his or her constitutional rights—for example, the right to assemble peacefully.

Criminal Case

A person accused of a crime is generally charged in a formal accusation called an indictment by a grand jury. Sometimes a grand jury is not needed and the prosecutor issues an accusation through what is called an **information**. The government, on behalf of the people of the United States, prosecutes the case through the United States Attorney's Office if the person is charged with a federal crime. A state's attorney's office prosecutes state crimes.

It is not the victim's responsibility to bring a criminal case. In a kidnapping case, for instance, the government would prosecute the kidnapper; the victim would not be a party to the action.

In some criminal cases, there may not be a specific victim. For example, state governments arrest and prosecute people accused of violating laws against driving while intoxicated because society regards that as a serious offense that can result in harm to others.

When a court determines that an individual committed a crime, that person will receive a sentence. The sentence may be an order to pay a monetary penalty (a fine and/or restitution to the victim), imprisonment, or supervision in the community (by a court employee called a U.S. probation officer if a federal crime), or some combination of these three things.

From: The Federal Judicial Center, Washington, D.C., on line at www.fjc.gov.

The first notice that a forensic scientist required for testimony will often be a subpoena describing the defendant's name, the jurisdiction, the date and time of the trial or hearing, and contact information for the requesting attorney.

Coordination with the requesting attorney is crucial to success as a witness on the stand. Most attorneys know very little about science and will need to discuss the intricacies of the scientists' expertise in order to be prepared for court. The scientist must explain what can and cannot be said about the scientific testimony. Insisting on a pre-trial conference, coming prepared, and being helpful are the best way to make testimony proceed as smoothly as possible.

Testifying

When a forensic scientist steps into a courtroom to testify, he is, in essence, entering a foreign realm where only some of the rules of science apply. As Lee Goff, a noted forensic entomologist, described it in his book, *A Fly for the Prosecution*:

> Academics and the legal system do not usually coexist in comfort. The laws of science and the rules of evidence have little in common. In theory, Academia functions on the principle of collegiality. In theory and reality, the American legal system is adversarial. The average academic entering the legal system is in for a tremendous culture shock (p. 5).

The legal arena has its own rules and most, but not all, apply to the scientist as expert and they must abide by the rules; experts, however, have leeway in the courtroom that no other witnesses have. It is a strange intersection between science and the law where even words have different meanings. Take, for example, the word "error." To a scientist, error is something that occurs naturally in all measurements and is accounted for in the statistics that are generated, such as "standard error of the mean." Errors in science cannot be avoided and are reported in due process. An attorney, on the other hand, hears the word "error" and thinks: Mistake! The scientist has just admitted to doing something wrong, in the lawyer's view, and has opened the door for further questioning. This "clash of cultures" does not always serve either side very well and may obscure what both parties seek.

Being a Witness and an Expert

Ordinary witnesses may testify only to what they have directly experienced through their five senses. This testimony must be factual in nature, and the witness, in nearly all cases, is barred from offering opinions, conjectures, or hypothetical information. Unlike other types of witnesses, however, expert witnesses are allowed to offer their opinions about evidence or situations that fall within their area of expertise. These opinions are allowed because the scientist is an expert in that area and knows more than anyone else in the courtroom about that topic; the scientist's opinion and expertise will assist the trier of fact in deciding the case.

Scientific evidence can be powerful. It can also be suspect. Judges and juries may ignore an expert's opinion evidence because it is just that: the expert's view on that issue. Often, however, those opinions and views are based on solid scientific data generated through valid analyses and therefore have a firm basis in fact. Expert witnesses must always tread a fine line between their science and the potential for advocacy in a case.

When a forensic scientist testifies, he or she does so as an **expert witness**—that is, someone who knows more about a topic or subject than the average person. The scientist is brought to court by either the prosecution or defense and offered as an expert in some area of study that will aid the judge or jury (generically referred to as **the trier of fact**) in reaching their verdict. The scientist then undergoes a process of establishing his or her education, training, experience, and expertise in that discipline. The scientist will often need to cite his or her educational degrees, work history, previous testimony experience, publications, professional associations, and other relevant information that will justify his or her expertise to the court. The attorney offering the scientist as an expert asks questions that will lay a foundation for the scientist's credentials; the opposing attorney may then ask questions in an attempt to weaken that foundation. This process is called **voir dire**, which is French for "speaking true" and is pronounced "vwa deer." It is important for the scientist to provide *relevant* qualifications to the court: Being coach of the local high school soccer team has no bearing on whether someone should be considered an expert in drug analysis, for example.

If the court rules that the scientist does possess sufficient credentials, then he or she may testify on that subject in the case at hand. The scientist must be careful to remain within the bounds of his or her expertise. It may be tempting for a forensic scientist to answer questions at the margin of his experience and offer speculative answers to be helpful or sound authoritative— but it shouldn't be done. The following fictitious example of over-extended testimony may clarify this idea:

Attorney: Dr. Medical Examiner, what type of wound was found on the victim's head?

Dr. ME: It was a contact gunshot wound.

Attorney: Would the perpetrator have had gunshot residue on his or her hands?

Dr. ME: Undoubtedly. Gunshot residue would have been on the perpetrator's hands.

Attorney: What is gunshot residue, Dr.?

Dr. ME: Small particles of material expelled from the weapon when it is discharged.

Attorney: What are those small particles composed of? What elements?

Dr. ME: Um . . .

Attorney: What's the best method to analyze gunshot residue?

Dr. ME: Scanning electron microscopy. We have one in our lab.

Attorney: How does it work?

Dr. ME: Well . . .

The medical examiner obviously has overstepped his bounds of knowledge and is now in danger of looking foolish or arrogant to the jury. Although he knows something about gunshot residue, he is clearly not an expert in this area and should have answered the attorney's second question with something like, "I'm a forensic pathologist and do not have specific expertise in the analysis of gunshot residue." It is better to answer truthfully with "I don't know" than to exceed your limits of knowledge, training, or experience.

CONSIDERATIONS FOR TESTIMONY

Preparation

Proper preparation for testimony is essential. Following are some "do's" and "don'ts". Review your paperwork and reports. Be familiar with the circumstances, times, dates, and names involved in the case. If possible, visit the courtroom in advance to get a feel of the room.

ALWAYS TELL THE TRUTH

As a witness, you have sworn to tell the truth to the best of your ability. Whatever the effects the facts may have on the case are solely the concern of the judge or jury. When you finish testifying, your part in the court proceedings is over.

PRIOR STATEMENTS

Anytime a person tells the same story twice, no matter how carefully, there are likely to be at least some differences. If there is an inconsistency with a prior statement you made, simply tell what you know to the best recollection you have. If there is an explanation for the inconsistency, give it ("If I said the evidence was returned on April 7, I misspoke. It was returned on April 17"). Your paperwork and notes should support your statements; be aware of this as you work.

DON'T DISCUSS THE CASE WITH ANYONE

It is possible that the defendant, his or her attorney, or someone on his or her behalf may try to talk with you about the case. You may speak with these people if you wish, but you don't have to discuss the case with anyone. It is not up to the prosecution or the defense to tell you whom to talk with. The only time you must answer questions is on the witness stand; that is the only time you are required to talk. If you do discuss the case prior to taking the stand, you may be asked about any alleged inconsistencies between your testimony and what you told whomever you spoke with. You will not have a court reporter's transcript to confirm or refute your claims. If the opposing attorney pulls you aside or wants to talk privately in the hallway, simply tell him or her that you'd be glad to do so with the other attorney also present. Otherwise, you may make a statement *ex parte* (away from one party in the case) that will then become part of the attorney's questioning in the courtroom ("Didn't you just tell me in the hallway . . .").

THE IMPORTANCE OF A PRETRIAL CONFERENCE

Arguably, the single most important part of your testimony experience is the pretrial conference that you have with the attorney. Both you and the attorney should be prepared to review all the significant aspects of the case and your testimony. The pretrial conference is like a dress rehearsal; it is critical that you thoroughly familiarize yourself with all the evidence, charts, and your paperwork. The pretrial conference is as important for the attorney as it is for the witness, and you should prepare a list of qualifying questions to aid the attorney in questioning you.

About Questions
LISTEN TO THE QUESTIONS CAREFULLY

One rule about questions: Answer as completely yet simply as possible. If you don't understand a question, ask for clarification; this is especially important if the question seems vague, value-laden ("Don't you feel that laboratory accreditation is important?"), or complicated.

DON'T VOLUNTEER INFORMATION

Confine your answers to what you are asked. Any information you volunteer may be inadmissible, irrelevant to the case, or may even open up a line of questioning that leads to confusion and trouble.

YOU CANNOT BE ASKED LEADING QUESTIONS ON DIRECT EXAMINATIONS

You cannot be asked leading questions, that is, questions that suggest an answer, on direct examination. For example, "Didn't you find marijuana in the sample submitted to you?" is leading; the question should be worded as, "What did you find in the sample submitted to you?" Take your time and answer the question completely. If you are asked, "Did anything else happen at that time?" or "Did you find anything else?" you may have omitted something you previously mentioned to the attorney.

BEWARE OF COMPOUND QUESTIONS

If you are asked several questions rolled into one, it will be difficult to answer them all accurately unless you do so one at a time. You could say, "I will answer your questions one by one," or you could ask, "Can you ask me those questions one at a time?"

About Answers
AVOID HEARSAY TESTIMONY

Unless you are specifically asked to testify about a conversation you had or to give your expert opinion, assume that every question calls solely for what you actually saw, heard, or did. Be careful of hearsay: Don't volunteer hearsay, such as, "Well, all the other examiners in my unit say. . . ."

OBJECTIONS

Lawyers have an absolute right and sometimes a duty to object, and you must give them that opportunity—it can be to your advantage. Don't answer too quickly; pause a second before every answer. If an attorney objects, *stop!* Don't answer! Wait until the judge rules and then either answer the question or stay silent.

REFERENCE TO DOCUMENTS

It is more effective if you can testify from memory without referring to your notes but if you must refresh your recollection, you are allowed to. Request permission from the judge: "If I had an opportunity to look at my notes, that would refresh my recollection as to the date."

DON'T GUESS

If you don't know the answer to a question, just say so. If you know most of the answer but not all of the details, just say so. No one remembers everything.

DON'T ARGUE WITH THE QUESTIONER

The cross-examiner is at a distinct advantage in being able to ask the questions. Argument or gamesmanship by a witness is not appreciated by a judge or jury. Good witnesses respond fairly and honestly and thereby retain their creditability and believability. Answer questions from the prosecutor and the defense attorney with the same tone, demeanor, and attitude.

NEVER GET ANGRY

When you are angry, you are least likely to do your duty as a witness, which is to give truthful answers. Your best reply is to remain calm, even-tempered, and to answer the questions. The more an attorney attempts to aggravate you, the more courteous and professional you should remain.

BEWARE OF YES OR NO

Some witnesses have the notion that all questions should be answered "yes" or "no." Many questions cannot be answered accurately with only "yes" or "no" because they are complicated or require additional qualification to not sound misleading. If the lawyer asks you to answer "yes or no," you are entitled to tell him or her that the question can't be answered "yes or no" without the answer being misleading. If he or she insists, you may respond, "I cannot answer 'yes' or 'no' without misleading the court." The judge normally will not direct you to answer "yes" or "no"; if he or she does, do so but expect additional questioning by the opposing attorney about your explanation.

REMAIN PROFESSIONAL ON THE STAND AT ALL TIMES

As a witness called on behalf of a party in a criminal case, you must remain professional on the stand at all times, from the moment you enter the courtroom and take the oath to when you leave the courtroom. Do not chew gum. Do not have things that you may have brought with you, other than necessary records, in your hands while testifying. Wear appropriate business clothing. Look at the jury when you answer questions. Follow the instructions of the judge. You represent forensic science, your laboratory, and yourself. Do so with honesty, integrity, and pride.

CHAPTER SUMMARY

Forensic science must operate in a legal context. The ultimate result of many scientific analyses is in a court room and the admissibility of this evidence is controlled by rules of evidence. Forensic science is part of the criminal investigation process which starts with the discovery of a crime. It is crucial that crime scene technicians properly recognize, collect and preserve evidence if it is to be effectively analyzed by forensic scientists. There are Constitutional and other legal constraints on how a criminal investigation can be carried out. These include discovery, search and seizure, protections against self incrimination, and due process.

The production of evidence at a trial is compelled by a subpoena; an order to appear in court. The admissibility of evidence is controlled by a set of rules that govern security of the evidence, authenticity, relevance and other issues. Scientific evidence is subject to all of these constraints as well as some that apply only to this type of evidence. These constraints arose in part from the decisions in *U.S. v. Frye* and *Daubert v. Merrill-Dow*. These cases set out validity and reliability rules for the admission of scientific evidence.

When forensic scientists examine evidence, they issue a scientific report. This report must be written to particular standards of accuracy and completeness. In some states the report is admissible by itself as proof of the facts it contains.

Although the same courts handle both civil and criminal cases in the U.S., they are different in their scope, rules and penalties. Discovery is much more liberal in civil cases. The only penalties for violation of civil laws is payment of money to the aggrieved party, whereas in criminal cases, life and liberty are at stake. In both cases, scientists act as expert witnesses, a status that enables them to offer expert opinions in matters within their expertise. Each time a scientist appears in court he must be qualified as an expert by the judge.

Test Your Knowledge

1. Should you answer questions with only "yes" or "no"?
2. What is an expert witness?
3. Do you have to respond to a subpoena?
4. What is a pretrial conference?
5. What should be included in a report?
6. How do forensic science laboratory casework reports differ from scientific papers?
7. List three things you should do when you testify.
8. List four things you should *not* do when you testify.
9. Why do some people describe all expert testimony as being opinion testimony?
10. Who is the trier of fact?
11. What parts of the Fifth Amendment of the Constitution affect evidence? How?
12. Who was James Frye? What was he accused of? What did he try to do in his defense?
13. The judge in *U.S. v. Frye* issued an opinion that affected the admissibility of scientific evidence for many years. What was the substance of his opinion?
14. Why is the practice of forensic science in crime labs considered to be part of the criminal investigation process?
15. What is an expert? Who decides if a witness is an expert? How is this done?
16. What privileges and responsibilities does an expert witness have that make him or her different from a non-expert (lay) witness?
17. How is evidence defined?
18. What is the difference between indirect and direct evidence?
19. What does voir dire mean? When is it used? What purpose does it fulfill?
20. What parts of the fourth Amendment of the Constitution affect evidence? How?

Consider This . . .

1. You are on the stand testifying in a case and the attorney asks you a complicated question, one with several questions entangled in it, and then demands that you answer it with either a "yes" or a "no." What are your options as an expert witness?

2. Part of the way through your testimony, you realize you misstated some statistics about the significance of your results (for example, you stated a frequency of 1 in 200,000 and you meant to say 1 in 200). What should you do?

3. What is the chain of custody? Explain the process that gives rise to the chain of custody and the document called the chain of custody. Why do expert witnesses often get more questions about the chain of custody in court than about the scientific testing they did? Give an example when a chain of custody may be broken. Why is evidence that is part of a broken chain of custody sometimes rendered inadmissible?

4. Explain what happened in *Daubert v. Merrill-Dow* that *led* to the trial? What type of case was this? What did the Supreme Court say about this case?

BIBLIOGRAPHY

Coulson, S.A., Buckleton, J.S., Gummer, A.B., & Triggs, C.M. (2001). Glass on clothing and shoes of members of the general population and people suspected of breaking crimes. *Science and Justice*, 41(1), 39–48.

Frye v. United States, 293F. 1013, 1014 [D.C. C/R. 1923].

Giannelli, P.C. (1996). *Giannelli Snyder rules of evidence handbook: Ohio practice 1996*. St. Paul, MN: West Information Pub. Group.

Goff, L. *A fly for the prosecution: How insect evidence helps solve crimes*. Cambridge, MA: Harvard University Press, 2000, p. 5.

Houck, M.M. (2005) The tyranny of numbers: Statistics and trace evidence. *Forensic Science Communications*, 1(1) Retrieved from **www.fbi.gov**.

Kieley, T.F. (2001). Forensic evidence: Science and the criminal law. Boca Raton, FL: CRC Press.

Moenssens, A.A., Starrs, J.E., Henderson, C.E., & Inbau, F.E. (1995). Scientific evidence in civil and criminal cases (4th ed.). Los Angeles: Foundation Press.

The Federal Judicial Center: **www.fjc.gov**

U.S. Federal Code, Title 28, Federal Rules of Civil Procedure.

www.fbi.gov/nq/lab/fsc/backissu/april2001/held.htm

INDEX